A NEW HISTORY OF INDIA

A New History of India

Fifth Edition

STANLEY WOLPERT

New York   Oxford
Oxford University Press
1997

Oxford University Press

Oxford   New York
Athens   Auckland   Bangkok   Bogotá   Bombay
Buenos Aires   Calcutta   Cape Town   Dar es Salaam   Delhi
Florence   Hong Kong   Istanbul   Karachi
Kuala Lumpur   Madras   Madrid   Melbourne
Mexico City   Nairobi   Paris   Singapore
Taipei   Tokyo   Toronto

and associated companies in
Berlin   Ibadan

Published by Oxford University Press, Inc.
198 Madison Avenue, New York, New York 10016

Oxford is a registered trademark of Oxford University Press

Library of Congress Cataloging-in-Publication Data
Wolpert, Stanley A., 1927–
A new history of India / Stanley Wolpert. — 5th ed.
p. cm.
Includes bibliographical references and index.
ISBN 0–19–510030–1 (cl). — ISBN 0–19–510031–X (pa)
1. India—History. I. Title.
DS436.W66 1997
954—dc20 96–19953

3 5 7 9 8 6 4 2
Printed in the United States of America
on acid-free paper

To the memory of
Professor W. Norman Brown
guru and friend
and
for my dear departed friends
Professor V. A. Narian
(1927–87)
Professor Holden Furber
(1904–94)
Madhu Mehta
(1926–95)
Raja Dinesh Singh
(1925–95)

PREFACE

A half century of freedom has tripled India's population (now some 960 million), more than quadrupled its gross domestic product, and lifted its economy in the last half decade of "globalization" into orbits of cyberspace, armed with nuclear ballistic missiles and satellite dishes. From booming Bangalore (South India's Silicon Valley) to smog-choked Delhi, from bustling Bombay (*Mumbai*) and Ahmedabad to revitalized Calcutta some 200 million urban-educated Indian business managers, skilled scientists and professionals, political leaders and government officials have in the last five years leapfrogged to afflu-ence, enjoying all the pleasures and pains of modern Western life, thanks to the "miracle" of economic privatization.

"Make our need your opportunity!"—New Delhi's modern man-tra to world capitalists has attracted more billions of private investor dollars to India since June 1991 than had been lured by its previous Soviet-inspired Five-Year Plans over the preceeding forty-five years. India's vast market—its ocean of cheap labor and fast-growing army of brilliant scientists and engineers, led by Harvard-, Oxford-, and UCLA-trained economists and freedom-thinking bureaucrats—has transformed Delhi's near bankrupt economy of 1991 into South Asia's only superpower. On the eve of the third millennium India thus seems ready, at long last, to "take off," wide awake and working harder than ever before to achieve all the proud hopes and glowing promises of its nationalist leaders' fondest futuristic dreams, or, as others might prefer to put it, to "return" to its ancient "Golden Age" of *Ram Rajya* ("Rama's Rule"). For to most Indians the past remains prologue, even as time itself is believed to be cyclical. History lives in India as in few other lands, where rebirth remains almost as universally axiomatic as death.

Yet for most Indians, even today, past poverty remains present reality. Indian male life expectancy continues to rise—five years longer now than the average Russian or Pakistani male; but for Indian women life remains perilous, often ending in felonious flames (whether from dowry-greed ignition or murderous *sati*-pyres) or suicide. Modernity has made youthful bride suicides "easier," tens of thousands of such tragic deaths occurring annually on India's railway tracks. With increased road traffic 160,000 Indians currently die annually in highway accidents, the world's highest rate of road accident fatalities. Urban crime and drug-related deaths have also recently multiplied, as has political corruption. Escalating "*hawala*" (money-laundering or "bribery") payoffs, revealed on the eve of India's recent national elections, have tarnished the reputations of many leaders in every major political party. So, like the electorates in the United States, Great Britain, and most other Western nations, many Indians have lost faith in all politicians and official probity. Coexistent with such modern cynicism, however, loom shadows of ancient India's pluralistic fears and hatreds, vows of vengeance, and the bitter *karma* of countless wars and ethno-religious conflicts, which continue to ripen in Ayodhya, Mumbai, Punjab, Tamil Nadu, and Kashmir.

I first visited India in February 1948, at the age of twenty, a maritime engineer. My ship reached Bombay on the day when one-seventh of Mahatma Gandhi's ashes were immersed in the waters of Back Bay. I witnessed that ceremony from Malabar Hill, watching hundreds of devout mourners swim after the beautifully painted and flagged ship that bore those ashes, hoping to touch some cinders as they blew away from the aft deck. I had never seen so many people before. I knew nothing of Gandhi's life, except that people called him India's "Saintly Father," and yet he was murdered by one of his fellow Hindus. The enormity of that crime and its paradox changed the course of my life. I abandoned marine engineering for Indian History, and have been a student of that uniquely fascinating subject ever since. In 1958 I returned to India as a Ford Fellow to conduct research on my doctoral dissertation, *Tilak and Gokhale*, which was first published in California in 1962. By then I had been teaching Indian History at UCLA for almost four years, and have remained there ever since, except for brief leaves or sabbaticals abroad.

This year, thanks to my good friend, inspiring historian-educator Professor Vartan Gregorian, president of Brown University, I have had the honor of launching Brown's Dr. E. S. P Das Chair in Modern

Indian History and Culture here in Providence. I thank Dr. Das for his generous support and flattering appreciation of my *New History*. I also must thank my brilliant students at Brown, especially Raj and Amit, Mohit and Ritesh, Shehriar, Piyush, Shoma, Sara, Sheela, Sonali, and Nimish, who have helped me "Rediscover India."

The greatness and glory of India's civilization, the beautifully rich wonder of its culture, the gentle wisdom and goodness of its best children, have over this past half century only increased my appreciation of Indian History, intensified my love of all that is good about India, and reaffirmed my devotion to its people. There are too many for me to thank, but I must name just a few members of my extended Indian family, whose love and generosity compel me to return Home, despite my longing at times for release: Madhukar and Tunni Shah, Bala and Bhanu Sar Desai, Feroze and Sillu Dordi, Doctorji and Mrs. Amarjit Singh, Inderji and Shiela Gujral, Khushwant Singh, Patwant Singh, Chhote Bharany, Pran and Prabha Chopra, and all their dear children.

For most kindly sharing their wisdom, and for generous support over many long years, I thank friends Vartan Gregorian, J. Kenneth Galbraith, Lloyd Cotsen, John Hawkins, Mimi Perloff, and Bob Dallek. I also thank my good editor Nancy Lane, and her fine assistant Thomas LeBien, and Oxford's excellent trade editor Paul Schlotthauer for having done so much to help publish two of my books this year.

Finally, dear Dorothy, to you and our youngest friends, Sam and Max, this book and its old author remain dedicated, with love.

Providence   S. W.
May 1996

CONTENTS

1. The Ecological Setting   *3*
2. Indus Culture (ca. 2500–1600 B.C.)   *14*
3. The Aryan Age (ca. 1500–1000 B.C.)   *24*
4. North Indian Conquest and Unification (ca. 1000–450 B.C.)   *37*
5. India's First Imperial Unification (326–184 B.C.)   *55*
6. Political Fragmentation and Economic and Cultural Enrichment (ca. 184 B.C.–A.D. 320)   *70*
7. The Classical Age (A.D. 320–ca. 700)   *88*
8. The Impact of Islam (ca. 711–1556)   *104*
9. Mughal Imperial Unification (1556–1605)   *126*
10. Western Europe's Vanguard (1498–1669)   *135*
11. Great Mughal Glory (1605–1707)   *149*
12. Twilight of the Mughal Empire (1707–64)   *168*
13. John Company Raj (1765–93)   *187*
14. The New Mughals (1793–1848)   *201*
15. Unification, Modernization, and Revolt (1848–58)   *226*
16. Crown Rule—A New Order (1858–77)   *239*
17. Indian Nationalism—The First Movement (1885–1905)   *250*
18. The Machine Solidifies (1885–1905)   *265*
19. Revolt, Repression, and Reform (1905–12)   *275*
20. The Impact of World War One (1914–19)   *286*
21. Toward Independence (1920–39)   *301*
22. The Impact of World War Two (1939–46)   *329*
23. The Nehru Era (1947–64)   *351*
24. From Collective Leadership to Indira Raj (1964–77)   *371*

25. From Janata Raj to Rajiv's Death   *407*
26. India Today   *441*
     Bibliography   *455*
     Glossary   *483*
     Index   *489*

MAPS

1. India—Physical Features   *2*
2. A.   Average Annual Precipitation   *7*
   B.   Population Density (1976)   *7*
3. Pre-British Indian Empires
   A.   Indus or Harappan "Empire" (ca. 2000 B.C.)   *64*
   B.   Mauryan Empire in the Reign of Ashoka
       (269–232 B.C.)   *64*
   C.   Gupta Empire under Chandra Gupta II
       (A.D. 375–415)   *65*
   D.   Delhi Sultanate in 1236   *65*
   E.   Mughal Empire at the Death of Akbar (1605)   *66*
4. Pre-Muslim India (ca. 1200)   *102*
5. India (ca. 1500)   *123*
6. British India (1797–1805)   *202*
7. South Asia Today   *350*
8. Pakistan   *392*
9. Bangladesh   *393*
10. India Today   *442*

A NEW HISTORY OF INDIA

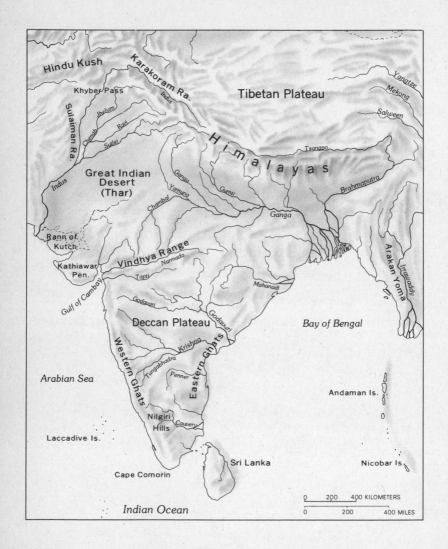

Hindu Kush

Karakoram Ra.

Khyber Pass

Tibetan Plateau

Yangtze

Mekong

Sulaiman Ra.

Jhelum

Indus

Salween

Chenab

Ravi

Sutlej

H i m a l a y a s

Indus

Great Indian
Desert
(Thar)

Ganga

Tsangpo

Yamuna

Gumti

Brahmaputra

Chambal

Ganga

Rann of
Kutch

Vindhya Range

Narmada

Arakan Yoma

Kathiawar
Pen.

Tapti

Mahanadi

Irrawaddy

Gulf of Cambay

Godavari

Deccan Plateau

Godavari

Bay of Bengal

Western Ghats

Krishna

Eastern Ghats

Arabian Sea

Tungabhadra

Penner

Andaman Is.

Laccadive Is.

Nilgiri
Hills

Cauvery

Sri Lanka

Nicobar Is.

Cape Comorin

Indian Ocean

0   200   400 KILOMETERS

0   200   400 MILES

India—Physical Features

ONE

THE ECOLOGICAL SETTING

India is named for the Indus River, along whose fecund banks a great urban civilization flourished more than four thousand years ago. That unique civilization, evolved in South Asia and sustained unbroken throughout four millennia, ranks with Western and Sinitic civilizations as one of the world's most brilliant cultural systems. The sophistication of Indian thought, the beauty of Indian art, and the power and wealth of Indian imperial unifications endow the pageant of Indian history with singular glory. Long before the dawn of the Christian era, India fired the imaginations of distant peoples and tempted conquerors from Macedonia to Central Asia to invade its subcontinent and seek to master its population and their arts. More recently, other invasions—whether launched by the zeal of Islam or Christianity or stimulated by the prospect of commercial profits or power—have brought fresh waves of migrants to South Asia. Each invasion has added diversity to India's vast population and complexity to its rich patterns of culture; nonetheless, many of the earliest seeds of the original civilization have survived in clearly recognizable form. Like the branches of a giant banyan tree that have long since surpassed the original trunk in girth, the great traditions of Indian civilization have spread from epoch to epoch across the subcontinent of their birth, drawing sustenance from countless local traditions, often stooping to conquer, and enduring essentially through change.

The story of Indian history must begin where Indic civilization itself emerged, in the ecological setting of South Asia. The subcontinent of South Asia encompasses an area of more than one and a half million square miles, from the Hindu Kush and Baluchi Hills on the west and the Great Himalayas on the north, to the Burmese mountains on the east and the Indian Ocean on the south. Within this kite-shaped subcontinent, whose north-south and east-west braces extend roughly 2,000 miles each, can be found virtually every sort of topography, climate, and geological formation: from sub-sea-level desert wastes to the world's highest peaks (Everest is 29,028 feet); from perennial drought to some of the earth's most heavily drenched terrain (Cherrapunji in Assam receives an average annual rainfall of 426 inches); from the ancient Precambrian granite of the peninsula (ca. 500 to 2,000 million years old) to a northern tier of relatively youthful Cenozoic mountains (ca. 60 million years old).

Geographically, the subcontinent may most simply be divided into three major horizontal zones: the northern mountain belt; its neighboring offspring of Indo-Gangetic alluvial plains; and the peninsular massif of the south, which may originally have been part of Africa. The northern mountains have served as a natural protective wall against both invading armies and Arctic winds, shielding South Asia from frost even more than from force, and providing its Indus plains above thirty degrees north latitude with January temperatures averaging fifty degrees Fahrenheit, while the rest of the subcontinent basks in mild warmth of up to eighty degrees. Though India's southernmost tip only dips to eight degrees latitude north of the equator, its climate throughout the year is subtropical, thanks to the northern shield. Heat is the most pervasive fact of India's ecological setting, and it is not surprising to find both sun and fire deified by Hinduism to this day. Though the impact of heat on Indian thought, work habits, and health cannot be measured, its significance should not be ignored. In contrasting the civilizations of China and India, and especially their modern achievements, the enervating effect of heat on Indian productivity is a factor that can hardly be underestimated.

Perhaps because of India's heat, water has always played an especially sacred role in Indian life and thought. The waters of the Indus River system, one of whose lesser tributaries is the Soan, were to become the cradle of North Indian culture; and, like the alluvial valleys of the Punjab ("Land of the Five Rivers") and Sind, whose silt was borne in their torrents, they are perennial gifts of Himalayan ice and snow.

The earliest traces of human habitation in South Asia survive as flakes of stone found scattered around the valley of the Soan River in what is now the northern part of Pakistan. These primitive tools or weapons are the only surviving signs of Paleolithic man's presence in North India. They appear to indicate that at some time during the second interglacial age, between 200,000 and 400,000 years ago, humans migrated to South Asia over the Hindu Kush Mountains of the northwest, or possibly climbed directly over the high Himalayas ("Abodes of Snow") from their original habitations in Central or East Asia, where Paleolithic skeletal remains, as well as flake tools, have been unearthed.

Fed by the glaciers of southern Tibet, the Indus flows almost a thousand miles north and west through Kashmir before it veers sharply to the south, cutting its gorge through Nanga Parbat, down the Malakand Pass, to capture the waters of Afghanistan's Kabul River. Both rivers join in the region of Gandhara, just north of the Khyber Pass, which was to become the historic highroad into India of invading armies from the west. North India's two other great river systems, the Yamuna-Ganga and Brahmaputra, originate in the same region of Tibetan ice, so close to the source of the Indus that they may once have belonged to a single mighty lake, whose prehistoric unity and tranquility was perhaps shattered by the titanic eruptive force of the birth of the Himalayas, driving their waters off in diverse directions. This ancient natural displacement of the waters that have brought tons of sedimentary earth to South Asia's northern plains and that continue daily to fructify that soil, is politically reflected at present in the threefold division of the subcontinent into Pakistan, India, and Bangladesh. These three nations depend most vitally upon the Indus, Ganga-Yamuna, and Brahmaputra rivers, respectively. Hindus have long worshiped "Mother Ganga" as a goddess. Hardwar—where the Ganga emerges with torrential force from her mountain wall to flow more placidly across the plains of Uttar Pradesh ("Northern Province")—is but the first of many sacred cities, including Allahabad and Banaras, that line the Ganga's crescent path of more than fifteen hundred miles to the Bay of Bengal in the east. At Bengal's delta, the mouth of the Ganga meets that of "Brahma's Son" (Brahmaputra), whose thousand-mile journey north of the Ladakh Range ends only after it has veered back upon itself to cut its way down between Bhutan and Burma into the "Land of Bengal" (Bangladesh).

South of the high, geologically youthful mountains of the north and their offspring alluvial plains lies the barren desert land (Rajasthan) and the rugged Vindhya and Satpura ranges of central India's ancient mountainous bedrock. This central mountain belt below the Tropic of Cancer has always presented a formidable natural barrier to easy communication between northern and southern India, encouraging the development of virtually independent cultures as well as empires to the north and south of the Vindhya-Satpura-Chota Nagpur divide throughout most of India's history.

The Deccan Plateau rises just south of the river Tapti, below the Satpura Range, and dips like a weatherworn old table to the east, obliging all of South India's major river systems—the Mahanadi ("Great River"), Godavari, Krishna, and Cauvery—to empty into the Bay of Bengal. The western edge of the Deccan Plateau is a spinelike wall of mountains called the Western Ghats ("Steps"), which average some three thousand feet in height and catch most of the annual rain that blows from the Arabian Sea during the southwest monsoon. The Deccan Plateau is, therefore, mostly parched and barren badlands, resembling the Southwest United States. The narrow coastal littoral of western India, however, is more like a tropical rain forest, enjoying from one to two hundred inches of rainfall annually. This area, the Malabar coast, is one of the world's best environments for growing pepper, nutmeg, and other spices that proved so potent a lure to Western appetites.

Denied the north's bounty of perennially snow-fed streams, South India has always depended on rain for its water. To this day, the annual advent of the June monsoon is greeted with ritual dance and ecstatic worship by southern India's peasantry. In Bengal and along the neighboring coastline of Orissa, the monsoon often arrives with hurricane force. Spending their fury on Assam and Burma, the rains then move west and are deflected by the Himalayas to water the Gangetic plain as far north as Delhi.

The winds that annually bring revitalizing rain to the south also probably brought the first humans to peninsular India by sea from East Africa, possibly at about the time East Asian migrants first wandered into the northern Soan River valley. Again, we have no skeletal remains, only tools, to inform us of human nomads in South India during the Paleolithic age, but here the tools are core stone implements, rather than flakes. Crude hand axes have been found in western, central, and eastern sites across the Deccan, but since most of

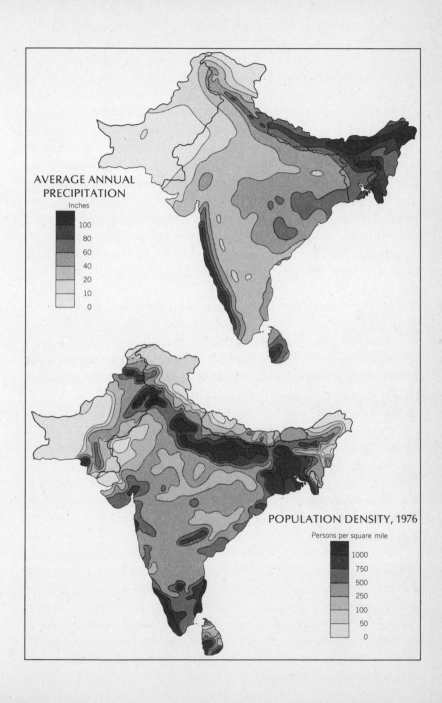

AVERAGE ANNUAL
PRECIPITATION
Inches

100
80
60
40
20
10
0

POPULATION DENSITY, 1976

Persons per square mile

1000
750
500
250
100
50
0

these early finds were located in Madras, on the peninsular's eastern (Coromandal) coast, this stone-core production is called Madras industry. Its techniques and products are almost identical with those found in South Africa and southern Europe.

Thus, for the first and longest era of Indian history, there were at least two distant, separate regions of human habitation and cultural development in South Asia. What led one group of people to save the flakes as their primary tools or weapons, while another kept the cores of stone? Did the one initially worship a god of the mountain, the other a mother goddess? How did their languages differ when they first fashioned the tools of speech? History offers no answers to such questions. We do know, however, that Dravidian, the linguistic family still dominant in South India, is a unique mode of communication, quite distinct from the Indo-European, Indo-Aryan languages of North India. There are several other great language families that have long been represented in South Asia, including Sino-Tibetan and Tibeto-Burman in the north and a wide variety of tribal languages spoken by isolated hill peoples, such as Munda, related to the Mon-Khmer languages of Southeast Asia. Historically, we can date with assurance only the relatively recent arrival of Indo-Europeans, however, and though much scholarly work has been done on all Dravidian languages, no proto-Dravidian grammar has yet been produced. Hence, we really have no way of fixing with any certainty the arrival of Dravidian speakers to Indian soil. Modern South Indian separatists demand an "autonomous" Dravidistan ("Land of Dravidians") on the basis of their claim that "Dravidian civilization" antedates the advent of the Northern Aryans, who only subsequently conquered and superimposed their "lesser" civilization upon the peoples of peninsular India, but history offers no positive evidence in support of such claims.

There seems to have been a second great wave of human migration to South Asia from East Africa or southern Europe sometime after the final receding of the glacial ice that had so long paralyzed human progress, during the Mesolithic Age, which began around 30,000 B.C. Numerous microliths have been found scattered across the face of the Deccan and up into central India, and even in the Punjab. These tiny stone weapons, called pygmy tools, so closely resemble those found in France, England, and East Africa that it would appear they were brought to South India by hunters and food-gatherers who were quite different from South Asia's paleolithic pioneers. However, recent finds in the Narmada River valley, which runs through central India between the Vindhya and Satpura mountains, appear to indicate an evolutionary line of development from early to middle Stone Age peoples in that region.[1]

In Mesopotamia, Egypt, and Persia, India's western neighbors made the transition from primitive hunting and food gathering to crop raising, a transition that marks the dawn of the New Stone Age and the advent of civilization, between the ninth and fifth millennia B.C. That Neolithic revolution seems to have occurred in South Asia only after 4000 B.C., the approximate date of the earliest Neolithic settlements found thus far in the hills of Baluchistan on the northwest frontier.[2] Why it should have taken some five thousand years for the concept of civilized settlement to spread from the region of Jericho to that of Baluchistan remains a mystery. The interlude is, indeed, so long compared to the distance (much of which was soon covered en route to Persia) that there is reason to hope that future archeological excavations in India and Pakistan may unearth more ancient Neolithic sites than any as yet discovered in South Asian soil.

Ecologically, the Indo-Iranian borderlands of South Asia may be viewed, as Dr. Agrawal[3] has done, as a single region precariously supporting a semiarid agricultural and herding economy. With less than ten inches of rain a year and all of it during the winter, this has long been a region of dry farming and transhumance; its multiplicity of village sites and cultures probably reflects a rather small number of seminomadic occupants. These Baluchi villagers constructed houses of mud brick, used tools of stone and bone, and appear to have domesticated sheep, goats, and oxen. No metal or pottery have been dis-

[1] H. D. Sankalia, "Pre- and Proto-History in India and Pakistan: New Discoveries and Fresh Interpretations," *Journal of Indian History* (University of Kerala, Trivandrum), 45, pt. 1 (April 1967), p. 103.

[2] From a dig near modern Quetta, Baluchistan's capital, with radiocarbon dates fixed at 3688 and 3712 B.C. Bridget and Raymond Allchin, *The Birth of Indian Civilization: India and Pakistan Before 500 B.C.* (Harmondsworth: Penguin Books, 1968), p. 101.

[3] D. P. Agrawal, *The Copper Bronze Age in India: An integrated archaeological study of the Copper Bronze Age in India in the light of chronological, technological and ecological factors, ca. 3000–500 B.C.* (New Delhi: Munshiram Manoharlal, 1971), pp. 209 ff.

covered at the earliest sites, though handmade pots of yellowish-red clay were produced somewhat later in this region. Less than a hundred miles north of Quetta, a number of sites in the Zhob River valley have yielded intriguing evidence of very early settlement on the frontier of South Asia. Mother goddess figurines from these excavations, though undated, suggest that India's most popular form of contemporary worship was possibly its oldest religious cult as well. Humped bulls made of terracotta have also been unearthed in the valley of the Zhob. The bull was later to be most intimately associated with Shiva in Hinduism, deified as Nandi, divine bearer of the great god. In addition, numerous stone phallic symbols were found scattered throughout this region, further indicating some type of ancient fertility worship similar to that later associated with Shiva, whose major icon remains the stone phallus. In southern Baluchistan, near the Makran coast, other early village sites have been discovered, the remains of which are generally classified as belonging to a Kulli culture, whose buffware pottery bears resemblance to that of its western neighbors but whose favored burial practice appears to have been cremation, which later emerged as the most common form of Indian burial. An abundance of clay "goddess" figurines as well as small bulls have been found at this culture's sites. Some of the geometric motifs of its pottery, such as the heart-shaped *pipal* ("fig") leaf, remain important religio-artistic symbols of Indian civilization. It was presumably while seated under a *pipal* tree that the Buddha attained enlightenment some three millennia later.

The neighboring Indus Valley plain is ecologically much the same as the Nile and Tigris-Euphrates valleys, cradles of neolithic civilizations. It is a semiarid region, requiring no iron tools to clear and settle, and with ample flood-borne supplies of silt to serve as natural fertilizer. Early archeological theories of a possible eastward shift of monsoon winds and the subsequent "progressive dessication" of the Indus plains[4] sound less plausible in the light of Fairservis' bril-

---

[4] Aurel Stein, *An Archaeological Tour in Gedrosia* (Calcutta: Government of India, Central Publications Branch, 1931), and *Archaeological Reconnaissances in N.W. India and S.E. Iran* (London: Macmillan and Co., Ltd., 1937); Stuart Piggott, *Prehistoric India to 1000 B.C.* (Harmondsworth: Penguin, 1950); R. E. M. Wheeler, *Early India and Pakistan* (London, 1959); and John Marshall, ed. *Mohenjo-Daro and the Indus Civilization* (London: Probsthain, 1931).

liant and painstaking analysis, which appears to prove that Indus riverine gallery forests sufficed to bake the bricks used in building and periodically rebuilding (about every 140 years) Indus cities.[5] It also seems clear that the larger and most problematic Indus animals, the tiger and rhinoceros, gradually driven away or destroyed by agriculture and grazing, were in fact the natural fauna of riverine long grass and open forest country. The elephant, it is true, was never found west of central India, but as Raikes and Dyson have pointed out, elephants might well have been imported to Sind.[6]

The great Indus urban civilization may have been preceded by a period of village culture roughly contemporaneous with that already noted in the Baluchi Hills. Amri, the first such village, discovered in 1929, is in Sind, barely a mile from the Indus, on the river plain just beyond the flood line. Exactly how long it took for India's early villagers to develop sufficient surpluses of wheat and barley and sufficient technological competence to allow them to make the extraordinary leap from village culture to urban civilization is impossible to say. Five hundred years would hardly seem too long a time for achieving a social transition of such historic magnitude. Amri, at any rate, reveals at least four phases of development, all of which antedated the higher stage of Indus civilization. From crude ditches in which handmade pottery and storage jars were found buried, the site also yielded later buffware with intricate geometric designs, which was made on wheels, and fragments of copper as well as bronze. The mud brick buildings of the final phases of this village culture are associated with beautiful painted pottery on which the humped Indian bull (*zebu*) and other domesticated animals are depicted. Several similar village settlements have been found at the same respectful distance from the river, most recently at Kot Diji, about twenty-five miles northeast of Mohenjo-daro, the great southern "capital" of the Indus civilization.

The Rajasthan desert region east of the Indus Valley may have enjoyed the same ecological environment as did Sind between four

[5] Walter A. Fairservis, *The Roots of Ancient India: the archaeology of early Indian civilization* (New York: Macmillan, 1971).

[6] R. L. Raikes and R. H. Dyson, Jr., "The Prehistoric Climate of Baluchistan and the Indus Valley," *American Anthropologist* 63 (1961): 265.

and five thousand years ago, when the now dry Saraswati "river" system flowed across its long barren sands to the Arabian Sea. At Kalibangan, 120 miles southeast of Harappa, the Indus' northern "capital," another pre-Harappan village, has recently been excavated. It is the first such site to be found in modern India rather than in Pakistan. Although it is not known what caused the Saraswati to shift course or dry up, such evidence of early community life proves that Rajasthan was not always desolate. Perhaps with Gujarat, whose shovellike jaw juts into the Arabian Sea to the south, Rajasthan belonged to a single ecosystem, the heartland of which was Sind and the Punjab, a region hospitable to Neolithic development and rapid growth and easily conquered by the copper, bronze, and stone tools and weapons of early Indian civilization.

The tough, monsoon-nourished *sal* forests of the Yamuna-Ganga plain east of the Rajasthan desert, however, posed more formidable barriers to human settlement, not to be overcome until iron ploughs drawn by oxen were developed well after 1000 B.C. With annual rainfall of up to forty inches and tropical steppe soil on older alluvium, the Doab region between the rivers Yamuna and Ganga has yielded no pre-Aryan traces of settlement, its jungles apparently too dense and foreboding for sustained Neolithic human penetration. Fragments of redware and ochre-color ware found in Doab subsoil may reflect the wanderings of early tribal peoples who inhabited the region, but the substantial hoards of painted greyware that were unearthed at Hastinapura near Delhi do not appear to be dateable earlier than 1000 B.C. The northern black polished ware also found at that site, which may have been the first great city of Aryan occupation in India, is clearly the product of a later era (ca. 500 B.C.), possibly brought to Hastinapura from the eastern Gangetic plain (around Bihar), where the iron that gave it its dark luster was found in abundance.

Near the confluence of the rivers Ganga and Yamuna, where the modern city of Allahabad stands in Uttar Pradesh, the average annual rainfall reaches forty inches; to the east, it increases to a hundred inches and more in Bengal and Assam, where it is possible to cultivate rice, sometimes two or three crops a year. Wheat, barley, and millets, on the other hand, are the basic crops of the western Gangetic plain and the northern Punjab, where annual rainfalls are closer to twenty than forty inches. Sugarcane will grow in both ecological zones and is one of India's most common and important crops. The region of relative historical isolation called Bengal or Bangladesh is the world's largest aggregate of deltaic land and must have been covered with *sal* forests when the great urban civilization of the Indus flourished more than four thousand years ago.

TWO

INDUS CULTURE

(ca. 2500–1600 B.C.)

The monumental excavations of the ancient Punjab city of Harappa (Hara is one of Shiva's names), begun in 1921, and of its mighty reflection south along the Indus at Mohenjo-daro ("Mound of the Dead"), started a year later, have transformed our historical understanding of ancient India. The discovery of these two Indus cities, some four hundred miles apart, served first of all to extend the roots of urban Indian civilization back at least a thousand years before the Aryan invasions. The archeological remains at these sites also emphatically reversed the relative cultural status of India's Aryan conquerors and her pre-Aryan peoples. The pre-Aryan *dāsas,* or "slaves," whose darker skins differentiated them from Aryan "color," were suddenly revealed as more advanced, sophisticated, and technologically precocious than the semibarbaric hoard of Aryan invaders from the west, whose only "civilized" advantages seem to have been some superior weaponry and the use of harnessed horses.

When Sir John Marshall sent D. R. Sahni to lead the dig at Harappa in January 1921, he had good reason to suspect that something significant lay buried beneath the enormous earth-covered mounds rising along the eastern bank of the river Ravi in the Punjab's Montgomery district. The site had been known to British officials in India for almost a hundred years. Sir Alexander Cunningham, a British general and amateur archeologist, visited it several times in 1853 and 1873, finding and recording small, indecipherable seals, which he picked up with other tantalizing objects near the mounds. There were also hundreds of thousands of burnt bricks scattered over the region, enough to provide ballast for a hundred miles of steel track on the Lahore to Multan Railroad, which practical-minded Indian contractors and British engineers used for precisely that purpose.

The richness of the archeological yield at Harappa has equated the name of that city with the Indus Valley civilization it represents. Radiocarbon dating now verifies that from at least 2300 to 1750 B.C. this great city, no less than three and a half miles in circumference, flourished behind massive ramparts of brick, forty feet thick at their base. The great wall of Harappa protected the citadel from the waters of the Ravi as well as from possible human invasion. The citadel itself rose almost fifty feet and was similar in size and geographic orientation to the one found at Mohenjo-daro, also standing to the west of the lower town, with its workers' quarters. The same central board of imperial urban planners seems to have designed both cities; even their bricks were baked to similar standard measurements. The extensive use of Harappan bricks for railway support, however, has made it impossible for archeologists to identify any clear building remains within the citadel walls, such as those discovered at Mohenjo-daro, but several granaries have been found north of the citadel, along the river bank. Constructed in two rows with six storage areas in each and ventilation ducts to circulate air freely through the thick-walled warehouses, these granaries could have been used to keep an ample surplus of wheat and barley at hand for Harappa's population (estimated to have been about 35,000); or they might have been used to store produce to be shipped downstream or imported goods brought from Mohenjo-daro or from distant Sumeria. These large and costly facilities attest to the wealth of the Harappan civilization, which, like Sumer, may have been ruled by a priest-king who was worshipped as divinity incarnate. A small stone statue that could perhaps represent such a monarch has been found at Mohenjo-daro. His elongated eyes, thick lips, and impassive, bearded face reflect almost supernatural serenity and power, while his jeweled head- and arm-bands, as well as the cloverleaf design on his robe, might have been insignia of exalted rank.

Between the granaries and the walled citadel of Harappa were discovered workers' quarters or barracks that were more sturdy and well drained than the slum dwellings in most later Indian cities, where lower caste or "outcaste" laborers hover in abject poverty beyond the town wall, clinging to the tattered fringes of Indian society. Was Indus society already polarized, as Indian society would later be, into a priestly-warrior elite, ruling over a mass of urban artisans, merchants, servants, and other "slaves of the gods"? No historic records have been found that name any priest-king or his bureaucratic assistants, and we can only assume that they somehow mastered the skills necessary to protect their subjects from floods and wild beasts. We do know, however, from the remarkable images carved on Indus seals, that tigers, elephants, and rhinoceros, as well as buffalo and bulls, all inhabited the forests of this now almost desert region. From the same seals, we also know that the people of the Indus had a written language.

The dig at Mohenjo-daro yielded a much clearer map of that ancient city on the west bank of the Indus, 250 miles north of the Arabian Sea. There were, in fact, no fewer than ten cities, constructed with singular conservatism one on top of the other over what must have been a period of many centuries. Marshall and his assistants were so excited by the lavish wealth of bricks, shards, seals, statues, beads, and other artifacts scattered in profusion in the sands of Sind that they sacrificed scientific precision to speed. We have, therefore, no clear idea of the chronological sequence of the finds or the longevity of Mohenjo-daro's many reincarnations, the earliest three of which are now submerged beneath the rising water table. We do know that the walled citadel contained many imposing buildings, one of which was a great hypocaust built of bricks coated with bitumen. It was complete with dressing rooms and was similar in size and complexity to the "bathing tanks" that still form an integral part of most Hindu temples. No great temple or center of worship has been found as yet at Mohenjo-daro, but that may be because the appropriate place for one, just east of the bath, has not been excavated. A sacred Buddhist *stupa* stands over that spot on the citadel, and permission has never been granted to move it. A large granary has been found to the west of the bath, and near that an even more spacious building (230 by 78 feet)—possibly a royal palace—has recently emerged.

Beyond the citadel's walls sprawled the humbler city of the mass of Mohenjo-daro's population. Planned along a north-south grid, the town was divided into a number of major blocks, which may have separated various occupational or kinship groups from one another, as in subsequent Indian cities castes would reside in their own enclaves. Some of the blocks contained spacious homes, whose sturdy brick foundations indicate multilevel structures. Domestic life was oriented around open interior courtyards that ensured privacy, much the way well-to-do Indian homes have been built and organized ever since. The covered brick drainage, both inside individual homes and on public streets, was technologically sophisticated and more sanitary than that found in many modern Indian towns. Mohenjo-daro's municipal administration, whether theocratic or not, was certainly conscientious. The standard sizes of streets and blocks, as well as of bricks, further attest to centralized authority. A block of small cells has been identified in one area of the low town as having been either a police station barracks or a monastery. There were public wells, and shops lined the main streets of Mohenjo-daro's cities during their half millennium at least of historic vitality.

Wheel-made pottery, much of it redware or buffware painted black, at times designed with animal as well as geometric motifs, has been found in profusion at all major Indus sites. The countless small shards of drinking cups appear to indicate that these people may have initiated the later Indian habit of using clay cups only once, possibly from fear of contamination or pollution. Many large sewer seepage pots and curious perforated vessels have also been found. In addition to potters, Mohenjo-daro's artisans included fine metallurgists, who were adept at fashioning tools and weapons of copper and bronze as well as stone.

The heavy brick walls and unadorned streets of Mohenjo-daro, Harappa, and the more recently discovered Indus sites of Kot Diji, Lothal, and Kalibangan leave an overall impression of ponderous utilitarianism, yet some smaller surviving artifacts offer tantalizing glimpses into the artistic sensibilities and beauty that must have flourished in Indus cities as well. A naked bronze "dancing girl" of Mohenjo-daro, for example, is most realistic, a figurine whose gaunt and boyish femininity still seems to mock mankind provocatively with her footless stance, haughty head, and petulantly poised arm. But the most exciting, yet frustrating, of all Indus artifacts are the small, square steatite seals found in such profusion at Mohenjo-daro. These magnificent seals, probably made for merchants who used them to "brand" their wares, provide brilliant portraits of Brahmani bulls, "unicorns," tigers, and other animals, whose realistic likenesses reappear two thousand years later on the capital plinths of the pillars of Ashoka. The seals also contain pictographic signs (usually above their

animal figures)—the still-undeciphered first writing of Indian civilization. Almost four hundred different pictographs have been identified
on the seals, far too many for the signs to have been phonetic, but too
few for the language to have been ideographic. We assume that the
pictographic words are the proper names of Indus merchants, written
from right to left. Many attempts have been made to crack the Indus
script and link it with the later Brahmi alphabet, the most recent by a
team of Russian Indologists, who used computers to help them analyze
the pictographs.

One seal found in the northern area of Mohenjo-daro depicts a
figure seated in yogic, ithyphallic posture, surrounded by a tiger, elephant, rhinoceros, water buffalo, and deer. He wears a horned head-
dress, appears to have more than one face (or possibly a tiger's
mask), and wears a tiger's skin on his torso and bangles on his arms.
This image may be the earliest artistic depiction of Shiva, India's
"Great God" (*Maheshvara*) as "Lord of Beasts" (*Pashupati*). From
the great number of phallic stones discovered at Indus sites, and from
the popularity of the bull as a device on the seals, we may postulate
that Shiva was worshipped in several of his subsequent forms by the
people of the Indus. In a later era, Nandi, Shiva's mount, would be
found patiently waiting, legs tucked beneath his body, at the entrance
to every Hindu Shaivite temple. Deep inside such a temple, however,
in its sanctum sanctorum or "womb-house" (*gārbha-griha*) usually
stands the icon that represents Shiva himself: a polished stone phallus,
sometimes encrusted with jewels or embellished with images of Shiva's
face, but generally unadorned. Shiva's dual role of yogic fertility deity
and tamer or destroyer of jungle beasts as Lord of the Hunt, helps to
explain the continuing duality of his image throughout the many muta-
tions of Indian civilization. He is at once the planter of seed, who
gives life, and the destroyer, whose wrathful power can strip the skin
from a tiger with the flick of his smallest finger. Another fascinating
seal from Mohenjo-daro also seems to depict a three-horned deity
(the trident is one of Shiva's symbols) standing in the middle of a
tree, from whose center he appears to have emerged. Another
figure outside the tree is bent in suppliant posture with arms raised as
though in worship to the "tree god." Behind him the bull stands in
waiting, while beneath them both stand seven girls, who may be dancing around the "sacred tree" in festive joy, celebrating the harvest,
perhaps, or some ancient rite of spring.

Many small clay figurines have also been unearthed at Mohenjo-daro, including crudely modeled images of women whose female organs are so prominently displayed that they appear to have served as ritual images of the mother goddess. Numerous yoni-shaped stones, moreover, have been found at Indus sites, strengthening the assumption that mother-goddess worship was an important element in the religious life of Harappan civilization, as it has remained throughout Indian history. The connection of Shiva with female "power" or *shakti*—embodied in the mother, whose names are many (Shri, Devi, Parvati, Kali, Durga) but whose essence is the earth and life-bearing nature—would thus appear to have been well established by the time Indian civilization emerged in full-blown urban form.

Indus civilization, now represented by no fewer than seventy unearthed sites, extended over almost half a million square miles of the Punjab and Sind, from the borderlands of Baluchistan to the desert wastes of Rajasthan, from the Himalayan foothills to the tip of Gujarat, probing the limits of its ecosystem during the half millennium of its mature survival. Recently discovered Harappan outposts along the Makran coast, including that of Sutkāgen Dor near the border of modern Iran, clearly attest to brisk and continuous trade with Sumer, especially during the reign of Sargon of Akkad (2334–2279 B.C.). Thanks to Indus seals found in the Sumerian dig at Ur in 1932, we know that merchants from Harappa and Mohenjo-daro were trading with their Sumerian counterparts between 2300 and 2000 B.C. The huge granaries beside the river at Harappa seem to indicate that Indus merchants exported their surplus grain to Sumeria, and possibly elsewhere.

By this time (ca. 2000 B.C.), the Indus people had begun to spin cotton into yarn and weave it into cloth, a dyed fragment of which has been found at Mohenjo-daro. The use of cotton for clothing is one of India's major gifts to world civilization, and the spinning and weaving of cotton was to remain India's premier industry, with expanding production for both the home market and export. It is impossible to determine how extensive cotton production was at this time, but some of this rare and remarkable produce was surely exported to Mesopotamia and may indeed have been a major item in that trade. Most Indian exports, however, were probably luxury items of relatively little bulk and weight, such as etched carnelian beads, shell and bone inlay goods, ivory combs, and possibly even peacock feathers and apes—commodities later imported from India by King Solomon—plus pearls and precious wood products. It is unclear whether spices were as yet used and traded, though from green amazon stones found at Indus sites, we do know that the merchants of Mohenjo-daro imported products from as far south as the Nilgiri Hills in southern India and may have established commercial relations with the Malabar coast. The presence of Tibetan jadeite at the sites indicates the probable northern limit of Indus commercial contact, while silver, turquoise, tin, and lapis lazuli were imported from Persia and Afghanistan.

By the time of Mohenjo-daro and Harappa, the nomadic hunting, food-gathering, and fishing subsistence economy of later Stone Age peoples and the marginal small-village agricultural economy of the Baluchi Hills had been clearly displaced by a sophisticated inundation-and-irrigation agricultural and commercial economy capable of supporting a large surplus urban population. Though wheat seems to have been the most important grain in the Indus Valley, rice (whether wild or cultivated) has been found at Lothal; other crops included field peas, dates, mustard seeds, and sesamum. The people of the Indus had domesticated a wide variety of animals, including the dog, cat, camel, sheep, pig, goat, water buffalo, *zebu,* elephant, and chicken. (One or two clay figurines that appear to depict horses have been found at a few Indus sites, but these may reflect foreign travel or imports, since it seems most probable that the horse first invaded India along with the Aryans.) The domestication of fowl obviously enriched the diet of Mohenjo-daro and Harappa and, along with cotton spinning, ranks among India's earliest gifts to world civilization. The *zebu* and water buffalo had already become India's leading draught animals, though the depiction of both on seals suggests they were worshipped as well as whipped. Much the same sort of ambivalence toward draught animals, who occupy the front room of a peasant's home all night but feel the wrath of his stick whenever they loiter in harness, remains characteristic of Indian behavior to the present. Such a seeming paradox of personality may only reflect the ambivalent natural environment, where lush harvests coexist with barren soil, drought with flood, feast with famine. It is hardly surprising that Hindu logicians would be the first to posit the coexistence of contradiction.

No graves have as yet been discovered at Mohenjo-daro, but cemeteries have been found at Harappa and more recently at the port of Lothal and in Kalibangan. Careful analysis of the skeletal remains of almost one hundred corpses indicates that the people of Harappa were a mixture of predominantly Proto-Australoid and Mediterranean physiques, as in modern peninsular India, ranging in height from five feet to five feet and nine inches, and averaging around thirty years of age at death. Most of the bodies were inhumed outstretched on their backs with their heads pointed toward the north. A number of pots were usually buried with the bodies, and some of the skeletons had been decorated with ornamental jewels. Though no monumental or "royal" tombs have been found, several coffins were placed inside substantial brick chambers, and at Kalibangan one cist two by four meters in size has recently been discovered. The most interesting burial find, however, comes from the port of Lothal in Gujarat, some 450 miles southeast of Mohenjo-daro, where three double graves, each with a male and female skeleton, have been discovered. This find is possibly the first indication of the well-known Hindu custom of *sati,* according to which widows were expected to follow their husbands in death as well as in life. The custom of *sati* as we know it was generally associated with the cremation pyre rather than the grave, but the idea of eternal wifely devotion may well have anticipated the practical change in disposal of the dead that was to become almost universal throughout India at a later date.

Sometime shortly after 1750 B.C., a number of factors began to transform the character of Harappan civilization, impairing its quality of life and disrupting its hitherto orderly urban environment to the extent that streets no longer followed any careful grid pattern, homes diminished in size, and pottery as well as drainage deteriorated or disappeared. South of Mohenjo-daro near the small Indus town of Chanhu-daro are two villages whose names have been bequeathed by archeologists to these decaying phases of Indus civilization, called the Jhukar and Jhangar cultures. It appears that the highly efficient, wealthy, and powerful empire of Harappa sustained some cataclysmic blows, which left its great cities and towns to the occupancy of squatters from neighboring villages or from more remote regions; people whose level of culture was more primitive than that of their predecessors. Their buffware was much cruder, their seals simpler and totally different in design from those of Mohenjo-daro in its heyday. (Some of the geometric motifs and crude sketches are, nonetheless, reminiscent of the Harappa and earlier Baluchi Hill cultures.)

The Jhukar people's use of faience beads may, as Piggot suggests, be explained by the "conscription" of "local craftsmen" who learned the art at Chanhu-daro and were forced by their "new rulers" to continue producing their wares;[1] or it might simply reflect the fragmented and disjointed continuity of tradition maintained by a remnant of the earlier urban culture who had somehow escaped natural catastrophe. Perhaps the homes of the faience craftsmen were further from the river than those of the seal engravers or potters. For we no longer assume that the mighty walled citadels of Harappa and Mohenjo-daro were scaled by conquering hordes of Aryan tribesmen, whose horse-drawn chariots and hafted axes sufficed to bring down an empire on whose distant ramparts the sun rarely set. Thanks to the recent careful work and creative archeological studies undertaken by George Dales and Robert Raikes, it is now generally recognized that, around 1700 B.C., a series of floods caused by tectonic earth movements brought an end to this once-glorious civilization.[2] For whatever reason, the Indus seems to have changed its course, spelling ruination to the delicately balanced system of Harappan agriculture. Hoards of jewelry and other precious objects, including copper tools, have been found in the highest strata at Mohenjo-daro, indicating the sense of impending doom and the widespread fear that must have gripped the populace as the flood waters rose. Many homes were hastily abandoned. Cooking pots were found strewn across kitchens; straw was found that had smoldered after a roof collapsed, as were fragments of burned wooden door jambs. But most revealing of all are the skeletons of people fleeing, a family of five caught in the debris as their walls and ceilings crumbled, as the earth moved, or as the river raced in. At least thirty skeletons have been found at Mohenjo-daro alone, not buried but trapped, killed by some terrible disaster. They were assumed until recently to have been massacred by an invading army,

---

[1] Stuart Piggott, *Prehistoric India* (Harmondsworth: Penguin, 1950), p. 226.
[2] George F. Dales, "New Investigations at Mohenjo-daro," *Archaeology* 18,2(1965):145–50; R. L. Raikes, "The End of the Ancient Cities of the Indus," *American Anthropologist* 66,2(1964):284–99.

though now it seems more likely that all were fleeing a combined earthquake and flood.

The chaos that characterized the last days of Mohenjo-daro apparently spread to Harappa in the north and may have reverberated as far as Lothal in the south as well, though our evidence from these sites is less clearly defined, and Lothal at least seems to have prospered long after the core of Indus civilization decayed. At Harappa, a cemetery (Cemetery H) was discovered in 1946 whose pottery resembles the work of the decadent Jhukar culture and whose burial practices differ significantly from those of an earlier, neighboring cemetery where inhumation was used. The method employed at Cemetery H was the fractional burial of bones in large urns, which marks the transition to cremation at Harappa and which may reflect the habit of new peoples, possibly recent conquerors, perhaps Aryans. The designs on the shoulders of the burial urns include such familiar motifs as peacocks and bulls (or possibly cows) and plant forms, among which is the Harappan *pipal* leaf that later appears in the iconography of Buddhism. In addition, however, there are stars, tiny human figures inside peacocks, and a man with long, wavy hair who is holding beasts, one of which is menaced by a dog. Pandit M. S. Vats, who directed the latter phase of the Harappan dig, suggested that perhaps the human figures represented dead souls, while the dog was the hound of death, called Yama by the Aryans, and the urns a product of early Indo-Iranian tribesmen who had come in the vanguard of invaders.[3] By the time of the Jhangar culture, at any rate, we find only poor grey-black burnished pottery, crudely incised in decorative patterns reminiscent of shards discovered in Gujarat dating from about 1000 B.C. The glory of Harappa had more than faded. The city empire with its wondrous citadels was gone, disappearing as dramatically, almost as inexplicably, as it had emerged, washed over by the silt of the Indus and its rushing tributaries of the Punjab, like a world of sand castles reclaimed by rising tide, its crumbling walls and vague outlines all that remain for the next wave of invading children to occupy.

[3] M. S. Vats, *Excavations at Harappa* (Delhi: Government of India, Manager of Publications, 1940).

THREE

THE ARYAN AGE

(ca. 1500–1000 B.C.)

Around 2000 B.C. the original Indo-European-speaking, seminomadic barbarians, who most probably lived in the region between the Caspian and the Black seas, were driven from their homeland by some natural disaster, possibly drought, prolonged frost, or plague. Whatever the cause of their dispersion—it may even have been a series of Mongol invasions from Central Asia—the ancestors of the Italic-, Greek-, Germanic-, English-, Celtic-, Iranian-, Sanskritic-, and modern Hindi-speaking peoples were forced to flee from southern Russia to survive. These tribes moved in every direction, splitting up into smaller, more cohesive units, driving their herds of cattle, sheep, goats, and domesticated horses with them, and opening a new chapter in the history of Europe, as well as of India. The Hittites were the first Indo-Europeans to settle in a new homeland, for we find traces of them just south of Caucasia in Cappadocia that date from approximately 2000 B.C. Other tribes pushed on, however; some to the west, across Anatolia, and some to the east, across Persia (now named Iran, a cognate of Aryan, for the Indo-Iranian language brought by Indo-Europeans to that region between 1800 and 1500 B.C.) The Indo-Iranians seem to have lived for some time in harmony following their long migration. By about 1500 B.C., however, they appear to have split once more, and pastoral tribes known to history as the Indo-Aryans, or simply Aryans, advanced still further east, across the perilous Hindu Kush Mountains, into India.

Our knowledge of the earliest history of our linguistic ancestors is derived from over a century of patient reconstruction of their "Urheim" by philologists such as Friedrich Max Müller (1823–1900), who developed the science of linguistic paleontology. By analyzing all the languages within this great linguistic family and extracting the geographic, climatic, botannical, and zoological terms that all have in common, it has been possible to chart an ecological map of the "original homeland," which most closely resembles Caucasia. The ingenious insight that sparked this magnificent labor of comparative linguistics and philology was publicized most effectively by Sir William Jones (1746–94), a judge of the British East India Company's High Court. Sir William studied Sanskrit after his arrival in Calcutta in 1783, and three years later he wote an illuminating and learned paper that noted the vocabulary and other linguistic bonds that make the Greek, Latin, Germanic, and Sanskrit tongues all relatives within one single Indo-European family of languages.

We have no archeological evidence for the first centuries of India's Aryan age (from about 1500 to 1000 B.C.), but we have been able to piece together some picture of the era from the Aryans' religious "Books of Knowledge," or Vedas, which were so sedulously preserved by the bards of each tribe through rigorous oral tradition. The oldest and most important of these sacred works, the Rig Veda (literally "Verses of Knowledge"), consists of 1,017 Sanskrit poems, most of which are addressed to various Aryan gods and solicit their bounty. It is the world's earliest surviving Indo-European literature.

Unlike the pre-Aryan peoples of Harappa, the Aryans lived in tribal villages with their migrant herds. Their houses, fashioned of bamboo or light wood, have not survived the ravages of time; they baked no bricks, built no elaborate baths or sewer systems, created no magnificent statues or even modest figurines; they had no seals or writing, no faience art, no splendid homes. Were these relatively primitive tribal peoples, in fact, capable of storming and conquering fortified Indus cities? Perhaps. They had harnessed their horses to chariots, and they seem to have wielded hafted bronze axes (a few of which were found in the highest strata of Indus cities), as well as long bows and arrows. They had been toughened by their trek, had endured blistering sun and crossed high passes deluged with sleet and snow. The Rig Veda itself, however, is unconscious of that journey and of the Aryan "invasion" of India, yet it does mention Aryan victories against "fortified places" (*pur*), within which dark-skinned people (*dasas*) had sought in vain to defend themselves against the fairer-skinned ("wheat-colored") Aryans. (The Sanskrit word *dasa* later came to mean "slave.")

Since the Rig Veda was not written down before about 600 B.C.
and the earliest surviving text dates back only as far as about 1200
A.D., we may ask, first of all, how we know that the Vedic hymns were
actually composed as early as 1500 B.C., when it is generally assumed
the Aryans first invaded India? Before 1909 the only way the age of
the Vedas could be approximated was by Max Müller's "dead-reck-
oning-backwards" technique. Müller analyzed the entire corpus of
Vedic literature, linguistically as well as ideologically, noting the various changes in Sanskrit case endings and word forms, as well as
syntax, vocabulary, and meaning. He determined how long it had
taken to effect comparable changes in ancient Greek and Latin and
framed a similar timetable for the evolution of Sanskrit. There are
three major stages in the evolution of those sacred Vedic texts that
are still considered "revealed" literature, or *shruti* (literally "heard"),
by Hindus. The initial stage includes the Rig Veda and three other
ancient "collections" (*Samhitas*) of hymns and magical incantations
or spells, the Sama, Yajur, and Atharva Vedas, all of which are ar-
chaic poetic texts. Next emerged a series of prose commentaries on
each of the Vedas, elaborating upon those often cryptic hymns and
describing in minute detail the procedures required for preparing the
Vedic sacrifices and properly propitiating the gods. Because they exalt
the significance and role of the Aryan priestly class, the *brāhmans*
(from the word "sacred utterance," or "those who chant sacred utter-
ances"), the commentaries are called the Brahmanas. Finally, a third
group of mystical philosophic works appeared, whose predominant
form, the poetic dialogue, and whose radical new religious messages
sharply differentiate them from the Brahmanas and Samhitas alike;
these are the Vedanta Upanishads, many of whose ideas are similar to
those in early Buddhism. Müller reasoned that they must have been
composed at about the time the Buddha lived, or somewhere during
the sixth century B.C. Then, by dead reckoning backwards, he estimated that within the 108 surviving Upanishadic texts there had been
significant ideological, if not linguistic, evolution, which probably took
several centuries. That would move the date of their composition back
to about the eighth century B.C., from whence it would have taken at
least two centuries for the Brahmanas to have been written. If the last
of the Vedic Samhitas was then completed and ready for commentary
by 1000 B.C., it seemed safe to infer that the oldest sections of the

Rig Veda must have been at least four centuries older, and hence the estimated date of around 1400 B.C. for the compilation of the Rig.

In 1909 excavations at the Hittite site of Boghaz-köi in Cappadocia yielded tablets containing a treaty concluded between the Hittite King Subililiuma and his Mitanni neighbor to the east, King Mattiwaza, who reigned in about 1400 B.C. Invoked as divine witnesses to this treaty were four gods, Indara, Uruvna, Mitira, and the Nasatiya, whose Sanskrit names in the Rig Veda were spelled virtually the same (*Indra, Varuna, Mitra,* and *Naksatras*), proving that by this date the Vedic pantheon had acquired its identity. This confirmation of Müller's estimates leads us to assume that, since the Rig Veda itself does not mention the Aryan invasions of India, the process must have begun at least a century earlier, or probably around 1500 B.C. The final wave of tribal invasions may have come centuries after the first Aryans started over the northwest passes. This was the most important invasion in all of India's history, since the Aryans brought with their Caucasian genes a new language—Sanskrit—and a new pantheon of gods, as well as the patriarchal, patrilineal family and the three-class social structure (priests, warriors, and commoners) into which their tribes were organized.

Limited as we are by Vedic sources, we naturally know more about ancient Aryan religion than other aspects of the culture. Before considering Rig Vedic religious beliefs and practices, however, let us see what can be gleaned from that work about Aryan social organization and other mundane matters. The term *aryan,* while primarily a linguistic family designation, had also the secondary meaning of "highborn" or "noble." The Aryan "commoners" or *vish* (the word later used to designate the largest class in Aryan society, *vaishyas*), were most broadly divided into "tribes," or *jana.* Though united by language and religion, as well as in warfare against their common non-Aryan, "dark" enemies, the *dasas,* these tribes appear generally to have been at war with one another. The foremost Aryan tribe was called Bharata, probably the name of its first *rājā* or "king"; it is honored to this day by the Republic of India, which adopted Bharat as its official Sanskrit name with the inauguration of its constitution in 1950. The later Aryan Epic, *Mahābhārata* ("Great Bharata") is the tale of many raja cousins and their interminable battles. Each Aryan tribe had its bards, who were priests, and they alone memorized the

Vedic hymns and officiated over the sacrifices. The raja, his brahmans, and *vish* settled in villages (*grāma*), keeping their herds of valuable cattle, horses, sheep, and goats in nearby pasture. Cows were so highly valued by Aryans that they came to be treated as currency and were paid to brahmans for performing religious services. The Vedic Aryans were, however, beef eaters and wine drinkers as well as warriors. It is not clear exactly when the Indians began to consider the cow divine; it must have been a later development, or perhaps a pre-Aryan concept resurrected in transmuted form, since we can assume that the pre-Aryans worshipped the bull. Indians may thus have been the earliest people, though they would not remain the only ones, to worship their money.

Just as each Aryan tribe was ruled by an autocratic male raja, each family was controlled by its father, whose dominant role over his wife and children was to become the standard pattern for subsequent Indian familial relationships, in which male supremacy and hierarchy dictated by age were to remain the rule. The joint or extended family, where the wives of all sons come to live and raise their children within the patriarchal household, seems to have developed early in the Aryan age. One of the most common prayers in the Vedas is for "manly, heroic" (*vira*) sons, who were not only needed to help care for the herds, but who would also bring honor to their fathers and tribes in battle and be able to perform the sacrifice to aid the souls of their fathers attain peace after death. The word *vira* in the Rig Veda was, in fact, hardly distinguishable from "son," since all young men were expected to be heroic fighters. Daughters, however, were little valued. Dowries would be required for them, and, although the status of women in Aryan society was probably higher than it was to remain throughout most of Indian history, they were forbidden to participate in any sacrifice to the gods since their presence was considered a source of pollution. Sons alone could inherit property, which was usually divided equally among them after the father's death. Primogeniture was reserved for royal families, unless the first-born was blind or otherwise seriously handicapped. We find no evidence of polygamy or child marriage in the Vedas, though both were subsequently practiced in most parts of India, which may again reflect the reemergence of pre-Aryan customs.

The country inhabited by the Aryans during the period in which the Rig Veda was composed was known as the "Land of the Seven Rivers" (*Sapta Sindhava*). It consisted primarily of the Punjab, whose five great rivers (the Jhelum, Chenab, Ravi, Beas, and Sutlej) flow into the Indus, which may then have captured the seventh river, called Sarasvati, now only a minor stream in the Rajasthan desert. The river Ganga was barely known to the Aryans by the end of the Rig Vedic era (it is mentioned only once in a late book), indicating that the conquering tribes expanded quite slowly toward the east, taking about five centuries to move from the Khyber to beyond the region of Delhi. Throughout that interlude, the process of Aryan and pre-Aryan conflict, cooperation, and assimilation must have effected major changes in the character of Aryan society and thought, as well as in the nature of the indigenous civilization.

The simple tribal structure grew more complex during this period as warfare and conquest brought new peoples and problems under the ruling rajas, who required the assistance of noble "warriors" (*kshatriyas*) and the advice of "councils" (*sabha* and *samiti*) of household elders to govern their burgeoning tribes. We know little more about these earliest political institutions of India than their Sanskrit names, but from what is known of tribal rule elsewhere, especially among other Indo-Europeans, it seems fair to assume that each king picked the heartiest soldiers to serve in his entourage and sought the advice of those patriarchs in his tribe who were most shrewd or powerful. Surely the *rishis* ("sages"), who are named in the Rig Veda, were among those first approached by rajas for assistance in practical as well as spiritual matters. The distinct separation between martial, royal, and priestly classes found among the Aryans appears to have diminished over time; if, indeed, priest-kings had ruled pre-Aryan Harappa, it may be that Aryan rajas learned from their slaves to rely more heavily upon the counsel of their own brahmans. At any rate, the hymn of the "Sacrifice of the Cosmic Man" (*Purusha-sukta*), which appears in the tenth and final "book" of the Rig Veda, explains that the four great "classes" (*varna*) of Aryan society emerged from different parts of the original cosmic man's anatomy: the *brahmans* issuing forth first, from the mouth; the *kshatriyas* second, from the arms; the *vaishyas* third, from the thighs; and the *shudras* last, from the feet. This "revelation," according to which all rajas, who were *kshatriyas* by birth, fell below all *brahmans,* who alone were asso-

ciated with the cosmic "head," may well have roused ancient royal
wrath, though not enough to delete the sacred hymn or alter a word
of it.

Gold, the metal most frequently mentioned in the Rig Veda,
must have been panned from the rivers of the northwest and used in
ritual sacrifices as well as for jewelry. The next most common metal
was *ayas,* which appears initially to have meant bronze, rather than
iron, since later, in the Atharva Veda, we find the distinction made
between "red" *ayas* and "dark" *ayas,* the former most likely referring
to bronze and the latter to iron. Some of the passages in the Atharva
Veda are known to have been added a few centuries after the Rig
Veda had been finished, and it is most probable that iron was not discovered in India until the Aryans had moved as far east as the modern
state of Bihar (where rich deposits of ore continue to be mined to
this day), which could not have been before 1000 B.C. By that time,
the use of iron had spread to Iran from its center of discovery by the
Hittites in the west: it may have entered India with fresh waves of invading Iranians, since its use was initially associated with pins and
other parts of horse harnesses, as well as weapons. B. B. Lal's excavations in the Doab east of Delhi, starting in 1950, unearthed traces of
the ancient city of Hastinapura, the epic capital of a great Aryan
tribe. Shards of painted greyware date its lowest level to about 1000
B.C., while from later levels iron weapons and tools have been unearthed, the oldest found as yet in India.

By the time the Rig Veda was written, the Aryans had apparently made the transition from a nomadic pastoral economy to a combined agricultural and pastoral one, for they reaped some variety of
"grain" (*yava*), which must have been barley or wheat. There is no
reference to rice, however, until the Atharva Veda. The lion was
known in Rig Vedic times, as was the elephant, whose Sanskrit name
means "beast with a hand," but neither the rhinoceros nor the tiger,
both so prominent on Indus Valley seals, is mentioned. The horse was
second only to the cow in importance, and chariot racing was one of
the Aryans' leading sports.

The Vedic Aryans appear to have loved music, wine, and gambling, as well as war and chariot racing. All their hymns were chanted,
but the Sama Veda was especially designed for song, and there were
lutes, flutes, and drums, which the gods and goddesses were said to

have played. From this era, at least, Indians continued to use song and dance as an integral part of their religious worship, and no sacred Hindu ceremony today, including the funeral procession, would be complete without its musicians, no temple without its dancers. Indian devotion to song and dance, however, predated the Aryan invasions, for the "dancing girl" of Mohenjo-daro survives to speak mutely to us of an art that is surely as old as Indian civilization itself. Another statue, found at Harappa though possibly of later vintage than that civilization, is a grey stone miniature torso of a dancer, whose distinctive three-way twist of shoulders and hips reminds us of Shiva in his most magnificent pose as "King of the Dance" (*Nātārāja*). As for drink, *soma* was daily imbibed by the Vedic warrior god Indra to help him overcome the terrible demon with whom he did battle, and we must assume that Aryan tribesmen also indulged in such libation. It is not quite clear whether this heavenly drink was alcoholic, psychodelic, or narcotic, though we assume that it was made from a plant that grows wild in the foothills of the Himalayas, which may have been hashish or peyote. The effects of this drink were so remarkably powerful that *soma* was deified, and the *soma* sacrifice became the foremost religious event of the Aryan year.

Since the game of dice, like chess, was invented in India, and many dice carved of nuts were found at Mohenjo-daro, it is hardly surprising to note that the Aryans were avid gamblers. "Cast on the board like magic bits of charcoal, though cold themselves, they burn the heart to ashes," lamented the Rig Veda's gambler in one of the rare secular hymns of that sacred collection. "The abandoned wife of the gambler mourns. . . . In debt, fear, and need of money, he wanders by night to the homes of others." Gambling continued to preoccupy Aryan Indians throughout the epic era as well, and in the *Mahabharata* we find the five virtuous and noble Pandava brothers losing their very kingdom and their single beloved wife (the most famous case of polyandry in Sanskrit literature) to the treacherously seductive roll of the dice.

Other secular hymns give us insight into the daily occupations and aspirations of Rig Vedic Indians. There were carpenters and wheelwrights, blacksmiths and tanners, weavers and spinners, as well as farmers and herders, among the tribesmen who settled down to the routine of village interdependence as they expanded toward Delhi and the Gangetic plain. The *shudras,* who did the menial labor, may originally have been pre-Aryan *dasas,* reduced to serfdom or slavery by captivity and easily kept in lowly status because of darker skin color. The Sanskrit word that came to mean "class" (*varna*) and that is still used with the modifiers *brahman, kshatriya, vaishya,* and *shudra* to identify the four broadest categories of Hindu caste society, originally meant "covering," associated with skin covering and its varying colors. Each *varna* had its distinguishing color; white for *brahmans,* red for *kshatriyas,* brown for *vaishyas,* and black for *shudras.* Acute color consciousness thus developed early during India's Aryan age and has since remained a significant factor in reinforcing the hierarchical social attitudes that are so deeply embedded in Indian civilization. There is no reference to "untouchables" in the Rig Veda, but fears of pollution became so pervasive in Indian society that it is difficult to believe they were not, in fact, pre-Aryan in origin. In all probability the subclass of untouchables emerged late in the Aryan age, recruited first perhaps from those *shudras* or *dasas* who performed tasks that were considered "unclean," such as the work of tanners, associated with animal carcasses, and that of sweepers, especially among the ashes of cremation grounds. While all *shudras* were, therefore, held to be fully a life below the three higher "twice-born" classes, whose sacred thread ceremony of "rebirth" marked their attainment of manhood and access to Vedic lore, some were considered so much less worthy than others that they were cast beyond the pale of recognized society.

The religion of the early Aryans centered around the worship of a pantheon of nature gods, to whom sacrificial offerings were periodically made for the good things of life and for repose thereafter. No one deity ruled over the pantheon, which included some thirty-three divinities named in the Rig Veda, but the most powerful gods, to whom many hymns were addressed, were Indra, Varuna, Agni, and Soma. Indra was the war god, youthful, heroic, ever victorious. Like Thor, he wields thunderbolts and hovers in atmospheric realms, assisted by an obscure storm god named Rudra, who comes to be identified only much later as the Rig Vedic form of Shiva. The association of Indra with the power to release waters as well as win wars helps explain his special significance, for he is hailed as "surpassing floods and rivers in his greatness." Perhaps he was the first great leader of the Aryan conquest, a historic figure whose youthful force overcame all obstacles, standing so tall and strong he seemed to hold Father Sky upon his shoulders, separating it from Mother Earth, as one simplistic myth of Vedic creation insisted. He required much nourishment and drank his *soma* greedily in three gulps every morning before going forth to defeat the demon Vritra, whose limbless body enclosed all creation, including the sun, waters, and cows, holding life in a state of inert suspension and darkness. With his "mighty and fatal weapon," the thunderbolt, Indra pierced the dark demon's covering and released the dawn (which is why Hindu prayers to Indra are chanted so early every morning, to help him defeat the night), leaving the demon "prostrate" while the waters, "like bellowing cows," rushed lowing toward the ocean. Indra then became "the lord of what moves and what remains rested." Vritra was the symbol of pre-Aryan power, "warder" of the *"dāsa* lord"; hence the hymn that tells of the battle between Indra and Vritra may be viewed as of historic as well as cosmogonic significance, conveying the essence of the Aryan victory. It has, indeed, been suggested that Vritra was no demon at all, but a dam constructed across the Indus by pre-Aryans to control the river for irrigation agriculture, and that by destroying that "barrier" or "cover" the Aryans flooded the region and its great cities, facilitating their conquest.

Once Indra's victory was achieved, however, Varuna, the King of Universal Order (initially *rita* and later *dharma*) stepped forward to take the central position of Aryan religious authority. Presiding over the sun-filled sky, Varuna was the divine lord of justice "who has spread out the earth, as the butcher does the hide, by way of a carpet for the sun . . . extended the air above the trees . . . put strength in horses, milk in cows, willpower in hearts, fire in waters, the sun in the heaven, and soma upon the mountain." Varuna was the divine judge of Aryan India, and appeals for mercy would be chanted to him by those who strayed from the path of virtue. Older and wiser than Indra, Varuna was most honored by the Aryans, the elder statesman on high, closely connected with the sun god Surya and with one of his lesser manifestations in the Rig Veda, Vishnu, who later shared with Shiva virtual monotheistic dominance over Hinduism.

Agni was god of fire, and as such had many forms, traversing the three realms of earth, atmosphere, and heaven. He was needed for every sacrifice; he mirrored the sun; and he had the power to heal, save, defend, or destroy. Hailed as "offspring of the (primeval) waters," he was also called "illuminator of darkness," and as the many-tongued deity of the sacred altar he presided over "ritual function." Soma was the god of immortality, the nectar whose "glorious drops" impart "freedom" and protect one's body from disease. "May we enjoy with an enlivened spirit the juice thou givest like ancestral riches," chanted the Aryans to divine Soma. "O Soma, King, prolong thou our existence . . . favour us and make us prosper. . . . For thou hast settled in each joint, O Soma."

Among the lesser personified powers of nature worshipped by the Vedic Aryans, the loveliest was Ushas, the dawn, "rosy-fingered" daughter of the sky. For her most beautiful poems were chanted. She brought of all "lights the fairest."

The seeming simplicity of the Aryan nature-worshipping religion was soon obscured by the Vedic quest for an understanding of cosmic origins and control over cosmic forces. The immediate purpose of a sacrifice was to secure some divine favor, whether fortune, longevity, or progeny, but it also had cosmic meaning in that its proper performance helped maintain the balance of order in the universe. The Aryan householder gave his gods soma, *ghi* (clarified butter), and other delicacies not simply in return for their favors, but because it was his duty to propitiate them so, just as they in turn were obliged to act in the appropriate fashion toward him. For gods as well as men had their individual duties, which were part of the cosmic scheme of things, and only when all behaved properly would the universe function as it was designed to do—in accord with the *rita,* the true order. Demons of falsehood were always trying to destroy that perfect balance, starting floods, bringing drought or famine, appearing in the guise of tigers or mad elephants; they were ever present as mosquitos and other evil creatures that buzzed, crawled, or walked upon the earth. The balance was tenuous at all times, which was why so many sacrifices were required, and why brahmans had to be employed day and night to chant the hymns they memorized. Truth (*rita*) could always be subverted by falsehood (*an-rita*), just as the "real" (*sat*) or existent world might always be disguised by imagined or "unreal" (*asat*) illusions, fantasies, and nonexistent fears and terrors. The word *sat,* which originally meant "existent," came thus to be equated with cosmic reality and its underlying ethical principle, truth. To Vedic man the universe was divided between earth's fair surface and the heavenly dome above it, the realm in which *sat* prevailed, and the demon-darkness beneath this world, where unreality and falsehood dominated all. Indra's daily battle renewed the wonder of creation, but speculation about this mighty hero soon led to profound questions: "Who ever saw him? Who is he that we should praise him?"

Before the Rig Veda was finished, such speculation was responsible for the creation of a number of superdeities, whose all-embracing qualities and impersonal characteristics more nearly resembled monotheistic than pantheistic gods. Prajāpati, whose name means "Lord of Creatures," emerged as a more comprehensive god than Indra, as did Visvakarman, the "Maker of All," and Brāhmanāspati, "Lord of the Sacred Utterance (brahman)." The introduction of the last name clearly connotes the growing power and presumption of the brahman priests, who further exalted their ritual chantings by deifying "speech" itself as the goddess Vāc. The evolution of a monistic principle of creation, however, came only at the very end of the Rig Veda (Book X, hymn 129), when we find a neuter pronoun and numeral, Tad Ekam, "That One," cited as the source of all creation, anticipating differentiation of any sort and all deities, self-existent, self-generating, unique. "There was not then either the nonexistent (asat) or the existent (sat). There was no sky nor heavenly vault beyond it. What covered all? Where? What was its protection? Was there a fathomless depth of the waters?"[1] begins this most remarkable and precocious of all Vedic hymns. It continues:

There was neither death nor immortality then.
There was the sheen neither of day nor of night.
That One [Tad Ekam] breathed [came to life], though uninspired by breath, by its own potentiality. Besides it nothing existed.

There was darkness hidden by darkness at the beginning.
This all was an unillumined flood. The first [with power of evolution] which was hidden by a shell, That One, was born through the power of its own [creative incubating] heat.

By about 1000 B.C., then, India's Aryans were asking questions and positing hypothetical solutions to problems that still remain un-

[1] W. Norman Brown, *Man in the Universe* (Berkeley and Los Angeles: University of California Press, 1966), pp. 29–30.

fathomable. In crediting self-incubating "heat" (*tapas*) with the origin of creation, the Rig Veda sounds so surprisingly scientific that we may find it difficult to reconcile such latter-day sophistication with much of the Rig Veda's earlier naiveté. The word *tapas,* however, was subsequently used in relation to yogic contemplation, and its use in the Rig Veda may reflect the reemergence of India's oldest form of religion as well as "science," the self-imposed rigor of isolated meditation that gave birth to so many illuminating insights throughout Indian history. "Desire" (*kāma*), which later came to mean "love," was the source of That One's stirring to life, the force behind creation, moving even a neuter spirit to sow the first "seed of mind," as it was so often to move India's noblest sages and gods from the austere depths of their contemplation to peaks of passionate bliss. Had pre-Aryan yogis (such as the one depicted on the seal from Mohenjo-daro) learned by now the language of their conquerors, teaching their own secrets in turn to brahman bards? Was India's "heat" and ancient wisdom starting already to take its toll of youthful Aryan energy and optimistic self-assurance?

FOUR

NORTH INDIAN
CONQUEST
AND UNIFICATION

(ca. 1000–450 B.C.)

The Aryan "conquest" of North India was thus a process of gradual institutional assimilation and sociocultural integration between invading barbaric hordes and their more civilized pre-Aryan "slaves." It would take more than a thousand years for that process of historic change to reach its peak of political unification under Mauryan rule in 326 B.C., by which time the center of North Indian power and civilized achievement had shifted more than a thousand miles east of the Indus to the region of modern Patna (then called Pataliputra) in the Gangetic plain. Our sources for reconstructing that slow and complex process of historic evolution are both literary and archeological; yet for many centuries, in most parts of India the record remains as blank as most Indian cave walls, and even when excitingly beautiful historic frescoes are found, they emerge only as fragments.

The *Mahabharata,* whose epic core probably reflects Indian life at about 1000 B.C., starts with King Santanu's intoxicated love for the beautiful goddess Ganga, whom he "marries," symbolizing the Aryan advance east of the Doab into the Gangetic plain. It was no simple romance or conquest, however, requiring heavy ploughs and other sturdy tools to clear the primitive forests of that lush region that even sophisticated Harappan technology had not been able to master. Recent discoveries of painted greyware at Hastinapura, Alamgirpur, and Kausambi in Uttar Pradesh indicate that, by about 1000 B.C., the Aryans had mastered the metallurgy of iron, which they may have learned from their Indo-European cousins who ruled the neighboring

Iranian plateau during Sialk VI.[1] Iron ploughs yoked to oxen and hafted iron axes were able to clear ground that had been impervious to tools of stone, copper, and bronze.

Although they were armed with force enough to overcome the natural barriers to their eastward advance and equipped with weaponry to subdue resistance from their *dasa* precursors, the Aryans could never seem to resolve their own intertribal conflicts. The *Mahabharata,* much like the *Iliad,* is drenched in the blood of endless warfare, echoing the cries of royal cousins who are locked in mortal combat over their legacy to King Santanu's domain. A residue of quite primitive, or perhaps non-Aryan, barbarism survives in the blood lust of the noble warrier Bhima, who howls like a wolf and dances wildly about the battlefield of Kurukshetra after drinking the heart's blood of his slaughtered cousin, Dushasana. Aryan chivalry had by now reached the level of Arthurian legend. Religious virtue was all too humanly united, however, with hopeless addiction to gambling in the hero Yudhishthira, *Dharmarājā* ("King of Religious Law") himself, who loses at dice not only his fortune and kingdom, but all that was owned by his noble Pandava brothers as well, and their lovely polyandrous wife, Draupadi. The epic struggle marks a transition at about this time from pastoral nomadism to the consolidation and confederation of Aryan tribes into royal territorial kingdoms with capitals like Hastinapura (near Delhi), dominating the surrounding agricultural and forest domain, which came to be called Arya-varta, "Land of the Aryans."

Several elaborate sacrifices, designed to consecrate royalty, appear in the Brahmana commentaries on the Vedas, composed from about 1000 to 700 B.C., further attesting to the growing significance of kingship. In the battle between gods and demons, we are told, the gods were losing until they decided to "make a king," choosing Indra, who then swiftly led them to victory. The "royal consecration" (*rājasūya*) sacrifice therefore refers to the raja as a "partner of the gods," inheriting some of the powers of Indra. Other sacrifices were performed to rejuvenate aging monarchs, who drank the revitalizing *soma* and raced in chariot meets, which they were invariably permitted to win. The "horse sacrifice" (*ashva medha*) was performed to

---

[1] N. R. Banerjee, *The Iron Age in India* (Delhi: Munshiram Manoharlal, 1965).

enhance a raja's domain and further prove his prowess. A great white stallion was turned loose and left to wander free for a year, followed by a troop of royal horsemen, who staked their king's claim to all the territory within boundaries surveyed by his stallion. At year's end the horse was driven home, first to be ritually mated with the king's wives, then killed and carved into quarters, symbolizing his universality and that of his monarch. As kings became more powerful, they were no longer content with the mere rank of raja, but assumed more pretentious titles, such as "great king (*maharaja*) and "ruler over all" (*samraja*). The earlier Rig Vedic tribal chieftains, little more than the first among many heads of families who met in regular council, bore as much resemblance to such monarchs as bamboo villages did to cities like Hastinapura.

The Satapatha Brahmana allegorically relates the eastward expansion of the Aryans as the spread of Agni's divine fire, consuming forests as he advanced, pausing only at broad rivers long enough for his devotees to learn to carry him across without destroying themselves in his flames. Videha Mathava was credited with having rowed Agni across the river Gandak, the natural border of what was then called Kosala (modern Uttar Pradesh); hence the region east of that wide river was named Videha (modern North Bihar), after its pioneer hero. We assume from this story that the Aryans used slash and burn techniques in clearing the jungle north of the river Ganga before applying their plows to the rich soil. Rice was by now cultivated in this alluvial monsoon region, and it is safe to conjecture that before long the Aryans would come to rely upon some system of irrigation in assisting its growth. Kosala and Videha were the kingdoms in which Rama and Sita, the hero and heroine of North India's shorter epic, the *Rāmāyana,* were born.

Less than one-quarter the size of the *Mahabharata,* the *Ramayana* is attributed to the poet-sage Valmiki. We assume it was composed sometime before 500 B.C., and its epic kernel actually seems to predate that of the *Mahabharata,* since no mention is made of any of the martial heroes of the latter in the *Ramayana,* though the story of Rama and Sita is recounted several times in the longer work. Opening at the Kosala capital of King Dasharatha, Ayodhya, from which the region of present day Uttar Pradesh derived its earlier name, Oudh, the *Ramayana* gives us many insights into the character of later Aryan court life. We learn of the machinations and endless intrigue of the aging king's three wives, who struggle and conspire to place their own sons in the position of heir apparent. We see how powerful a force religious law, or *dharma,* has become in dictating "proper" behavior, even for a monarch. As guardians and interpreters of that sacred lore, officiators of the royal sacrifice, the brahman priesthood maintained its special privileges and courtly influence.

In sharp contrast to the luxurious decadence of the court, however, we see the austere natural life of the hermit-sage, whose primitive forest home is a major alternative setting for the *Ramayana's* heroic "exiled" couple. The spiritual as well as economic interdependence between the settled city and the untamed forest regions of the still predominantly wild Gangetic plain helps us understand better the process through which the gradual socioeconomic transition of North India was effected in this epic era. The great sages, who seek only to practice their yoga undisturbed in the forests, are constantly attacked by "demons," whose evil king, Ravana, is as much the prototypical villain of Indian literature as Rama is its divine hero-king. Though Ravana's insular kingdom was named Lanka, it is doubtful that the island of Sri Lanka (Ceylon) had as yet been discovered or conquered by the Aryans. The demon king and his domain were probably based on pre-Aryan marauders closer to Oudh, the neighboring southern badlands region of Malwa perhaps, or possibly the peninsular Deccan. We know from the *Mahabharata* that the Aryans had by now probed south beyond the barrier of the Vindhya Mountains (which are from 1,500 to 3,000 feet high), since the tale is told of how a brahman sage named Agastya "humbled" that central Indian range. Jealous of high Himalaya, Vindhya had grown so big with pride that it blocked the very path of the sun, rousing the gods on their Himalayan abode to send their messenger, Agastya, who was Vindhya's guru, on a trip to the south. When Vindhya saw his guru approach, he naturally bowed low, and Agastya shrewdly asked him to remain that way until he returned—but never went back.

The entire *Ramayana* may be read as an allegory of Aryan and pre-Aryan conflict, culminating in the Aryan "conquest" of the south. While Rama and his lovely wife Sita, daughter of King Janaka of Videha, are in forest exile, Ravana kidnaps Sita and carries her off to Lanka. During the ensuing search for his wife, Rama seeks allies among the monkeys of the forest, whose leader, Hanuman, helps him find Ravana's fortress. The monkey army then link tails, improvising a bridge that Rama can cross to kill his demonic enemy and liberate his wife. Obviously, the Aryans lacked force enough to "conquer" South India by the martial invasion of armies launched from the north. There were no adequate communication or supply lines at this time, and the process seems, in fact, to have been closer to the one recounted in epic tale. The "turbulent frontier" of the jungle must have brought repeated raids to Aryan settlements, with robbery, rape, kidnap, or murder luring Aryan stalwarts to venture beyond the comfortable security of their homesteads in reprisal. Whether alone or in small bands, such heroes doubtless won both loyal allegiance and much-needed support from many of the tribesmen they encountered along the way, who soon learned from them the secrets of Aryan "civilization," including the manufacture of iron tools and weapons and the use of ploughs, seed, and water to ensure a grain surplus the year round. Hostile tribes were fought and ultimately subdued by the superior weaponry and self-confidence of the northern invaders, who naturally believed that the gods were on their side and fought with all the fury of Indra himself, hurling "thunderbolts" of iron at intractable "demons."

Such a process of expansion, settled agricultural production, and pluralistic integration of new peoples led to the development of India's uniquely complex system of social organization, which was mistakenly labeled the caste system by the Portuguese. For what the Portuguese called "caste" in the sixteenth century was, in fact, the ideal Rig Vedic "class" (*varna*) system of *brahmans, kshatriyas, vaishyas,* and *shudras,* whereas what Indians mean by caste is really a much more narrowly limited, endogamous group related by "birth" (*jati*). The actual social pattern that began to emerge in the later Vedic period, however, was a combination of both *varna* and *jati* systems. The classic threefold Indo-European tribal division into priests, warriors, and commons quickly proved inadequate, as already noted; while those three classes became the "twice-born" *brahman, kshatriya,* and *vaishya varnas,* the conquered Harappans were included quite early into the system as only once-born *shudras.* Or at least so we assume, for the origins of the *shudras* remain obscure and can only be inferred from their second-class legal status and from the fact that

they were defined by the brahmans as "servants" of the twice-born who
could be "exiled at will" or "slain at will." *Shudras* were not permitted
to hear or study, the Vedic hymns, which were considered potent
magic and reserved for twice-born ears only; indeed, later legal texts
prescribed "pouring molten lead" into the ears of any *shudra* caught
secretly listening to Vedic mantras. As more and different tribal peoples were absorbed within the spreading boundaries of Aryan society,
however, it soon became necessary to add still a lower class, one
whose habits or occupations were so strange or "unclean" that *shudras*
did not wish to "touch" them. Hence the emergence of those beyond
the pale of the four-*varna* system: the untouchables, also called
"fifths" (*panchamas*), or outcastes. So much for what can be inferred
as to the origins of the *varna* system. The roots of the more complex
jati system seem to be buried far deeper in India's pre-Aryan soil.

From the varying sizes of Mohenjo-daro floor plans in different
quarters of that ancient city, we can hypothesize a social hierarchy
based, perhaps, on a combination of occupational status and atavistic
fears of pollution through miscegenation or commensality. By the
later Aryan era, however, we do find detailed listings of specialized
occupations in Brahmana texts. These listings indicate more than a
mere proclivity to catalogue job descriptions, for these are sacred
books, and there seems to be a special sanctity as well as significance
to the type of work performed by various groups of people in Indian
society. We will subsequently note the emergence of occupational
guilds and the appearance of entire villages within which members of
a single occupation, whether weavers or fisherfolk, reside, geographically insulated from "polluting" neighbors. Not that occupation alone
can explain the origins of a caste system, whose reflexive ritual roots
of totem and taboo were surely at least as important as the type of
work performed or the nature of tools employed. Primitive fears of
losing power, whether through poisoning or castration, naturally dictated marriage only within the limits of a trusted group, whose *jati*
was close enough to one's own family to assure friendship and support. Tribal fears of losing "identity" or racial fears of losing "purity"
further complicated the causal pattern of development of a *jati* system
that was eventually to subdivide each *varna* into hundreds, and in
some cases thousands, of "castes" within India's subcontinent. The
actual pattern or patchwork quilt of social hierarchy that emerged
varied greatly from region to region since it was, in fact, a synthesis of

Aryan and pre-Aryan interaction. In South India and Bengal, for example, we find virtually no intermediary *varnas* between the conquering brahmans and their *shudra* converts to Brahmanism and the Aryan way of agricultural life.

Plough and irrigation agriculture greatly increased the food supply available to Aryan settlers, permitting rapid expansion of India's population as a whole and the growth of extended family units within villages as well as towns. Bonds of kinship and marriage alliances as well as economic interdependence linked the villages within each territorial kingdom, adding social strength and wealth to the most powerful monarch's domain. So the eastward and southern expansion and cultural synthesis continued, a constant blend of the "great" Sanskritic Aryan and "little" pre-Aryan traditions, of conquest and assimilation, or withdrawal behind walls of primitive magic fortified by fears of death, or worse still, of life devoid of tribal mana. Each Aryan household kept its sacred hearth, where sacrificial offerings were supposed to be made no less than five times each day by the brahman householder or his brahman priest or cook (cooking was the second most popular profession for a brahman, since no one feared taking food prepared by such sacred hands). Annually, the great *soma* altar was built, a painstaking ritual process that took the entire year to complete, and one that only the great and wealthy could afford, hiring dozens of brahmans to pour the libations and chant the mantras that were, by this Brahmana age, exalted above the very gods themselves. The fire sacrifice was believed the magic key to cosmic power, holding within its sacred ritual the utterance of syllables pregnant with meaning as mighty as *dharma,* capable of prolonging life by inducing rainfall. The *soma* sacrifice was timed to end just before the monsoon usually began.

There were years, however, when the rains came before the ritual altar was quite ready, or, more disconcertingly, failed to come for days or weeks after the fires had been fed and all the brahmans had been paid their herds of cattle. At such times the mighty rajas and *kshatriyas* and the wealthy *vaishyas,* who had expended so much on sacrifices that failed, must have questioned their validity, or at least questioned the powers of the brahmans. The optimism and self-confidence of the early Aryans were, in fact, gradually eroded as settlement brought little salvation and suffering overshadowed joy, even as death lingered along the brightest high roads of life. During

Rig Vedic times the dead had been buried in their "little clay houses,"
from which they either ascended to Yama's heaven to join the shining
solar deities, or fell to the unreal (*asat*) darkness of demonic chaos
below. By the Brahmana age the idea of hell was elaborated upon, as
was the concept of heaven, but whether in emulation of the sacrificial
fire ritual or to purge the spirit of its bodily dross, even as iron or
gold were purged, cremation became more common as centers of
Aryan settlement moved east. A new theory also emerged at this
juncture (new at least to Vedic thought, for it may well have been
pre-Aryan in origin), positing that one might experience such excru-
ciating pain and torments in hell after "death" that he might reawaken,
only to suffer again. Another Brahmana text of this time notes that
the "food" consumed by a man in this world will "consume him" in
the next, anticipating, it would seem, the ideas of *karma* (action) and
samsāra (metempsychosis) that were later to become axiomatic to
Hinduism. Thus, while the cities and villages of Aryavarta reverberated with the froglike chants of brahmans and the air above the
Ganga filled with smoke from funeral pyres and altars, some wise
men watched in silence and wondered.

Was it the wastefulness of resources that appalled them most?
The banality of time-worn ritual and uninspired mantras? Or the
venality of those all-too-worldly priests who officiated? Perhaps it
was the futility of it all, for where was the raja who never died? Or
the rich man who never fell ill? Or the strong man powerful enough
to survive the blow of a tiger's claw, or a mad elephant's tusk? By
about 700 B.C. such disenchantment with Brahmanic ritual and with
the established system it represented had alienated at least some of the
Gangetic plain's most brilliant minds, who wandered in the forests as
hermits, meditating and teaching their wisdom to worthy disciples,
seeking solutions to problems exacerbated by recent change, yet as
timeless as Indian history itself:

> From the unreal [*asat*] lead me to the real [*sat*]!
> From darkness lead me to light!
> From death lead me to immortality!

Such was the threefold quest of the Upanishadic mystics, 108 of
whose poetic-philosophic dialogues survive as the final fruit of re-

vealed Vedic scripture, representing the orthodox intellectual revolt against Brahmanism that emerged in the eastern Gangetic plain in the eighth century B.C. *Upanishad,* which literally means "to sit down in front of," indicates how these speculative messages were transmitted by gurus (predominantly *kshatriya,* it would seem) to student disciples in forest "seminars." The *kshatriya-brahman* struggle for *varna* primacy may, therefore, be seen as a social counterpart to the religio-philosophic "revolt" against orthodoxy that the Upanishads represent. Without rejecting Rig Vedic mantras or sacrifices as possible aids to salvation, the authors of the Upanishads stressed a different path toward the attainment of *moksha* ("release"), which hereafter becomes the ultimate goal of Vedantic meditation. When the sage Yājñavalkya is ready to leave his household to become a mendicant forest wanderer and asks his wife what "final settlement" she would like, all Maitreyi requests of her husband is the "secret of immortality." Then he expounds upon the "reality" of *ātman* ("soul"), for "with the understanding of *atman,* this world-all is known." Yet paradoxically, that real essence is invisible, "as, when a drum is being beaten, one would not be able to grasp the external sounds, but by grasping the drum or the beater of the drum the sound is grasped." In the Rig Veda *atman* simply meant "breath," and it must have been the association of breath with life, and its cessation with death, that led to exalting its status, first to mean "self," then "soul," the animating universal element. Many students, however, found it difficult to accept anything they could not see as the cosmic essence. Young Svetaketu, a "conceited" youth who thought himself wise after having studied with conventional brahmans for twelve years, was such a skeptic, but his father instructed him in the "reality" of *atman* by asking him to "bring hither a fig." The dialogue continued thus:

"Here it is, sir."
"Divide it."
"It is divided, sir."
"What do you see there?"
"These rather fine seeds, sir."
"Of these, please, divide one."
"It is divided, sir."
"What do you see there?"

"Nothing at all, sir."

"Verily, my dear, that finest essence which you do not perceive —verily, my dear, from that finest essence this great Nyagradha [sacred fig] tree thus arises. Believe me, my dear, that which is the finest essence—this whole world has that as its soul. That is reality. That is *atman*. That art thou, Svetaketu."

The last line equates the cosmic soul with the individual soul, offering the Upanishadic key to mastering reality and controlling cosmic forces and events: learning to "understand," that is, master and control our real selves. Such mystic knowledge and understanding could be guaranteed no one, however; it was intuitive wisdom that might come in a flash of "blinding" illumination or might never come at all, for learning its mystery was as hard as walking barefoot across "a razor's edge." Yoga exercises, postures, breath, and diet, helped prepare the mind for receiving such wisdom, just as studying Upanishadic dialogues did, but nothing tangible assured release, or "freedom from desire," and "liberation" from all aspiration, self-conceit, and rebirth, the quiescent, mindless, motionless inactivity of *moksha*. Women as well as men joined in this strenuous struggle to comprehend the meaning of reality before it was too late. Some of the most difficult questions are asked by Upanishadic heroines like Gargi Vācaknāvi, whose ever curious mind drove the sage Yājñavalkya to lose his temper. "On what, pray, is all this world woven, warp and woof?" asked Gargi. The wise man began by answering merely "wind," but Gargi only asked the same question of wind, and so on through "the worlds of the sun," "worlds of the moon," "worlds of the stars," and "worlds of the gods," till finally Yajnavalkya was driven to admit that all was based upon "the worlds of Brahma."

"On what then, pray, are the worlds of Brahma woven, warp and woof?"

"Gargi," the sage replied, "do not question too much, lest your head fall off."

From the primacy of "sacred utterance" and the divinity of Brahma emerges the neuter monistic principle, Brahman, whose transcendental spirit pervades the cosmos, and is equated in the

Upanishads with an individual's *ātman*. Truly to understand reality, therefore, one must appreciate the identity of that cosmic equation whose distilled Vedantic essence is the Sanskrit formula *tat tvam asi,* "thou art that one." *Tat,* the third person singular pronoun ("that one") stands for Brahman; *tvam,* the second person singular ("thou") represents *atman;* and *asi* is merely the present singular form of "to be." For thousands of years Hindus would continue to ponder that seemingly simple equation, trying to grasp its mystic meaning. Whether we call it pessimism or realism, the Upanishadic view of this world and its inhabitants was ultimately negative. As old King Brihadratha, the hermit sage, put it, "in this ill-smelling, unsubstantial body, which is a conglomerate of bone, skin, muscle, marrow, flesh, semen, blood, mucus, tears, rheum, feces, urine, wind, bile, and phlegm, what is the good of enjoyment of desires?" That very problem was infinitely compounded now by the fully developed belief that there was an endless cycle of existence" (*samsara*), of rebirth, re-death, and rebirth, whereby man found himself "like a frog in a waterless well," while the whole world decayed, great oceans dried up, and mountain peaks fell away, even as gnats and mosquitos were born to perish. The one hope, the only "way of escape" was through knowledge of the mystic identity of self and all. To "control" the universe, then, one merely had to control oneself; to "escape" universal pain and suffering, one would merely need release from the snares that entangled one's *atman,* even as the vines and weeds of a forest entangle a tree. Desires, deeds, "action" (*karma*) of any sort came now to be considered hindrances, snares, delusory traps in the soul's search for *moksha,* and the Law of Karma emerges, linked to the concept of *samsara,* as a distinguishing axiom of Indic civilization.

That "law" posited that every action, good and evil, had repercussions, bore fruit in kind at some future time. The sum of our past individual *karma* thus determined our present lives, even as our current behavior would dictate our future conditions of existence, and not only in this lifetime. *Karma* alone could tell how many lives we might be condemned to relive on the chain of cyclical suffering to which we were bound. If one were evil enough, he might be reborn as a mosquito! Good *karma* was naturally better than bad, and simply to be born human was considered a great advantage—far superior to the world of demons, insects, and most animals—and the higher one's *jati* and *varna,* the better. But virtue, like wealth and power, had its own problematic rewards, and the yogis of the Gangetic plain looked beyond such "pleasures" to the pure bliss of eluding all mortal pain. That bliss was described in the Upanishads as "a deep dreamless sleep," and the perfect release "beyond," even as Brahman was described negatively as "not this, not this" (*Neti, neti*). Like human life, time was deemed cyclical, passing through *kalpas* ("epochs") and *yugas* ("ages") of millions of years, each of which is measured as but a moment in the life of Brahman, whose years are unlimited. What we "saw" was but "illusion" (*māyā*), to which we gave "name and form" (*nāma rūpa*) because of our ignorance of its unreality. One Upanishad called the "illusion-maker" the "Mighty Lord" (*Maheshvara*), a name later used for Shiva, whose yogic powers and creative force may have been equated by this time with control over life and death. Punjabi Aryan optimism and faith in the shining *sat* (real or true) seems thus to have been displaced in the woods of Bihar by a pessimistic belief in the pre-Aryan darkness, chaos, or nonbeing, the *asat* of demonic Vritra, which now more closely resembled the ultimate goal of Indian reality than did either the world of mortals or gods.

"Verily, freedom from desire is like the choicest extract from the choicest treasure," notes the Maitri Upanishad. "In thinking 'This is I' and 'That is mine' one binds himself with himself, as does a bird with a snare! Hence a person who has the marks of determination, conception, and self-conceit is bound. Hence, in being the opposite of that, he is liberated. Therefore one should stand free from determination, free from conception, free from self-conceit. This is the mark of liberation [*moksha*]. This is the pathway to Brahma here in this world. This is the opening of the door here in this world. By it one will go to the farther shore of this darkness, for therein all desires are contained."[2]

By the sixth century B.C., Buddhist sources named sixteen major kingdoms and tribal oligarchies in North India, from Kamboja in Afghanistan to Anga in Bengal. The most powerful of these *mahājanapadas* ("great tribal regions") were Magadha and Kosala, the former commanding the eastern Gangetic plain south of the Ganga, the

[2] *Maitri Up.* VI. 30 in Robert Ernest Hume, *The Thirteen Principal Upanishads* (London: Oxford University Press, 1931), pp. 442–43.

latter controlling the domain slightly west of Magadha and north of the great river artery of Aryan settlement and trade. Kosala was the region that contained the former epic capital of Ayodhya and absorbed the once independent *janapada* of the thriving river city of Kāsī (later called Vārānāsi and then Banaras, the "capital" of Hindu worship). With its capital at Srāvastī near the Himalayan foothills, Kosala had its central base of power just west of the Sākya tribe, whose most famous member, *Sākyamuni* ("Sage of the Sākyas") Siddhartha Gautama, the *Buddha* ("Enlightened One"), was born in Kapilavastu around 563 B.C. Many hill tribes like the Sākyas came within the ever expanding orbit of Aryan suzerainty at this time, retaining their tribal identity and independence for the price of tribute, or what would later be called taxation. Tribal princes like Siddhartha, at once fascinated yet repelled by the "civilized" trappings of Aryan ritual and rule, were driven to search for solutions to the same sort of problems tackled by Upanishadic teachers. When he was about thirty, Gautama left his sumptuous life of ease and wandered for the next six years in the woods of Kosala and south through the kingdom of Magadha, before attaining the enlightenment that changed his name and bequeathed to the world one of its great religious philosophies.

Magadha, with its capital at Rājgīr ("The King's House") commanded not only the eastern Gangetic trade, but the rich mineral resources of the Barābar Hills, where iron lay encrusted on easily accessible outcroppings. Expanding east as well as south, Magadha soon absorbed the Bengali *janapada* of Anga into its regal orbit, and by the reign of Bimbisara (ca. 540–490 B.C.), the Buddha's great patron, it was the wealthiest as well as the most powerful kingdom in North India. Magadhan princes, ironsmiths, and wealthy merchants must obviously have felt less indebted to brahman prayers and sacrifices for their power and riches than Punjabi farmers had, and Bimbisara himself found the Buddha's heterodox rationalism a much more attractive doctrine than Brahmanism for reasons of logic as well as state. One story preserved in the Buddhist Pali Canon tells how the Buddha won Bimbisara's favor by asking a pompous court brahman, who had just insisted the king should sacrifice fifty of his finest goats without concern since "whatever he sacrificed went directly to heaven," whether his father was alive. The brahman replied that he was, whereupon the Buddha inquired, "Then why not sacrifice him?",

saving the king's favorite goats and driving the brahman from court.
The Buddha rejected the brahmans' pretentions to inherited piety, insisting that only a person who "behaved as a brahman should" de-
served to be treated like one. Substituting his monastic order of virtuous conduct, nonviolence, and poverty for Brahmanism's priestly
monopoly of magic and wealth, the Buddha bolstered rising *kshatriya*
and *vaishya* expectations and launched his great peaceful revolution
in Kosala as well as Magadha.

In a deer park at Sarnath, on the outskirts of Kāsī, the Buddha
was believed to have set his "wheel of the law" (*dhamma;* Pali for
dharma) in motion about 527 B.C. by preaching his first sermon after
achieving enlightenment. That sermon on the four noble truths embodied his message and was to become the philosophic core of
Theravada ("Teachings of the Elders") Buddhism, or what would
later be called *Hināyana* ("Lesser Vehicle") Buddhism by the post-
Christian era *Mahāyāna* ("Greater Vehicle") Buddhists. The first
noble truth was "suffering" (*dukkha*), and how all existence was in-
exorably bound up with it. From birth to death, through sickness and
old age, sorrow was everywhere, gaining poignance in separation
from those we love, intensified by proximity to those we hate; no facet
of ordinary life could escape it. The second noble truth was "ignorance" (*avidyā*), the basic cause of suffering. It was no lesser ignorance than that of the fundamental nature of reality, and here, as
in the pessimistic tone of his first truth, it seems that the Buddha spent
some of his wandering years listening to Upanishadic sages. His
threefold definition of the nature of reality is, however, significantly
different from any orthodox or other major contemporary heterodox
school, for he posits a "sorrowful," "transient" (*anicca*), and "soulless" (*anatta*) world. The latter most sharply differentiates Theravada
Buddhism from either idealistic Upanishadic Brahmanism or Jainism,
and while we know of several materialistic schools prevalent in North
India at that time (the *Ajīvikas,* "Non-Soul," and *Chārvāka* or
Lokāyata, "People's"), virtually none of their teachings have survived
the ravages of time or religious opponents.

Had we the wisdom to understand reality's soulless, transient
misery, the Buddha went on to explain in his sermon at Sarnath, we
would be able to elude suffering, or at least diminish it, since our "will
to exist" would weaken, as would the passions of our sense organs

and our eagerness for "contact," "sensation," "craving," "clinging," "becoming," and "birth," all of which chain us to the wheel of cyclical suffering, reborn and redying. In what may have been borrowed from India's precocious medical thought of this time, the Buddha prescribed as his third noble truth that any *"ill"* which was understood could, in fact, be cured. (We know of at least one great Magadhan physician, Jīvaka Komārabhacca, who journeyed more than six hundred miles from Rājgir to Ujjain in the sixth century B.C. to cure the Avanti king, Pradyota, and was handsomely rewarded for his success.) The fourth and final of the Buddha's truths was the noble eightfold path to the elimination of suffering: to hold, practice, and follow right views, right aspirations, right speech, right conduct, right livelihood, right effort, right mindfulness, and right meditation. The difficulty, of course, was in properly defining "right," but if one trod the eightfold path without misstep, the goal of *nirvana* (which literally meant "the blowing out," as of a candle's flame), could be achieved, and the pain of suffering would finally be overcome. *Nirvana* was thus the Buddhist equivalent of *moksha,* a "paradise" of escape rather than pleasure.

The Buddha spent the next forty-five years of his life teaching these four noble truths to disciples who gathered round him in such numbers that he soon established a monastic "order" (*sangha*), which continued to grow and was destined to spread throughout the world after his death. Initially, only men could join the *sangha,* and the vow of chastity (*brahmacārya*) was as important as those of nonviolence (*ahimsa*) and poverty (*apārigraha*)—three vows that would become integral to Hindu concepts of piety (all three were taken by Gandhi during the latter half of his life). Nuns were admitted to the *sangha* shortly before the Buddha died, but he was always doubtful about the nature of female influence upon his monks. When Ananda, his foremost disciple, asked how best to behave with women, the Buddha advised "not to see them." Yet what if one *happened* to see them? Then it was best "not to speak to them." But suppose it was impossible to avoid speaking with them? "Then keep alert, O Ananda!" the Buddha warned.

All members of the *sangha* pursued a rigorous course of "right discipline" (*sila*), yogic concentration, and thoughtful study in their search for *nirvana.* Not only did they have to abandon all family

bonds and prospects of progeny, but they were daily enjoined to beg
for their food, bestowing "merit" upon those who placed rice in their
bowls, which would hereafter remain an Indian symbol of virtue
rather than shame. With heads shaved, the saffron-robed, barefoot
disciples of the Buddha marched the length and breadth of the Gangetic plain, teaching his message of moderation, nonviolence, and love
for all creatures. The idea of monasticism achieved such popularity
that it attracted religious leaders in other parts of the world, spreading west to the Near East and thence to Europe, wandering north and
east to China and Japan. Though Buddhist monastic orders in India
never acquired martial power or wealth approaching that of their
counterparts in China and Japan, they did become a formidable
ideological force against Brahmanism and attained great political
significance in Magadha.

One Buddhist text tells how King Ajātasatru, Bimbisara's son
and heir, went to call upon the Buddha, who was "lodging" in a
Magadhan mango grove with no fewer than 1,250 disciples. Just before his elephant reached the grove, however, the king became fright-
ened, fearing a trap, since he heard "no sound," and found it almost
impossible to believe that 1,250 people could assemble anywhere
without a "sneeze or cough." Yet such, supposedly, was the disciplined strength of the sangha. Before he died at the age of eighty (ca.
483 B.C.) in Kusinagara, the Buddha was virtually worshipped by his
disciples, who prayed to him daily for "refuge." Yet, when Ananda
sought his advice about how the sangha should be administered after
he was gone, the Buddha's final message was, "You must be your
own lamps, be your own refuges. Take refuge in nothing outside
yourselves. Hold firm to the truth. . . . Whoever among my monks
does this will reach the summit. All composite things must pass
away. Strive onward vigilantly."

Another kshatriya prince who founded an order of monks and
rejected Vedic and Brahmanic authority was Vardhamāna Mahāvira
(ca. 540–468 B.C.). Born in Besarh, near the modern city of Patna,
Mahavira (literally "Great Hero") was a son of the chief of the Jñā-
trika tribe, and like the Buddha, he abandoned his hedonistic life to
become a wandering ascetic when he was about thirty years old. He
seems first to have joined a sect of nudist ascetics called the Nirgranthas ("Free from Bonds") with whom he remained for about ten
years. The founder of the Nirgranthas was named Pārshva and may

have been Mahavira's guru. After his death Mahavira and his own followers split from the parent group to establish a new sect, the Jainas. *Jaina* means "follower of *jina* (conqueror)," the honorific title bestowed upon Mahāvira for what must have been his remarkable powers of self-control. He also came to be known as the twenty-fourth *tirthankara* ("ford maker") after the Jain canon was finally recorded at the Council of Vallabhi in 454 A.D. The *tirthankaras,* who have "crossed over the waters" to preach the faith, are the Jain equivalents of gods, and Mahavira was the last of them; Pārshva was the twenty-third. Mahavira adhered to a more extreme form of asceticism than did the Buddha. He not only went naked, but also advocated and practiced self-torture and death by starvation as the surest paths to salvation. Though it took him thirteen years from the time he resolved to starve himself to death before he finally succeeded in doing so, Mahavira was supposedly the last Jain to attain the pure and perfect peace of *tirthankara* paradise, the topmost level of release from mundane matters, which resembles *moksha* and *nirvana* in its total lack of karmic activity.

The central doctrine of Jainism is that all of nature is alive: everything from rocks and earthworms to gods has some form of "soul," called *jiva.* Like *atman,* all *jiva* are eternal, but in contrast to Upanishadic idealism, there is no Jain equivalent to the infinite cosmic *atman,* only a finite number (millions of billions) of various degrees of *jiva,* some much more powerful than others. The *jiva* of a rock, for example, has only one "sense," that of touch, and is thus far weaker than the *jiva* of a man, god, or hell-dweller, which commands the multisensory capability of touch, taste, smell, sight, hearing, and mind. According to Jainism, *jiva* are not themselves created by any divinity, but have always existed as an eternal cosmic pool of souls. At the dawn of the two-*kalpa* cycle of Jain time, all *jiva* are quiescent and free, unencumbered by "particles of matter" (*pudgala*) that ensnare them with invisible karmic nets and thereafter burden and tarnish them through the trillions of years it takes for the cycle to run itself back to quiescence.

Almost as important a Jain doctrine as the concept of *jiva* is that of "nonviolence," *ahimsa.* It is probably thanks to Jainism that *ahimsa* became so significant an aspect of later Hinduism, though Buddhism also prohibited the killing of any living creature, and Buddhist monks, like their Jain counterparts, were required to take vows of nonviolence. Mahavira's "pure unchanging eternal law" was that "all things breathing, all things existing, all things living, all beings whatever, should not be slain, or treated with violence, or insulted, or tortured, or driven away." The Jain prohibition against destroying life was so complete that to this day devout followers of Mahavira's religion wear masks to guard them against inhaling invisible organisms and carry dusters to whisk other invisible *jiva* from chairs or floor mats before they sit down. Like Buddhism, Jain philosophy soon acquired all the characteristics of a religious faith, practiced first by an order of male monks, joined later by nuns and a supportive lay community. An important economic result of Jain nonviolence was that even lay members of the community rejected agriculture for fear of ploughing under living things and turned instead to commerce and banking, nonviolent occupations, and often lucrative ones. The Jain community, centered primarily in Gujarat, soon became quite wealthy and remains one of the great mercantile communities of modern India. Paradoxically, the only living being a devout Jain was encouraged to "kill" was himself, through starvation, though such a death would be viewed as the liberated "birth" of one's hitherto entrapped *jiva*. More than two thousand years after Mahavira's suicide, Gandhi was to revive the fast-unto-death as a political weapon.

The emergence of Gangetic kingdoms, the growth of cities and towns, the development of North Indian communications and trade routes, as well as the founding of new schools of heterodox religious philosophy, were thus all products of the impact of later Aryanization upon the indigenous tribes and mores of the eastern Gangetic plain. Time and distance served not only to weaken the original fiber of Aryan ritual and tribal pastoralism, but they transformed the very meaning of the word *Aryan* itself from a basically linguistic family to a fundamentally socioeconomic pattern of cultural settlement. By the end of this era of dynamic change, Brahmanism had bogged down in the flood plains of Kosala and Magadha, which themselves were rapidly stripped of their timber to wall mighty cities that almost seemed reincarnations of the great urban monuments that once lined the Indus. By the fourth century B.C., the warring kingdoms and tribes along the Ganga were to be unified under a single imperial umbrella, one far grander than any Harappan empire had ever been.

FIVE

INDIA'S FIRST IMPERIAL UNIFICATION

(326–184 B.C.)

Although, by the sixth century B.C., Magadha had emerged as first among many competing kingdoms and tribal confederacies of the Gangetic plain, it took two centuries after the reign of Bimbisara for that richly endowed region to assert control over the subcontinent. Punch-marked Magadhan silver coins found at Taxila in the northwest proved that trans-Indo-Gangetic trade continued steadily from the reign of Ajātasatru, who died about 459 B.C., through the succeeding Magadhan dynasties of Sisunāga and Nanda. The wealth of the latter dynasty, whose monarchs were supposedly "usurpers" of *shudra* origin, was reported to have been far greater than that of its predecessors, attesting to Magadha's growing economic capacity for asserting imperial dominance under properly inspired leadership. The monarch who provided that leadership was Chandragupta Maurya (reign, 324–301 B.C.), whose North Indian unification came in the wake of Alexander the Great's catalytic invasion of the Indus in 326 B.C. Just as the Achaemenid Empire of Cyrus the Great (558–530 B.C.) seems to have inspired his contemporary, Bimbisara, to establish the kingdom of Magadha, Alexander's dream of world unification might well have sparked India's first great unifier to transform Magadha into the Macedonia of South Asia.

The Gandharan region of India's northwest, whose capital was Taxila (near Islamabad, the capital of Pakistan), fell under Persian control in 518 B.C. As the twentieth satrapy of Darius' Achaemenid Empire, India paid an annual tribute of no fewer than 360 talents in gold dust, acording to Herodotus, who told many fabulous tales of India in his account of *The Persian Wars,* including descriptions of giant gold-digging ants who labored in gold-strewn deserts. Such tales must have stimulated Alexander's appetite for venturing beyond the Indus. The "ivory, apes, and peacocks" of "Ophir" had, by Biblical authority, been brought to King Solomon's temple by sea and were further evidence to the Western world of exotic riches in that alluring land beyond the Persian East. The glory, power, and wealth to be won by conquering so vast and distant an empire thus inspired the world's greatest young general to shatter Persian power and hurl his mighty army across the river Indus, north of Attock, in the spring of 326 B.C. The raja of Taxila, a wise ruler named Ambhi, welcomed Alexander's irresistible force without firing an arrow and opened the gates of his border city to the invading horde, which was estimated at from 25,000 to 30,000 Macedonian cavalry and Asiatic Greek foot soldiers.

Advancing east with his army, Alexander was soon confronted with another river, the Jhelum (Hydaspes), beyond whose banks lay the domain of the great Aryan Raja Porus (Puru), who vainly tried to defend his homeland against the invaders. Porus had two hundred war elephants, India's natural tank corps, but Alexander's undefeated cavalry charged around their flanks, while flaming Macedonian arrows drove them to turn in terror and crush Porus' infantry, which was massed behind them. (The combination of infantry and elephant wall has traditionally served as India's two-edged first line of martial defense.) Following this crushing defeat, no Indian raja seriously contested Alexander's advance, but as he moved east across the Punjab, rebellion threatened his garrisons to the rear, first in Kandahar and then in Swat. By the time Alexander reached his fifth great river, the Beas (Hyphasis), he had apparently heard of the wealth and power of distant Magadha and was anxious to venture on to the "Eastern Sea," which may have been the Ganga. According to Justin and Plutarch, Alexander met a "young stripling" at this time named "Sandrocottus," who has since been equated with Chandragupta and who may, in fact, have been the future founder of the Mauryan Empire. At the Beas, Alexander's army resisted any further march to the east; the mighty general finally bowed to the wishes of his troops, and, late in July of 326, he turned back toward the homeland that he never lived to see again. As the tide of Macedonian power receded, that of a revitalized Magadha spread west under Mauryan rule.

Though we know surprisingly little about the origins of India's first imperial family, the name Maurya (in Pali, *Moriya*) was probably derived from the word for peacock, which may have been the clan's original, pre-Aryan totem. Some accounts say that Chandragupta was the son of herdsmen, however, and others that his mother was in the royal harem of the Nandas, but whatever his family stock may have been, it was obviously exceptional, since the dynasty he founded ruled over most of India for 140 years. It was long believed that the genius behind young Chandragupta's remarkable rise to power was an older brahman prime minister, Kautilya (or Chānakya), who has also been credited with authorship of the *Arthāshastra* ("Science of Material Gain"), a textbook in realpolitik that is reminiscent of Machiavelli's *Prince*. The Indian ideal, at least, of sage brahmanic "control" to temper and guide the impulsive actions of a youthful conqueror is thus traditionally established, whether or not it was actually practiced during Chandragupta's reign.

Thanks to Thomas Trautmann's recent painstaking statistical analysis of the *Arthashastra* text and its most common Sanskrit words,[1] we know that Kautilya could not possibly have written the entire, many-layered work, which was probably completed around A.D. 250. He may, however, have written an early part of the text, contributing his ideas and talent—as doubtless other brahman ministers or lesser bureaucrats in the service of many Indian monarchs did—to this summary statement concerning the nature, politics, and economy of the ancient Indian state. While the *Arthashastra* must not, therefore, be used alone as a primary source of early Mauryan polity, it does provide invaluable evidence concerning the practical operations as well as the long-range goals and ideals animating India's kingmakers and kings for many centuries. The work begins with a chapter on the education and training of a king, enjoining him to be "energetic" and "ever wakeful." When at court, the raja is advised never to keep his petitioners waiting at the door, for a king who makes himself "inaccessible to his people" is sure to "create confusion in business, and

[1] Thomas R. Trautmann, *Kautilya and the Arthāshastra; A Statistical Investigation of the Authorship and Evolution of the Text* (Leiden: E. J. Brill, 1971).

cause public disaffection." The monarch must learn to control his senses, especially those "six enemies," lust, anger, greed, vanity, hautiness, and exuberance. He must also control his subjects—particularly powerful ministers, wealthy merchants, wise brahmans, and beautiful queens—and to help him in such a difficult task, he must hire an army of spies. Spying seems, indeed, to have been a major occupation of early Indian bureaucracy, and the *Arthashastra* specified that spies are to function in various "guises," such as pseudo-students, priests, householders, saints practicing renunciation, merchants, poisoners, female mendicants, and other such "artful persons."

To sustain his army of spies, soldiers, and civil bureaucrats, which at the peak of Mauryan power probably totaled more than a million men, the Indian monarch claimed a share, usually one-fourth, though sometimes as high as one-half, of the value of all crops raised throughout his domain. Trade, gold, herds, and other forms of wealth were also taxed, but since by this time most Indians worked as agriculturists, settled in villages "of not less than a hundred and not more than five hundred families," land revenue became and was to remain the mainstay of all Indian kingdoms and empires, including the British. In fact, it had probably already provided the Nanda monarchs of Magadha with the wealth that helped them expand their boundaries before Chandragupta ousted the last of their line from his throne at Pātaliputra (modern Patna) sometime between 324 and 322 B.C. Pātaliputra seems to have been the largest and greatest city in the world during Mauryan rule. The great Magadhan capital, commanding the south bank of the Ganga just east of its confluence with the Son, was no less than eight miles long and a mile and a half wide, surrounded by a timber wall with 570 towers and a moat 600 "cubits" (900 feet) wide and 30 deep. The municipality was administered by six boards of five men each, the traditional *panchayat* ("five-member") council of elders that governed India's villages, guilds, and castes, as well as towns and cities. We get a tantalizing glimpse of the sophistication and complex vitality of the Mauryan capital from the names of its governing boards, all that is preserved from fragments of the diary of the Greek ambassador, Megasthenes: industrial arts, trade and commerce, tax collection, foreigners, vital statistics, and the maintenance of public places, including markets and temples.

From the *Arthashastra* we can appreciate how shrewdly Mauryan officials were chosen and how carefully they must have been observed and supervised. Government servants "should be constantly kept under vigilance in their duties" that textbook on bureaucratic management advised, for men are "by nature fickle and temperamental." Officials should work without "either dispute or unity among themselves," since if they work united, "they consume state revenues," but if they work in discord, "they damage the work." Bureaucrats should work only "as directed" and do "nothing" without the knowledge and approval of superiors. Bureaucracy, with all its mixed blessings, was thus obviously no recent Western import to Indian soil, and may have had indigenous roots in Harappan society. Megasthenes observed seven "classes" in Mauryan India, the highest being the royal councillors, whom he ranked above brahmans; the others were agriculturists, herdsmen, soldiers, artisans, and spies.

Chandragupta spent the last quarter century of his life in consolidating his grip over all of North India, extending Magadha's power to the Indus and beyond. In 305 B.C. we know that he concluded a treaty with Seleucus Nikator, Alexander's Greek heir to western Asia, fixing the western border of the Mauryan Empire along the spine of the Hindu Kush Mountains. In return for the Greek withdrawal of all forces in the Punjab to the Iranian plateau, Seleucus received five hundred war elephants from Chandragupta and exchanged ambassadors with Pataliputra. There was a cryptic "marriage" clause in that treaty as well, and though it is not clear whether Seleucus sent one of his daughters to Chandragupta's court, later reference to *Yāvani* (Greek) women serving as an "elite guard" over Chandragupta's bedchambers, indicates that possibility.

The Mauryan Empire was divided into *janapada* "districts," which reflected earlier tribal boundaries and were administered by the emperor's closest relatives or most trusted generals. The Mauryan army was organized in four major corps, whose strengths were reported to have been six hundred thousand infantry, thirty thousand cavalry, eight thousand chariots, and nine thousand elephants. Even if those figures were exaggerated or valid only during peaks of imperial power, the support of so huge a force attests to the state's remarkable size and centralized administration. Estimates of India's population at this time are at best conjecture, yet based on so large a military and civil bureaucracy, it seems fair to assume that there were close to fifty million people in South Asia by the third century B.C.

To encourage the clearing of forest lands, tax remissions were

granted, and tribal areas were settled with state support, mostly by
shudras, whose status was little higher than slaves. The *Arthashastra*
cautioned that a "denuded treasury" posed a "grave threat to the
security of the state," advising the monarch against too many remissions of land revenue. The Mauryan state owned and operated all
mines, as well as such vital industries as shipbuilding and armament
factories. Large centers of spinning and weaving were also owned by
the state, by which was meant, at least theoretically, the king, for the
polity described in the *Arthashastra* was a socialized monarchy, with
strict enforcement of working regulations imposed on artisans and
professionals alike. Workers who "failed" to fulfill "contracts" were
subject to be fined one-quarter of their wages and could further be
"taxed for damages" up to twice their wage. Washermen (*dhobis*)
who beat the laundry they washed on any but "smooth stones" could
be punished, as could physicians (*vaidyas*) whose patients died of
neglect or careless treatment. Weights and measures, as well as currency, were all state controlled. The standard coin of the Mauryan
realm was the silver *pana,* minted at 3.5 grams. A king's councillor
received 48,000 *panas* as his annual salary, after which the scale fell
precipitously to 1,000 for engineers, mining superintendents, and
military officers; 500 for soldiers of the line and spies; 120 for carpenters and other skilled craftsmen; and 60 for the unskilled laborers
at the bottom of the economic ladder. We assume that the last-named
wage barely sufficed to feed and clothe a man and his immediate
dependents.

The Mauryan state and its king had thus evolved dramatically
from the rustic monarchy of preceding centuries. While in theory the
king owned all land and wealth, there were, in fact, large tracts of
tax-free property, either temporarily exempted from revenue, as in
the case of newly cleared domain; or permanently relieved, as was
religious property, brahmanic as well as Buddhist and Jain, or gifts
bestowed upon valiant servants of the crown and their heirs. Many
artisan and merchant guilds (*shreni*) were also privately owned
corporate bodies. *Shreni* exercised judicial autonomy over their members much the way *jati* did, establishing enclaves in towns and cities,
the way castes had within villages, ruled by their councils of elders
and headman, even as villages were. An interdependent private enterprise and state-controlled economy has thus existed for more than two

thousand years in India. *Shreni* are also mentioned as military units, and the term appears to have been used generally to differentiate any substantial, cohesive Mauryan "group," somewhere between a tribe and a *jati* in size, though it was later confined to its primary meaning of "guild." Some *shreni* were used to colonize wild or waste lands; sturdy paramilitary peasant craftsmen, their social bonds and manual talents made them excellent guardian-settlers of the Mauryan frontiers.

The *Arthashastra* also expounded India's classic *mandala* ("circle") theory of foreign policy, according to which the domain of the "conqueror king" was viewed as the center of twelve concentric rings, with the territory immediately adjoining the king's called that of "the enemy," and that just beyond, the land of "the friend." Such realpolitik definitions continued with "friends of the friend," and so on, to the last two rings, where the "intermediate king" and the "neutral king" resided. It was most important for the conqueror to prevent the intermediate monarch from allying with his enemies, or, ideally, to convince him of the wisdom of joining his side. The neutral king was generally remote enough to remain aloof, though he controlled sufficient force to prove dangerous should he ever be provoked or tempted to join the fray, as was distant Alexander.

According to Jain tradition, Chandragupta abdicated his throne in 301 B.C. to become a Jain monk in South India, where he fasted until his death, while his son, Bindusara, took control of Pātaliputra. We know surprisingly little of Bindusara's thirty-two-year reign, though he expanded the limits of Mauryan power south of the Vindhyas and continued to maintain diplomatic relations with his Greek neighbors to the west. He is perhaps best remembered for his curious request of Seleucus' successor, Antiochus I, to whom he wrote once asking for some Greek wine, figs, and a Sophist. Antiochus sent the figs and wine, but politely explained that there was as yet no market in philosophers. Bindusara's greatest contribution to Indian history was his son, Ashoka ("Sorrowless"), who reigned from 269 to 232 B.C. We have a clearer image of the policies and personality of Ashoka than of any other monarch of ancient India, thanks to the edicts he had carved onto the great rocks and polished pillars of sandstone that he erected Persian-style throughout the limits of his enormous empire, like mighty ribs supporting the sky. James Prinsep, an amateur epigraphist who worked in the British mint in Calcutta, first deciphered the *Brāhmi* script of Ashoka's edicts in 1837, and since then many Indologists have devoted their lives to reconstructing the enduring message of India's most powerful, and possibly most enlightened, emperor. Some 5,000 words were carved by Ashoka's orders on at least 18 rocks and 30 pillars (though only 10 pillars remain standing in good condition). Most of the inscriptions were in the *Brahmi script,* which has been identified as the antecedent of the *Devanāgari* ("City of the Gods") script of Sanskrit and modern Hindi, phonetically the world's most scientific script at that early date. Although Ashoka's inscriptions are the earliest Indian writing as yet to be deciphered, their epigraphy clearly attests to several centuries of historical evolution. A few of the edicts found in Northwest India were carved in a different script, *Karoshthi,* which has, in fact, been identified as a variant of Aramaic, written from right to left, obviously transmitted to India through Achaemenid Persia, where it was then in wide use.

For the first eight years of his reign, Ashoka behaved as most ancient monarchs usually did, consolidating and expanding his power in as ruthlessly expeditious a manner as possible. Then, following the classical prescription of the *Arthashastra,* which advised that "any power superior in might to another should launch into war," Ashoka invaded the frontier tribal kingdom of Kalinga to his south (modern Orissa), subduing it after the bloodiest war of his era. In the longest of his edicts, Ashoka told of how many people were "slain," how many more "died," and how many others were taken "captive" from that conquered and settled land. With its last major tribal opposition in South Asia annihilated, the Mauryan administration could now afford officially to abandon its policy of conquest in favor of the more enlightened advocacy of peace and nonviolence (*ahimsa*). Pillar edicts proclaimed Ashoka's revulsion at past carnage and "remorse at having conquered Kalinga," which, it was said, made him resolve thereafter to reject violence in favor of the Buddha's law of nonviolence. Ashoka may actually have converted to Buddhism in the tenth year of his reign, though whether or not he did is less important than the policy of pacification to which his newly unified empire was "converted." "If one hundredth part or one thousandth of those who died in Kalinga or were taken captive should now suffer similar fate," proclaimed the imperial edicts—which, though written in the name of a monarch whose epithets were "Beloved of the Gods" (*Devanāmpiya*) and "He of Gentle Visage" (*Piyadāssi*), must be read as state propaganda issued by the bureaucratic machine—that "would be a matter of pain to His Majesty."

The Mauryan state, in the wake of the Kalinga conquest, thus proclaimed its emperor's resolve to bear "wrong" so far as "it can possibly be borne" without resorting to violent retribution, to "look kindly" upon all subjects, including the "forest tribes," though these were advised to "reform." Those who had not as yet surrendered to Mauryan rule, however, were warned to remember that the emperor was not only compassionate but powerful, using stick as well as carrot in completing the Mauryan settlement of South Asia, determined to ensure the "safety, self-control, peace of mind, and happiness" of all "animate beings" in the realm. Before the end of Ashoka's reign, Mauryan rule claimed revenue from Kashmir to Mysore, from Bangladesh to the heart of Afghanistan. Only three Dravidian "kingdoms" (Kerala, Chola, and Pandya) remained independent to its south, as did Ceylon. Mauryan India maintained diplomatic relations with all of its neighboring states, as well as with Antiochus II of Syria, Ptolemy II of Egypt, Antigonus Gonatas of Macedonia, and Alexander of Epirus. Ashoka was hailed as the first true *chakravartin* ("he for whom the wheel of the law turns"), universal emperor of India. He addressed all Indians as "my children" and carved in stone his paternalistic administration's express desire "that they may obtain every kind of welfare and happiness both in this world and the next." Ashoka was said to have informed his subordinates, doubtless to reassure them, that no matter where he was, whether eating or in his harem, he was always "on duty" to carry out the "business" of state. He may well have enjoyed the great power he wielded and pursued his imperial labors with religious dedication, but he could not administer so vast an empire alone. He appointed many special "overseers of the law" (*dhamma-mahāmāttas*) to tour the empire as his emissaries to local governments, supervising local officials in the performance of their duties, which, given the great distances involved and the equally vast differences in customs, laws, and languages among India's diverse regions, must have been an almost impossible task. It was, nonetheless, the beginning of an attempt to enforce central bureaucratic control over what had hitherto been fiercely autonomous or

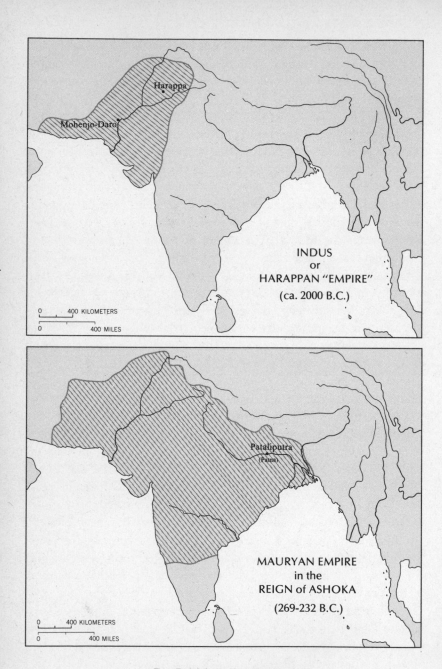

INDUS
or
HARAPPAN "EMPIRE"
(ca. 2000 B.C.)

Harappa

Mohenjo-Daro

0   400 KILOMETERS

0   400 MILES

MAURYAN EMPIRE
in the
REIGN of ASHOKA
(269-232 B.C.)

Pataliputra
(Patna)

0   400 KILOMETERS

0   400 MILES

Pre-British Indian Empires

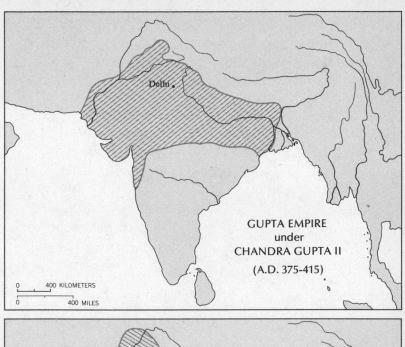

GUPTA EMPIRE
under
CHANDRA GUPTA II
(A.D. 375–415)

Delhi

400 KILOMETERS
400 MILES

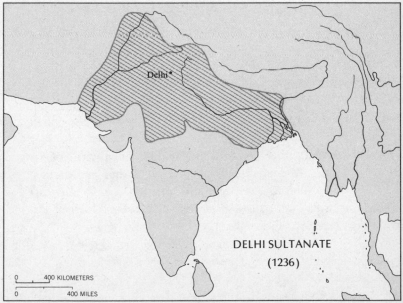

DELHI SULTANATE
(1236)

Delhi

400 KILOMETERS
400 MILES

Pre-British Indian Empires (cont.)

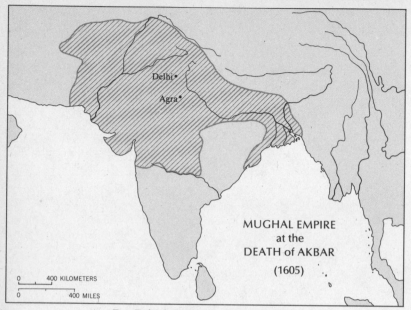

MUGHAL EMPIRE
at the
DEATH of AKBAR
(1605)

0   400 KILOMETERS

0   400 MILES

Pre-British Indian Empires (cont.)

virtually unsettled areas. We may safely assume that the regions most remote from Pātaliputra were least profoundly affected by the periodic visits from central headquarters, but Ashoka's *mahamattas,* like later British collectors, showed the flag in the remotest corners of the realm, and that in itself was significant.

The surviving rock edicts of Ashoka are filled with good moral advice; they urge the Mauryans to "listen to father and mother," to practice "liberality to friends, relations, brahmans, and ascetics," and to abstain "from the slaughter of living creatures." Other key admonitions were toleration for all "sects," "compassion," and "truthfulness." *Dharma,* that unique word which means religion, law, duty, and responsibility, was used more than any other term by Ashoka, who deemed it "most excellent." In his twenty-sixth year, the emperor inscribed the following message: "Both this world and the other are hard to reach, except by great love of the law, great self-examination, great obedience, great respect, great energy . . . this is my rule: government by the law, administration according to the law, gratifi-

cation of my subjects under the law, and protection through the law."
Ashoka abandoned the traditional annual royal hunt in favor of a
"pilgrimage of religious law" (*dharma yātra*), which allowed him to
visit distant corners of his empire personally, the living symbol of imperial unity to his people, who must have viewed him as a divinity
incarnate. To facilitate communication throughout the empire and accelerate the process of integration, Ashoka had shade trees planted
along the roads over which he journeyed, and sponsored from his
treasury such public works as the digging of wells and the erection of
rest houses on major highways built in his reign. Thanks to Ashoka's
conversion to *ahimsa,* many more Indians became vegetarians at this
time. The emperor's personal example was used, moreover, to help
inspire millions beyond his domain, for his emissaries were sent to
convert the peoples of Ceylon and Burma, and possibly of more distant regions of Southeast Asia, to Buddhism, and through Buddhism
to Mauryan pacification and Indian civilization.

Sometime between 250 and 240 B.C., Ashoka hosted the Third
Great Council of Buddhism at Pātaliputra, which had by then become
Asia's foremost center of art and culture. The Ashokan pillars were
topped by capitals decorated with animal sculpture, the most famous
of which are the four lions of Sarnath, three of whom have become
the national symbol of modern India. The lions supported an enormous stone "wheel of the law" (*dharma chakra*), commemorating the
Buddha's first sermon at Sarnath, and they rose above an abacus
around which were carved four smaller wheels and four animals—an
elephant, a horse, a lion, and a bull. The horse and bull are extraordi-
narily vital and beautifully realistic works of art. So perfectly do they
mirror the ingeniously executed figures incised two thousand years
earlier on the seals of Mohenjo-daro that we would be tempted to say
they were carved by the same craftsman. The artistic gap in Indian
history during the era of early Aryan rule would thus seem to be a
gap in the survival of resources, rather than any testimony to the demise of India's great tradition of artistic craftsmanship. That tradition, passed in unchanging form from generation to generation, as it
has continued to do in the floodlight of more recent history, defied all
the turmoil of battle and the erosive impact of time. Ashoka's royal
artists could as easily have worked for the ruler of Harappa as they
did for the master of Pātaliputra.

Tradition credits Ashoka with having built no fewer than 84,000 *stupas* (literally "gathered"), or Buddhist reliquary mounds, among which the ashes of the Buddha were supposedly subdivided from the eight original *stupas* erected earlier. These hemispherical mounds of solid stone, the largest and most famous of which was erected at Sanchi in central India, were subsequently worshipped by Buddhists, who would circumambulate the *stupa* in a clockwise direction. Centuries after Ashoka's death the *stupas* were embellished with elaborately designed stone railings and "gates" (*tornas*), topped with many tiers of umbrellas above the "little square house" (*hārmika*), set over the egg-shaped mound, through which the Buddha's spirit was said to pass from terrestrial to celestial release and eternal repose. The multitiered umbrellas are seen by some art historians as the architectural prototype of the East Asian pagoda, later introduced into China by those indefatigable cultural couriers of early Asian history, the Buddhist missionary monks. *Stupas* were further evidence of the vigorous revival of Indian interest in and patronage for monumental art, and they may in fact be a stone version of the earlier, less permanent Aryan sacrificial mounds. The *stupa* seems, at any rate, to have symbolized the universe, and it may, therefore, be viewed as a primitive ancestor of the Hindu temple, which was later to share with it such microcosmic meaning. As sacred monuments, *stupas* were also called *chaityas* and were carved within the patiently excavated caves of solid rock that were used by monks throughout Magadha as their places of worship. Some of the most beautiful sculpture in India is found in the *chaitya* hall caves at Ajanta and Ellora, which were completed centuries after Ashoka, but others that were begun in his reign are still remarkably well preserved. The monks themselves lived and studied in rock-cut caves called *vihāras*. So many thousands of them were dug in this era that Magadha came to be known as "the land of *viharas*," or what is now called Bihar.

Ashoka seems to have withdrawn almost entirely from public life during the last years of his reign, which ended with his death in 232 B.C. Following Ashoka, Mauryan rule lost much of its vitality, falling into economic as well as spiritual decline. The coins were soon debased and the ramparts of empire attacked and eroded in north and south alike. Many sons contested the throne, and there is no consensus among Indian sources as to whom it went, though Kunala, his son Samprati, and Ashoka's grandson Dasharatha are highest on the list of immediate successors. The technological obstacles to integration and the cost of so enormous a bureaucracy soon made a fiction of centralized Mauryan imperial rule, however. With power so attenuated by the attempt to maintain that rule, it is hardly surprising to find that fragmentation, local reassertion of independence, and interregional rivalries and invasions swiftly followed the demise of the God-King-Father Ashoka. Nonetheless, the first great dynasty of Indian history continued to rule over Magadha at least until 184 B.C., when Brihadratha, last of the Mauryans, was killed by his brahman general, Pushyamitra Shunga, who started a new line of rajas in central India that ruled a much diminished "empire" until 72 B.C. India's first great unification thus lasted 140 years; it was won by the swords of Chandragupta and Bindusara, ruled in accord with the shrewd pragmatism of the *Arthashastra,* and consolidated under the royal paternalism of Ashoka, whose toleration for peoples of every faith, tongue, and stage of development accepted the realities of the Indian subcontinent's regional pluralism and extended over all a single system's "white umbrella" of righteous law. Less surprising than its disintegration in the wake of northern invasions, southern defections, and constant bickerings over succession is the fact that the idea of unity survived so long at so early a state of technological integration of the subcontinent— for, after all, the Mauryans ruled India roughly as long as the British would more than two thousand years later.

SIX

POLITICAL FRAGMENTATION AND ECONOMIC AND CULTURAL ENRICHMENT

(ca. 184 B.C.–A.D. 320)

For five centuries, from the collapse of Mauryan suzerainty to the reassertion of a single dynasty's claim to central authority over all of north India by the imperial Guptas in A.D. 320, India was fragmented politically. A series of Central Asian invasions coincided with the growth of new regional monarchies in the south to reduce Magadhan power to its pre-Mauryan status as one of several competing regions of quasi-feudal independence. Throughout this era of political disunity, however, India enjoyed lucrative profits from greatly expanded overseas trade, with many merchant and artisan *jati* and *shreni* prospering from the flow of wealth and with ever-increasing demands for Indian products from both the Roman and the Chinese empires. Intellectually and artistically, this was a time of important growth and cultural syncretism.

The northwest region of Bactria had declared its independence from Seleucid rule around 250 B.C. By 190 B.C., Greco-Bactrian invaders had recaptured Peshawar, and a decade later, with the collapse of Mauryan rule, they took the entire Punjab under their control. These heirs to Alexander's conquest struck magnificent coins with images of Heracles, Apollo, and Zeus on the obverse of the heads of their kings—Euthydemus, Demetrius, and Menander. We know that the last of those monarchs ruled the Punjab from his capital at Sagala (modern Sialkot) around 150 B.C. and was converted to Buddhism by the monk Nāgasena, whose dialogues with the king are preserved in the *Milindapañho,* "Questions of Milinda," which was Menander's Indianized name. A number of inscribed columns commemorating the Bactrian conquests of northwestern India have survived, the most famous of which is the Garuda pillar at Besnagar (Bhilsa), which was erected by Heliodorus, a Greek ambassador of King Antialkidas to the fifth Shunga monarch of Magadha, Bhagabhadra. Heliodorus proclaimed himself "a worshipper of Vāsudeva," identified with Krishna ("Black"), a pre-Aryan local god of northwest India. Krishna was Aryanized as an "earthly emanation" (*avatāra*) of Vishnu, the Vedic sun god who gradually emerged as one of the two great gods of Hindu theism. Krishna was incorporated into a later portion of the *Mahabharata* sometime around 200 B.C. as divine charioteer for the heroic Arjuna, whose fallen spirits he revitalizes in the theistic dialogue the *Bhagavad Gītā,* or "Song of the Blessed One." Vishnu would eventually have ten *avataras,* including not only Krishna, but Rama and the Buddha as well—cloquent testimony to Hinduism's capacity to assimilate diverse regional cult heroes and heterodox divines into its elastic fold. Perhaps it was just such flexibility of religious outlook that attracted Greeks like Heliodorus, who was not alone among India's foreign conquerors to succumb to her cultural appeal. The dialogue by which the monk Nagasena converted King Menander to Buddhism was heard, we are told, by no fewer than 500 Greeks and 80,000 monks. Even if the latter number is exaggerated, we may safely assume that many, if not most, of the former followed their king's lead in thought as well as action. Indian acculturation seems to have weakened the grip of kings like Menander over their home base in Bactria, however, where revolts led by Eucratides, at about 175 B.C., brought a second line of Greco-Bactrian monarchs to power.

Eucratides and his heirs, who stamped coins with the equestrian figures of Castor and Pollux on their obverse, moved from the Kabul Valley to Taxila, dominating the region of Gandhara for more than a century. The distinctive classical Buddhist art, which has come to be known by the name of that region, was the most enduring legacy of

the process of Indo-Greek syncretism generated in Gandhara, where
many other currents of Indian and Western thought flowed with the
growing volume of international trade. Medicine, astronomy, and
astrology were all influenced by such exchange of ideas, as were
Indian and Western religions.

The Bactrian bridge between east and west, erected and maintained during the crucial centuries prior to the dawn of the Christian
era, may have been a vital catalyst in the emergence of Christianity,
as well as we know it was in the transformation of Theravada Buddhism's "lesser vehicle" to the "great vehicle" of Mahayana Buddhism, whose many sects were to capture the allegiance of hundreds
of millions of devotees throughout Asia. The central tenet of Mahayana Buddhism is the concept of the *Bodhisattva* ("he who has the
essence of Buddhahood"), a compassionate and loving savior who,
rather than selfishly abandoning the world, pauses at the threshold of
nirvana to reach down to help all mankind attain liberation from sorrow and rebirth through his grace. Was the idea of the Bodhisattva
inspired by reports of the life of Christ that may have reached India
through Persia and Bactria, or was the process of cultural diffusion
one that flowed from east to west? Or did such similar concepts
emerge totally independent of one another in such distant regions at
virtually the same time? No definitive historical evidence has as yet
been discovered to answer these questions.

At around 50 B.C., Hermaeus, last of the Greco-Bactrian kings,
sought in vain to defend his beleaguered land from a two-pronged
attack: Scythian invaders moving down from the north, and Parthians
moving in from the west. The Scythians, also called Shakas, had been
driven from their original homeland in Central Asia by the Kushans
(Yueh-chih), Indo-European-speaking nomads who were themselves
forced to flee from the depredations of the predatory Hsiung-nu (later
known as Huns in the west). Following its great imperial unification
under the Han dynasty (founded in 202 B.C), China expanded its
borders, which obliged the "barbarians" beyond its walls to migrate
west, eventually triggering a series of invasions of India and later of
Rome and Western Europe. The Scythians were thus but first among
a number of Central Asian nomads who poured into the South Asian
subcontinent during this half millennium of fragmentation. For the
century or more of Scythian invasions and domination over the Pun-

jab, in fact, we find that Persian Pahlavas appear to join them in looting and attacking the Indus Valley, which may indicate that before advancing into India, the Shakas had moved south of Bactria to conquer Parthia. There may, however, have been two independent and virtually simultaneous lines of attack, one by Scythians over the Kabul Valley in the North, the other by Pahlavas over the Bolan Pass further south. We find, in any case, a number of coins bearing the Greek title "king of kings" (*basileos basileon*), used by Parthian rulers, with the names of Scythian rulers such as Maues and Vonones on their other sides. We also find Sanskrit titles such as *"rajatiraja"* or *"maharajasa rajarajasa"* ("king of kings" or "great king and king of kings"), indicating Persian royal pretentiousness. Maues seems to have been the first Scythian conqueror of the Punjab, and Vonones his younger contemporary. There were several Shaka kings named Azes, but the most famous of the Parthian rulers of this era was Gondophernes, whose name has long been connected with that of the Apostle Thomas. According to the account in the apocryphal version of the Acts, when the apostles divided the world among themselves for preaching the word of Christ, India fell by lot to Thomas, who protested that he was too weak to go so far away. Yet even as he was protesting, a "certain merchant" appeared before them, sent by King *Gudnaphar* (Gondophernes), who said that he needed "a skillful carpenter." That is supposed to have convinced St. Thomas to go, and legend insists that he died in South India, where many Christian converts still consider him the martyr-saint who brought Christianity to India.

The Kushan invasions of India began about the middle of the first century of the Christian era, when the last of the Pahlava kings was defeated by these more powerful Central Asian nomads, while the Shakas were driven deeper into India, settling in the region of Malwa around Ujjain. The process of Indianization was by now well under way for the displaced Scythian warriors, who adopted *kshatriya* names and had their horoscopes cast by brahmans in keeping with Vedic lore. Such periodic infusions of Central Asian blood helped generate a particularly vital, romantic, and variegated regional culture in Rajasthan and Central India's "Middle Province," Madhya Pradesh. Many of the Hindu warriors of Rajasthan (Rajputs), who were among the fiercest opponents of the Muslim invasions a thousand years later, were descendants of these Shakas, Kushans, and later Hunas. Kujula Kadphises, the first Kushan king to rule over India's northwest region, was succeeded by his son, Wima, after his death in ca. A.D. 64.

A silver scroll found at Taxila, which is dated A.D. 78, records the enshrinement of some Buddhist relics and mentions a *"Maharaja Rajatiraja Devaputra Kushāna,"* which indicates that Kushan monarchs not only called themselves "great king" and "king of kings," but also "son of heaven," in emulation of that Chinese imperial title. The king who may have used this exalted title was probably the greatest of the Kushan monarchs, Kanishka, who reigned for more than two decades around A.D. 100. Ruling from his capital at Purushapura (modern Peshawar), which commands the Pass, Kanishka expanded his empire from Bactria to Banaras, including Kashmir as well as the Punjab and Sind, Delhi, Mathura, and Sanchi. Like Ashoka and Menander, Kanishka converted to Buddhism, and he hosted the Fourth Great Buddhist Council in Kashmir. The authoritative texts of the Hinayana canon were engraved on copper plates at this council and supposedly deposited in a *stupa* especially built to protect them, but the location of this *stupa* is unknown. The council also gave impetus to the development of Mahāyāna Buddhism and its emigration to China. It was, furthermore, during Kanishka's reign that Buddhist statues began to be carved in stone and bronze throughout Gandhara and in Mathura.

At the hub of trade routes linking India, China, and the West, Kanishka's Kushan Empire prospered. The gold coins struck by Kushan monarchs who followed Kanishka matched the weight of the Roman *denarius,* and reveal mighty, burly figures dressed in the heavy coats and padded boots that were necessary to withstand the chill winds of Central Asia. A headless stone statue found at Mathura may have been Kanishka himself, for the solid stance reflects great power and confidence, while the royal sword the figure holds looks almost too heavy for any mortal to wield. Kanishka's wealth and wisdom attracted artists, poets, and musicians to his court, as well as monks and merchants from all parts of Asia. The greatest of these luminaries was the Sanskrit poet-dramatist Ashvaghosha, a Buddhist who is credited with having converted the emperor. His "Life of the Buddha," *Buddha Chārita,* is one of the earliest examples of classical Sanskrit poetic literature. We have reports from visiting Chinese

Buddhists of the magnificent architecture at Kanishka's capital, including a fourteen-storied wooden reliquary tower, more than six hundred feet high, which was, unfortunately, burned down sometime after the sixth century. A beautiful silverwork casket, however, only eight inches high, was unearthed in 1908 at Peshawar; it bears an inscription with Kanishka's name. What is more, it bears the earliest images of the Buddha and Bodhisattvas found to date, visually marking the transition from Hinayana to Mahayana Buddhism. Kanishka's empire was administered through a number of satraps and must have been defended by a vast army. Despite his conversion to the religion of nonviolence, the mighty sovereign from Central Asia is said to have remained addicted to warfare until the bitter end of his life when, according to legend, he was smothered in his field camp by his own disgruntled lieutenants. The Kushan dynasty endured for more than a century after the death of Kanishka, only to be overthrown after about A.D. 240 by the expanding power of the Sassanians from the West.

While wave after wave of Central Asian and Persian invasions thus removed India's northwest from indigenous control, the aftermath of Mauryan political collapse further fragmented the northeastern, central, and southern regions of the subcontinent. Magadha remained under Shunga rule till 72 B.C., and included Bengal as well as Bhopal and Malwa within its boundaries. Despite the orthodox Brahmanic faith of its founder, later Shunga monarchs clearly became generous patrons of Buddhism, as the elaborate *stupas* built at Sanchi and Bhārhut under Shunga rule demonstrate. The last of the Shungas was deposed from Magadhan overlordship by his "servant," Vasudeva Kanva, founder of the Kanvāyana dynasty, which lasted almost half a century. In 27 B.C. Magadha was conquered by the explosive power of the mighty Andhra (or Sātavāhana) dynasty of South India. Apparently originating somewhere between the peninsular rivers Godavari and Kistna (Krishna), homeland of the Dravidian Telugu-speaking peoples whose descendants now live in a state called Andhra, the great Andhra dynasty spread across much of South and central India from the second century B.C. till the second century A.D. Conquering the northwest Deccan region of Mahārāshtra ("The Great Country"), the Andhras later established their capital at Paithan on the Godavari, about a hundred miles northeast of modern Poona (ancient Puné), the center of subsequent Maratha power. The

Sanskrit word *sātavāhana* most probably meant "seven mounts," referring to the seven-horse chariot of Vishnu, whose mounts each represent one day of the week; it thus signifies Aryanization of the Dravidian Andhras. The elder Pliny wrote of the "andarae" as a "powerful race," controlling numerous villages and at least thirty walled towns plus an army of 100,000 infantry, 2,000 cavalry, and 1,000 elephants. For four and a half centuries after 230 B.C., this mighty dynasty ruled India's midland, ranging north of the Vindhyas to wrest Malwa from the Kanvāyanas in the first century B.C. and south to the Tungabhadra and Kistna rivers which divided them from Tamilnad ("Land of the Tamils"). Amarāvati on the banks of the Kistna, which was later the southeast capital of the Sātavāhanas, flourished on its trade with Rome, Ceylon, and southeast Asia and may well have been the most prosperous city in India during the second century of the Christian era. The Sātavāhanas fought the Shakas, both in Malwa and Gujarat, from which they were ultimately driven by the latter invaders. To the east they came into conflict with a rejuvenated Kalinga under Maharaja Khāravela, whose rise to power prior to the dawn of the Christian era is primarily known from the lengthy "Elephant Cave" (*Hāthīgumphā*) inscription found near Bhuvaneshvara in Orissa. Though the Hathigumpha inscription is undated, it was carved sometime between 200 and 25 B.C. It refers to three successful invasions of North India by the great Kalinga monarch, one of which brought the king of Magadha to his knees. Magadhan loot brought back to Orissa after Khāravela's great victories supposedly provided the wealth required to build the first temples at Bhuvaneshvara, which subsequently became the center of the distinctive Orissan style of Indian art and architecture.

South of the Andhras and Kalingas were three ancient Tamil "kingdoms," the Cheras (or Keralas) in the west, the Pandyas in the center, and the Cholas on the east coast ("Coromandal" literally means Chola circle). References to the "pearls of Pandya" date back to Megasthenes and the *Arthashastra,* evidence that the fame of this southernmost realm had reached Pātaliputra by at least the fourth century B.C. Pandya may have been a flourishing political reality during the age of King Solomon's temple building as well. The name Pandya seems to be derived from *Pāndu,* the righteous royal family of the epic *Mahabharata,* indicative perhaps of the earliest source of the

Aryanization of Tamilnad. Inscriptions of royal Pandyan grants to favored subjects have been found in South India, dating from as early as the second century B.C. Madura, the Pandyan capital, still one of South India's greatest temple cities, was the center of Tamil literary culture, and several *sangams* ("college" or "academy"), reputed to have included more than 500 poets, flourished there from at least the second century of the Christian era. Over 2,000 *sangam* poems have survived, collected in nine anthologies, and a Tamil grammar, the *Tolkāppiyam*. From the latter we learn not only grammar but much about the social life of early Tamils, who were divided into "castes" based on geographic domain; i.e., hills people, plains people, and forest, coastal, and desert folk. Within each of these five basic groups were occupational subdivisions, including pearl divers, fishermen, boatmakers, and boatmen. The early Tamil kingdoms appear to have been matriarchal, and matrilineal succession survived in Kerala and some other pockets of Tamilnad till the nineteenth century. The Dravidian kinship system was based on cross-cousin marriage, strikingly different from the prevailing Indo-Aryan system of the north, where exogamy led to the marriage of "strangers" rather than relatives. The *sangam* literature names several very early Chera kings, one of whom was supposed to have attacked *Yavana* (Greek or Roman) ships. There is also early reference to Chola monarchs, whose trade was predominantly with Southeast Asia. Ancient Tamilnad, enriched by the discovery of gold and precious jewels and by an expanding foreign trade, was a fabled region of fine feasts and fortune. *Sangam* poets sang of "succulent chops of meat" and "triple water" (probably coconut milk, palm fruit juice, and sugarcane juice) imbibed with fiery "toddy" (the intoxicating fermented beverage of the palmyra palm) at the royal feasts that lasted for weeks.

Court life was also enhanced by troops of actors, who sang, danced, and played their own musical accompaniment, much the way bands of *bhajan* ("devotional") musicians still continue to offer peripatetic entertainment throughout South India. Women actively participated in these dramatic road companies, which, with the later development of the classical *bhārat nātyam* dance form and the evolution of temple dancer-prostitutes (*devadāsis*), were vital facets of culture contributed to Indian civilization by the Dravidian south. The epics of Tamilnad, written several centuries after the *sangam* era, tell of the adventures of two courtesan-dancers: Madavi, the *femme fatal* of the first epic, *Silappadikāram* ("The Jewelled Anklet"); and her daughter, Manimegalai, whose name is also the title of the second poem. After a series of passionate loves, both women eventually abandon their occupations to become Buddhist nuns.

The three Tamil kingdoms of the south seem initially to have alternated between political alliances designed to ward off their northern neighbors, and warfare among themselves. From an inscription dated about 165 B.C., we know that all three formed a "confederacy" against the invading army of Khāravela of Kalinga, who nonetheless defeated them. Perhaps their common defeat convinced the Tamil kings of the greater wisdom of selfish action, for more common than united defense was the pattern of conflict among the Cheras, Cholas, and Pandyas that developed after the middle of the first century B.C. and continued until the fourth century of the Christian era. Centrally situated as they were, the Pandyas alternately allied themselves with Cheras against Cholas and vice versa. This debilitating warfare left the kingdoms vulnerable to conquest once again early in the fourth century, when a new dynasty, the Pallavas ("robbers"), appeared at the former Chola capital of Kanchipuram (Conjeevaram). The Pallavas were firmly entrenched in Tamilnad by A.D. 325, but the mystery of their origins remains one of the more fascinating problems of South Indian history. Though it is possible that the founders of this dynasty were originally Pahlavas or Parthians, descended from the north, there seems better reason to believe that they were southeastern feudatories of the Sātavāhanas, whose collapse in the second century led to their independent assertion of power around Kanchipuram, from which they managed to oust the Chola monarch. From copperplate and stone inscriptions and coins, the history of this dynasty's southern dominance can be traced between A.D. 325 and at least 800, making it one of the longest dynastic lines in all of Indian history.

The hoards of Roman coins found in South Indian ports attest to the brisk and lucrative volume of overseas trade, which was also reported in surviving fragments of the *Periplus* ("Marine Geography") *of the Erythrean Sea,* an anonymous guide to Indian Ocean commerce, probably written by an Arabian sea captain during the first century of the Christian era. Ivory, onyx, cotton goods, silks, pepper and other spices, and jewels were shipped from peninsular

India's ports, the most prosperous of which were Broach on the river Narmada, Sūrpāraka near Bombay, and Arikamedu near Pondicherry. The Romans paid for Indian produce with various items, including copper, tin, lead, antimony, and wine, but the number of gold Roman coins of the first few centuries A.D. found in Tamilnad indicate, as Pliny complained, that the balance of trade favored India. The international "drain" of specie thus flowed initially from west to east. Trade routes crisscrossed the subcontinent by now, and caravans of camels, oxen, and donkeys carried the seeds of urban culture with coins and produce throughout India and beyond its geographic limits. The great overland trade route from Taxila to Kabul branched off to the Central Asian north and China, as well as to Kandahar and the Persian Gulf in the south and west, making India the economic center of Sino-Roman trade in this era of growing commercial intercourse. *Shreni* prospered to such an extent that their wealthy members donated fortunes to religious orders, especially Buddhist and Jain, the traditional religious offspring of mercantile enterprise. In the Deccan, magnificently carved caves, the most famous of which are at Ajanta and Ellora, survive to this day as evidence of the affluence of merchant guilds, whose leaders are sometimes depicted outside entranceways to religious shrines to commemorate their patronage of the order. Throughout India, bustling new cities emerged at caravan stops as well as ports, where *shreni* assumed responsibility for the maintenance of municipal order and evolved legal regulations governing the social behavior of guild members, as well as their commercial conduct.

The increased use of coins and the growth of commerce and wealth led to the emergence of Indian bankers and financiers (*shreshthins* or *seths*), who helped support failing guilds as well as impecunious monarchs and lesser landowners. With the risk of travel and trade very high, interest rates were exorbitant, varying from 15 to 240 percent per annum, but profits sufficed to make the demand for money ever greater. Indian banking and commercial families established branches at as many of the great urban centers of enterprise as possible, both at home and abroad, thus early developing kinship networks of wealth that secured growing fortunes within *shreshthin* "houses" of regal resource and power. Gold coins based on the weight of the Roman *denarius* (124 grains) and silver coins, which had been introduced during Mauryan times, were in regular circulation, as were coppers and cowry shells. Village and rural regional economies, however, remained interdependent and nonmonetary, with traditional services rendered among *jati* according to hierarchical *jajmāni* ("patronage") patterns of barter. Every village servant and craftsman—sweeper, barber, laundryman, blacksmith, tanner—had his peasant patron (*jajmān*), who provided his family with rice enough to survive each year in return for an annual "retainer" on his services.

Increased trade and wealth stimulated Indic thought and arts in many ways. Not only did Buddhism evolve into a full-blown religion with Mahayana worship of the Bodhisattva, but Brahmanism also emerged as a faith whose central form of worship now shifted from Vedic fire sacrifices to personal devotion for some form of a sectarian deity, either Vishnu or Shiva. This period thus marks the emergence of Hinduism out of the earlier Vedic-Brahmanic ritual faith, and from the dawn of the Christian era, that Hinduism continued to grow in popular appeal throughout the subcontinent, though Buddhism and Jainism initially remained vigorous competitors. Hindu legal codes, like the *Mānava Dharmashāstra* ("Law Code of Manu"), were compiled at this time, as were the many "Ancient Tales" (*Purānas*), myths and fables about the various Hindu gods, especially Vishnu, Shiva, and Brahma. Vishnu came to be exalted as the divine savior of mankind, whose *avatāras* always appear on earth when demons threaten to destroy the cosmic *dharma*. To date there have been nine *avatāras:* the Fish (*Matsya*), the Tortoise (*Kurma*), the Boar (*Varāha*), the Man-Lion (*Nārasimha*), the Dwarf (*Vāmana*), Rama with an Axe (*Parāsurāma*), Rama the hero of the *Ramayana,* Krishna, and the Buddha. A tenth *avatāra,* a savior on horseback called *Kalkin,* was scheduled to come in the future whenever the world would need Vishnu's help to survive. By bringing the Buddha within Vishnu's devotional cult, Hinduism sought to recapture millions of those who had abandoned its ritual orthodoxy for Buddhism's promise of salvation.

Krishna, who shares with Rama the distinction of being Vishnu's most popular incarnation, was the fluting lover of India's milkmaids, most beautiful of whom was Radha. As divine lovers, Krishna and Radha were later magnificently portrayed by Rajput artists. Krishna also taught the message of Upanishadic philosophy and elucidated Hinduism's paths to salvation in his dialogue with Arjuna in the

Bhagavad Gita. As Arjuna's charioteer, Krishna was able to console that great warrior, who lost heart and lay down his weapons just before the final battle on the field of Kuru was about to begin. Arjuna had seen his aged guru and uncle among the ranks of the enemy (his Kaurava cousins) and could not bear the prospect of killing his own kin. He wanted to abandon the epic struggle before it started, but Krishna reminded him that as a warrior, it was his primary duty (*dharma*) to fight. The godly charioteer also explained the true nature of "reality," teaching Arjuna that the faces and bodies he saw at the far end of the field were but illusory figures, while the imperishable "souls" (*atman*) hidden within each of them could not be slain by sword or arrow. "He (*atman*) slays not, is not slain," Krishna said. "He is not born, nor does he ever die."

The central teaching of the *Gītā* is Krishna's explication of the "discipline of action" (*karma yoga*) path to salvation. No longer was any Vedic sacrifice required of those who would seek eternal release; nor was the virtually impossible withdrawal of all contact with the world of material possessions. Krishna taught Arjuna that so long as he acted without concern for the fruits of his action, no *karma* would attach itself to anything he did. "Indifference" to success or failure was the key to this new method, thanks to which one could escape the imperative of rebirth while remaining strapped to the wheel of life's activity. The elitist, razor's edge goal of Upanishadic salvation was made available to all who had sufficient discipline to restrain their passions while engaged in actions of every kind, performing deeds without selfish motive. "To whom loved and unloved are equal," that person would "transcend" death and rebirth, attain the perfect calm of *moksha*. Another major Hindu path to salvation—*bhakti* ("devotion")—was introduced in the *Gita* and later became even more popular than *karma yoga*. To convince Arjuna, Krishna finally revealed himself as the divine Vishnu, explaining that "Those who revere me with devotion (*bhakti*), they are in me and I too am in them," developing this concept of *bhakti* in much the manner of Christian grace. "Even if a very evil doer reveres me with single devotion, he must be regarded as righteous in spite of all. . . . Even those who may be of base origin, women, men of the artisan caste, and serfs too. Even they go to the highest goal."

Hindu salvation was thus made accessible to the peasant masses of India as well as the Brahman elite. Hinduism itself, however, re-

tained inequality as a central concept of its faith, inextricably bound
as it was to the "caste" system. The *dharma* (religion or duty) of a
brahman was quite different from that of a *kshatriya,* and each was a
world removed from the lowly *shudra,* yet he too came within Hinduism's fold. Untouchables, though "outcastes," were also Hindus, and
like *shudras,* they were promised ultimate elevation through rebirth
only if they devoted this lifetime to doing their duty (*dharma*). If the
lower classes were foolish enough to rebel against their servile status,
they would accrue only evil *karma.* Uncomplaining acceptance of
one's position at birth (*jati*) was, therefore, the surest method of
eventual advancement up the ladder of Hindu *varna,* in keeping with
the "laws" of *karma* and *samsara.*

The obscure Vedic rain god Rudra emerged now as Lord Shiva,
Hinduism's "Great God" (*Maheshvara*), no less exalted in the primacy of his divine status by sectarian followers than was Vishnu by
his devotees. Divine yogi and fertility god, Shiva was at once the aus-
picious creator of life, Lord of Beasts (*Pashupati*), and King of the
Dance (*Nataraja*) and the dark destroyer, haunting burial grounds
and consuming poison; fiercer than the tiger; death and time incarnate. Shiva epitomizes Hinduism's reconciliation of extremes: erotic
passion and ascetic renunciation, frenzied motion and unmoving calm,
violence and passivity. Though his modern sectarian followers pre-
dominate in the south, Shiva's mythic mountain home was Kailasa in
the Himalayas, and the wild rapids of the river Ganga were supposedly reduced to relative calm by the matted locks of his hair,
through which they flowed before falling to earth. Shiva's consort was
one or another form of the mother goddess—the benevolent Pārvatī,
the virtuous Satī, the dark and malevolent Kālī, or the demoness
Durgā. (All gods of Hinduism have female counterparts, whose potent
earthly "power" or *shakti* was believed to be the active life force.
Vishnu's consort Lakshmī was goddess of good fortune, while
Brahma's bride Sarasvatī was divine patron of the arts.) In addition
to many brides, Shiva had a number of children, including the war
god Kārtikeya (or Skanda) and the elephant-headed Ganesh, who
were worshipped in the north and south. Hindus continued to accept
the lesser divinity of many minor deities as well, including Brahma,
Indra, Agni, and demi-gods and goddesses, cows, and brahmans.

For a good Hindu, however, caste laws and mundane duties
were more important than deities. What really mattered was to eat

properly, drink properly, marry the right person, and act "correctly"; that is, in keeping with the law (*dharma*) of one's *varna, jati,* and *āshrama* ("stage") of life. According to Hindu legal texts (*Dharmashastras*), there were four *āshramas,* each of which applied only to the three twice-born *varna,* and all of which obviously reflected ideal rather than real stages of transition in the average Hindu's life. The first stage, called *brahmachārin,* was celibate studenthood and began after a boy's investiture with the sacred thread (usually between the ages of six and twelve), when he left his paternal home and went to live with his guru. The *Dharmashastras* emphasized the importance of celibacy during this stage of life, especially with regard to the guru's wife, who may often have posed a challenge to the young student's capacity for concentrated study. Since the ideal age for a Hindu wife was one-third that of her husband, many wives must have been closer in years and interests to the young pupils than to their brahman gurus. At any rate, the importance of celibacy during the *brahmacharin ashrama* of study was such that the word *brahmacharya* later comes to mean simply that.

After memorizing his Vedic mantras and possibly learning one or more adjunct Vedic "limbs" (*vedangas*), which included such subjects as phonetics, grammar, prosody, astrology, and etymology, the graduate returned home, ritually to bathe and then marry. With marriage, usually arranged by parental and astrological agreement, the good Hindu entered into his second *ashrama,* that of the householder, *gryhastha.* As a *gryhastha* one's *dharma* was radically different from that of prior or subsequent stages, for it was now a man's primary duty to start a family, prosper, and enjoy life. The pleasures and virtues of *kama* ("love"), including sexual enjoyment (*rati*), were now ritually prescribed, and the classic textbook of love, Vatsyayana's *Kama Sutra,* was probably written around the second century of the Christian era. A good householder was obliged to keep himself and his wife happy, to have as many children as possible, and to accumulate as much material wealth as he could handle, dispensing an appropriate portion of it to brahmans during every ritual and holiday of the busy religious calendar. After viewing the face of his grandson, who assured the continuity of his line and prayers for his *atman* after death, the twice-born Hindu male was ready to pass into the third stage of life, that of forest-dweller (*vāna-prastha*).

The third and fourth *ashramas* of Hinduism seem to have been

inspired by the asceticism of Buddhist and Jainist monastic ideals, and
they may reflect once more the assimilative powers and wisdom of the
brahmans, who preferred to conquer their opposition by absorption.
Relatively few wealthy Hindus actually abandoned their *gryhasta*
affluence to don a forest hermit's robe of poverty, yet such was the
ideal that was early instilled into their consciousness. Entering the
third *ashrama* meant leaving home and possessions behind (though a
man's wife was permitted to join him if she so desired) and going off
to live in humble simplicity in the forest, without wealth or work to
distract one from the leisure needed to prepare for the fourth and final
stage of ascetic wanderer (*sannyāsin* or *bhikshu*). Few men ever attain the wisdom or courage to embark upon that final *ashrama* of
homeless, bondless, isolated wandering. In this final stage, even the
willing wife must be left behind. All ties with the past, all links with
the world of action, accumulation, karmic reality, and humanity
should be severed in preparation for death. Only the calm and puri-
fied soul, liberated from every bond and worldly tie, could hope to
enjoy the bliss of ultimate release, *moksha*.

Christianity may have entered India at this time. A small but influential group of Syrian Christians in Kerala persist in claiming that
their sect was founded by St. Thomas, who may have sailed to
Malabar in the first century and who was supposedly martyred at
Mylapore, a suburb of modern Madras, in A.D. 68. The tiny Jewish
community of Cochin also claims to have been founded in the first
century, but no clear historical evidence of such early Jewish settlement in Malabar has as yet been discovered. Like Buddhism and
Hinduism, Jainism now became subdivided into sects, the naked
Digambaras ("Sky-clad"), who migrated south and settled in modern Mysore, and the white-robed *Svetāmbaras* ("White-clad"), who
established their base in Gujarat and Kathiawar in the west.

Virtually all of the great artistic and architectural remains from
this period are Buddhist. The Buddha image, which first evolved in
the north under Kushana rule, is one of the most exciting artistic
legacies of this era. It was at Mathura that the image developed in its
distinctly Indian style, and throughout Gandhara in its Hellenistic
Roman style. With the Buddha's deification by Mahayana sects, his
depiction (both as Buddha and Bodhisattva) became an important
challenge and a stimulus to religious worship. The yogic calm and

strength of the Mathura Buddha's visage and monumental form, which eventually incorporated no fewer than thirty-two major signs (*lakshanās*) that symbolically represent his wisdom and universal virtue and power, is one of India's finest contributions to world art. The prolifically produced Gandharan Buddhas, on the other hand, often more nearly resembled Roman emperors and senators than yogis. But a third style of Buddha figures, produced further east in Orissa, were, like the Mathura form, totally Indian in inspiration and artistic execution. The ingenius artists who first brought the Enlightened One to visual life in stone and metal devised many poses with varying hand gestures (*mudrās*) to depict the different messages of the Buddha, including his most famous blessing of reassurance, the "forget fear" (*abhāya mudrā*) gesture, which Gandhi would later adopt as one of his favorite symbols.

The gateways and railings of the great *stupa* at Sanchi were beautifully decorated during this time with voluptuous *yakshis* (mother-goddess figures), dancing elephants, and other carved figures designed to depict stories of the Buddha's previous lives and births, recounted in literary form as *jātaka* ("birth") tales. Architecturally, the most important work was done in rock-hewn caves, mostly in the Deccan, under Andhra rule. The *chaitya* ("sacred spot shrine") cave at Karla, near Poona, is the most beautiful example of early Deccan architecture of this era and remains one of the best preserved caves in India. The handsome patrons carved outside the entrance were probably a wealthy merchant and his wife, who dedicated the funds for constructing this cave as their meritorious service to the Buddhist order. Deep within the cool stone of the cave, monks could comfortably worship the Buddha by walking around the *stupa* excavated from the far end of the long *chaitya* hall, beyond its many columns carved from the living rock. *Viharas,* multicell living quarters for the monks, were also carved at this time; they usually reflected the austere life of their residents, however, and were made without sculpture. Similar *chaityas* and *viharas* survive at Nasik, Bhaja, Kondane, Junnar, and Ajanta, as well as Ellora, but the last two sites became the most extensive and impressive centers of Indian cave art. The twenty-seven caves at Ajanta, which stretch across an entire crescent-shaped mountainside, were rediscovered by British travellers in the late nineteenth century after having been buried under a landslide for centuries. The most beautiful paintings of Ajanta appear to have been produced at least three centuries after the fall of the Andhra dynasty, probably during the sixth and seventh centuries. The thirty-four caves at Ellora also date from about the fifth through the eighth centuries of the Christian era. At Amarāvati, furthermore, a great *stupa* was built, but only its reflection survives today in the miniatures carved in bas-relief around the *stupa*'s railings. Some of the Amaravati reliefs depicting scenes from the *jatakas* are among the most beautifully composed gems of Indian sculpture.

Science as well as the arts flourished in this period of prosperity and increased contact with the world beyond India's borders. Medicine and astronomy both benefited from Bactrian rule, with Hellenistic ideas flowing into North India even as Indian concepts in these fields of study helped fructify Western thinking. The precocity of Indian medical thought may be attributable to early yoga practice, which, thanks to Buddhism and Jainism, gained imperial patronage from at least Mauryan times. Indian interest in herbal medication and the use of magic potions dates back much further, to the Atharva Veda (ca. 1000 B.C.), where *ayurvedic* (Hindu medical science of longevity) prescriptions are first found. It was in the second century of the Christian era, however, that Chāraka compiled the oldest surviving Indian medical textbook, *Chāraka Samhita,* which in some ways resembles the works of Galen and Hippocrates. While yoga taught Indians to focus on their spinal columns, they did not as yet appreciate the powers of the brain, believing the heart to be the organ of human intelligence. They also believed that health depended primarily upon maintaining a proper balance among the three bodily "humors," which were associated with the three "strands" (*gunas*) of ancient Indic philosophy: *sattva* ("truth"), *rajas* (passion), and *tamas* (darkness).

Indian interest in astronomy dates back at least to Vedic times, when proper building of the sacrificial altar required astronomical information. Prior to contact with the Hellenistic world, the Indians used a lunar calendar, dividing each month into "bright" and "dark" fortnights of fifteen *tithis* (lunar days). Every two months formed an Indian season, of which there were six, and every thirty months a leap month was added to catch up to the solar year. Western astronomy seems to have introduced the solar calendar, the seven-day week, and the hour to India, together with our zodiac. Indian astrology had earlier divided the heavens primarily into twenty-seven "lunar mansions" (*nākshatras*). In some fields, however—notably mathematics and grammar—Indians were far more advanced than Western thinkers. Pānini's brilliant physiological and morphological analysis of the Sanskrit language in his "Eight Chapters" (*Ashtādhyayī*) was probably completed before the close of the fourth century B.C., and Patanjali's "Great Commentary" (*Mahābhāshya*) on that masterwork was written in the second century B.C., thus making Sanskrit the first language to be scientifically analyzed. Unfortunately, the weighty authority of these ingenious works contributed to the standardization of Sanskrit and the development of long, unwieldy compounds, which diminished its utility as a "living" language. Other, less "refined" popular tongues, called *prākrits,* emerged as the dominant dialects of India's various regions, and from these were subsequently to evolve the many modern Indo-Aryan languages of North India, each twice removed from its classical Sanskrit ancestry.

SEVEN

THE CLASSICAL AGE

(A.D. 320–ca. 700)

The reunification of North India under the imperial Guptas (ca. A.D. 320–550) and the reign of Harsha Vardhana (606–47) of Kanauj may be studied as classical prototypes of the Hindu state, their era comprising India's classical age. New popular forms of Hinduism emerged at this time, together with monuments of brilliant temple art and Sanskrit literature, which have remained sources of inspiration throughout Indian history.

The Guptas, like the Mauryas, established their base of imperial power in Magadha, where they controlled rich veins of iron from the Barābar Hills. Chandra Gupta I, who was no relation to his namesake, the founder of Mauryan rule, struck coin to commemorate his coronation at Pataliputra in February of 320, and he assumed the exalted Sanskrit title "Great King of Kings" (*Mahārājādhirājā*). By marrying the daughter of the king of the ancient and powerful Lichavi clan, which ruled the neighboring domain of Vaishali, north of the river Ganga, Chandra Gupta secured his grip over that vital Gangetic artery, which carried the major flow of North Indian commerce. From that powerful nucleus athwart the east-central Gangetic plain, the Guptas expanded their frontiers of empire to the Punjab on the west and Bengal on the east under Chandra's son and heir, Samudra, who reigned from 335 to about 375. Kashmir was also brought within the mighty new dynasty's domain, which soon extended south to the Deccan. A detailed eulogy to the intrepid warrior Samudra Gupta was engraved on one of Ashoka's pillars at Allahabad, thanks to which we have a clear record of the martial conquests of this Napoleon of ancient India, who was urged by his dying father to "rule the whole world" and who came close to achieving that elusive goal within South Asia. "Skilled in a hundred battles," Samudra Gupta was credited with having "forcibly uprooted" nine kings of North India, humbling eleven more in the south, while compelling another five on the outskirts of his empire to pay "tribute" as his "feudatories." Even such mighty monarchs as the last of the Kushanas and Shakas were listed as his vassal tributaries, as was the king of Ceylon. Samudra Gupta thus had good reason to crown his victorious reign with the performance of the *ashvamedha* ("horse sacrifice"), marking the historic revival of royal patronage to Brahmanic ritual.

The peak of Guptan power and cultural glory, however, was attained during the reign of Samudra's son and successor, Chandra Gupta II (c. 375–415). We get some insight into the bold and lively personality of Chandra Gupta II from a popular Sanskrit drama, "The Queen and Chandra Gupta" (*Devīcandragupta*), written by Vishākhadatta several centuries later (probably in the sixth century). That courtly play tells of how Chandra's older brother Rama inherited their father's throne but proved himself weak and treacherous by promising to surrender his wife to a barbaric Shaka ruler who had defeated him in battle. Dressing as a woman, Chandra Gupta valiantly volunteered to take the queen's place, murdered the Shaka monarch as soon as he reached the enemy's harem, and returned a hero to his brother's court. Vishākhadatta then has Chandra Gupta murder Rama and marry his widow. If the tale was based on truth, it certainly reveals a resourceful individual and helps explain how his reign could become the most exciting one in the Guptan era. We know more about India during Chandra Gupta's reign than at most other early interludes because of the works of his ingenious court poet-playwright, Kalidasa, and the diary kept by the visiting Chinese Buddhist pilgrim Fa-hsien, who traveled around India for six years at the beginning of the fifth century. Seven of Kalidasa's brilliant Sanskrit classics have survived, yet virtually nothing is known of the life of this "Shakespeare of India" except that his name (which literally means "Slave of the Goddess Kali") might indicate South Indian *shudra* birth; internal literary evidence, on the other hand, suggests he was a brahman from Mandasor or Ujjain. Whatever his origins may have been, there is no question of the inspired quality of Kalidasa's writing and little doubt that his great royal patron, "Whose Splendor Equalled that of the Sun" (*Vikramāditya*), and who, tradition tells us, drove the Shakas from the lovely city of Ujjain, was Chandra Gupta II.

Numismatic evidence attests to the final defeat of the Shakas by the Guptas in 409, after which Chandra Gupta's empire had direct control over the ports of the Arabian Sea and the riches of Western trade. By marrying his daughter, Prabhāvatī, to the Deccan's Vākātaka King Rudrasena II, Chandra Gupta II extended the influence of his empire south of the Vindhyas. Chandra Gupta himself married Kuvera, a queen of the Nagas, thus further consolidating his power over the eastern wing of his empire by political alliance. Fahsien's account depicts a state of internal peace and remarkable personal freedom and tolerance enjoyed by common people throughout North India during this enlightened reign, where "the king (Fa-hsien never mentions Chandra Gupta by name) governs without decapitation or corporal punishments." As a Buddhist monk, Fa-hsien naturally took great interest in such evidence of nonviolent behavior, which he perhaps reported for didactic reasons, also noting that the "well-bred" Indians were vegetarians. He found Pātaliputra a city of palaces and such widespread affluence that hospitals were provided free of charge to which "the poor of all countries, the destitute, crippled, and diseased, may repair." Despite the Brahmanic revival and establishment of Hinduism over North India, Buddhism continued to flourish, with thousands of monks residing in Mathura alone and many hundreds in Pātaliputra. Nonetheless, Fa-hsien noted "untouchables" hovering beyond the pale of Hindu society, carrying gongs to warn passing upper-class people of their polluting presence.

Of Kalidasa's seven works, *Shakuntala* is the most beautiful, a poignant drama named for its heroine, an orphaned forest nymph who captures the heart of King Dushyanta during his hunt near her hermitage. Shakuntala's "bewitching youth" so enthralls the king that he forgets his wife and court responsibilities entirely. He marries the lovely girl by the *kshatriya*'s simple "love-match" (*gandhārva*) rite, living with her at the hermitage long enough to leave her pregnant. After returning to his court, the busy king forgets Shakuntala and refuses at first to acknowledge their child, only much later recognizing the youth as his true son and heir. The richness of Kalidasa's imagery, the universal and timeless reality of his dialogue, his sensitivity to natural beauty of every sort, and the humanity of his characters suffice to overcome any weakness of plot or dramatic convention, endowing *Shakuntala* with a classic brilliance. Neither of Kalidasa's other plays match *Shakuntala*'s beauty, but at least one of his poems, *Meghadūta,* "The Cloud Messenger," is a literary gem, capturing the longing of distantly separated lovers in the panoramic imagery of India as seen from a fleeting cloud, bearing love's message from Vindhyas to Himalayas. Kalidasa's genius was unique, but he was not the only author in this era who used Sanskrit to create enduring secular literature. Shudraka's contemporary "Little Clay Cart" (*Mrichakatika*) is the realistic story of a poor brahman, Carudatta, who falls hopelessly in love with the virtuous courtesan, Vasantasena. It is the only Sanskrit drama to include a legal trial scene.

During the Gupta era, royal support was lavished on Hindu, Buddhist, and Jain faiths, and the Hindu temple emerged as India's classic architectural form. Like its Grecian counterpart, the Hindu temple (whose finest examples from this era survive at Deogarh in Central India and Aihole in the south) emerged as the home of a deity, where devotees went to offer gifts and to pray to the icon inside it's "womb-house" or sanctuary. Every temple was erected to a specific deity: Vishnu, Shiva, or a mother goddess (at Aihole, in the name of Durga). The early Hindu temples were modest in form as well as size, initially consisting of little more than the sanctuary, entered through a much larger, square hall (*mandapa*), where worshippers could assemble. Outside was a porch, often beautifully decorated with sculptured reliefs depicting the deity in various forms or mythological scenes. The roofs were flat, but later had imposing towers. Stone blocks were fitted together without mortar, the entire temple usually raised on a special platform within a paved rectangle, representing the cosmos in miniature. Temple entrances, like the entranceways to Buddhist and Jain caves, were generally decorated with images of the deity and his consort in godlike yet human form. Hindu temples would evolve from such humble origins into extravagantly ornate structures and even entire cities, especially in the south, after the eighth century.

The Guptan era also marked the apogee of cave art and sculpture. The magnificent *fresco secco* fragments that still adorn Ajanta's dark walls and cave ceilings depict graceful human figures whose opulent jewelry and lavish hair styling reflect fashions of the Guptan and Vākātakan courts, earthly paradises worthy of preservation as scenes from Mahayana Buddhist "heaven." Yashodhara's fifth-century

commentary on the *Kama Sutra* includes a detailed discussion of the
"six limbs" of painting, providing this relatively new art form for
India with classical prescriptions. Guptan figure sculpture was characterized by a classical serenity and simplicity, reflected in the Buddha
images from Sarnath and Mathura that were produced at this time. We
also find voluptuous pink fragments of mother-goddess-like women,
whose fecundity of form and plasticity of flesh seem to belie the stone
substance from which they were carved. Even the gold coins minted
by Guptan monarchs were works of art.

Commerce as well as Buddhism stimulated Indian intercourse
with China and Southeast Asia at this time. From the port of Tam-
ralipti in Bengal (then called Vanga), Indian vessels sailed south to
Ceylon and then east through the Straits of Malacca to Southeast
Asian entrepôts, from which Chinese merchant ships carried Indian
cottons, ivory, brassware, monkeys, parrots, and elephants to the
Middle Kingdom. China's exports to India consisted primarily of
musk, raw and woven silk, tung oil, and amber. Neither opium nor
tea had as yet become factors in the trade between Asia's two greatest
powers. The Hindu kingdom of Fu-nan (South Vietnam) was by now
a prosperous state ruled by the Brahman Kaundinya, one of many
Indian voyagers to Southeast Asia, who introduced Sanskritic culture
to most of the peninsula, as well as the archipelago. The island of
Bali still reflects that early, potent infusion of Indian influences. The
kingdoms of Sri-Vijāya in Sumatra and Cho-po in Java were also
islands of Hindu, and later Buddhist, power within China's southern
orbit. While Indian trade with China passed by camel caravan over
the great northern silk routes of Central Asia, as well as across the
more precarious sea routes of the south, the opening of Southeast
Asian markets to Indian cloth and the exploitation of Southeast Asian
spice resources by Indian merchants made the southern route increasingly profitable. India's trade with Rome declined after the third
century as the Asian drain of Roman gold undermined the economy
of that crumbling empire, but contemporaneous Roman coins have
been found in Madura and other South Indian cities for most of the
first century of the Guptan era. We can appreciate the value of India's
exports to Rome by noting that Alaric's ransom for sparing that city
in A.D. 410 included some three thousand pounds of Indian pepper.
Precious Indian jewels, ivory, perfumes, woods, spices, and cloth re-

mained in demand throughout the civilized Western world. All that Indian monarchs and merchants really wanted in return was specie and Arabian horses.

The Guptan Empire, like every other Indian empire, was supported primarily by the land revenue "share" that India's peasant villages provided from every harvest to the royal treasury. In addition to the traditional one-fourth of their harvests demanded by the state, peasants were obliged to pay water taxes if their land was irrigated and as a rule would be taxed one-fiftieth of their cattle and gold and one-sixth of their fortune in trees, meat, honey, perfumes, herbs, and fruit. Forced labor was usually demanded one day each month to maintain roads, wells, and irrigation systems. Despite such high taxes, Indian agriculture seems to have blossomed in this period, providing a rich variety of succulent produce enjoyed by foreign visitors, in addition to such staples as rice, wheat, and sugar cane. Fruits included the mango, melons of several kinds, plantain, coconut, pears, plums, peaches, apricots, grapes, pomegranates, and oranges. Ginger and mustard have been added to the more common spices, and Indian cows naturally provided all forms of milk products. Though meat was no longer widely consumed, fish continued to play a vital role in the diets of Bengal and the south. Many varieties of coin were stamped, a money economy flourishing in the major cities thanks to trade. *Shreni* by now had their own insignia—colored banners and yak tails (*chauris*)—which would be used in parade on festive holidays, and payments within guilds were made according to seniority by "shares" of the total income. Most artisan guilds had four ranks: apprentice, advanced student, expert, and teacher, qualifying for one, two, three, and four shares, respectively, of the profits. In 465–66, a guild of oilmen received a perpetual endowment of land from a neighboring sun temple in return for the promise of always providing that temple with the oil it needed for worship and cooking, attesting to the *jajmani* relations that were established between some *shreni* and religious institutions. India's expanding overseas trade and high interest rates led to inflation and widespread bankruptcy, and considerable attention was devoted to "debtors" and punishments for failure to pay debts in the growing corpus of Hindu legal literature.

The Guptan state, somewhat like its more centralized Mauryan model, owned all salt and metal mineral mines, as well as crown lands in the environs of the capital, and operated various industrial enterprises of special use to the king, including the royal mint, munitions factory, gold and silver workshops, and weaving and spinning mills to clothe and decorate the royal family and harem. The major difference between the Guptan and Mauryan monarchies was that of religious affiliation. The Gupta bureaucracy was perhaps less pervasive than Ashoka's had been, but its elite corps of spies continued to maintain the administrative machine of state in secure daily operations, while the elephantine military establishment served both to augment imperial real estate holdings and to dissuade foreign as well as domestic opposition from attempting to topple the monarch, who claimed universal dominion as his birthright.

When Chandra Gupta II died in 415, his son Kumara Gupta succeeded to the throne and reigned for forty years. To demonstrate his prowess, he performed the sacred horse sacrifice and struck gold coins on which the six-headed, ten-armed war god Kārtikeya, mounted on a peacock, appeared as the king's personal symbol of power. Kumara's reign was peaceful, though in its later years invasions threatened from the west, where the nomadic Central Asian Hsiung-nu began their first fierce knocking at the door of India's Khyber Pass. Kumara handed over his domain intact after 455 to his son, Skanda Gupta, last of the great Guptan monarchs. The Hunas (Huns) soon advanced in force against Skanda Gupta's northwest border, and for the duration of his twelve-year reign Skanda was obliged to ward off their predatory assaults. The drain on the treasury so weakened the empire, however, that shortly after Skanda's death in 467 rapid decline ensued, leading to the empire's fall by the close of the fifth century. Toramana, the Huna leader who conquered Persia in 484, headed an invasion of India that wrested the Punjab from Guptan control just before 500. His son, Mihirakula, expanded that power base to absorb Kashmir and much of the Gangetic plain after 515. Guptan resistance was shattered, and the imperial glory of that dynasty faded by the middle of the sixth century.

For half a century following the collapse of the Guptan Empire, North India reverted to much the same pattern of political fragmentation that had preceded Chandra Gupta's rise to power. In the Kathiawar Peninsula on the west, the kingdom of Valabhi emerged independent of Magadha, which was ruled by the Maukharis; while in the east, Bengal, Assam, and Orissa all broke free, as did Nepal and Kashmir in the north. In Rajputana a new kingdom was founded by the Gurjaras at Jodhpur, and another branch of that same clan, which probably originated in Central Asia, established its independent sway over Broach. A line of later Guptas, not related to the imperial Guptas, ruled in Malwa. In A.D. 606 a remarkable young monarch named Harsha Vardhana came to the throne of Thanesar, north of Delhi, and sought once again to unify all of North India. We know a great deal about Harsha thanks to the eulogistic "Life of Harsha" (*Harsha Cārita*) written by his brahman courtier, Bāna, as well as the record of his reign preserved for us by China's most brilliant visiting Buddhist pilgrim, "The Master of the Law," Hsuan Tsang, whose journey "In the Footsteps of the Buddha"[1] brought him to Harsha's India from 630 to 644. Only sixteen when he came to power, Harsha eventually ruled most of North India, from Kathiawar to Bengal, during the forty-one years of his reign. Deeply devoted to his sister, whose life he saved as she was about to follow her husband upon the pyre, Harsha was as supportive of Buddhism as Ashoka had been, and he was a poet as well as a warrior-king. As Harsha's domain expanded eastward, he shifted his capital to the Gangetic town of Kanauj, which became one of the major political centers of North India. During Harsha's enlightened reign, Hinduism grew in popularity, developing its classical forms of "worship" (*puja*), which entail devotees bringing their offerings of fruits, sweets, and other delicacies to the icons of gods, who were also worshipped with "devotion" (*bhakti*), as well as by the performance of a number of secret rituals associated with female "power" (*shakti*) that have come to be called Tantric.

The esoteric nature of Tantrism[2] obscures its roots and rituals, though it clearly seems to antedate Brahmanic Aryan religious concepts, harking back to ancient mother-goddess worship and Shaivite forms of worship. The orgiastic character of Tantric ritual—where group intercourse often occurs at "polluted" cremation grounds and is associated with eating meat and drinking alcoholic beverages—is so

[1] The title of René Grousset's account of Harsha's pilgrimage. *In the Footsteps of the Buddha* (London: George Routledge & Sons, Ltd., 1932).

[2] Kees W. Bolle, *The Persistence of Religion* (Leiden: E. J. Brill, 1965).

antipathetical to Brahmanic ethics and the norms of other types of Hindu behavior that we can hardly explain the popularity and persistence of such actions as reflecting anything but the most deeprooted aspects of Indian cultural consciousness. Eastern India, possibly Bengal, appears to have been the regional center from which Tantrism spread throughout the subcontinent, transforming Buddhism as well as Hinduism with its forms of worship. *Vajrāyana* ("Vehicle of the Thunderbolt") Buddhism also emerged here in the seventh century, introducing as its most powerful divinities a number of female saviors, called *tārās,* the consorts of weaker Buddhas and Bodhisattvas. This third form of Buddhism subsequently became most important in Nepal and Tibet. Tantric Hinduism and Buddhism both exalted female power, the generative earth power associated with the mother, as the highest form of divine strength. True worship of the mother through sexual intercourse (*maithuna*) was extolled as divine ritual, but the mores of Brahmanism and earlier forms of Buddhism were by now too deeply entrenched to permit open and public practice of Tantric orgies, which were thus relegated to secret nocturnal sites that would, as a rule, be avoided by all others. Tantric ritual is intimately connected with the practice of yoga, since control of the body and breath are central to the proper performance of *maithuna.* The study of yoga as spiritual discipline was pursued first of all along the eightfold path of "royal yoga" (*rāja yoga*), through self-control and the observance of proper conduct, practice of posture (*āsana*) as well as breath (*prāna*) exercises, and organic restraint and mind steadying, to the perfect achievement of deep meditation (*samādhi*) and the absolute freedom of *kaivalya.* The "yoga of force" (*hātha yoga*), which stressed acrobatics and *maithuna* as paths to salvation, and the "yoga of dissolution" (*laya yoga*), were even more significant for the pursuit of Tantrism.

Yoga was only one of six schools of classical Hindu philosophy that emerged in this era and that continue to be studied throughout India to this day. The *Samkhya* ("numbers") school, closely related to yoga, may also antedate the Aryan conquest, though its oldest surviving text, Isvara Krishna's "*Samkhya* Verses," dates only to about the second century of the Christian era. This philosophy analyzes the world as consisting of twenty-five basic principles, twenty-four of which are "matter" (*prakriti*), the other being "spirit" or "self" (*purusa;* literally "man"). There is no divine creator in this system;

all matter is eternal, uncaused, but basically threefold in its qualities or "strands" (*gunas*): truthful (*sattva*), passionate (*rajas*), and dark (*tamas*). Which of these qualities dominates will determine the "nature" of things and people, virtuous and noble, strong and bright, or inert, dull, mean, cruel, and so forth. The twenty-four forms of matter evolve from *prakriti,* which brings forth intelligence (*buddhi*), giving birth to ego-sense (*āham kāra*), thence to mind (*manas*), from which the five senses emerge, and then the five sense organs, the five organs of action, and finally the five gross elements (ether, air, light, water, earth). *Purusa* stands alone, however, and there are an infinite number of such "men," all equal, each of which unites with the feminine gender *prakriti,* much the way *jiva* was ensnared by matter in Jainism. The ultimate salvation of *purusa* lies in his recognition of separateness and distinction from *prakriti,* allowing the spirit or soul to cease suffering and attain the freedom of true release.

The four other schools or "visions" (*darshana*) of classical Hindu philosophy are generally also coupled: *Nyāya* with *Vaisesika,* and *Pūrva-mīmānsa* with *Vedānta. Nyaya* means "analysis" and is the Hindu system of logic, which teaches salvation through knowledge of some sixteen categories of reasoning and analysis, including syllogism, debate, refutations, quibbling, disputations, and argument of every sort. The *Nyaya* syllogism is more elaborate than the Greek, its typical five-part example being: (1) the hill is on fire; (2) because it is smoky; (3) whatever is smoky is on fire, as in a kitchen; (4) so with the hill; (5) therefore, the hill is on fire. Not only do Hindu logicians insist on a middle example, they also caution against no fewer than five kinds of fallacious middles, making *Nyaya* the world's most intricate and elaborate system of logical analysis. The *Vaisesika* ("individual characteristics") school of Hindu philosophy is sometimes called India's "atomic" system, for its basic premise is the unique character of each element of nature. The material universe emerges from the molecular interaction of the atoms that make up earth, water, fire, and air. There are, however, nonatomic "substances" (*dravyas*) as well, such as soul and mind or time and space, to help explain whatever the atomic theory may leave inexplicable. Salvation in this school, as in the others already noted, is achieved through perfect knowledge, following which the "self" is released from matter and rebirth.

The *Purva-mimansa* ("early inquiry") school was based entirely

on the study of Rig Vedic ritual and sacred texts. For these Hindu
fundamentalists, salvation was equated with the precise performance
of the Soma sacrifice, since everything prescribed in the Vedas must
be taken literally as eternal truth. Less a school of philosophic inquiry
than a remnant of Brahmanic orthodoxy in its most extreme form,
this system attracted fewer and fewer adherents over time. *Vedanta*
("end of the vedas") derives its inspiration from Upanishadic speculation rather than Rig Vedic sacrifice, and is alternately called *Uttara-
mimansa* ("later inquiry"). It has been Hinduism's most influential
philosophic system, developing many subsidiary branches and appealing widely throughout India to leading intellectuals of all ages and
beyond India's borders the world over. Through the monistic principle of Brahman, *Vedanta* philosophy seeks a reconciliation of all
seeming differences and conflicts in Hindu scripture. The greatest
Vedanta teacher was Shankara (ca. 780–820), a South Indian brahman whose school of unqualified monism developed the idea of our
world as illusion (*maya*), the one reality being Brahman, whose name
was also Atman. During his brief but brilliant career, Shankara wandered from his Kerala home to the Himalayas, establishing many religious schools (*māthas*) in his wake, becoming one of the most revered individuals in all of Indian history, a saintly teacher second only
to the Buddha. Several other schools of *Vedanta* later departed from
Shankara's strict monism. The most popular of them was founded by
the great Vaishnava teacher Rāmānuja (ca. 1025–1137), who viewed
Brahman as a divine being, with the world and finite humanity consti-
tuting real parts of his infinitely perfect body. Rejecting *maya,* Ramanuja stressed the importance of *bhakti,* by which he meant intense and
loving meditation and devotion to god, as the surest path to *moksha.*

The *bhakti* movement became most important in South India
during this era, and may be thought of as reflecting popular Tamil reaction to Brahmanic Aryanization, which, though imposed by royal
edict, hardly changed the spirit of common people. Tamil poet-saints,
whether Shaivite *Nāyanārs* or Vaishnavite *Alvārs,* sang of their fervent
devotion to god, of their adoration and love for his spirit and blessed
form, writing with the passion of intoxicated lovers consumed by bliss.
Whether because of South India's climate or because of the fire of
Dravidian language and culture, these poets of Tamilnad added new
vitality to Hinduism that would soon sweep north, weaning masses

from Buddhism and Jainism to the revitalized Hindu fold, whose central tenet became love of god. Throughout this era, the passionate devotional poems of Tamil saints were sung to accompanying music and dance in South India's temples. While the process of Aryanization continued to transform the south, Dravidian ideas and patterns of behavior moved north, and the Deccan bridge between these two polar centers of Indian culture was itself significantly modified by this accelerating ideological traffic.

The political system of South India should not be thought of as a group of competing, centrally developed bureaucratic states, as was the case in the north, but rather as "a multicentered system of power."[3] There were, on the one hand, nuclear areas of village-based agricultural power centered around the drainage basins of the major peninsular rivers, and coexistent centers of relatively isolated upland and forest peoples, who retained their tribal organization and were antagonistic to the villagers. The great warrior families who asserted their authority over these areas, ruling from Kanchipuram as the Pallavas did, or from Tanjore as had the Pandyas and Cholas, were brahman and *sat-shudra* (especially of the *vellala* caste) transmitters of Aryan culture and Hindu civilization. The so-called imperial Pallavas, who were believed to have dominated much of peninsular India by A.D. 600, were, in fact, little more than plunderers, using their urban fortresses to store wealth they looted from most of Tamilnad. Their Chola rivals were equally adept at plunder, and thanks to their seafaring capability, they expanded the horizons of their predatory raids to Southeast Asia as well. The Pallava-Chola era of dominance over South India, which lasted until the twelfth century, was characterized, however, by the continuing struggle between agricultural and forest peoples. Unlike the north with its snow-fed rivers, irrigated agriculture in the south was primarily dependent upon tank and reservoir storage, only secondarily on river channels. Two agrarian institutions that helped integrate nuclear areas were the *brahmadeya* (Brahman-controlled village circle) and the larger *periyanadu* ("great country" provincial assembly). While brahmans controlled the former, each

[3] Burton Stein, "Integration of the Agrarian System of South India," in *Land Control and Social Structure in Indian History,* ed. Robert E. Frykenberg (Madison: University of Wisconsin Press, 1969), pp. 179–88.

region's dominant peasant caste directed the latter. Contemporary inscriptions also refer to itinerant merchant guilds, *nanadesi,* and artisan guilds resembling *shreni,* which seem to fall under the jurisdiction of the *periyanadu,* or at least share with it responsibility for maintaining order within the nuclear zone of settlement. Nuclear areas were relatively self-sufficient economically, and rules of caste as well as other traditional Hindu social processes prevailed within them, though the basic South Indian division of brahman, nonbrahman, and untouchables of either the left-hand (*idangai*) or right-hand (*valangai*) varieties applied. The latter divisions seem to have originated with occupational differences associated on the right hand with agriculture and on the left hand with animal husbandry or artisan labor.

One of the greatest Pallava kings was Mahendra Vikrama Varman I, who ruled Kanchipuram for the first three decades of the seventh century, while Harsha reigned over the north. "Great Indra" Varman was also a poet, credited with authorship of a ribald one-act play called "Sport of Drunkards" (*Mattavilāsa Prahasana*). The rock-cut Pallava temples (*mandapas*) of this period reflect Mahendra's pretensions to imperial power. He adopted the lion as his heraldic symbol, and it would thereafter remain the Pallava dynasty's beast. During the reign of his successor, Narasimha Varman I (ca. 640–68), beautiful monolithic *rathas* ("chariots") were carved at the Pallava seaport of Mahabalipuram, south of Madras and near the mouth of the Palar River, some forty miles up which stood the capital, Kanchipuram. Hewn from granite hillocks, the "seven pagodas" of Mahabalipuram (only one of which remains) are among the earliest and finest examples of Dravidian architecture and South Indian rock sculpture. Though drifting sands and shifting tides have long since altered the coastline, leaving Mahabalipuram a virtually deserted port village, remains of an elaborate tank and canal system also attest to the once-flourishing prosperity of this town, whose merchant princes must have sent richly laden fleets to Southeast Asia and China. Indeed, much of the art and architecture of Cambodia and Java, including Angkor Wat and Barabudur, were certainly inspired, if not produced, by Pallava artists and craftsmen. The subsequent reigns of Rajasimha and Mahendra Varman III (ca. 700) saw the development of temple building both at Mahabalipuram and Kanchipuram. The Shiva and Vaikuntha Perumal temples at the capital were the highpoints of Pallava art and the beginnings of the Dravidian temple style, with its truncated pyramidal tower (*shikhara*) rising over the central shrine.

The Vākātakas of Bundelkhand, whose origins may possibly be traced to Scythian Shakas, had emerged by the Guptan era as rulers over the western Deccan, establishing their center of power around Nagpur in the region of Vidarbha. The rugged tableland of the Deccan, whose rivers all flow east to the Bay of Bengal, remained the scene, nonetheless, of many competing bands of roving warriors and of continuing conflict with those who ruled the more fertile plains of Tamilnad and the Andhra coast. Lack of rain was the spur to Deccan militancy, the barren, serrated hills of this "Great Country" (Maharashtra) serving as a natural training ground for many of India's toughest foot soldiers and horsemen. In the mid-sixth century, a mighty new dynasty, the Chālukyas, seized control of the southwestern Deccan. From their capital just south of the river Krishna at Badami, they consolidated their grip over the Bijapur district under Pulakeshin I (r. 535–66). In the next century, under the mighty Pulakeshin II (610–42), they expanded to the Arabian Sea, conquering the island of Elephanta (*Ghārāpuri*) off Bombay; they marched north to defeat Harsha's army, possibly when the latter sought to invade the Deccan, and east to subdue Kalinga, then south to defeat a Pallava force under Mahendra Varman I. With good reason, then, did Pulakeshin II assume the regal title "Lord of the Eastern and Western Waters." In 641, when Hsuan Tsang visited Maharashtra under Pulakeshin's rule, he reported that the land was peaceful and the people hard working, proud, and totally submissive to their maharaja's will. By this time Badami was some five miles in circumference, a capital of temples as well as forts, and Pulakeshin had hundreds of war elephants under his command, ready if need be to "trample everything down, so that no enemy can stand before them." Yet such were the vicissitudes of Indian power politics that only a year later the great Pulakeshin was killed in battle by the vigorous young Pallava monarch, Narasimha Varman I. After the Pallava occupation of Badami in 642, another Chalukya dynasty emerged in the eastern Deccan, with its capital north of the river Godavari at Pishtapura on the Bay of Bengal. The Chalukyas of Badami later returned and survived in a much weakened reincarnation till the middle

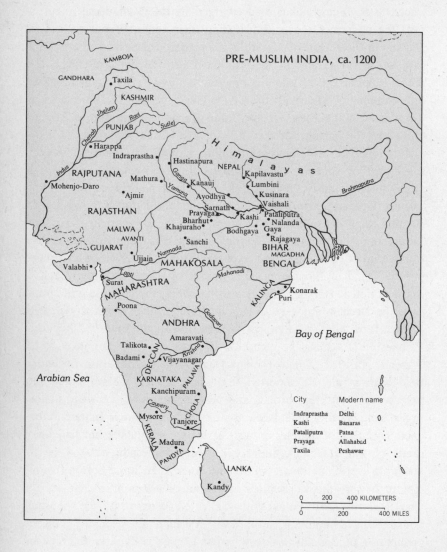

PRE-MUSLIM INDIA, ca. 1200

KAMBOJA

GANDHARA
Taxila

KASHMIR

Jhelum *Ravi*

PUNJAB *Sutlej*

Chenab

Indus Harappa

Indraprastha Hastinapura

H i m a l a y a s

NEPAL

RAJPUTANA Mathura *Ganga* Kanauj Kapilavastu

Mohenjo-Daro *Yamuna* Ayodhya Lumbini

Ajmir Kusinara

RAJASTHAN Sarnath Vaishali

Prayaga Kashi Pataliputra

Bharhut Nalanda

MALWA Khajuraho Bodhgaya Gaya

AVANTI Sanchi Rajagaya

GUJARAT *Narmada* BIHAR

Ujjain MAHAKOSALA MAGADHA

Valabhi *Tapti* BENGAL

Surat *Mahanadi* *Brahmaputra*

MAHARASHTRA

Poona KALINGA Konarak

Godavari Puri

ANDHRA

Amaravati Bay of Bengal

Talikota *Krishna*

Badami DECCAN Vijayanagar

KARNATAKA PALLAVA

Kanchipuram

Cauvery

Mysore CHOLA

Tanjore

KERALA Madura

PANDYA

Arabian Sea

LANKA

Kandy

City	Modern name
Indraprastha	Delhi
Kashi	Banaras
Pataliputra	Patna
Prayaga	Allahabad
Taxila	Peshawar

0

0	200	400	KILOMETERS
0	200	400	MILES

of the eighth century. They were finally defeated and superseded by one of their own feudatories, the Rashtrakutas ("country lords").

Dantidurga, the first of the Rashtrakuta kings, seized power over Badami in 752. The Rashtrakuta capital, however, was established at Ellora, near the modern city of Aurangabad, north of the Godavari, where Dantidurga's successor, Krishna I (r. 756–75), subsidized the excavation of India's foremost rock-cut temple, one of the architectural wonders of the world: the Kailasanatha ("Mount Kailasa") temple to Lord Shiva. Larger than the Parthenon, this unique evocation of Shiva's paradise was carved out of a mountain of solid rock, behind whose facade it still stands hidden till the visitor steps within a narrow entranceway and suddenly finds himself surrounded by monumental figures of elephants, gods, and goddesses under open sky. The genius of Indian stone carving has never been so brilliantly demonstrated, and though the erosive force of time has taken its toll of the intricate lacework designs on many of that temple's portals and pillars, Kailasanatha stands as the single most awesome tribute to Rashtrakuta power. It is one of the finest examples of the Hindu impulse to express worshipful devotion through the medium of visual art. No wonder that the master architect was reputed to have exclaimed after stepping back to view his finished temple, "Oh, *how* did I do it?" There are many Buddhist, Jain, and other Hindu caves at Ellora, some quite beautiful, but there is only one Kailasanatha.

Rashtrakuta power remained ascendant throughout the ninth and early tenth centuries, but by the end of the tenth, a new Chālukya dynasty rose to recapture the glory of its ancestral line in western India, basing itself this time in Kalyāni, northwest of Hyderabad city. By now, however, the classical age was at an end, for India and its civilization were challenged by a new force from the West, a series of stunning invasions that brought with them a new religious system of values and ideas in many ways antithetical to the basic concepts, the very warp and woof, of Hinduism itself.

EIGHT

THE IMPACT OF ISLAM
(ca. 711–1556)

The birth of Islam in the sands of Saudi Arabia in the year 622 was destined to divert fundamentally the course of Indian history. Not since the Aryan dispersion more than two thousand years earlier had any series of invasions had so profound an impact on South Asia as would those that brought the religion of the Prophet Muhammad (570–632) to Indian soil. The historic legacy of that impact may be seen not only in Pakistan and Bangladesh, but also among some hundred million Muslims still living in India.

It is difficult to imagine two religious ways of life more different than Islam and Hinduism. The central tenet of Islam, which literally means "surrender," is that there is but one God, Allah. Every Muslim must at all times surrender himself to Allah's will, revealed to Allah's most perfect prophet, Muhammad, through the archangel Gabriel, and recorded in the sacred Qur'an (or Koran). When he was about forty, the Prophet started to receive revelations, but his preaching won more ridicule and torment than converts in Mecca, finally leading to his "flight" (*hijrat*) north to Medina, after that city invited him to become its temporal and spiritual leader in 622, which is reckoned as the first year of the Muslim calendar. The unequivocal monotheism of Islam served to unite all Muslims (those who surrendered to Allah's will) into a brotherhood (*umma*) that was at once a mighty social as well as military force. To become a Muslim, one needed first of all to affirm the credo: "There is but one God, Allah, and Muhammad is his prophet." For the love of Allah, all Muslims were obliged to give alms (*zakat*) to the poor; to pray five times a day facing Mecca, the center of Arab power, toward whose conquest Muhammad directed his people's united energies after consolidating his grip over Medina; to fast during the ninth lunar month (*Rama-*

dan); and to make at least one pilgrimage (*hajj*) to Mecca. "Holy war" (*jihad*) was waged vigorously against those who failed peacefully to submit to God's will, though Jews and Christians were, from the birth of Islam, given special status as "protected peoples of the book" (*dhimmis*), since their scripture was believed to be based on the "partial revelation" of lesser prophets. As *dhimmis* they could continue to follow their own faiths, providing they paid a special "head tax" (*jizya*), about 6 percent of each individual's total monetary worth, to their Muslim rulers. Pagans, however, were offered only the options of Islam or death. The conquest of Mecca in the Prophet's own lifetime gave Muslims a solid core of Arab power around which to expand their brotherhood in the decades following Muhammad's death. Within a single century, Islam burst explosively across North Africa and Spain and over the Tigris and Euphrates to Persia and India. Never before in world history had an idea proved so contagious and politically potent. Martial fervor, combined with the ethic of social unity that replaced Arab intertribal conflict, made Muslim forces virtually invincible during the first century of their zealous expansion. The last judgment day, when all dead would be raised to hear God's eternal decisions, was a concept vividly articulated by Muhammad, and none were promised better prospects of residing in the cool, fountain-filled gardens of Allah's paradise than those valiant warriors who died in righteous battle.

India remained blissfully oblivious to Islam's existence during the first two decades of that new faith's vigorous growth. Arab merchants, however, brought home enough South Asian wealth to whet the appetites of Muslim warriors, soon to be infuriated as well by Sindi attacks upon Muslim shipping. The Arab commander of the first Islamic force to reach India reported from Sind to his caliph in 644 that "water is scarce, the fruits are poor, and the robbers are bold; if a few troops are sent they will be slain, if many, they will starve."[1] This pessimistic assessment postponed further attempted Muslim conquest until A.D. 711, when the piratic plundering of a richly laden Arab ship as it passed the mouth of the Indus so enraged

[1] "Futu'hu-l Bulda'n" of al'Bila'duri in H. M. Elliot and John Dowson, eds., *The History of India, As Told By Its Own Historians: The Muhammadan Period: Historians of Sind, I* (Calcutta: Susil Gupta Ltd., first ed. 1867, second ed. 1955), vol. 25, p. 17.

the Ummayad governor of Iraq that he launched an expedition of six thousand Syrian horses and an equal number of Iraqi camels against the rajas of Sind. The Arab force swiftly conquered Brahmanabad, and its leader saw to it that "the infidels" (*kaffirs*) he found there either converted to Islam or perished. *Dhimmi* status was, however, soon added to those extreme alternatives, as Muslim scholars learned of Hindu religious books and Muslim monarchs wisely recognized that there were too many Hindus to exterminate. The concept of protected peoples had, in fact, been expanded by now to accommodate Zoroastrians (known as Parsis when they fled Muslim persecution to Western India) during the Islamic penetration of Persia, so it was hardly surprising to find Hindus offered the same option of second-class subjectship in return for special tax payments. What was surprising, however, was how long it took Islam to spread beyond the narrow confines of Sind to other regions of the subcontinent. Nor should we think of Sind as the staging ground for Islam's expansion across South Asia, since it was not from that Arabian Sea backwater, but rather from the Afghan high board of the Khyber Pass that Islam's major invasions would be launched, and then not before the end of the tenth century.

By the tenth century, Islam had changed in many ways, transformed since the founding of the Abbasid caliphate in the mid-eighth century, when it was ruled from Baghdad rather than Damascus or Medina, into an empire embellished by Persian civilization, protected by Turkish armed slaves. Flourishing commerce and industry brought the world's riches to Baghdad, but the empire soon became too vast, too diverse for any one *caliph* ("deputy" of God) to control, and from its fringes, east and west, independent kingdoms emerged under regional rulers, who by the eleventh century assumed the secular royal title, sultan. Central Asian-Turkish slaves (*Mamluks*) had, since the ninth century, been purchased and imported in considerable numbers to bolster the armies of sagging Abbasid caliphs, but by the tenth century entire tribes of Turkish nomads were driven into Afghanistan and Persia by China's expansion to the west. The first independent Turkish Islamic kingdom was founded by a Sāmānid warrior slave named Alptigīn, who seized the Afghan fortress of Ghazni in 962, and from that capital established a dynasty that was to endure almost two hundred years. It was his grandson, Mahmud

(971–1030), the "Sword of Islam," who led no fewer than seventeen bloody annual forays into India from his Ghazni perch, waging his *jihads* at least as much for plunder as for the promise of paradise.

Mahmud of Ghazni began his raids in 997, leaving his frostbitten capital each winter to descend into the Punjab's plains, zealously smashing countless Hindu temple idols, which he viewed with Islamic iconoclastic fervor as abominations to Allah, and looting India's cities of as many of their jewels, specie, and women as he and his horde of Turkish cavalry could carry back across the Afghan passes. Thanesar, Mathura, Kanauj, Nagarkot, and finally even the fabled Kathiawar temple city of Somnath were singled out as targets to be razed by Mahmud, who used their fabled wealth to convert remote Ghazni into one of the world's greatest centers of Islamic culture in the eleventh century. The brilliant physician, astronomer, philosopher, and historian al-Biruni (b. 973), and the great Persian poet Firdawsi, author of the *Shāh Nāma,* were but two of Islam's luminaries lured to Ghazni by Mahmud. Though the court chronicler Utbi clearly exaggerated his sultan's prowess when he claimed that ten thousand Hindu temples were destroyed in Kanauj alone by Mahmud's sword, it is not difficult to appreciate the legacy of bitter Hindu-Muslim antipathy left by raids that may have taken even 1 percent of that toll. In 1025, for example, the Hindu inhabitants of Somnath were reported to have stood calmly watching the advance of Mahmud's fierce army toward their temple city's walls, confident that Shiva, whose "miraculous" iron *lingam* hung suspended within a magnetic field inside Somnath's "womb-house," would surely protect his worshippers from harm. Here, too, the chronicler probably exaggerated, for he wrote that more than fifty thousand Hindus were slain that day and that over two million dinars' worth of gold and jewels were taken from the hollow *lingam* shattered by Mahmud's sword. Yet the bitter shock of such attacks, whatever the factual sum of their deadly impact, was even more painfully amplified in the memories of those who had watched helplessly as friends and family were slain or enslaved by invaders who claimed to kill, rape, and rob in the name of God. Before his death, Mahmud annexed the Punjab as the Easternmost province of his empire.

The Ghaznavids were but the first in a series of Turko-Afghan Muslims to invade North India and shatter autonomous Hindu power, initially in the Punjab, then further south and east. A century and a half after Mahmud's death, Ghazni itself was seized by rival Turkish Ghurids, also fierce Central Asian nomads in origin. Sultan Muhammad of Ghur and his slave lieutenant Qutb-ud-din Aybak led their first raid into India in 1175, destroying the Ghaznavid garrison at Peshawar in 1179, capturing Lahore in 1186 and Delhi in 1193. Muhammad returned to Ghazni, leaving his lieutenant to consolidate the Ghurid grip over North India from Delhi. For talented Turkish soldiers like Qutb-ud-din, martial slavery proved the highroad to power, and as an institution in eastern Islam, it was an important means for permitting bright, courageous, and loyal young men to rise rapidly above their status at birth. The Turks' Central Asian horses and crossbows fired at the gallop (thanks to the iron stirrups with which Ghurid marksmen could brace themselves as they rode) tipped the martial balance of power in favor of Qutb-ud-din's forces, but Hindu India did not surrender its homeland without a struggle.

The Rajputs (literally "king's sons") waged the staunchest battle, keeping the Turko-Afghans busy for years seeking in vain to drive these ferocious *kshatriyas* from their desert domain. Though claiming direct descent from either the Aryan sun (*sūrya*) or moon (*chandra*), all four of the major Rajput dynasties(Pratihāra, Paramāra, Chauhān, and Chalukya) probably originated in Central Asia themselves, which may help to explain their success at warfare and their singular tenacity. They stood, at any event, as the vanguard of Hindu India's spirited opposition to the Muslim conquest, and even when defeated in battle or driven from one desert fortress after another, they never completely surrendered. Bengal (including Bihar, Assam, and Orissa) had prospered under a series of independent dynasties, first the Pāla, then the Varman, and finally the Sena, whose capital at Nadia was conquered by Turko-Afghan power in 1202. India's major centers of Buddhism, including the great university at Nalanda, where more than ten thousand monks lived and studied, were sacked at this time, driving thousands of Buddhists to flee toward Nepal and Tibet and killing uncounted others who weren't swift enough to escape. The severity of Turko-Afghan persecution directed against centers of Buddhist monasticism was so unrelenting that the religion of the Buddha was now sent into exile from the land of its birth, never to return again in any significant numbers until 1954 when B. R. Ambedkar, India's learned leader of Hindu untouchables, publically converted to Buddhism with some fifty thousand of his followers as a political protest. Though Buddhism flourished in Nepal, Tibet, China, Japan, and most of Southeast Asia after its diaspora, the *sangha* found no sanctuary on Indian soil for some seven and a half centuries. Perhaps Vaishnavite adoption of the Buddha as an avatar, coupled with the growing popularity of the *bhakti* movement, Hindu asceticism, and subsequent Muslim mysticism, sufficed to satisfy the spiritual needs of those Indian Buddhists who survived by conversion.

In 1206, Muhammad of Ghur was assassinated in Lahore, following which Qutb-ud-din Aybak proclaimed himself sultan of Delhi, initiating a new phase in Muslim relationships toward India and the dawn of a series of Islamic dynasties in South Asia. The founding of Qutb-ud-din's "Slave" (Mamluk) dynasty transformed North India into *Dar-ul-Islam* ("Land of Submission") from *Dar-ul-Harb* ("Land of War"). The sultanate at Delhi lasted 320 years, including five successive Turko-Afghan dynasties. Qutb-ud-din died when he fell from his polo pony in 1210 and was succeeded by his capable son-in-law, Shams-ud-din Iletmish (Iltutmish), who reigned from Delhi for a quarter of a century (1211–36). Faced when he took the throne with a number of political challengers in the west, Iletmish wisely exerted himself first in consolidating his strength around Delhi and over the Gangetic plain, winning support from most of the Turkish military bureaucrats (*iqtā'dārs*) by confirming grants of revenue from landed areas (*iqtā'*), which had been assigned by Qutb-ud-din. By diplomacy he kept the armies of Ghengiz Khan from invading Delhi. He assured all Hindus the status of *dhimmis* and left many local Hindu chiefs and petty rajas in control of their domain as long as they paid the requisite revenue to his treasury, which never was empty. Relatively mild in manner, Iletmish was hailed by the chroniclers and poets of his reign as a devoutly religious monarch who preferred the tactics of persuasion and tolerance to those of naked power. He could, nonetheless, vigorously suppress opposition, as he did in Bengal, where rebellion flared from 1225 until his victory there in 1229. His campaigns against the Rajputs were less decisive, though he did capture Gwalior in 1231 and Ujjain in 1235. By his death the following year, the Delhi sultanate was clearly the most powerful state in North India, and his vigorous daughter Raziyya succeeded her father to the throne, which she managed to hold for three years. Raziyya was the only Muslim woman ever to rule on Indian soil, but she would not be the last remarkably astute woman in Indian history to succeed her father to the pinnacle of political power in Delhi. Raziyya was murdered in 1240, and her father's palace guard of Mamluks, "the Forty" as they were collectively called, ruled conjointly for the next six years.

The shrewdest and most unscrupulous of the Forty was Balban, who had served as Sultana Raziyya's chief huntsman. He seized effective power in 1246 by becoming grand chamberlain (*amir-i-hajib*) to the puppet king Bahram, and he ruled as such until 1266, when he assumed the title as well as the power of sultan, which he retained for twenty more years. The continuing struggle for power and political intrigue among the Forty at Delhi left provincial Turkish governors, east and west, to develop what were virtually independent regional power bases, several of which, notably Oudh and Sind, sought to challenge Balban's central authority in 1255. Balban's leadership prevailed, however, and he not only defeated his own governors, but he also kept the Mongols at bay by a combination of firmness and diplomatic wooing. While there is good reason to believe, as Professor Nizami has argued,[2] that Balban seized his throne by regicide, he assiduously labored to exalt the position of sultan, once he occupied it, to divine status. Always surrounded by his elite corps of palace guards, Balban spoke to none but his leading officials and was as impartially ruthless in disposing of his own relatives as he was in dismissing strangers. He never hesitated to stamp out any opposition in the environs of his court by calling for his elephants, and he used the glittering facade of Persian ritual and royal pomp to overawe his populace, becoming more of a grand monarch in the traditional Hindu mould than the "first among equals" ideal of orthodox Islam. Prostrating and toe kissing were both insisted upon by this slave who became sultan. Employing assassins to poison all of his former comrades among the Forty, Balban ended his long reign with no friends to remind him of his former status. In his constant dependence upon spies and his enhancement of the espionage arm of the sultanate, Balban

---

[2] Khaliq Ahmad Nizami, "Balban, the Regicide," in *Studies in Medieval Indian History and Culture* (Allahabad: Kitab Mahal, 1966), pp. 41–53.

was employing traditional Indian techniques, reaching back at least to the *Arthashatra* for prescriptions of how to retain power. He did, however, restrict the highest posts in his service to pure-born Turks, but as his bureaucracy grew, he found it necessary to admit Afghans and Indian-born Muslims to its middle ranks, especially in the army, which was expanded and greatly strengthened. With Balban's death in 1287 his dynasty also expired, though it continued nominally for three years longer, inept and competing grandsons standing as puppets for rival court factions while the bureaucratic machine managed to maintain its own momentum.

Jalāl-ud-dīn Fīrūz Khaljī used his position as general of Balban's army to catapult himself to the sultan's throne in 1290, launching a coup that inaugurated the second Delhi dynasty. Originally Turkish, the Khaljis had moved to Afghanistan and then settled in India after the Ghaznavid and Ghurid invasions. Jalal-ud-din was in his seventies when he seized power, and could only retain it for six years, but the transition from Slave to Khalji dynasties marks still another step in the Indianization of the sultanate, which now more freely relied upon the support of indigenous peoples. Jalal-ud-din himself did little more than repress Balban's supporters; the short-lived yet significant dynasty that he began left its mark on Indian history, however, due to the reign of his nephew-successor, Alā-ud-dīn, who had his uncle murdered in 1296. Shortly before usurping the throne, Ala-ud-din Khalji ventured into the Deccan to raid and loot the wealthy capital of Devagiri, home of the Yadavas, and with the gold he carried north from that plundering venture bought the loyalty of those who took his uncle's life. South India thus served involuntarily to bolster the Delhi sultanate, whose fortunes were now linked in some measure with the Deccan and Tamilnad, two regions hitherto oblivious to the impact of the Muslim invasions.

The northern Deccan Yadavas (descendants of the Yadu tribe) were feudatories of the later Chalukyas who had been weakened in struggling against neighboring Hindu feudal states to their south, especially the Hoysalas of the Mysore region and the Kākatiyas of Andhra to the east. These medieval Hindu monarchies were essentially feudal in origin, usually starting with the grant of tax-free land to a warrior by his overlord. The feudatory commonly referred to himself as *sāmanta* ("tributary" or "high feudatory"), until he felt powerful

enough to adopt the royal title of *maharaja* instead or in addition to
mahasamanta, which was at times added to the former. As long as the
feudatory acknowledged his overlord's suzerainty, he was obliged to
maintain troops for his lord's defense, remit some portion of his revenues, and offer the lord his daughters in marriage. Though Indian
feudalism differed significantly from that phenomenon as developed in
Western Europe, including neither large-scale demesne farming nor a
widespread manorial system, it bore sufficient resemblance to validate
use of the term, which primarily connotes political decentralization
and local economic self-sufficiency, as well as the conditional holding
of land by lords in return for the promise of martial service. The
monarchal granting of tax-free estates, most commonly called *jāgīrs,*
as reward to soldiers would remain a common feature of later imperial
as well as feudal systems throughout Indian history until the late nineteenth century, when such alienated revenue was to be reclaimed by
the British. Tax-free domains were also often granted to brahmans
and their devout followers by Hindu monarchs, but no single church
hierarchy ever emerged under Indian feudalism as it did in the West.

Tamilnad also remained fragmented at this time, though Chola
attained greater prominence and wealth from its growing trade with
and expansion into Southeast Asia. The Chola monarch, Vijayālaya
(r. 846–71), managed while still a feudatory of the Pallavas to capture Tanjore in 850, providing his heirs with a rich, temple-city center
of power. During the following reign, Kānchipuram fell under Chola
control, and after that Parantaka I (907–55) virtually established a
Chola kingdom by ousting the Pandyans from their capital at Madura.
During the later, vigorously expansionist era of Chola King Rājārāja
(r. 985–1016), Kerala was conquered, as was Ceylon. Under Rajaraja's son, Rājendra I (r. 1016–44), Chola armies even marched
north of the Tungabhadra to wreck havoc upon the western Chālukyas
of Kalyani. The Chola navy was by this time the mightiest in the
Indian Ocean; not only did it conquer the Maldives, but in 1025 it
defeated the fleet of Sri Vijāya, the great maritime empire of Southeast Asia that spanned Malaysia and modern Indonesia.

Many Chola bronzes of Buddhist as well as Shaivite divinities
have been found in the archipelago dating from this era of South
Indian domination. The art of bronze casting reached its peak of perfection in the Chola temple cities of Tanjore, Kānchipuram, and Chi-

dambaram, where the anonymous tenth-century genius labored who produced Shiva as *Nataraja* ("Lord of the Dance"), casting bronze the way Benvenuto Cellini poured gold, with inspired brilliance. Poised with one foot crushing the back of the demon dwarf Muyalaka, Shiva stands ready to begin his cosmic dance of life, to restore vitality to a world in ashes, raising his mighty left leg high above the burial ground of forgotten dreams, holding the little drum of awakening in the fingers of one hand, the fire of creation as well as destruction in another, and gesturing with one of his two remaining hands to "forget fear" (the *abhaya mudra* of the Buddha). Rarely has an artist achieved such perfect balance and harmony in any medium as in this metal statue, whose patina of green oxide only enriches its beauty, whose symbolism embraces all of Hindu civilization in its mythic power, and whose exquisite form has led to its almost infinite reproduction throughout India, though none of the later versions have ever approached the Chola craftsmen's quality of perfection. Whether their subject was Shiva or the goddess Lakshmi, whether it was Rama or Krishna with his consorts and *vāhana* (Garuda), the worksmanship of the imperial Chola artists was flawless in its beauty, magically transmuting metal to fleshlike texture, imparting the breath of life to their subjects. In temple architecture, the Cholas built upon foundations inherited from the Pallavas, and the Shiva temple in Tanjore, completed in about 1000, is a monument worthy of the greatness of Rajaraja. It was the most ambitious and enormous Hindu temple of its time, the forerunner of the giant Dravidian temples that still distinguish the landscape of Tamilnad. A massive pyramid of stone, the *shikhara* ("tower") rises 190 feet above the inner shrine, a pillar of worship visible for miles across the countryside. A second great Chola temple, the Gangaikondacholapuram, erected by Rājendra I in about 1025, was famed for its huge gateway "cow fortresses" (*gopuram*) and is virtually all that remains of the once-flourishing Chola city of Kumbakonam. Chola maritime power continued to exploit the resources of the east coast of India, while Muslim Arab merchants seized control of almost all Indian trade on the west coast and across the Arabian Sea.

During the reign of Ala-ud-din Khalji (1296–1316), the sultanate reached its peak of centralized power and acquired imperial dimensions. Resuming many of the tax-free *jagirs* that had previously been assigned to Muslim nobles, Ala-ud-din paid his officers in cash and kept personal control over his mighty army. To maintain a full treasury, the sultan raised the land tax to 50 percent of each crop and strictly enforced its collection from all Hindu subjects in his realm. He also introduced two new taxes, one on milch cattle, the other on houses. His network of spies and loyal courtiers was efficient enough to make him more feared than hated, and his homosexual relationship with the convert slave, Malik Kafur, the second most powerful figure in the sultanate, perhaps helps account for the singular intrigue at his court. He was a cruel but remarkably capable monarch, the first to introduce a successful system of wage and price controls in Delhi. Private hoarding of gold and silver, common throughout much of Indian history, was temporarily ended during Ala-ud-din's reign. The prices of food, grains, and cloth were kept low enough to permit soldiers and average workers to survive without high salaries. Merchants were licensed, and their profits were kept under strict state control; peasants were obliged to sell their grains only to registered food merchants at fixed prices. Hoarding was forbidden and, if discovered, severely punished.

Such economic controls were beneficial to the majority of Delhi's population, but they could not be enforced much beyond its environs, and they aroused opposition from merchants as well as neighboring peasants, who felt discriminated against. More serious than such internal opposition, however, were the fierce Mongol invasions Ala-ud-din faced from 1303–6, threatening to destroy most of Northwest India's cities. Thanks to his economic reforms, however, Ala-ud-din managed to raise sufficiently powerful armed forces to confront the Mongol horsemen, eventually driving them back to Afghanistan. With his northern marches secured, Ala-ud-din turned south once again in search of more gold. The Yadava capital of Devagiri was taken in 1307 and used as a springboard for still deeper Khalji penetration into the Deccan. Many Rajput chiefs were defeated by Ala-ud-din's army, as were those of Gujarat. In 1309, the South Indian kingdom of Kākatiya was captured, and a year later Khalji armies raided into Pandyan territory in Tamilnad, bringing the Muslim faith to peninsular India's southernmost tip. Ala-ud-din's imperial conquests were never consolidated by his successors, however, for with his death in 1316, his line collapsed. Malik Kafur sought unsuccessfully to retain control of both court and army, but he was murdered by his own soldiers. One of Ala-ud-din's sons, Qutb-ud-din Mubarak, survived his father's death by four years, retaining tenuous hold of the reins at Delhi by abandoning all attempts to keep commodity prices down, lowering revenue demands, and generally relaxing administrative authority while he enjoyed his transvestite pleasures at court.

For most of the fourteenth century, the Tughluqs ruled Delhi. The founder of this third Muslim dynasty in India was Ghiyas-ud-din Tughluq, the son of a court Turkish slave and a Hindu Jat woman, who ruled mildly for five years. In 1325, Ghiyas-ud-din and his favorite son were both killed when a victory pavilion erected by his other son and successor, Muhammad (r. 1325–51), suddenly collapsed. Catapulted to power over the corpses of his father and brother, Muhammad Tughluq's long reign may be read as a search for religious expiation of this guilt by a man of not inconsiderable sensitivity and talent. The Muslim world traveler Ibn Battuta, who journeyed through Asia and Africa from 1325 to 1354,[3] served for a while as chief judge (qazi) at Muhammad's court and recalled how strict the sultan was about compelling "the people to master the ordinances for ablutions, prayers, and the principles of Islam." He could not, however, succeed in teaching others to do what he personally ignored, and rebellion was widespread throughout this era. In a grandiose attempt to subdue southern resistance to his rule by establishing a second capital in the Deccan, Muhammad Tughluq forced many nobles and officials to abandon their homes in Delhi in 1327 and journey south over five hundred miles, across the Vindhya-Satpura divide to Daulatabad (Devagiri). Many died in that arduous journey. For those who survived, moreover, Deccan heat and lack of perennial river water made Daulatabad an inhospitable site for the sultanate's secondary capital.

In 1329–30, Muhammad Tughluq attempted another surprising innovation, the issue of new currency. Possibly hoping to emulate the imperial Chinese, whose use of paper currency was quite successful, the sultan issued brass or copper tokens that were proclaimed to be the equivalent of the increasingly rare silver tanka (140 grains). The scheme was apparently as well motivated as any paper currency issue

[3] *Ibn Battuta's Travels In Asia and Africa, 1325–1354,* trans. and selected by H. A. R. Gibb (New York: Robert M. McBride & Co., 1929).

and might have worked were it not for the fact that foreign merchants refused to take tokens for their goods, even though they used them readily within India. What is more, Indians were permitted to turn in their coppers at the royal mint for silver or gold, and they did so more vigorously each month. Indeed, most Indians seem to have become kitchen coppersmiths in the wake of this "reform," and mountains of Tughluq copper tokens were subsequently said to have remained lying for a century outside the sultanate's treasuries. Within three or four years, the sultan was obliged to withdraw his special coins because of the heavy loss of public treasure. From 1335 to 1342, India suffered one of its most severe and prolonged periods of drought and famine. Although all accounts of foreign travelers make note of the plaintive suffering at that time, the sultanate made no concerted effort to assist its subjects by general tax relief or food distribution. Widespread economic discontent led to rebellions throughout the sultanate, which proved most effective in the south. In 1335, an independent Sultanate of Madura was established by the Tughluq governor Ahsan Shah. When Muhammad's army moved south to subdue that fire, others raged in Lahore and Delhi, forcing the sultan to return north. Hindu chiefs, noting the success of Muslim officers in their Tamilnad rebellion, raised similar banners of independence from Delhi. Muslim intrusion into the Deccan had driven many Hindu warriors south of the Tungabhadra River, where a new Hindu kingdom arose in 1336, named for its capital, *Vijāya-nagar,* "City of Victory."

Harihāra I (r. 1336–57), founder of Vijayanagar, had converted to Islam in order to serve the Tughluqs as a governor in his southern homeland, but he reconverted to Hinduism and quickly became leading tributary overlord of the southern Deccan. Winning the support and allegiance of the most powerful local landholders (*nāyakas*), Harihara conquered the domain of the Hoysala overlord, Ballāla III, and developed his urban capital as the most "important place" in the region of peninsular India. By 1343 Vijayanagar claimed sovereignty over all lands hitherto tributary to the Hoysalas, but was less successful later in seeking to extend its control north of the Tungabhadra. Rebellion against Muhammad Tughluq's reign broke out in Daulatabad and other Deccan cities in 1345, led by discontented Muslim nobles, among whom was Hāsan Gangu, who proclaimed himself sultan of the Deccan in 1347. Taking as his title-name

Ala-ud-din Bahmān Shah, Hasan founded the Bahmāni dynasty, the mightiest and longest-lived Muslim dynasty of the Deccan, which remained united for some two hundred years and survived in fragmented form for a century after that. For the last half of the fourteenth century and all of the fifteenth, the Bahmani dynasty fought a series of ten indecisive but bloody wars against Vijayanagar over the fertile *doab* soil between the rivers Krishna and Tungabhadra, which marked their disrupted boundary. South India was thus effectively removed from the control of Delhi's sultanate, which had reached its peak of power and entered a phase of rapid deterioration following the death of Muhammad Tughluq.

Bengal declared its independence from Delhi in 1338. Following a deadly struggle among the region's leading *māliks* ("nobles"), Mālik Hāji Ilyās emerged victorious, assuming the title of Sultan Shāms-ud-din (r. 1339–59) and founding the Ilyās Shāhi dynasty, which ruled over Bengal from its capital at Lakhnawati for almost a century. Bengali independence from West Indian Muslim rule was thus asserted early and maintained until the sixteenth-century Mughal conquest. Nor was it language and political independence alone that differentiated the region of Bengal from the rest of North India in this era, for even the form of Islam predominantly practiced there was peculiarly attuned to its cultural character and ancient heritage. Sufism, Islam's mystic thread, which evolved primarily as a legacy of Persian influence upon Islamic orthodoxy, struck a responsive chord in the mass of Bengal's population, especially among the lowest class of Hindu outcastes and former Buddhists, who were left without a priesthood to turn to for spiritual guidance after 1202; it appealed as well to many Muslims, for whom it revitalized the message of Islam.

Three orders of Sufi saints had appeared in India by the thirteenth century: the Chishtī, Suhrawardi, and Firdawsi, all of which appealed to the same mystical yearning for union with God, experienced by so many Hindus as well as Muslims and other "God-intoxicated" seekers the world over. The fervent love of God, which played so important a role in the passionate yearnings for "the mother goddess" that had characterized Bengali religious consciousness from time immemorial, may well have its roots in the same lush alluvium, fed by torrential monsoons which gave birth to "golden" jute, the premier crop of Bangladesh. The ineffable blend of love and sorrow,

of human frailty and poverty amid the awesome forces of nature's
lush wealth, helped fashion the uniquely sensitive, passionate, and
mercurial temperament characteristic of Bengal; it seems also to have
predisposed millions of Bengalis to the ardent appeal of Islamic mysticism. The wandering *pirs* (Sufi preachers), who went into Bengal's
remote villages to bring their message of divine love to impoverished
peasants, sounded much the same as Hindu *bhakti* ("devotional")
saints offering salvation through worship of the mother goddess, or
Mahayanist Buddhists bearing the promise of divine salvation by
grace of a Bodhisattva's blessing. Bengali translations of epic Sanskrit
literature were encouraged by court patronage during this era of independent Muslim rule, serving at once both to weaken the grip of Brahman pundits over the people and to provide the national mortar of a
literary vernacular language to this culturally fecund region. Under a
succession of independent Muslim dynasties, Bengal was to retain its
sovereign status, free of interference from Delhi, till the peak of great
Mughal power and Akbar's conquest in 1576.

When Muhammad Tughluq was killed fighting rebellion in Sind
in 1351, his orthodox cousin, Firuz, ascended the throne of Delhi,
from which he reigned for thirty-seven years (1351–88). Famed for
his abolition of torture, his passion for building, and his lifelong ad-
herence to the tenets of Islamic orthodoxy, Firuz was blessed with a
reign of good monsoons and abundant harvests. He sought in vain to
reconquer Bengal, but soon contented himself with building a new
capital instead of carrying on futile wars. The new Delhi constructed
in his name, Firuzabad, was replete with gardens as well as mosques
and colleges. His generous patronage of the *ulama* gave Firuz the best
"historic press" of all the sultans of Delhi, and he is credited with having constructed no fewer than forty mosques, thirty colleges, a hundred hospitals, and two hundred new towns in the environs of the
capital. He was also attracted by various irrigation schemes, supporting the construction of some fifty dams and reservoirs and bringing
hitherto barren land into productive use. Less suspicious, perhaps,
than his predecessors, Firuz cut down on the royal spy corps, investing more heavily in productive enterprises. He seems to have been one
of the most intelligent, if not enlightened, monarchs of the sultanate,
although his orthodoxy led him to a number of hostile actions against
the Hindus, which ultimately alienated the majority of his subjects.

Brahmans felt particularly persecuted, since they had hitherto generally been exempt from paying the *jizya* (as they were from the payment of all taxes) but were now obliged to do so. Many nonbrahmans
converted to Islam during Firuz's reign, thus escaping the poll tax
entirely, though they were still faced with the obligation of paying the
less onerous Muslim *zakat* (alms).

Firuz was the last of the strong sultans of Delhi. Within a decade
of his death, half a dozen transient monarchs held tenuous sway over
the precarious fortunes of that fast-declining kingdom. While rival
factions fought among themselves in Delhi, the Central Asian armies
of Timur the Lame (Tamerlane), waiting beyond India's northwestern gates, swooped down through the passes to plunder the Punjab
and in 1398 entered Delhi itself. The death and plunder of Old Delhi
by Timur's forces left "towers built high" with the heads and ravaged
bodies of Hindus. Tens of thousands of slaves were dragged away as
living booty, and the great mosque at Samarkand was later to be rebuilt by the stonemasons of Delhi. Timur abandoned the Indian plains
before the hot weather in 1399, leaving a trail of blood and torture
behind him as he moved north to his homeland; Delhi was so spent in
the wake of his orgiastic attack that for months the city lay in the
death throes of famine and pestilence, "not a bird moving." Slightly
more than a century later, Timur's great grandson, Bābur, would return to found the Mughal dynasty on that same site.

The immediate aftermath of Timur's invasion was accelerated
fragmentation of the sultanate, with Gujarat declaring its independence under Zafar Khan in 1401. Gujarat flourished on its trade from
the West, and its capital at Ahmadabad came to be known as the
"Venice of India" by the mid-fifteenth century. The magnificent
mosques and sumptuous palaces constructed in that era remain to this
day in the midst of modern cotton mills, Jain palaces, and British
"bungalows" as ample testimony of the wealth that trade and political
independence poured into this most enterprising region of India. The
Hindu Gujarati merchant caste, called *bania,* soon aroused the envy
of other Indians and foreigners alike, thanks to their singular good
fortune. To the east of Gujarat, a separate Sultanate of Malwa was
established in 1401 by Dilavar Khan, an Afghan noble, who also
found himself liberated from monarchic control in Delhi. Though
barren Malwa lacked the affluence and power enjoyed by Gujarat, its

strategic position between Gujarat and the Gangetic plain gave it access to the rich caravans obliged to traverse its domain. Through their plunder, and ultimately by the imposition of transit tolls on all merchandise, Malwa's sultanate managed to amass both a fortune and a power base of its own.

Two dynasties claimed the mantle of Delhi's tattered sultanate after the searing wounds inflicted by Timur began to heal. The first, whose surname, Sayyid, should indicate descent from the Prophet, was founded by the Turkish Khizr Khan in 1414 and retained control over Delhi until 1450. The Sayyids were deposed by the Lodis, a clan of Afghans, who were raised to "noble" status for their success at horse breeding and given control over much of the Punjab by their predecessors in Delhi. Buhlul Lodi reigned over Delhi and the Punjab for nearly forty years (1451–89), during which time significant numbers of Afghan peasant-herdsmen settled in North India. Buhlul Lodi's son, Sikāndār (r. 1489–1517), has been hailed by the historians of his court as the wisest and most dedicated, hard-working, and far-sighted sultan ever to sit upon Delhi's throne. He wrote poetry himself and invited scholars of every sort to his side, encouraging the compilation of books on medicine (*Ma'dan-ul-Shifa*) as well as music (*Lahjat-i-Sikāndār Shāhi*). The stimulus to learning and cultural creativity provided by Sikāndār's patronage seem, moreover, to have helped several important syncretisms of Islam and Hinduism to emerge at this time, though Sikandar himself was surprisingly orthodox. Perhaps the fact that his mother was Hindu and that he had early fallen in love with a Hindu princess drove him to seek by periodic displays of temple-razing and iconoclastic fervor to prove himself more orthodox than the *pirs* and *mullas* of his realm. At any rate, several new sects of mystic toleration, blending the best of both great schools of religious thought, came to life before Sikāndār expired.

Before the end of the fourteenth century, the wave of *bhakti* Hinduism, born in South India, had reached the Ganges, along whose banks at Banaras Ramanuja's greatest disciple, Rāmānanda, settled to preach his doctrine of divine love. Among the many disciples of Ramananda was an illiterate Muslim weaver named Kābir (1440–1518), whose devotional poems and dedicated life inspired millions of followers to abandon their sectarian perceptions of Islam and Hinduism in favor of his syncretic path of simple love of God. In the

Punjab, Nānak (1469–1538), born a Hindu but reared on the democratic doctrines of Islam, rejected caste and became the first *Guru* ("divine teacher") of the *Sikh* ("disciple") faith, which he founded. Conceived by Nanak as a doctrine of loving devotion to the "one God, the Creator," whose name was Truth—*Sat*—this religion only later became a martial one, with subsequent gurus driven to take up the sword by Mughal persecution. In Bengal, the most popular of all *bhakti* Hindu preachers was the teacher Chaitānya (1485–1533), whose intoxicated devotional frenzy was such that his disciples believed him to be a reincarnation in one body of the divine lovers, Krishna and Radha. The syncretic sect founded by Chaitanya was centered around *Satya-Pir* ("Truth-Saint") worship, which proved singularly successful in attracting Bengalis to its fold.

By the dawn of the sixteenth century, India was thus not only fragmented politically, but divided spiritually into many religio-philosophic camps. The last of the Lodis, Ibrāhim (1517–26), was unable to command the allegiance of those who had served his father, and soon found himself confronted with rebellion in Bihar, then Lahore, and finally from the Hindu Rajput confederacy, led by Rānā Sāngā of Mewar. So desperate was the political struggle for power in the north that Western Europe's vanguard, the Portuguese, who had reached the Malabar coast in 1498 and since then returned in ever-greater force to secure a tochold of trade, went unnoticed, undiscussed in Delhi. For Ibrahim, the mortal and immediate threat came not a thousand miles away by sea, but galloping over the flat plain from nearby Lahore, whose city gates had been flung wide in welcome to the king of Kabul, Babur (1483–1530), the "Tiger," descended from Timur the Barlas Turk on his father's side and Ghengiz Khan the great Mongol on his mother's. Daulat Khan, the governor of Lahore, had invited Babur to "save" him from Ibrāhim, and the Tiger, who wrote in his *Memoirs,*[4] that for the past twenty years he had "never ceased to think of the conquest of Hindustan," readily accepted the invitation. He came to India, however, not as an ally of the viceroy of the Punjab, but as founder of the greatest Muslim dynasty in Indian history, as first *pādishāh* ("emperor") of the Mughals.

---

[4] *Memoirs of Zehir-ed-Din Muhammed Babur,* trans. J. Leyden and W. Erskine, rev. by L. King (London: Oxford University Press, 1921), 2 vols.

"I placed my foot in the stirrup of resolution and my hands on the reins of confidence in God," wrote Babur, "and marched against Sultan Ibrahim . . . whose army in the field was said to amount to a hundred thousand men and who . . . had nearly a thousand elephants." They met on the field of Panipat, just north and to the west of Delhi. Babur had some ten thousand of his own Chagatai Turks and Afghans at his command, including Central Asia's finest horsemen, and brought matchlock *feringi* ("foreign") cannon, which he lashed together with rawhide and positioned in front of Ibrahim's advancing wall of elephants. "The sun had mounted spear-high when the onset began," Babur recorded in his *Memoirs,* "and the battle lasted till mid-day, when the enemy were completely broken and routed. By the grace and mercy of Almighty God this difficult affair was made easy to me, and that mighty army, in the space of half a day, was laid in the dust."

The date was April 21, 1526, dawn of the Mughal Empire, Babur seized the treasure stored in Delhi and distributed it liberally among his troops, who advanced without rest to Agra, some one hundred miles south along the river Yamuna (Jumna), conquering that city, which was to become the twin capital of the Mughals, the next day. All resistance to Mughal power was not as yet defeated, however, for Ibrāhim's brother Mahmud escaped from the carnage at Panipat to raise an army in Bengal, and the Rajput Confederacy was marshaling its formidable forces round the banner of Rana Sanga (r. 1509–28). Worshipped by his people as heir to Rama's throne and "child of the sun," the Rana of Mewar stood first among the thirty-six royal tribes of the Rajputs.[5] Almost alone among the confederacy, Mewar's dynasty had survived the centuries of Muslim invasions and rule, thus enhancing its primacy and sacrosanct status, and Sanga was the greatest Rana of his noble line. Bearing scars from eighty wounds won in battle, the one-eyed, one-armed leader of Hindu India's united stand against Mughal power almost won his claim to divine descent on the field at Khānuā, less than forty miles west of Agra, in March 1527. His much larger force had Babur's army surrounded. To bolster the demoralized spirits of his besieged men,

[5] For a history of Mewar, see James Tod, *Annals and Antiquities of Rajast'han,* new impression (London: Routledge & Kegan Paul, first pub. 1829, reprinted in 2 vols. 1957), vol. I, pp. 173–401.

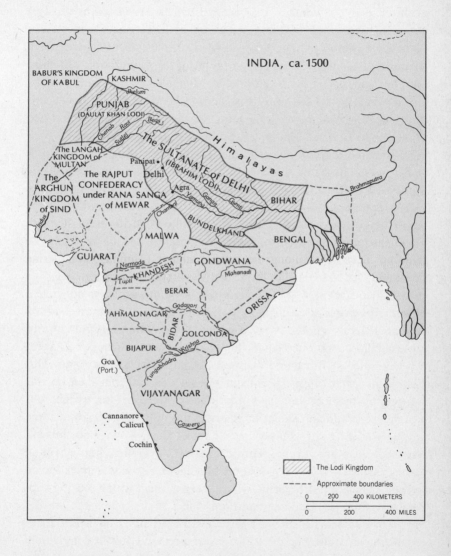

INDIA, ca. 1500

BABUR'S KINGDOM
OF KABUL

KASHMIR

Jhelum

PUNJAB
(DAULAT KHAN LODI)

Chenab *Ravi* *Beas*

Sutlej

The SULTANATE of DELHI

H i m a l a y a s

The LANGAH
KINGDOM of
MULTAN

Panipat

The SULTANATE of DELHI
(IBRAHIM LODI)

Brahmaputra

The
ARGHUN
KINGDOM
of SIND

The RAJPUT
CONFEDERACY
under RANA SANGA
of MEWAR

Delhi

Agra

Yamuna *Ganga* *Gumti*

BIHAR

Indus

Chambal

BUNDELKHAND

BENGAL

MALWA

GUJARAT

Narmada

KHANDESH

GONDWANA

Mahanadi

Tapti

BERAR

AHMADNAGAR

Godavari

BIDAR

GOLCONDA

ORISSA

BIJAPUR

Krishna

Goa
(Port.)

Tungabhadra

VIJAYANAGAR

Cannanore
Calicut

Cawery

Cochin

The Lodi Kingdom

- - - - - Approximate boundaries

0   200   400 KILOMETERS

0   200   400 MILES

Babur shattered his golden wine goblets and distributed their fragments among the poor. Then he ordered all of his supply of wine
poured into the ground, vowing never to drink prohibited liquor
again. Rana Sanga waited too long. By the time he attacked the Mughals, his own confederacy, undermined by caste rivalries, had lost its
cohesion, and some of his allies deserted him. Babur's victory at
Khanua on March 15, 1527, forced Mewar to flee, mortally wounded,
shattering Rajput hopes of ever recapturing Agra or Delhi. The third
and final attempt to stem the tide of Mughal might was crushed by
Babur when he led his army east across the river Gogra in what is
now Uttar Pradesh to defeat the joint Afghan and Bengali force raised
by Mahmud Lodi on May 6, 1529, the sultanate's last stand.

In three martial victories, the Tiger thus secured all of North
India as his household domain, but little more than a year following
the last of his battles, on December 26, 1530, he died in Agra, after
calling upon God to take his life in exchange for the recovery of his
son, Humāyūn (1508–56), who was deathly ill at the time. Though
his dynastic line survived on Indian soil, Babur had never ceased to
long for the cool air and fine muskmelons of his Central Asian
homeland. And Humayun almost lost the empire his father fought so
hard to bequeath him. More interested in opium and astrology than
power, Humayun was challenged by his younger brothers and by
Afghan generals who had served under Babur. The most powerful of
these Afghans was Sher Khan Sur, who established himself as independent ruler in Bihar and claimed Bengal as well after 1536. Humayun managed to oust him from Bengal's capital, but then he was
trapped by the monsoon and lost part of his army to treacherous
weather before facing defeat by Sher Khan's shrewdly deployed force
at Chausa in 1539. Following Chausa, Sher Khan had himself
crowned *shah* and chased Humayun west, defeating him again at
Kanauj in April 1540. The second Mughal emperor was then forced
to flee across Sind into Persia, where he spent his fortune seeking support from Shah Tamasp to recapture Delhi.

For five years Sher Shah Sur ruled North India from Delhi, reorganizing the revenue system to ensure greater equity to individual
cultivators and a more reliable flow of funds to his treasury. He revitalized every aspect of administration by taking personal interest in
all appointments to positions of power, consolidating central authority

over outlying districts, and designing a blueprint for more efficient imperial rule, which would serve to inspire many subsequent reforms. He was the most talented Afghan sultan, and he might have established a dynasty of almost Mughal dimensions had he not been killed in battle in May 1545. Even with only five years to consolidate his power, Sher Shah managed to bequeath his kingdom to son and grandson, but neither had his wisdom or energy. The last of his line, a mere boy who was killed by his own uncle in 1554, vacated the throne of Delhi to rival claimants on the eve of Humayun's return to India. With a Persian army at his command, Babur's son recaptured the Punjab, Delhi, and Agra early in 1555. Within a year of his return, however, he tripped, light headed from a pipe or two of opium, cracking his head on the stone stairs of his private astronomical observatory, and as one chronicler put it, "stumbled out of life as he had stumbled through it" in January 1556. The legacy of Babur now passed to his thirteen-year-old grandson Akbar, who would soon prove himself fully worthy of a name that meant "Great."

NINE

MUGHAL IMPERIAL UNIFICATION

(1556–1605)

Jalal-ud-din Muhammad Akbar was born on October 15, 1542, at Amarkot in the Sind desert, while his father was fleeing toward Persian exile. The infant Akbar was left with his Persian mother, Hamīda Bānū Begum, and his head nurse, Māham Anaga, at the Afghan fortress of Kandahar, under guard of his father's trusted lieutenant, Bayram Khan, who remained at his side after he ascended the throne. Reared in rugged Afghan exile rather than Delhi's palatial ease, Akbar spent his youth learning to hunt, run, and fight and never found time to read or write. He was the only great Mughal who was illiterate, but he later noted that the prophets were illiterate and advised believers to keep at least one son in that unlettered condition. Perhaps illiteracy predisposed him toward mysticism, as did his love of animals and nature and the early hardships he endured.

Akbar was thirteen when Humayun died, and Bayram Khan, knowing there would be rival claimants to Delhi's throne, quickly crowned the boy at Kalānaur in the Punjab, where Akbar had been serving as his father's governor. Several heirs to the Afghan Sur dynasty did, in fact, claim to inherit control over North India, but none of them proved formidable obstacles to Akbar's force, led by his chief minister and general, Bayram Khan. A Hindu named Hemū, however, who had served as *vazir* of the Sur dynasty, marched against Agra and Delhi at the time of Humayun's death, seeking to restore Hindu power. He seized both cities and proclaimed himself *Rājā Vikramāditya* ("King Whose Effulgence is Equal to the Sun's"), winning enough Afghan support to muster 100,000 horses in his army. Hemu also kept his elephant corps well fed, even though each of his 1,500 war elephants ate an average of 600 pounds of grain a day while the populace of Delhi starved through one of India's worst famine years. But despite Hemu's military strength, the Mughal army won the prize of North India once again on the battlefield of Panipat on November 5, 1556.

For the first five years of his reign, Akbar remained subservient to Bayram Khan, until the regent was deposed at the instigation of Akbar's nurse, who hoped to run the empire herself. The faithful old regent, hustled off on a pilgrimage to Mecca, was stabbed to death by a band of Pathans in Gujarat in 1561. For a few years, the ambitious nurse, Maham Anaga, enjoyed the prerogatives of court patronage and petty power, but in 1562, the seventh year of his reign, Akbar liberated himself from harem rule and took firm personal command of his court and its policy. Akbar's unique achievement was based on his recognition of the pluralistic character of Indian society and his acceptance of the imperative of winning Hindu cooperation if he hoped to rule this elephantine empire for any length of time. First of all, he decided to woo the Rajputs, marrying the daughter of Raja Bhārmal of Amber in 1562, thus luring that Hindu chief with his son and grandson as well to his capital at Agra, the start of four generations of loyal service by that Rajput house in the Mughal army. That same year, Akbar showed his capacity for wise as well as generous rule by abolishing the practice of enslaving prisoners of war and their families, no longer even forcibly converting them to Islam. The following year (1563), he abolished the tax that from time immemorial had been exacted by kings from Hindu pilgrims traveling to worship at sacred spots throughout India. Akbar had been tiger hunting around Mathura when he chanced to learn of the pilgrim tax and, insisting it was contrary to God's will, ordered it abolished immediately. In 1564 he remitted the hated *jizya* (non-Muslim poll tax), which was not reimposed for more than a century, and with that single stroke of royal generosity won more support from the majority of India's population than all other Mughal emperors combined managed to muster by their conquests.

Not that Akbar was adverse to using force as well as conciliation in unifying India. When the *rana* of Mewar refused to follow Amber's example of joining the Mughal army and Akbar realized that repeated attempts at diplomacy simply inflated the pride of the Rajput heirs to Rana Sanga, he personally led the seige of Chitor in

October 1567 and ordered the massacre of some thirty thousand of its defendants when it fell in February 1568. No prisoners were taken; only the massive regalia, Mewar's kettledrums and candelabra, were dragged back to Agra as symbols of the Mughal victory, which broke the back of remaining Rajput resistance. Ranthambhor surrendered in March 1569, and Kālinjar that August. By November 1570, virtually all the chiefs of Rajasthan—except for the *rana* of Mewar, who had fled to the hills—had sworn allegiance to Akbar.

The picture of Akbar preserved by a number of contemporary historians, the best of whom was Abu-l Fazl (1551–1602), and by foreign visitors, including several Jesuits at court, is of a most energetic and powerful, yet singularly sensitive, often melancholy man, whose fits of depression were as prolonged and profound as his flights of manic celebration were frequent. He seems to have been epileptic, and until the age of twenty-seven, he remained childless. (Twin sons born to his wife in 1564 died after only a month.) Anxiety that he might leave no heir to his empire weighed so heavily upon his consciousness that Akbar went to seek help from a Sufi saint, Shaikh Salīm of the Chishti order, who lived at a spot called Sīkri, some twenty-three miles from Agra. Repeated sessions with the shaikh seemed to soothe the emperor's troubled spirit, and a year later, in 1569, his first son and heir, Salim, was born to the daughter of Raja Bharmal. The next year a second son was born, and two years later, a third. Akbar's gratitude to his sufi guide went far beyond the mere naming of his heir after the shaikh, for in 1571 at the village sight of Sikri he erected a magnificent red sandstone palace-fort, moving his entire household and court to that "sacred" spot, thereafter called Fatehpur ("Fortress of Victory") Sikri. To accommodate the sudden influx of people, an artificial lake had to be dug, but by 1585 that supply of water proved so inadequate or polluted that the splendid new capital city had to be abandoned. It stands to this day as a ghostly reminder of Akbar's reign, at once a tribute to the impulsive grandeur of his spirit and a symbol of the evanescence of power.

Akbar invaded wealthy Gujarat in November 1572, personally marching at the head of his army into Ahmadabad, then capturing Surat in February 1573. With great Mughal power thus securely anchored in the Arabian Sea, Akbar could turn his attention east to Bengal, where he marched in 1574, finally integrating that hitherto independent region into his imperial scheme in 1576. The northwestern limits of Akbar's power were firmly fixed by August of 1581, when he marched triumphantly into Kabul with no fewer than 50,000 cavalry and 500 elephants at his command. By pacifying Afghanistan for the remaining quarter century of his rule, Akbar managed to achieve more than British arms would ever be able to command. In 1592 he added Orissa, and in 1595 Baluchistan, to his imperial domain. His effective control over northern and central India was, in fact, greater than that of either the Mauryas or the British, and after conquering those regions he established stable administrations within them, creating a pattern followed by his Mughal descendants as well as by early British administrators.

The *mansabdari* (*mansab* is Persian for "office" and *mansabdars* were "officeholders") system of administration developed by Akbar divided the higher echelons of Mughal officialdom into thirty-three ranks. Each "rank" was classified on the basis of the number of cavalry an official of that *mansab* was expected to raise and lead in the emperor's service at a time of martial emergency. The princes of royal blood were given the highest *mansabs,* ranging from five to ten thousand, while lesser nobility would be assigned *mansabs* of from five hundred to five thousand. The lowest *mansabdar* was a "commander of ten," whose appointment would come from the emperor through nomination by some court *amir* (noble). All of the higher ranks were hand-picked officials drawn mostly (70 percent) from Muslim soldiers born outside of India, but they also included many major ethnic, regional, and religious groups within India, about fifteen percent Hindus, most of whom were Rajputs. Akbar's administration thus relied, as would the British Raj, primarily upon foreign administrators, but it also contained some of the most talented as well as vigorously ambitious young Indians, many of whom might have raised armies against Mughal power had they not been invited to share in it. Twenty-one Hindus held *mansabs* of five thousand and above during Akbar's reign, and Raja Todar Mal, who was Hindu by birth, held the second most powerful post in the bureaucracy—*diwan,* or minister of revenue—and was for a brief interlude elevated to the premiership, as *vakil,* the emperor's bureaucratic right hand.

Akbar's empire was divided into twelve provinces (*subas*) and subdivided into districts (*sarkars*), which were in turn usually broken

down into *parganas* (subdistricts). Each province was ruled by its
subadar ("governor"), whose lavish court was a miniature reflection
of the great Mughal's. Municipalities were separately administered by
a *kotwal* ("city governor"), who supervised the various bureaucratic
boards. Cities supported their own police force, while the countryside
was secured by district military commanders (*faujdars*), who assisted
the district revenue collectors (*amalguzars*) in gathering the emperor's
"share" at each harvest. Though the decennial revenue "settlement"
reached with each peasant (*ryot*) or landed overlord (*zamindar*)
in Akbar's time varied from region to region, depending upon the
fertility of the soil, actual crops grown, availability of irrigation water,
and other factors, the average overall assessment was about one-third
of the total annual harvest, or one-third of its cash value, somewhat
lower than it had been before or was to be later. To help peasants survive seasons of drought or crop failure, moreover, Akbar's revenue
collectors were ordered to remit taxes in afflicted districts, and thanks
to the sympathetic administration of that ministry by Hindus, there
was far less coercion or brutality used in the general collection of
revenue during Akbar's reign than at any other era of Muslim rule.
Islamic law, codified in the orthodox *Shari'at* ("Highways") and
interpreted by learned scholars (*ulama*), was enforced by judges
(*qazis*), whose decisions could always be appealed to the emperor.
After 1579, Akbar asserted that as God's earthly representative he
would always judge correctly. Few prisons were maintained in Mughal
India. Whipping, public humiliation and display, banishment, and
death were the usual forms of punishment. At the local level, where-
ever a legal dispute involved two Hindus, Hindu law (*Dharmashastra*) was applied, and Brahmanic opinion, or the decision of the
village *panchayat* would generally be accepted as final, unless appealed to the emperor.

A total population of about a hundred million, less than one-
tenth the density of modern South Asia, and a total revenue demand
of little more than one-third the yield of the land, meant that the
average inhabitant of Akbar's India was economically better off than
his peasant heirs have subsequently been. As for the elite courtiers,
indeed all *mansabdars,* their descendants would never again attain
such affluence, enjoy so much power, or live in such luxurious splendor. A mere *mansabdar* of one thousand received the rupee equivalent

in salary of a British lieutenant governor, and his pay permitted him
to live in thoroughly regal luxury. Mughal grandees knew that their
wealth and property would revert to the emperor when they died, so
their level of conspicuous consumption was even greater than their
exorbitant salaries alone dictated. There was very little opportunity
for capital investment, hence lavish spending on high living became
the order of daily Mughal society. Grandees kept their stables full of
Arabian horses, their harems replete with Indian and African dancing
girls, their servants' quarters crammed with slaves, and their jewel
boxes overflowing. India's premier industry in Akbar's day was the
production of textiles for export, as well as for domestic sale. Gujarati cotton goods clothed most of Africa and Asia at this time, and
Indian peasant weavers did not yet face competition from Lancashire
mills for the home market. Impoverished Bengali peasants wore sack-
cloth woven of jute; the wealthy covered themselves in homespun
silk; and Northerners who could afford to wore wool during the brief
but often bitter winters. In addition, Kashmiri shawls and carpets
were prized both at home and abroad for their soft warmth and the
brilliance of their artistic design. Indigo and opium were other important Mughal exports, but they ranked far below cotton cloth and
spices in total value.

Not only did Akbar's efficient administrative system help stimulate and expand India's economic development and trade, foreign as
well as domestic, but it also resurrected Ashoka's imperial idea of
bringing the entire subcontinent under a single "white umbrella." Like
the ancient Guptan and Mauryan emperors, Akbar endowed his high
office with trappings of divinity, which he may have based more on
contemporary Persian models than on early Indian ones, but which
were clearly popular with most of his subjects. For it was not as an
orthodox Muslim monarch that Akbar ruled, but rather as a divine
Indian emperor, the spiritual as well as the secular father of all his
people. Perhaps, like the predisposition to a social system based on a
hierarchy of "castes" where birth alone confers high or low status for
life, something in the "soil" or "climate" of India made its populace
more amenable to royal rule (raj) of a "divinely" imperial variety
than to other forms of government. Akbar, at least, seems to have perceived such a predilection. He made the most of his imperial position
not merely by issuing his "infallibility decree," which raised the deci-

sions of "a most God-fearing king . . . for the benefit of the nation" above those of sectarian Islamic law, but also by founding a "Divine Faith" (*Din-i-Ilahi*) at his court in 1581. The motto of that court religion, used as a salutation by its devotees, was *Allahu Akbar,* which could mean either "God is great" or "Akbar is God." It was doubtless interpreted both ways, depending on whether the person using it was more of a devout Muslim or an imperial Mughal. Akbar himself had by then abandoned orthodox Islam for its mystic Sufi form, and much of the ritual associated with his court religion was derived from the practice of Sufi orders. Some of his ideas came from Hinduism and Jainism, however, and others were borrowed from Parsis, Sikhs, and Christians, who were regularly invited to Fatehpur Sikri to discuss their beliefs with the ever-curious emperor. One of the most interesting buildings at Akbar's capital was his octagonal "Hall of Private Audience" (*Diwan-i-Khas*), a small chamber whose central support is a stout pillar from which catwalk spokes emerge about eight feet above floor level. Akbar would stand on the catwalk and, looking down at the gathering of learned leaders of every religion attending below, throw out provocative questions to one and then another wise man, vigorously engaging them in debate and inciting arguments among them while he listened and paced overhead.

Orthodox Muslim leaders like the *mulla* of Jaunpur came to fear that the emperor had abandoned Islam entirely, and called upon their congregations of the faithful to rise in revolt. In 1581 rebellion flared in Bengal and Kabul, but Akbar managed easily to suppress such opposition to his popular "national" policy. Akbar went so far as to forbid cow slaughter by imperial decree, making that offense against Hinduism and his Divine Faith punishable by death. During the final decade of his reign, orthodox Muslim opposition mounted in the north, while much of his time and fortune was lavished on expeditions to overwhelm Shi'ite Muslim sultanates of the Deccan. The sultanates of Ahmadnagar, Khandesh, and Berar, heirs to the northern half of the Bahmani sultanate, were all defeated by Akbar's forces and incorporated (nominally, at least) within the Mughal Empire. Lacking a modern railroad or telegraph system, however, the Mughals were to find it much more difficult to retain effective control south of the Vindhya-Satpura divide than to win martial victories there. Though Akbar and his successors repeatedly tried to consolidate their tenuous grip over the south, their invasions served more to stimulate local opposition than to unite the subcontinent permanently. Indeed, it proved easier for Akbar to expand his hold over Afghanistan than over the Deccan, weaning Kandahar from Persia and adding it to Mughal domain in 1595.

The importance of Persian cultural influence on the Mughal Empire and court can hardly be exaggerated; it was found not only in Akbar's Sufism, but in the reintroduction of Persian as the official language of Mughal administration and law (Persian had been used by the Tughluqs, but not by the Lodis). The elegant decadence of Mughal dress, decor, manners, and morals all reflected Persian court life and custom. Mughal culture was, however, more than an import; by Akbar's era, it had acquired something of a "national" patina, the cultural equivalent of the Mughal-Rajput alliance. That new syncretism, which has come to be called "Mughlai," is exemplified by Akbar's encouragement of Hindi literature and its development. While the Persian and Urdu languages and literatures received the most royal patronage and noble as well as martial attention, the emperor also appointed a poet laurcate for Hindi. Raja Birbal (1528–83) was the first poet to hold that honored title, thanks to which many other young men of the sixteenth century were induced to study the northern vernacular that has now become India's national tongue, helping to popularize it through their poetry and translations of Persian classics. Most popular and famous of the Hindi works in this era was the translation of the epic *Ramayana* by Tulsi Das (1532–1623).

In architecture as well as painting, Akbar's era reflects a blend of Perso-Islamic and Rajput-Hindu styles and motifs. The buildings at Fatehpur Sikri are a unique synthesis of Indian craftsmanship and design employed in the service of one of Islam's most liberal monarchs. The multitiered roofs topped by howdahlike nests in sandstone resemble the palaces of Rajasthan more than those of Isfahan or Shiraz, and the pillared porticos suggest the Hindu temple in its classical form, rather than the mosque. In painting, even more vividly than in architecture, the central theme of Akbar's policy of fashioning a Mughal-Rajput alliance reached its peak of artistic expression. More than a hundred painters were employed at court, honored with *mansabdar* rank, and constantly encouraged by the emperor himself to improve their magnificent pictures, which would be exhibited each

week before him. Their portraiture, book illumination, and naturalistic animal and bird paintings remain among the most beautiful treasures of Indian civilization. The Persian calligrapher par excellence of
Akbar's court was Khwaja Abdul Samad, but most of the great artists
were Hindus: Daswanath, Basawan, Lala Kesu, Haribans, and others
whose names were recorded by Abu-l Fazl, though much of their
work has perished or been lost. We do, however, have a number of
portraits and miniatures dating from this era, the finest fruit of the
union of Hinduism and Islam. When challenged by orthodox Sunnis,
who reminded him that Islam prohibited depiction of the human
form, Akbar replied that he could not believe God, the "Giver of
Life," would be repelled by the human beauty portrayed in works of
true art.

The last four years of Akbar's life were plagued by his eldest
son's rebellion, the curse of an inherited pattern of Central Asian succession struggles that haunted the Mughals and proved so debilitating
a drain on Indian resources. Prince Salim proclaimed himself *padishah*
in Allahabad in 1601, while his father was preoccupied with Deccan
warfare. Akbar deputed his most trusted lieutenant, Abu-l Fazl to
"take care" of his wayward son, but the loyal Fazl was murdered on
his way back to Agra by an assassin hired by Salim. Akbar reasserted
his paternal power briefly, but the following year, on October 17,
1605, the great Mughal emperor died of what seems to have been a
dose of poison administered by his son. Salim now assumed his Persian title name Jahāngīr ("World Seizer"), starting his twenty-two
year reign at the age of thirty-six. The empire he inherited was probably the most powerful in the world at the time, and certainly the
strongest in Indian history to date.

TEN

WESTERN EUROPE'S VANGUARD

(1498–1669)

When Vasco da Gama sailed into the Malabar coast port of Calicut on May 27, 1498, his voyage round the Cape of Good Hope marked the culmination of more than half a century of Portuguese nautical persistence. Prince Henry's school of navigation had been motivated almost as much by Christian zeal as by the promise of profits from the spice trade, since finding a southern direct passage to the East would strike a blow for Christianity against "Moorish" Arabs, Turks, and Mongols. Vasco's historic arrival in India initiated an era of Western European imperial penetration and conquest that would last four and a half centuries. But in 1498, neither Vasco nor the petty Hindu ruler of Calicut, whose title was *zamorin,* dreamed that the small but sturdy flagship *San Gabriel* was to be the vanguard of some eight hundred Portuguese galleons destined to dominate the Indian Ocean during most of the next century. For the ensuing three centuries, in fact, tens of thousands of still larger freighters and men-of-war from Holland, England, and France would follow the Portuguese route, forming a chain of floating power that would bind India to Western Europe, as a giant elephant might be bound by a chain secured to one of its ankles.

When Vasco arrived at Calicut, it was a thriving port familiar to Arab, Hindu, and Chinese merchants and seamen, who came from all parts of Asia and Eastern Africa to purchase the pepper and ginger of Malabar with gold, ivory, silks, and jewels. The *zamorin* was one of the wealthiest petty princes along the Malabar coast, but the one thing he and all of the pre-Portuguese traders in the Indian Ocean area lacked was mounted cannon on their ships. Vasco's flagship alone carried twenty guns, but he did not use them at first, for, having reached India with only two of his original fleet of four ships, he felt uncertain of his power. Admiral da Gama ordered his men to pay whatever they were asked for anything the "natives" had to sell—even ginger loaded with red clay and cinnamon of the poorest quality. Despite the fact that the Portuguese were thus "tricked" into paying almost twice the price charged to more experienced merchant visitors like the Arabs for the cargo they loaded at Calicut, Vasco returned home to sell his spices for some sixty times the total cost of his entire two-year expedition. When word of this 3,000 percent profit spread through Lisbon, eleven new Portuguese ships were immediately assembled and outfitted, setting sail for India under the command of Pedro Alvarez Cabral in March 1500. Only six of Cabral's vessels reached Calicut, but that was a sufficient show of strength to persuade the *zamorin* to sign a treaty of peace and friendship with Portugal, allowing Cabral to purchase a warehouse, in which he left fifty-four Portuguese merchants, called factors, to buy spices when prices were low and keep them stored near the dock until the next fleet could return to Calicut.

The transition from the initial "fleet" to the more permanent and potentially profitable "factory" stage of Western European enterprise in India proved disastrous, and the era of friendly Indo-Portuguese collaboration was followed by a longer period of bitter conflict and hatred. For, soon after leaving Calicut, Cabral plundered a Muslim ship laden with spices, and in retaliation the "Moors" attacked the Portuguese factory and killed all the Westerners they found living there. Vasco da Gama returned with a fleet of fifteen heavily armed ships in 1502 and blasted the port until it was all but reduced to rubble. Then he captured several Muslim vessels and cut off the hands, ears, and noses of some eight hundred "Moorish" seamen, sending the lot to the *zamorin*'s palace for his highness's "curry." Such orgies of piracy and plunder served to secure Portugal's direct route to the East and assured an uninterrupted supply of pepper and cloves. With less than a million population, Portugal never attempted to establish any large land bases in Asia, but did fortify whatever factories were thereafter founded at strategic positions in the Indian Ocean to prevent further massacres of their men. The bulk of Portuguese power remained at sea, however, and until Portugal itself was brought by "personal union" under the control of Philip II of Spain, the Indian Ocean remained virtually a Portuguese monopoly. In keeping with the Treaty of Tordesillas of 1494 and its subsequent confirmation by papal bulls, Spain held the monopoly of New World gold, leaving India (and Brazil) to Portugal. The Protestant sea powers of England and Holland were as yet too weak to crack the Iberian Catholic monopoly over the "undiscovered" worlds, East or West; hence Portugal's remarkable interlude of uncontested plunder in Asia throughout most of the sixteenth century.

Dom Affonso d'Albuquerque, viceroy of Portugal in the East from 1509 to 1515, was master architect of Portugal's Indian empire, a strategist of no mean vision and a religious fanatic whose hatred of Islam was almost as great as his knowledge of the Indian Ocean. Albuquerque realized that to command the Indian Ocean, it would be necessary for Portugal to build fortresses at the mouths of the Persian Gulf, the Red Sea, and the Bay of Cambay, as well as somewhere along the Malabar coast and athwart the Straits of Malacca. He decided upon centrally situated Goa as the best Malabar base for his headquarters and in 1510 seized control of that island, which was to become Portugal's capital on Indian soil. Goa had previously been under Muslim control, ruled by the sultan of Bijapur, who abandoned his allegiance to the Bāhmani sultan of the Deccan in 1490, carving out for himself a kingdom that included the Konkani coast and Goa. Most of the population of Goa were Konkani-speaking Hindus, and since Albuquerque had the good sense to cut their taxes in half after doing no less to the bodies of every Muslim soldier and official or overlord he could find, Portuguese power was quickly and securely established on the island that soon came to be known as the Golden Babylon of the East. Western Europe's first toehold on India would be the last spot reluctantly relinquished over four and a half centuries later. With Goa, Viceroy Albuquerque wrote his king, Dom Manoel: "I do not believe that in all Christendom there will be so rich a King as Your Highness, and therefore, do I urge you, Senhore, to strenuously support this affair of India with men and arms, and strengthen your hold in her, and securely establish your dealings and your factories; and wrest the wealth of India and business from the hands of the Moors."[1] Albuquerque dreamed of diverting the Nile in order to dry up Egypt and of stealing the remains of the Prophet from Mecca,

[1] *The Commentaries of the Great Affonso d'Albuquerque, Second Viceroy of India*, trans. from the Portuguese ed. of 1774 by W. de Gray Birch (London: Hakluyt Society, 1875–84), nos. 53, 55, 62).

so fierce was his hatred of Islam. No Muslim was permitted to hold any office, no matter how humble, in Albuquerque's imperial regime, though Hindu "police" (*sipahi,* later *sepoy*) were readily employed. Western Europeans learned long before the British arrived how best to exploit the communal conflicts and social divisions within India's fragmented, pluralistic society.

Portuguese spice imports rose from less than a quarter of a million pounds in 1501 to more than 2.3 million pounds per year by 1505, when Venetian merchants found that they could buy barely one million pounds of spice in Alexandria, though their annual purchase in 1495 had been 3.5 million pounds. Arab and Venetian merchants remained in the spice trade throughout the century of Portuguese power in Asia, but the balance of trade had shifted dramatically, and Western European nations persisted in short-circuiting Arab middlemen carriers as the European demand for spices continued to increase. Before his death in 1515, Albuquerque also secured Malacca, Ormuz, Diu, and Socotra, keeping his promise to his king, who was fast becoming the wealthiest monarch in Europe.

As Portugal's wealth and power in Asia grew, however, Jesuits back home feared that the mire of "Eastern sin" might undermine the empire built in such great measure by the zealous spirit of Christian crusading. In 1542 the first Jesuit missionaries arrived in Goa, zealously seeking to convert India's "heathen" population to Catholicism, but in 1548 Francis Xavier wrote Ignatius Loyola to report pessimistically that "all these Indian nations are very barbarous, vicious, and without inclination to virtue, no constancy of character, no frankness." It was to be a familiar lament of Westerners during the centuries ahead, by no means unique to the Portuguese, nor to the Jesuit fathers. Xavier himself soon abandoned India for Japan, finding Hindu converts less interested in biblical instruction and "spiritual food" than free rice. There were enough converts, nonetheless, combined with the offspring of officially encouraged intermarriages (Luso-Indians), most of whom became Catholics, to warrant introducing the Inquisition to Goa in 1560. Soon thereafter Portuguese trade and power began to decline rapidly, following the terrible defeat of Vijayanagar at the battle of Talikot in 1565. The mighty Hindu kingdom of South India had, from the time of the viceroyalty of Albuquerque, maintained friendly diplomatic relations and brisk trade with the Portuguese, who supplied Krishna Deva Raya (r. 1509–29) with all the horses he needed for his vigorous royal consolidation. Albuquerque readily accepted any enemy of Muslim powers as a friend, and since Vijayanagar was in continuous conflict with the Muslim sultanates of the Deccan, to its north, he warmly solicited the friendship of its "God King" (*Deva Raya*). While Vijayanagar prospered, Goa remained both prosperous and secure, but by 1564 at least four (Ahmadnagar, Bijapur, Bidar, and Golconda) of the five Deccani sultanates that had emerged as independent territorial powers from the Bahmani sultanate united in declaring *jihad* against Vijayanagar. Though wealthy Vijayanagar was reported to have armed nearly a million soldiers in its defense, they failed to ward off the fury of the united Muslim force; not only was the Hindu army at Talikot slaughtered, but the great "City of Victory" south of the Tungabhadra was plundered and destroyed. The Portuguese were now left to languish for lack of trade, wondering how long it would take before the newly enriched and emboldened Muslim sultans of the Deccan attacked their precarious ports. Though the sultans did not drive the Portuguese from India, Philip II of Spain took possession of all Portuguese domains after 1580, personally unifying Iberia under his rule, and closing the port of Lisbon to Dutch merchants. The immediate aftermath of that hostile Spanish policy was to intensify Protestant European competition in its search for alternate direct passages to India.

In the fifteenth century, the English had begun exploration for a northern passage to India, which would have been "legal" even by the Treaty of Tordesillas, since that agreement divided the undiscovered world between Spain and Portugal along a "line" 370 leagues "west and *south* of the Cape Verde Islands," but for Britain, the years 1494 to 1600 proved only to be a century of failures. The first English chartered company organized to seek a northeast passage to India was formed in 1551 by the cartographer Sebastian Cabot, its governor, and twelve councilors who contributed the capital of £6,000 to discover "Cathay, and divers other regions, dominions, islands, and places unknown." Sir Hugh Willoughby commanded the London Company's first expedition in 1553, but was found two years later with all seventy of his crew frozen dead in Russia's Siberian ice. In 1554 England's Muscovy Company was launched with a royal charter promising it a monopoly of trade through Russia to Persia, and it sur-

vived as a rival to England's East India Company until late in the
eighteenth century. Intrepid English sea captains like Drake and Cav-
endish defied the Spanish and Portuguese monopoly by voyaging
around the world, looting, shooting, and pirating their way to knighthood and historic fame. But when nonbuccaneering merchants like
Osborne and Staper subsidized the outfitting of a boat named *Tiger* in
1583 and sent her east to run the Portuguese blockade, their ship was
captured off Ormuz and its merchant "factors" were taken as prisoners to Goa. Among those merchants was Ralph Fitch, who survived
his ordeal and wrote a number of letters from India that served to
stimulate British interest and enterprise in the eastern seas.

"Here be many Moors and Gentiles," Fitch reported in the first
Englishman's letter "home" from India. "They have a very strange
order among them—they worship a cow and esteem much of the cow's
dung to paint the walls of their houses. They will kill nothing, not so
much as a louse, for they hold it a sin to kill anything. They eat no
flesh, but live by roots and rice and milk. And when the husband dieth
his wife is burned with him if she be alive . . . In the town they have
hospitals to keep lame dogs and cats, and for birds. . . . Goa is the
most principal city which the Portugals have in India . . . Here be
many merchants of all nations."[2] Fitch told of how wealthy the inhabitants of Goa were, how richly varied their commodities appeared,
how sumptuous their palatial homes seemed, and how busy and bustling other Indian cities were as well. In 1585 he visited Agra and
Fatehpur Sikri, and he estimated the population of each of Akbar's
capitals at twice the size of London, whose population then was only
about a hundred thousand. When Fitch returned to London in 1591,
he had many more tales to tell and Indian jewels and other produce to
show, further whetting the appetites of his merchant comrades to risk
all on the passage east.

After 1588, when the Spanish Armada fought its way to the bottom of the English Channel, there was less real risk involved in that
passage, and Dutch as well as English sea captains hoisted sail to join
the race around the Cape of Good Hope. For the Dutch, the attack
against Catholic Spain's eastern monopoly was no less than a national

[2] Letter of Ralph Fitch, "Merchant of London," 1583, in J. C.
Locke, *The First Englishmen in India: Letters and Narratives of Sundry
Elizabethans* (London: G. Routledge and Sons, 1930).

movement, a cause to which every patriotic Dutch burgher readily subscribed. A great school of cartography grew up at Antwerp during the latter half of the sixteenth century and Amsterdam, Haarlem, Rotterdam, and half a dozen other cities of the States-General bustled with merchants and sea captains ready to risk their fortunes against Spanish "tyranny." Two years before Ralph Fitch was brought as a prisoner to Goa, Haarlem-born Jan Huygen van Linschoten arrived at the Portuguese capital in the east to serve as secretary to the archbishop. Linschoten remained in Goa from 1583 to 1589, returning home in 1592 with more detailed information about India than Fitch had amassed, plus priceless Portuguese navigation maps of the Indian Ocean, which taught the Dutch how to use the monsoon winds to their best advantage. The Portuguese always guarded such maps as jealously as they had defended their monopoly trade, but with the Dutch publication of Linschoten's *Reysgeschrift* ("Sailing Guide" to the Eastern Seas) and *Itinerario* in 1595, and their subsequent English translation, the top-secret maps became public knowledge throughout Northern Europe. Linschoten's descriptions of "the heathenish Indians" of Goa, whom he called "very subtill and cunning," were to become clichés of Western characterization during the next three centuries. In 1595 the Dutch launched their first fleet, four ships commanded by Cornelius de Houtman, to run the southern blockade to the Indies. Following Linschoten's directions and charts, de Houtman reached Java in a year and was back home after little more than two. Only one-third of his men returned to tell that tale, but their cargo brought a profit of 80,000 florins and stimulated the launching of no fewer than twenty-two fresh ships, financed as five separate voyages, to risk the blockade in 1598.

Determined to let no force on sea or land deter them any longer, the Dutch organized themselves into a nation of blockade runners and spice traders, sending sixty-five ships to Java, Sumatra, and the Molucca ("Spice") Islands between 1595 and 1601. On March 20, 1602, the States-General issued a charter to the Vereenigde Oostindische Compagnie (United East India Company, hereafter called the VOC), granting its seventy-six directors—who had invested some 6.5 million guilders (ten times the total capital raised for the English East India Company two years earlier)—a monopoly of all trade between the capes of Good Hope and Magellan for twenty-one years. Within three years of its birth, the VOC sent thirty-eight ships into the Indian Ocean, defeating the Portuguese fleet off Johore and seizing their Spice Island fortress at Amboina.

While the Dutch raced so effectively to enter the Eastern trade, establishing their first bases in the archipelago of Southeast Asia, the merchant "adventurers" of London had not been altogether unenterprising. On December 31, 1600, Queen Elizabeth I granted her royal charter to "The Governor and Company of Merchants of London," promising them monopoly privileges on all trade with the "Indies" (that is, between Good Hope and Magellan), though only for fifteen years and with an export limit of £30,000 in specie fixed on each voyage. It was clearly less than a united national effort for England at this juncture. The company's first governor, Thomas Smythe, and his twenty-four "committees" (directors) raised £68,373 in capital from some 217 initial subscribers and outfitted four ships, which were launched in February 1601 under James Lancaster's command. This first expedition reached Sumatra in June 1602, starting a factory there, and stealing nine hundred tons of calicoes and spices from a Portuguese ship pirated in the Straits of Malacca. As Protestant cohorts joined in combat against the same Iberian Catholic "tyranny," the English initially worked in cooperation and harmony with the Dutch, settling together at Bantam and Amboina to reap a fertile harvest of cloves, nutmeg, and pepper. In 1603 alone one million pounds of raw pepper were shipped to England from Southeast Asia. The London adventurers' investment paid off handsomely, with an average profit of 170 percent on each of their first eight voyages.

The third voyage launched by the merchants of London left in March 1607, three ships commanded by Captain William Hawkins, whose flagship, *Hector,* dropped anchor off Surat at the mouth of the River Tapti on August 24, 1608. It was the English East India Company's first visit to India. Surat was the Mughal Empire's principal port; after 1619 it became the site of England's first factory in India and remained the west coast headquarters for John Company (East India Company) until that key role would be taken over by Bombay in 1687. When "ambassador" Hawkins arrived, armed with 25,000 pieces of gold and a letter from James I to the Mughal Emperor Jahangir, Surat was a bustling city. This western gateway to India was packed with Muslim pilgrims waiting for the annual ship to Mecca; its indigo and cotton cloth warehouses bulged with merchandise ready for export; its Mughal grandees were carried on their palanquins by African slaves; its brothels were filled with nubile beauties from all parts of India, bedecked in pearls and diamonds; its bazaars were jammed with merchants from most of Asia, peddling everything from peacock feathers to white elephants, from coarse grain to opium, from palm leaves to gold. Neither Hawkins nor his ship were eagerly awaited by the governor or merchants of this busy seventeenth-century port, and Hawkins would soon learn to his despair that India had no more need to trade with England at the dawn of that century than she had interest in the produce of any Western power. Nothing England made at this time was really desired by Indian merchants or officials except for specie, the commodity that prevailing mercantile economic philosophy taught Englishmen to covet most.

Hawkins was first ignored, next humiliated, then robbed by Mughal officials (his ship and most of his crew were captured by Portuguese pirates while he was ashore at Surat). Later, his life was daily endangered by the "machinations of Jesuits," who constantly "conspired" against him (so at least he wrote home) throughout his more than two years in India, trying in vain to negotiate a treaty of trade with Jahāngīr, who merely appointed him a minor *mansabdar* at court. England's second envoy, Paul Canning, who arrived at Agra in 1612, fared much worse, however, and was sent packing for home in a matter of months. Portuguese Jesuits at Agra had helped convert Indian apathy and neutrality toward Englishmen into positive aversion. But on November 29, 1612, British stock rose swiftly in the minds of the Indians (those of Surat, at least), when they watched Captain Best's proud ship *Red Dragon* sail into and disperse with well-aimed cannon fire no fewer than four Portuguese galleons and "a whole fleet of frigates."

Best's victory at sea shifted the balance of Anglo-Portuguese power in the Indian Ocean and effectively neutralized the influence of Portuguese Jesuits at Agra's court. When King James's ambassador, Sir Thomas Roe, visited Jahangir in 1616 to present his gifts and credentials, there was a new mood, a more cordial reception, awaiting him. Mighty as the Mughal army was, Jahangir had no fleet, and the Mughals had come to depend on Portuguese frigates for the costly protection of their annual pilgrim ship to Mecca; they now looked instead toward the more seaworthy, less bigoted British for this religiously vital service. Though it took Roe some two years and nine months of haggling, he managed to win permission in 1619 for the English East India Company to build a factory at Surat, but he cautioned the company against seeking territorial acquisition in India, advising his countrymen to rest content with profits derived from "quiet trade."

Only four years after the British factory at Surat was founded, the company was obliged to rely more heavily on it than Roe or any of his predecessors could have imagined. On February 23, 1623, the Anglo-Dutch honeymoon of cooperation in the Indies came to a violent end with the Dutch star-chamber "massacre" at Amboina of ten English merchants of that port for "conspiring" with some Japanese and Portuguese merchants (also executed) to "seize" the Dutch fortress. Jan Pieterszoon Coen, governor-general of the VOC and architect of the Dutch plantation system of exploitation of the Spice Islands, thus dramatically inaugurated his new anti-British policy, all but forcing John Company to abandon Southeast Asia and fall back on India as a second-best base of operations. Had the Dutch been less selfishly avaricious in fastening their monopolistic grip over the Indies, the British might never have established their empire over India. It was thus less in a "fit of absentmindedness" than as the fallout of failure elsewhere that the world's most powerful modern empire came to be born. While Holland busied itself with profiting so lavishly from its Spice Islands monopoly that by 1650 it had become the world's foremost power, artistically and scientifically as well as commercially, English merchants and sea captains quietly built up their tripod base on the fringes of Mughal imperial might.

From their premier base at Surat, the British soon gained control over the Arabian Sea and Persian Gulf, destroying Portuguese power at the mouth of the latter in 1622 by seizing Ormuz. Thereafter, Persian silk competed with Gujarati calico as England's favorite textiles from the East. (Calico was still used mostly for household linens, tablecloths, and towels in Britain and Western Europe, becoming popular for apparel only after 1660.) English annual imports of Indian calico "pieces" (twelve to fifteen yards in length) jumped from fourteen thousand in 1619 to over two hundred thousand in 1625; the demand for Persian silk grew less swiftly. Indigo and saltpeter were the other major imports from India, and the fact that both products were produced in the eastern Gangetic plain, especially in Bihar, stimulated British efforts to establish factories on the east coast as well as the west coast of the subcontinent. The terrible famine that paralyzed Surat for over three years after 1630 also helped convince the company's servants of the wisdom of diversifying their bases of operation in South Asia. Surat remained the company's west coast headquarters, but Bombay, which had been given to Charles II as part of Catherine of Braganza's dowry in 1661 and turned over to John Company in 1668 for £10 annual rent, would soon displace it.

The first English ship to reach eastern India stopped at the Coromandal coast port of Pulicat in 1611, only to be driven off by a Dutch fleet that had been based there since 1609. The Dutch used southern India as a major source of slaves for their Spice Islands plantations, as well as for the purchase of cotton cloth. Dutch investments in Coromandal cloth, which would then be sold for spices in the Moluccas and Bandas, proved a most profitable way of diminishing the specie "drain" from home. This technique of "triangular trade" was quickly learned and followed by the English, who were equally anxious to reduce the eastern flow of specie. London merchants soon learned that by investing their gold in South Indian weavers, whose products could easily be sold in Bantam for spices, they were able to buy four times the value of pepper and cloves for the same amount of specie. Small wonder that their interest in establishing a factory along the Coromandal coast quickly intensified. This region was then nominally controlled by the Muslim sultan of Golconda, though all along the coast, Hindu tributary princes ruled local clusters of fishing and weaving villages. One of these rajas, a descendant of the ruling dynasty of Vijayanagar, was overlord of the village of Mandaraz, some thirty miles south of Pulicat, where Francis Day purchased land on which to build an English fort in 1639. The fort was christened St. George in 1642, and the village came to be called Madras, the company's headquarters at Coromandal, soon to grow into one of the three great urban ports of Britain's Indian empire.

From ports in southeastern India, British merchants soon sought more immediate access to the mainstream of produce flowing down from the Gangetic plain to the Bay of Bengal. In 1633 chief merchant

Ralph Cartwright ventured as far north as Orissa, leaving his boat at
Cuttack on the Mahanadi ("Great River"), while he marched inland
to see if he could persuade the Mughal governor to grant him permission to trade. Governor Aga Muhammad Zaman, deputy to the *nawab*
("viceroy") of Bengal, ran his court in true Persian fashion, first offering his bare toe for Cartwright to kiss before listening to his pleas.
Englishmen were not too proud as yet to kneel before Indians, and
Ralph Cartwright did what Mughal custom required of him, winning
for his company immediate official permission to trade anywhere in
Orissa, free of customs, and the right to purchase land for factories,
as well as to provision and repair ships at any Orissan harbor. Those
concessions, however, hardly proved a blessing to the English factors
and seamen who settled on the rugged coast of that malaria-ridden,
hurricane-washed region. Before the year was over, five of the six
English merchants in Orissa were dead, and most of the crews of ships
that put into port at either of the company's factories at Hariharpur
or Balasor were stricken with a strange "fever" from which few recovered. By 1641 Orissa had to be abandoned by the company, and
not until 1650 would Englishmen work up the courage to move further north, into Bengal.

　　While Englishmen thus found it possible to trade with greater
liberty around India, the constant drain of bullion required to support
such commerce made the company a target of jealous competitors
back in London. To retain its monopoly and bullion-export privileges,
the company was obliged to pay more exorbitant "gifts" to courtiers
of the British crown than were expected at Agra or Surat. In 1628
the directors appealed to Parliament in their first Petition of Right,
asking for a public declaration of support, but Parliament saw no
need as yet to concern itself with the affairs of a crown company. In
1635 the crown, still poised perilously on the head of Charles I, abandoned its ward by issuing a license for East Indian trade to yet another company, Courten's Association. Sir William Courten and his
merchant friends, joined by Endymion Porter, then groom of the royal
bedchamber, had "lent" King Charles £200,000 for the favor of
poaching on the grounds of the old company's monopoly. In 1636 the
first of Courten's ships to reach Surat plundered a Mughal vessel of its
entire cargo, leading to the retaliatory detainment of the East India
Company's president and council at Surat for two months and a fine

of £18,000. The association, which did not expire until its royal patron was beheaded in 1649, remained a source of continuing embarrassment and irritation, as well as commercial competition, to the company. By 1640 East India Company stock, whose face value was £100, sold in London for £60. In desperation the company appealed again to Parliament the following year, but the Commons expressed no concern for its diminishing fortunes. Then in 1646, a fresh bill of appeal asked that the company be reincorporated under "Parliamentary license," which the Commons found inducement enough to warn Courten's association to "withdraw from India" within three years. The Lords, however, rejected that bill, and the company found itself unable to sell a new joint stock subscription, leading the governor to advise his fellow shareholders to "draw home their factors and estate." But the court of directors at Leadenhall Street in London voted to hold on a bit longer, obliged though they were to abandon seven of their weaker Indian factories in 1648.

Under Oliver Cromwell's Commonwealth, the company finally received the national support it had earlier sought in vain. Cromwell's Charter of 1657 inaugurated the first permanent joint stock subscription, which became the capital base of a newly revitalized company that thus embarked upon its modern phase of corporate immortality. But for Cromwell's Protectorate, the royal-born company would have died ignominiously. In 1654, Cromwell's treaty with Portugal gave English ships full rights of trade in any Portuguese possession in Asia, finally shattering the *de jure* monopoly Portugal had held since the Treaty of Tordesillas. Though Cromwell's charter was burned by royalists after the Restoration, its commitment was honored by Charles II, whose era became a true golden age for the revitalized East India Company. New royal charters granted the company rights it had never previously possessed, including permission to coin money, to exercise full jurisdiction over all English subjects residing at its factories or forts, and to make war or peace with "non-Christian powers" in India. The merchant adventurers of London thus became a virtual state unto themselves, and acted accordingly whenever east of Good Hope. More than a hundred English factors resided in India after 1660 and, for the remaining four decades of the century, they received enough specie from home to purchase whatever goods they required to assure the continuing flow of an average annual profit of 25 percent. Learning well their lessons from the experience of their European predecessors, the English invested in cloth woven by the peasants of Coromandal and Gujarat, bought their pepper, indigo, and silk where it was cheapest, and left the souls of Indians to be cared for by their own priests and pundits. In 1658 the newly solvent company was based at Hughli in Bengal, inheriting the former Portuguese factory more than a hundred miles north of the Bay of Bengal up the Hughli River, Mother Ganga's main tributary to the sea. All factors in Bengal at this time remained, however, under the official control of the governor of Fort St. George. The acquisition of Bombay similarly strengthened the British position in the west, for after 1669, when Gerald Aungier took over as governor of that island and started its fortification, Bombay became the company's premier port and an impervious British bastion.

ELEVEN

GREAT MUGHAL GLORY
(1605–1707)

The Great Mughals, whose reigns span the entire seventeenth century, have with good reason become universal symbols of power and affluence, of tenderness and cruelty, of ferocity and sensitivity; luxury loving, licentious, sentimental, brutal, and poetic, they were the embodiment of all those extremes characteristic of the Indian life-style known as Mughlai. Jahangir, Shah Jahan, and Aurangzeb each in his own way epitomized some aspects of the complex cultural syncretism within which they lived and over which they presided. The courts they maintained, the courtiers they chose, reflected a new syncretic patina of civilization that was a blend of Indian, Persian, and Central Asian manners and mores.

Jahangir's Rajput mother can hardly be credited with having converted her son to Hindu ways, but his remarkable Persian wife, whom he renamed Nur Jahan ("Light of the World") after marrying her in 1611, firmly entrenched Persian culture at Agra's court. A thirty-four-year-old widow when she married the emperor, this ingenius woman was, in fact, ruling empress of India long before her uxorious husband died. She raised her Khurasan-born father, Mirza Beg (renamed Itimad ud-Dawlah by Akbar) to the premiership and brought her brother, Asaf Khan, into position to serve the next emperor in that office by arranging the marriage of his lovely daughter, Mumtaz Mahal ("Exalted of the Palace"), to Jahangir's third son, Khurram ("Joyous"), whom she effectively supported as imperial successor to her husband. The combination of beauty, brilliance, and ambition that allowed Nur Jahan first to capture Jahangir's heart and soon virtually to usurp his throne, may not be uniquely Persian, but following her ascent to power, Persian was no longer simply the language of Agra's court, it set the tone and direction of North Indian administration and cultural life at both provincial and imperial capitals. Agra itself had grown by this time to twice the size of Isfahan, with an estimated population of over half a million. The city was dominated by Akbar's mighty Red Fort on the Yamuna, whose heavily guarded battlements were viewed with awe and terror. Persian poets and artists, architects and musicians flocked to this great capital, which Akbar had found more congenial than Delhi, home of so many hostile Afghans, and which Jahangir, his empress, and their successor, Shah Jahan, sought to make a model of Safavid elegance, luxury, and grandeur. From the glazed tile that covered the inner walls and walks of the palace to the enclosed formal garden, stone lace walls, and inlaid tiles of the domed tomb Nur Jahan had built for her father, precursor of the Taj Mahal, Agra mirrored the architecture of its contemporary Persian capital. The Taj itself, which was said to have taken twenty thousand workers over twenty years to build after the death of Mumtaz Mahal in 1631, was designed by two Persian architects and has often been called the greatest single work of Safavid art ever constructed; in its dependence on Indian materials and craftsmanship, however, not to mention its use of such motifs as the four Rajput canopies built around the base of its dome, it emerges as an excellent example of Mughal cultural syncretism rather than as a Persian import. In many ways, moreover, the love of silks and perfumes, the custom of draping both male and female figures with jewels, diaphanous veils, and peacock feathers, and the delight in song and dance, intoxicating drink, and the pleasures of the harem are habits and traditions at least as deeply rooted in Indian as in Persian soil. Safavid approval served to validate such behavior for Muslims as well as Hindus, and in a real sense the historic significance of the Great Mughal Persian impact may be said to have been that it helped to Indianize Muslim culture, which is in part why the rule of the Great Mughals proved so stable a unifying force over some two centuries.

What could be more traditionally Indian, after all, than a Mughal procession of silver-tusked, silk-caparisoned elephants bearing bejeweled nobles in tassled howdahs? Jahangir's love of wine, women, and dancing girls was in the best of maharaja traditions, as was his lavish expenditure of countless rupees on the month-long festivities

celebrating the marriages of his sons. What did most people care if
mullas rather than brahmans presided over the ceremony? Agra's impoverished masses must have at least derived some vicarious pleasure
from seeing and hearing how richly their royal neighbors lived. They
could even forget that these rulers were "foreign conquerors," for as
any casual observer of Mughal miniature portraits may note, there
was no sharp distinction in either dress or appearance between the
Great Mughal emperors and princes and their leading Rajput nobles
or other contemporary Hindu chiefs. Koranic caligraphy and geometric designs continued to decorate the fringes of Mughal miniature
paintings, but portraits of the hallowed Muslim monarchs were now
painted without prohibition by Indo-Persian artists of rare genius, as
were the virtually naked figures of servant girls, princesses, and embracing couples in the classic Rajput, and earlier Hindu, tradition.
Jahangir's own interest in gardens and natural beauty helped stimulate the emergence of a distinctively naturalistic style in Mughal painting that was again more traditionally Indian than Persian in character.
Among the great Persian artists at his court was Aqā Ridā of Herat,
his son Abū-l-Hasan, and Mansur, whose vivid paintings of animals
won unique acclaim. Govardhan and Manohar were Hindu artists of
almost equal fame who worked at court, and we know of another
Hindu, Bishandas, of whose portraiture work Jahangir thought so
highly that he was sent to Isfahan to paint the Persian emperor,
'Abbās I. Jahangir prided himself on his Persian poetry and artistic
skill and wrote memoirs (*Tūzuk-i-Jahāngīrī*) covering most of his
reign, which were completed by the Persian Muhammad Hādī. Many
other historians resided at court, and they were happy to follow the
emperor to the Himalayas for the hot season, an annual royal exodus
initiated by Jahangir that the British later loved to emulate.

Jahangir placed his son Khurram in command of his army in
1613, and therefore the prince, who was soon renamed Shah Jahan
("Emperor of the World"), led a number of campaigns against the
Rajput forces in Mewar and Kangra and the Deccani sultanates of
Ahmadnagar, Bijapur, and Golconda. The only threat to Mughal
power in this era came from Persia, when in 1622 Shah Abbas
wrested Kandahar from Agra's control. Jahangir was too preoccupied
with his gardens, wine, poetry, and women to lead an army over the
Baluchistan passes, and Shah Jahan refused Nur Jahan's order that

he do so, rightly sensing that the empress no longer favored him (her own daughter by a former marriage had just married his younger brother, Shahriyar) and was trying to get him away from Agra and Delhi. The year before, Shah Jahan's elder brother, the rebellious Khusrau, had died, probably poisoned by one of his brothers. With Khusrau's death, Shah Jahan was the leading contender for his father's mantle; his real opponent was Nur Jahan, who wished to continue to rule the empire no matter who reigned.

In 1623 Shah Jahan marched in open rebellion toward Agra but was driven off by imperial forces under the command of the mighty general Mahabat Khan, whom Nur Jahan had recalled from exile in Kabul. The rebellious Shah Jahan was chased around southeast India for three years before finally agreeing to return to his father's fold. By then, however, Mahabat Khan was so powerful and popular that Nur Jahan considered him a potential threat and banished him to Bengal, subsequently charging him with embezzlement of imperial funds. Mahabat responded by staging a coup in 1626, taking both Jahangir and Nur Jahan prisoner. He held them captive for several months without harming either, then lost his nerve and let them escape. Nur Jahan pardoned the simpleminded general and sent him off to hunt Shah Jahan, who was advancing north again with an army at his back. The wily empress now appointed her equally shrewd and politically adroit brother, Asaf Khan, premier of the realm. On October 29, 1627, Jahangir died, and Nur Jahan tried to bolster the sagging spirits of her sick son-in-law Shahriyar, who was then at the Punjab provincial capital of Lahore, hoping to imbue him with courage to fight for Agra's treasure. She sought to win her brother's support, of course, but Asaf Khan backed his own son-in-law, Shah Jahan, whom he informed by courier of his father's death. Shah Jahan rushed north to claim his throne, reaching Agra early in 1628. All of his closest relatives, who were potential rivals, were put to death in fine Mughal fashion, and then Shah Jahan enjoyed three weeks of lavish coronation celebrations. Nur Jahan was pensioned off and went to live in solitude at Lahore till her death in 1645.

Shah Jahan ruled for three decades (1628–58). The most lavish spender of all the Mughals, he was addicted to monumental architecture inlaid with jewels and semiprecious stones and to a harem, whose total population numbered five thousand. His beloved wife, Mumtaz

Mahal, bore him fourteen children, only half of whom survived to adulthood, before she herself died at the age of thirty-nine. The eldest son was Dara Shikoh (1615–58), whose deep interest in art, humanity, and eclectic philosophy led many courtiers and foreign visitors to believe that India would soon be blessed with another ruler as wise and liberal as Akbar, but Dara's austere, orthodox brother, Aurangzeb, had other ideas. Shah Jahan's first years as emperor were preoccupied with seeking to subdue rebellion in the Deccan and Bundelkhand. Ahmadnagar and Bijapur were both defeated and promised to pay annual tribute to their Mughal suzerain by 1635, and the following year Golconda also agreed, but the Deccan refused to remain permanently subordinate to northern power. While Shah Jahan's army waged its costly wars in the south, one of India's worst recorded famines desolated the Deccan's peasant population, and only five thousand rupees of imperial funds were spent each week to help relieve the widespread misery and starvation by order of the ruler who soon would lavish billions on a peacock throne and his wife's tomb. It was at Burhanpur, the Mughal provincial fortress in the Deccan, in the midst of that famine- and plague-racked region in 1631, that Mumtaz Mahal died in childbirth. "Empire has no sweetness, life itself has no relish left for me now," Shah Jahan was supposed to have wailed when he heard the news, yet he lived another thirty-five years. Maharashtra had claimed the first of its Mughal royalty, but she would not be the last. The rugged Deccan was to remain a continuing political trap, seductively elusive, a bottomless pit of expenditure and martial loss for the Great Mughals.

Shah Jahan continued to rely on his grandfather's *mansabdari* system, but while Akbar had eighteen hundred *mansabdars,* the number of higher bureaucrats dropped to only eight hundred under Shah Jahan, though the upper limit was now escalated to the rank of sixty thousand for Prince Dara Shikoh (Shah Jahan himself held a *mansab* of thirty thousand before inheriting the throne). The other three princes, Aurangzeb, Shuja, and Murad, were granted much lower *mansabs,* yet all were above ten thousand. If Shah Jahan was more than generous to his children, however, he was less than expansive to others, for only four of his courtiers held the rank of seven thousand, six were given *mansabs* of six thousand, and fifteen had *mansabs* of five thousand. (Aurangzeb was to elevate many more to the higher

echelons of service, primarily because he needed many more generals
to wage his expensive Deccan wars.) Soon after his mother's death,
Dara Shikoh married his cousin, Nadira, to whom he remained singularly devoted, and who bore his eight children. Aurangzeb (1618–
1707) was only fourteen at the time, but he seems already to have
hated his elder brother and coveted his father's throne. Apparently in
the hope of cooling this brooding prince's ambition by distance, his
father appointed Aurangzeb viceroy (*nawab*) of the Deccan in 1636,
but remoteness from Agra only fired his appetite for power, and after
eight years in exile Aurangzeb quit his post and returned to the northern center of empire. By that time, Shah Jahan had begun construction of a magnificent new capital in Delhi, to which his court would
move in 1648.

Not that Agra was ever totally abandoned by the builder of the
Taj, in whose reign the white marble Pearl Mosque (Moti Masjid)
was also erected like a giant jewel inside the Agra fort. Perhaps it was
the memory of Mumtaz Mahal that troubled the emperor; or the heat,
which is always more intense at Agra than Delhi; or perhaps it was
merely the desire to build his own city, his own palace, for building
became his primary passion now. That passion drove him back to the
rusty plain on which no fewer than six earlier Indian capital cities had
previously been erected, there to design and order the construction of
the seventh, Shah Jahanabad, or what would later be called Old Delhi,
after the British began building a New Delhi in 1911. Many of the
blocks and bricks required to erect the new walls and massive gateways to Shah Jahan's city were taken from the rubble of Firozabad,
the Tughluq's Delhi, whose Purana Qila ("Old Fort") still stands, a
shattered fragment of jagged stone and blackened brick, a wretched
remnant of the once proud sultanate. Shah Jahan built on a grander
scale, though he, too, used the red sandstone of neighboring hills,
erecting a Red Fort (*Lal Qila*) even larger than Agra's palace, a city
within the city that enclosed almost five million square feet within its
towering ramparts. Inside were royal apartments, harems, a secretariat, factories, storehouses, military barracks, a treasury, a mint, and
stables, a home for tens of thousands of servants, slaves, courtiers,
eunuchs, princesses, and a king mightier than England's monarch,
richer than China's, as strong as Persia's. After the fort was finished,
a beautiful mosque—the Jama Masjid—was built facing its main en-

trance. India's largest place of worship, its central courtyard alone is over a hundred thousand square feet, permitting tens of thousands of Muslims to gather there on Friday afternoons for united prayer. It remains India's noblest monument to Islamic culture, even as the Red Fort is still its mightiest.

In 1639 Prince Shuja, Shah Jahan's second eldest son, had been sent to Bengal, the Mughal Empire's province of "peace, plenty, and pestilence," and he presided over its wealthy destiny for eighteen years, remote from Delhi and Agra. Prince Murad, the youngest brother, on the other hand, was the black sheep of Shah Jahan's sons. He deserted his army in Central Asia, after he had led them to Balkh and Badakshan in 1646 on what was Shah Jahan's most wasteful military venture, a vain attempt to recapture his dynasty's ancestral home of Samarkand. Perhaps, like Babur, Shah Jahan missed the melons of Central Asia, but his youngest son missed only the high life back at court, and for the peasantry who paid to sustain such imperial folly, the cost of the two-year Central Asian expedition totaled some forty million rupees. Small wonder that Shah Jahan felt obliged to raise his revenue demands to an average of one-half of all crops, rather than Akbar's one-third. Between his monumental buildings and his martial misadventures—which included three more fruitless expeditions against the Persians at Kandahar—even the fabled surplus exacted from the perspiring backs of one hundred million people swiftly disappeared. Still, the emperor sat upon his Peacock Throne (thanks to which it was said that the world had run "short of gold"), encrusted with the largest diamonds, rubies, sapphires, emeralds, and pearls ever found, all but blinding distracted courtiers and visitors, who faced a firmament of unmatched brilliance and wealth whenever they had the good fortune of approaching the imperial presence. Unless required elsewhere on urgent business, all principal *mansabdars* mustered twice daily before the emperor at his Hall of Public Audience (*Diwan-i Am*), while lesser officials stood somewhat more remote, yet still within call should they be needed. The virtues of humility, obedience, patience, and loyalty were thus instilled in all of the mighty generals and civil administrators, at the cost of intellectual initiative, independence of mind, self-sufficiency, integrity, and courage. Bullied and treated like children by their emperor, it was hardly surprising to find such "nobles" behaving in turn as petulant petty tyrants to their servants, bearers, soldiers, and peasants. The whole system was a pyramid of power designed to perpetuate its imperial pinnacle, whether through ruthless violence, extortion, harem intrigue, bribery, or sheer terror. The formal gardens, marble mausoleums, and Persian miniatures were as nectar squeezed from a subcontinent crushed into obedience, milked of its riches by the few, who had reason to lyricize in Persian couplets carved into the ornate walls of Delhi's Hall of Private Audience (*Diwan-i Khas*), "If there be Paradise on earth, It is Here, It is Here, It is Here!"

Dara Shikoh's intellectual curiosity and religious tolerance was more than counterbalanced toward the end of Shah Jahan's reign by Aurangzeb's Islamic orthodoxy and militant zeal. While Shah Jahan remained healthy, the tension between these polar princes was held in balance, and though factions or parties partial to one or the other emerged at court, the emperor kept them from open warfare. Shah Jahan sent Aurangzeb back to the Deccan in 1652, but did not give him full rein to conquer Golconda and Bijapur by invasion, as Aurangzeb wanted to do, afraid, it seems, of the growing power and boundless ambition of his "prayer-monger" son. Dara, though viceroy of the Punjab, stayed close to his father at court, but he cared little for mundane glory and took less interest in harem intrigue or court politics than he did in mysticism, Hindu as well as Sufi, and in the translation of such Sanskrit classics as the *Bhagavad Gita* into Persian, to which he devoted himself. Thus, while Aurangzeb ventured forth from his capital of Aurangabad to temper his sword in dusty Deccan campaigns that frustrated as well as whetted his appetite for greater conflicts and clearer victories, Dara languished and grew weary in the Delhi *maya*-world of material glitter. Shuja remained in Bengal, and Murad was sent to Gujarat, where he was given as little to do as possible, left to idle his time with a busy cycle of provincial feasts and ceremonies, interspersed by royal hunts and holiday visits to neighboring cities and shrines, arranged by his shrewd "guardian" minister, Ali Naqi Khan. Late in 1657, with the beautiful mausoleum that would immortalize his wife completed, the aged Shah Jahan fell victim to his own insatiable harem lust and all but expired of strangury, throwing Delhi's court into a state of panic over the succession that soon reverberated to the distant corners of princely Mughal provincial habitation.

Aurangzeb lured simple Murad into alliance against "our disgraceful brother," as he called Dara Shikoh, first by denying that he had any regal ambitions, claiming it was only to save Islam that he stood ready to "defend" the empire against his elder brother, and later by promising Murad the Punjab and Sind for his support. Shuja was first to proclaim himself emperor, minting coin in his own name before the end of 1657 in Bengal, stimulating Murad to do the same in Gujarat. Aurangzeb was more cautious, waiting till July 1658 before he would claim the thrones of Delhi and Agra. By early 1658 Shah Jahan had regained his health, and he tried desperately to help Dara cope with his brothers' lust for power, but the mystic heir had spent too much of his time worrying about otherworldly matters to save his life or retain the empire bequeathed to him. The united forces of Aurangzeb and Murad marched north to defeat Jodhpur Rajput Jaswant Singh's formidable imperial force at Dharmat in April. Agra was mobilized in desperate preparations for the impending onslaught, and Dara set forth atop the mightiest royal elephant to lead his great army of about a hundred thousand troops to Dholpur, on the river Chambal, where he planned to block the advancing rebel force. Aurangzeb, however, was too shrewd a strategist for his brother. He eluded the blockade, fording the river upstream in the blazing heat of May and forcing Dara to fall back in hasty disarray to Agra's abandoned fortress. They met at Samugarh, several miles east of Agra, two elephant-led armies, Dara's riddled with disloyal Muslim cavalry like that led by the Uzbek Khalilullah Khan. Aurangzeb was clearly the stronger general. Rajputs in the thousands died for Dara Shikoh, who fled the field toward Agra, a fallen and hunted shade of his former self; his head was soon to be delivered in a box from the victorious Aurangzeb to their imprisoned father. Murad was at first flattered by his pious elder brother, while his army was still of use to Aurangzeb, who beseiged Agra's Fort till he forced Shah Jahan to surrender early in June; soon after that, Murad was taken prisoner by his brother, and three years later he, too, was decapitated.

Aurangzeb ascended the throne of Akbar as Alamgir ("World Conqueror") the First in July 1658 and reigned until his death in March 1707. He was at once the most pious and the most ruthless of the Great Mughals, a single-minded leader of brilliant administrative capacity and as cunning a statesman as ever mounted an Indian throne. He was hailed by Sunni Islam as India's only caliph and reviled by Hindudom even more than the Ghaznavids and Ghurs had been for the suffering he self-righteously inflicted on non-Muslim subjects in peacetime as well as in war. He had no friends, only worshipping admirers and enemies whose hatred was surpassed only by their fear of his cold fury. He spent more time at prayer than in his harem, and was most methodical in his relentless pursuit of power. Dara, who had fled to Lahore and from thence to Multan, posed no real military challenge after Samugarh. Shuja, who still retained control over Bengal and Bihar, hoped to "liberate" Agra and the imprisoned Shah Jahan while Aurangzeb was busy chasing Dara around the Punjab, but Emperor Alamgir proved too swift, too ubiquitous a force for any and all of his rivals. Defeated at Khajuha in January of 1659, Shuja retreated to his Bengali capital of Rajmahal on the Ganga, but was forced soon to abandon that stronghold for East Bengal's delta and Dacca. The next year he was driven out of India entirely, seeking refuge in Burma, where he reportedly met a violent end. Once the war of succession was over, Alamgir turned his energies to problems of an administrative and financial nature. Securing his hotly contested throne had cost the imperial treasury much of the remaining fortune accumulated during Shah Jahan's reign, but Alamgir's taste for building and court luxuries was the austere opposite of his father's. Monumental construction now came to an end, as did most royal expenditure on nonreligious celebrations, especially those including wine, song, and dance, all banished as infidel habits unfit for courtly patronage. Personally puritanical in dress and taste, Alamgir set a tone of zealous sobriety for his courtiers, which amply attested to his strength of character and helped impress his piety on those who had watched his bloody rise to power with misgivings.

The era of religious tolerance and Hindu-Muslim equality of treatment that had been initiated by Akbar was now abandoned. Alamgir appointed "censors of public morals" (*muhtasibs*) to every provincial capital and other large city in his realm, ordering them to be sure that Islamic law was obeyed and the proper prayers performed. In 1668 Hindu religious fairs were outlawed, and no new Hindu temple permits were issued, nor would royal permission be granted to repair rundown temples. Alamgir even tried to abolish gambling and "illicit sex" from the land that had virtually invented the former and practiced the latter as one means of worship. Hard-pressed to raise sufficient revenue to support the army he needed to keep himself in power, Alamgir reimposed the hated *jizya* poll tax on Hindus in 1679, after having more than doubled the duty Hindu merchants were obliged to pay on the same produce bought and sold by Muslims. When crowds gathered near his Red Fort to protest such inequity and blatant discrimination, imperial elephants were ordered out to crush them. Alamgir received many warnings that rebellion would spread across India if he persisted in his policy of strict partiality to the religious beliefs of the ruling minority, but he stayed firmly in power and only expanded his imperial domain.

The primary cause of the widespread revolt that occurred in the latter part of Alamgir's reign, however, was economic, not religious. The intolerable increases in revenue demands drove more and more *zamindars* as well as peasants to risk death from rebellion rather than accept inevitable starvation. With Mughal imperial power heavily dependent on its cavalry, administrative *mansabdars* were assigned *jagirs* of land to sustain themselves and the horsemen they were expected to lead into battle at the emperor's call. Greed tempted many *mansabdars* to raise the revenue demands imposed on their peasants, despite central government regulations to the contrary, and to support far fewer cavalry than their rank required, keeping the grain that would have been used to feed those men and horses for their own profit. Such inflation became quite common toward the end of Alamgir's reign and in the period of more rapid Mughal decline that followed. Even during the early years of the reign, however, the harshness of *jagirdar* demands led peasants in many parts of the empire to flee from their Mughal villages and rally their support behind regional *zamindars*, Jat, Maratha, Sikh, and Rajput.

Hindu Jat peasants under a *zamindar* named Gokulā revolted in Mathura in 1669, and three years later the Satnamis ("Truth Namers"), a sect of Hindu peasants in the Punjab, rebelled, marching toward Delhi until they were blown to shreds by Mughal artillery. The Sikhs by now had twice supported losing aspirants to Mughal power, once in Jahangir's reign, when they backed his rebel son, Khusrau, and again when they favored Dara's claim; this experience left a bitter residue of anti-Mughal passion in the Punjab, to which Alamgir added an intolerable potion of anguish. The Marathas also emerged throughout the latter part of his reign as bitter thorns in his Deccan side, and the Rajputs rose time and again, yet the "prayer-monger" held his

throne, convinced that he did so by the will of Allah, assisted by a
larger and better paid army than any previous Indian monarch had
ever sustained. Despite the "bad press" that has haunted Alamgir's
reputation since his death, his reign may be read as a lesson in the
efficacy of unrelenting force and untiring dictatorial dominance as
keys to the control of India's preponderantly docile, hard-working,
apolitical peasant population.

Akbar had hoped to win control over all of India by his enlightened policy of "love"—forging a multireligious alliance, reducing
taxes, and encouraging tolerance for all sects and ideas. Alamgir
achieved greater success through his policy of terror and naked power
—if we would measure imperial success, that is, by the conquest and
retention of real estate alone. It is, of course, harder to gauge the less
tangible, though at least equally important things, such as the "general welfare" or "happiness" of the public at large, especially since
there are so few surviving records of the lives and feelings of India's
people, other than imperial courtiers. We do know, however, of at
least one anonymous letter of eloquent complaint that reached the
emperor after he had reimposed the *jizya,* stating in part: "Your subjects are trampled underfoot; every province of your Empire is impoverished; depopulation spreads and difficulties accumulate . . .
If Your Majesty places any faith in those books by distinction called
divine, you will be there instructed that God is the God of all mankind, not the God of Mussalmans alone."[1] Such sentiment may have
prevailed throughout much of India during the latter part of the
seventeenth century, and in several regions it led to violent and sustained revolt. The Punjab has been mentioned as the first such major
region, perhaps because it had so long been "loyal" to its viceroy,
Dara, and his claims to Agra's throne. While Dara still had Lahore's
gold to dispense, he retained a substantial following of cavalry, Rajput
as well as Muslim, but that quickly melted as Alamgir's sun rose
higher, leaving, nonetheless, a hard core of Sikh opposition in the
Land of the Five Rivers.

Since the founding of the Sikh[2] faith by its saintly guru, Nanak,

[1] Jadunath Sarkar, *History of Aurangzeb* (Calcutta: M. C. Sarkar
and Sons, 1952), vol. 3, p. 34.
[2] The best history of this faith is Khushwant Singh, *A History of the
Sikhs,* vol. 1, *1469–1839* (Princeton, N.J.: Princeton University Press,
1963).

in the early sixteenth century, this popular new liberal community flourished in the Punjab, drawing its recruits mostly from hard-working peasantry of Hindu as well as Muslim birth. Recording the sacred sayings of Guru Nanak in a specially devised script, called *Gurumukhī* ("from the Guru's mouth"), his chosen successor, Guru Angad (1504–52), gave the community greater cohesion and a clearer sense of its own identity. The third guru, Amar Das, was patronized by Akbar, further inducing converts to the faith, which stressed community eating as well as prayer and abolished female *purdāh* ("seclusion") together with caste exclusiveness and untouchability. Ram Das, the fourth guru, had served at Akbar's court and was granted some land by the emperor between the rivers Sutlej and Ravi in the Punjab, which was to become the site of the Sikh's sacred capital. Ram Das's son and successor, Arjun (1563–1606), completed the great Sikh temple at this spot, naming the city Amritsar ("Pool of Immortal Nectar") for its tank filled with "sacred" well water. Under Arjun's careful guidance the Sikh scripture, Granth Sāhib, was compiled and deposited in Armitsar's temple. "In this vessel," wrote Guru Arjun of his book, "you will find three things—truth, peace, and contemplation." Jahangir charged Arjun with treason, however, and had him tortured to death for supposedly aiding the emperor's rebel son Khusrau, then refusing to admit his "guilt" or abandon his faith. Arjun's martyrdom inspired his own son, Hargobind, to arm his comrades, who stood ready to defend their religion with their lives, converting the pacifist faith of Guru Nanak into a militant new order pitted against Mughal tyranny. Though forced to retreat to Kiratpur in the Himalayan foothills, Guru Hargobind and his band held out against Mughal arms until the guru's peaceful death in 1644.

The seventh guru, Har Rai, was hounded even farther into the mountains, from which he emerged only after 1658 to support Dara Shikoh's claim to the throne. Following Alamgir's victory, however, Har Rai was obliged to send his son, Ram Rai, as hostage to Delhi's court, where the young man became a loyal follower of the emperor, losing his father's support. Shortly before his death in 1661, Har Rai designated his younger son, Hari Krishen, as his successor, but this son, too, was forced to seek Alamgir's favor in Delhi, and he expired there of smallpox in 1664. The ninth guru was Har Rai's granduncle, Tegh Bahadur (1621–75), who was arrested by Alamgir's soldiers in

Agra and brought to Delhi, where he was subsequently beheaded after refusing to convert to Islam. Guru Gobind Rai (1666–1708), Tegh Bahadur's son, was the tenth and last of the Sikh Gurus. He vowed to avenge his father's murder and to devote his life to ceaseless struggle against Alamgir's tyranny. Gobind Rai forged his community into an "army of the pure" (*khālsa*), taking as his new surname Singh, which means "Lion," and baptizing his closest followers with the same name. From this time at least, the Sikhs emerged as a close-knit force of toughened fighters who recognized one another by the sartorial symbols of their faith, for they vowed never to cut their hair or beards, always to carry a saber, to wear a steel bracelet on their right wrists and knee-length soldier's shorts, and to keep a comb for their hair. The Guru now merged with his community, since all were to become hunted men and full-time fighters, but the majority will of the *khalsa* would hereafter represent the Guru who was thus immortalized. Gobind was said to have had more than twenty thousand loyal supporters in his retinue at one time, but Mughal armies outnumbered and harrassed him till his last years of struggle and hardship. At least he had the ultimate satisfaction of outliving his hated rival, Alamgir, by a year and a half.

In Maharashtra[3] an equally fierce Hindu opposition to Mughal rule emerged under the leadership of Shivaji Bhonsle (1627–80), who was hailed by his followers as the founding father of the Maratha "nation" but reviled by the Mughals as a Deccan "mountain rat." Shivaji's father, Shāhji, had served the sultan of Ahmadnagar, northernmost of the five successor-kingdoms to the once mighty Bahmani sultanate. As the sultan's *jagirdar* in Poona, Shahji was an important local leader and soon found himself wooed by the sultan of neighboring Bijapur, for his was a borderland *jagir*. Shortly before Shivaji's birth, his father shifted allegiance to Bijapur and took a second wife as well as a new sultan. Emotionally abandoned by her husband, Shivaji's devout mother, Jiji Bai, lavished all of her attention and affection upon her son. After Shah Jahan came to power at Agra and advanced into the Deccan, Shahji shifted allegiance once more, accepting a Mughal *mansab* from late 1630 until Shah Jahan returned

[3] The best history of the Marathas remains G. S. Sardesai, *New History of the Marathas,* 3 vols., 2nd impression (Bombay: Phoenix Publications, 1957).

north in 1632, when the adroit Maratha statesman decided to return to his original overlord, the sultan of Ahmadnagar. The following year, however, Daulatabad fell, and the sultan was taken prisoner by the Mughal army. Shahji now sought to rally the remnant of Ahmadnagar forces, using guerrilla warfare tactics against far superior Mughal might in the hostile Deccan terrain he knew so well. With some twelve thousand followers, Shahji retained his independence of Muslim rule for a few years, but when Bijapur concluded a treaty with Shah Jahan in 1636, the fate of Hindu resistance was sealed, and Shahji himself surrendered to a joint Mughal-Bijapur army. He was banished thereafter from Poona, but Jiji Bai raised her son Shivaji in that city, which was to become the capital of Maratha power.

Shivaji was a fiercer fighter than his father, and he grew to manhood imbued with his mother's love of Hinduism and antipathy to Muslim Bijapuri as well as Mughal rule. That the Deccani sultanates were heterodox Shi'ites, while the Mughals were orthodox Sunnis made no difference to Shivaji's yearning for independence of all overlords, especially those of a "foreign" faith. He wanted "self-rule" (*sva-rāj*) and the full freedom to practice his own religion (*sva-dharma*) in the land of his birth, the Great Country—Mahā-Rāshtra. Shivaji left Poona at the age of twenty, leading a band of young Maratha followers who had been reared in the hills of their rugged region and could most effectively use guerrilla tactics in fighting both Mughal and Bijapur powers. These "mountain rats" would wait for caravans to wend their way into the hill country and then swoop down to plunder whatever they could use to strengthen their band, which soon had the arms, money, and horses to pose a formidable challenge to Muslim garrisons. Securing a number of key mountaintop plateaus as his fortresses, Shivaji soon commanded a substantial portion of Maharashtrian terrain, and though Bijapur tried to coerce him by holding his father hostage, nothing could intimidate this tenacious young Hindu warrior. His first stronghold, Sinhagārh ("The Fortress of the Lion"), dominates the plain southwest of Poona, its sheer rock face rising so nearly vertical that no human was believed capable of scaling its height. Maharashtrian lore claims that Shivaji used a giant Deccan lizard to assist him in this "superhuman" task, tying a rope to the lizard's rough tail and hurling the creature up against the wall so that its suction-cup feet fastened firmly to the sheer cliff and Shivaji and his men could climb over it in the dead of night to surprise the Muslim garrison there.

Whatever his secrets may have been, Shivaji clearly used an intimate knowledge of his homeland to considerable martial advantage, and he well deserves to be called one of the founding fathers of modern guerrilla warfare, a method he learned in part from Shahji. By 1659 Shivaji's daring exploits roused sufficient Bijapuri concern to launch the powerful general Afzal Khan and an army of several thousands troops, who completely surrounded Shivaji in his "Fortress of Valor" (Pratāpgārh), where he was trapped with insufficient food or water to survive a lengthy siege. Shivaji then offered to "surrender," insisting, however, upon meeting Afzal Khan personally, man to man, to "discuss final terms." The Muslim general was a giant bull of a man, and Shivaji barely more than five feet tall. Afzal advanced unafraid to a level spot just below the high wall of the fortress, and Shivaji left his lair wearing an innocuous-looking, loose-fitting cloth shirt with sleeves that covered his hands, each of which was lethally armed. He carried a "scorpion-tail" dagger in one hand, and had the fingers of the other sheathed in razor-edged iron "tigers' claws." Shivaji rushed to embrace Afzal Khan, who collapsed with a death cry that signaled the Hindu troops, hidden along the road Afzal had taken, to spring from ambush and kill the general's attendants. The leaderless army at the base of the mountain was easily dispersed in panic and despair, once the troops realized that the "invincible" Afzal was dead and there would be no prospect of pay, nor hope of plunder. Shivaji's murder of Afzal Khan marked the real birth of Maratha power; thereafter, no South Indian force would be able to challenge this intrepid Hindu leader, and only the full weight of Mughal imperial might could humble him. Shivaji was strong enough to raid Surat in 1664, plundering much of its wealth and even trying to loot the British factory there, which was stoutly defended by its small garrison. When such embarrassing exploits reached Alamgir's ears, he sent a huge army, led by the Rajput Jai Singh, to deal with this Maratha "upstart." Shivaji was now trapped at another of his hilltop forts, Purandhar; in 1665 he sued for peace, but this time he was obliged to surrender twenty-three of his twenty-five fortresses, in return for a *mansab* in the imperial Mughal service.

Shivaji actually appeared as a *mansabdar* of five thousand horse at Alamgir's court, but he was so outraged to find himself facing the backs of generals he considered his inferiors that he fussed and fumed and finally fainted, had to be carried off, and was placed under house arrest. Once again proving his ingenious talent for "miraculous" escape, Shivaji eluded his guards by hiding in a laundry or food basket until he was outside, then made his way back to the Deccan, where he was welcomed as a returning monarch. By 1670 he had recaptured most of his fortress perches and could launch a second, even more lucrative attack against Surat. In 1674 he felt sufficiently powerful to have himself crowned Chatrapati ("Lord of the Universe") in a traditional Hindu coronation at Rajgarh ("The Fortress of Rule"), where eleven thousand brahmans gathered to chant the sacred Vedic mantras, while fifty thousand loyal Maharashtrian friends and supporters swore undying allegiance to this reincarnation of Lord Shiva. Shivaji Maharaj thus publicly staked his claim to *svaraj* over the land of his birth, as countless Hindu voices throughout the Deccan took up the cry of *"Shivaji Maharaj-ki-jai!"* ("Victory to Great King Shivaji"). Alamgir was at the time preoccupied with Afghan conflicts and would not be able to focus his personal attention upon the Deccan for some years to come; not, in fact, until after Shivaji's early death in 1680. The death of Shivaji did not mean the end of Maharashtra's struggle for independence, however, since he bequeathed to his sons and countrymen his fierce spirit of Hindu nationalism—at least in this Maharashtrian regional form—and they continued his battle against Mughal power.

The year Shivaji died, the Rajputs of Jodhpur and Mewar rose in united opposition to Alamgir's rule. The emperor sent his son Akbar at the head of a powerful Mughal army to subdue Rajput resistance in 1680, but the young prince decided instead to join forces with the Hindus against his father's tyranny and soon declared himself emperor. The second Akbar could not, however, fully emulate his namesake; falling victim to Alamgir's greater shrewdness and martial experience, he was forced to flee with a few of his followers to the Deccan. Akbar sought help from Shivaji's elder son, Sambhaji (1657–89), who had also just assumed his father's royal title at Rajgarh. Alamgir now saw no alternative but to march south himself in order to subdue this potentially most dangerous of all rebellions he had faced. By November 1681 he reached Burhanpur, and in March 1682 he established his camp at Aurangabad, the Deccan capital he had built when still a young prince. But it was much easier for Alamgir to invade the Deccan than it was for him to find the Maratha forces he had come to crush. Sambhaji, like his father, used guerrilla-warfare strategy and scorched-earth tactics to harass and hide from the Mughal armies, retaining control of the treacherous hills and their forts, posing a constant source of annoyance and insult to Alamgir's far superior power. Akbar sought in vain to lure Sambhaji north with him, hoping to march upon Agra at the head of an allied force of Marathas, Rajputs, and those Mughals who rallied round his banner. Shivaji's son never really trusted the Muslim prince, however, and Indian regional differences once again served to undermine dreams of unification, for even though the Marathas and the Rajputs were both Hindus, they spoke different languages and hardly knew one another. The theme of regional Indian discord would often be replayed in the face of growing British power, and not until the latter part of the nineteenth century would Indians learn that if they hoped ever to rule themselves they would have to subordinate regional ambitions, prejudices, and fears to a common, united national effort. After watching his father's invincible force demolish Bijapur's defenses and bring that Deccan sultanate under the ever-expanding imperial umbrella, Akbar fled to Persia in 1686, where he died in exile before Alamgir. Sambhaji continued to evade direct confrontation with Alamgir's army, which moved like a swarm of locusts over the Deccan.

In 1687, soon after Bijapur fell, Golconda, the last of the Shi'ite sultanates, surrendered and was absorbed within the Mughal system, leaving only the Marathas as diehard opponents to the "World Conqueror." In 1689 Sambhaji was captured, tortured, and butchered to death, firing his people's spirit of resistance even more through the courage of his last anguished hours than he had during the latter years of his lifetime. His younger brother, Rājā Rām, now took the sacred title of Chatrapati and kept the banner of Maratha independence waving for another full decade, till his own death in 1700. Raja Ram's widow, Tārā Bāi, continued her husband's relentless struggle, heroically refusing to bow to the Mughal yoke. Alamgir had taken Sambhaji's son Shāhu and his mother into his camp, however, and raised the boy as a *mansabdar*, determined to incorporate Maharashtra into the empire that now extended beyond the limits reached even by the great Ashoka. Never before nor since would so much of the South Asian subcontinent fall within a single ruler's domain. Great Mughal glory thus reached its pinnacle of power at the end of the seventeenth century, under the fiercely fanatical leadership of an octogenerian despot whose will was obeyed from Kashmir to Hyderabad, from Kabul to Assam.

Yet the conquest of the Deccan, to which Alamgir devoted the last twenty-six years of his life, was in many ways a Pyrrhic victory, costing an estimated hundred thousand lives a year during its last decades of fruitless, chess-game warfare, in which Maratha fortresses would be taken one week and lost the next, only to be recaptured and relost a month later. The expense in gold and rupees can hardly be imagined or accurately estimated. Alamgir's moving capital alone—a city of tents thirty miles in circumference, some two hundred and fifty bazaars, with half a million camp followers, fifty thousand camels, and thirty thousand elephants, all of whom had to be fed, stripped peninsular India of any and all of its surplus grain and wealth throughout the quarter century of its intrusion. Not only famine, but bubonic plague arose to take countless more lives during this era of tragic conflict and waste. The macabre dimension of the drama seems somehow too great for seventeenth-century warfare, sounding more modern in its unyielding butchery, its senseless massacre of human and animal life. Even Alamgir had ceased to understand the purpose for it all by the time he decided to stop firing his guns and turn his army world back toward the north in 1705. The emperor was nearing ninety by then. He spent most of his days reading and copying the Koran, preparing himself for the final reckoning, while at the retreating fringes of his camp, Maratha horsemen rode roughshod over the rear guard, looting, plundering, and picking off Mughal stragglers, gaining strength from the bloated imperial presence that had come to plague the Deccan. "I came alone and I go as a stranger. I do not know who I am, nor what I have been doing," the dying old man confessed to his son in February 1707. "I have sinned terribly, and I do not know what punishment awaits me."[4] He expired on a Friday and was buried in a village near Aurangabad.

TWELVE

TWILIGHT OF THE MUGHAL EMPIRE

(1707–64)

The last decades of Alamgir's reign imposed too heavy a drain upon imperial Mughal resources and aroused too much opposition to Mughal rule to permit the empire that Alamgir's indefatigable energy and will had dominated to flourish long after its master's demise. Not that so vast and complex a machine as the Mughal bureaucracy depended upon the presence of a single individual alone, as Alamgir's own absence from Delhi throughout the final quarter century of his reign demonstrated, but fear of an energetic despot's wrath seems to have sufficed to inspire even lazy and mediocre officials to labor beyond their normal capacities. It certainly kept Aurangzeb's three living sons (his eldest had died in prison) hard at work in their respective provincial posts, though none of them inherited their father's appetite for power or his obsession with religion and duty. The first half of the eighteenth century was the twilight era of Mughal imperial rule.

Muazzam, Alamgir's eldest living son, was sixty-three when his father died, and left Lahore for Delhi, to claim the peacock throne as Bahādur Shāh ("Great King") the First (Shāh Alam). His brothers Azam and Kam Bakhsh, also laid claim to the throne, however, so that the first half of Bahadur Shah's brief reign of five years was consumed in the traditional Mughal war of succession. Bahadur Shah tried to calm the troubled waters of Hindu antagonism by making his peace with the Rajputs and leaving young Shahu, who had "escaped" Mughal supervision to return to Maharashtra as raja of Satara, to fight his own civil war against Maratha rebel diehards. The new emperor had the wisdom to recognize his limitations and those of his realm. He even won the support of Guru Gobind, who went to Agra to attend his court. Following Gobind's death in October 1708, however, the Sikhs resumed more militant action against Mughal garrisons in the Punjab, under the leadership of Bandā Bahādur (1708–16). Promising land to the landless and sharing his wealth with all who joined his force, Banda (the name means "slave" of the guru; he was born Laksman Das) gathered his army as he marched, storming the Mughal fort at Sirhind in 1710, and becoming virtual king of the Punjab. (He even introduced his own calendar and struck coin to commemorate his "reign.") Lahore alone withstood Banda's peasant uprising, and when Bahadur Shah's army marched against him, he was forced to flee to the hills. The emperor found himself obliged to remain in Lahore, faced with the sort of guerrilla warfare and peasant uprising that had plagued his father in the Deccan. There was, it seemed, little satisfaction or glory now in Mughal imperial power, only martial conflict and the challenge of rebel forces, north and south. Spiritless and depressed, Bahadur Shah died in February 1712.

A new storm of fraternal warfare now broke loose, driving all four of Bahadur Shah's sons scurrying about so desperately seeking their own support that none of them bothered to bury their father for nearly a month. Jahandar Shah, the eldest son, finally won the scramble for the throne, but he proved so incompetent an administrator that he lost the support of his own *vazir,* Zulfiqar Khan, and he squandered his treasure as profligately as he did his power, which he managed to hold less than a year. The emperor's thirty-year-old nephew, Farrukh-siyar, seized Delhi and Agra with the help of two powerful courtiers, the Sayyid brothers, who became the real rulers of Delhi. Shortly after the young emperor was enthroned, however, he lost confidence in his benefactors, and he wasted much of the six years of his reign in spying on and intriguing against them. He also launched a powerful army against Bandā, who held out in costly rearguard fighting until the end of 1715, when he was taken in chains and paraded through the streets of Delhi and was then slowly tortured before being executed the following year. Factional rivalry at court, which was stimulated by the emperor's own insecurity and ineptitude, led to a general breakdown in discipline and pay. The erosion of imperial authority ended when Farrukh-siyar was dragged from his harem, blinded by his own courtiers, and poisoned in 1719.

Husain Ali, the more powerful Sayyid brother, had returned from his viceroyalty over the Deccan at the head of an army that included some eleven thousand loyal Marathas, intending to depose the inept Farrukh-siyar. The Maratha civil war had ended, thanks in great measure to the shrewdness of a Chitpavan brahman, Bālājī Vishvanāth (ca. 1660–1720), who became Shahu's *peshwa* ("premier") over Poona in 1714; it was the most important post of Maharashtrian administrative power, and one that he passed on to his heirs for more than a century. In return for their loyal support of Husain Ali and the Mughal imperial faction he represented, the Marathas were officially "recognized" and given "tax-collecting power" (*diwāni*) over the land they actually controlled. Shahu thus became a member of the Mughal bureaucracy, a *mansabdar* of fifteen thousand, invested with command of all of Shivaji's forts and lands, including territory conquered from the neighboring central provinces of Khandesh and Berar and pockets of Maratha land in Mysore and along the coasts around Tanjore and the Carnatic. The total paper value (actual value was always less) of the annual revenue of the six Mughal provinces thus placed under Maratha rule was supposed to have been eighteen *crores* of rupees (180 million rupees), which was an estimated 35 percent of the total crop value of the region. The figure of 35 percent was agreed upon as a result of Balaji's tough negotiating and Mughal ignorance of earlier Maratha tax-collecting practices. Shivaji had devised two types of tax: one was for the *svarajya* or "home territory" of the Marathas, which was called *sardeshmukhi* and was only 10 percent of the annual yield; the other was for conquered lands beyond the *svarajya,* which were forced to pay a higher tribute of 25 percent (*chauth*) for Maratha "protection." Now the Marathas were given permission to collect both *chauth* and *sardeshmukhi* from all their domains in return for the promise to feed and keep ready fifteen thousand cavalry to serve the Mughal emperor upon call, plus an annual cash payment of ten *lakhs* of rupees (one million rupees) to the Mughal treasury. From the Maratha viewpoint it was a great bargain, supporting Maratha soldiers of fortune and giving them new license to raid almost at will and conquer more of central India, which could then be "officially" taxed—for who ever knew exactly where one Mughal province ended and the next began? It was the dawn of a century of unprecedented Maratha power and expansion at the expense of neighboring Mughal provinces in Gujarat, Malwa, Rajputana, central India, and Orissa, and ultimately of Agra and Delhi as well. Poona flourished with more revenue than it had ever before demanded or received, and it soon had the martial power to ignore entirely the nominal annual payments it owed Delhi. After all, what was the point in paying a monarch whom no one respected or feared?

The Sayyid brothers again selected the new "emperor," but he proved to be consumptive and had to be replaced in a matter of weeks by his elder brother, who was so sick himself that he died just three months after being crowned Shah Jahan the Second. Delhi's court was a sorry state of despair and confusion by the end of 1719, when the hard-working kingmakers, the Sayyids, came up with still another imperial front man, a young grandson of Bahadur Shah, who was crowned with the title Muhammad Shah, and was to surprise everyone by reigning for thirty years. The problem of having to find so many new monarchs in so short a time, however, was that rival claimants and their supporters were more swiftly embittered, rousing more intense hatred of the Sayyids at court, especially among the rival Irani and Turani factions, which now joined forces in opposing the domineering brothers. The strongest Turani noble, Nizam-ul-Mulk (1669–1748), had been sent as viceroy to the Deccan, where he led a successful assault against the Sayyids' nephews in Khandesh, luring Husain Ali from Delhi at the head of an imperial force. But the mighty Sayyid general was poisoned en route and his elder brother imprisoned and left to die in chains by Muhammad Shah, who had turned against his mentors and who now appointed Nizam-ul-Mulk his *vazir*. As premier, the Nizam tried to institute effective reforms in a court that had virtually abandoned administrative routine, but Muhammad Shah took less interest in appointing efficient *mansabdars* and punishing those who failed to remit revenue to his imperial coffers than he did in his harem and hookah. There were sycophants enough at court to satisfy the young emperor's ego, and the Nizam soon realized that he could reform Delhi only by removing most of his noble rivals from its Red Fort, a thankless task at best, and a perilous one. He chose instead to live as "king" in Hyderabad, the city he had come to love in South India. Late in 1723 he abandoned Delhi and, with his army, moved back to the south, where he founded a "dynasty" that would long outlive the Mughal Empire.

The Nizam and the Marathas now made competing claims to the same Deccan territory. Bājī Rāo (1700–1740), who succeeded his father as *peshwa,* soon proved to be a vigorous defender of Maratha fortunes. The death of the Sayyids had left the Marathas without support in Delhi, and the Nizam at first refused to honor the 1719 imperial settlement that gave Poona its claim to revenue from much of his domain. Baji Rao himself, though a brahman, took the lead in confronting the Nizam's army, which was also challenged by a Mughal force sent from Delhi with the emperor's newly appointed viceroy of the Deccan, Mubariz Khan. Rather than risk fighting both armies, the Nizam chose to conciliate the Marathas first, and he personally negotiated a settlement of "mutual friendship" with Baji Rao, near Dhar, in 1724. He then advanced to destroy Mubariz and his army, effectively severing South India from Delhi's control; he subsequently received Baji Rao at Aurangabad and rewarded him with ceremonial robes and gifts for having withheld his cavalry from the bloody battle. After the Nizam had firmly established his grip over Hyderabad, he wrote the emperor to "beg his forgiveness" and received in response formal title to the viceroyalty of the Deccan, which would prove useful to him in his ensuing struggles against his immediate rivals, the Marathas. While maintaining a facade of loyal imperial service, the Nizam became, in fact, an independent monarch, for he remitted no revenue to Delhi and he declared wars and concluded them to suit his own convenience; he did not, however, pose the formal challenge of open rebellion by proclaiming a new dynasty or striking coin in his own name. The eighteenth-century pattern of dismantling the Mughal Empire was thus established, and would soon be followed in other provinces as well, including Bengal, and Oudh, where powerful Mughal *nawabs* became independent kings in all but title and trivia.

While South India emerged independent of Mughal power, the Safavid dynasty was attacked by Afghan invaders, Isfahan falling in 1722. A powerful Persian general, Nadir Quli, managed eventually to drive the Afghans back to their homeland and ascended the Persian throne as Nadir Shah in 1736. He then called upon Delhi's Muhammad Shah to join him in a united effort to crush Afghanistan. Poor Muhammad Shah could hardly manage to defend his own Delhi from Baji Rao's army, however, when rough-riding Maratha light horse dashed to the very environs of the Mughal capital in 1737. The emperor was obliged to appeal in desperation to the Nizam for help, and in addition to exalting him with the unprecedentedly glorious title of *Asaf-Jah,* promised him five new provinces and ten million rupees in cash to remove the Maratha blight from central India. The Nizam couldn't resist so tempting an offer, and he left Delhi with thirty thousand crack troops, all the artillery he needed, and full discretion to do as he pleased in exterminating the Marathas. He ordered his elder son, Nasir Jang, to march up from the south, hoping to trap Baji Rao in their pincers. But the *peshwa* proved the more brilliant general, not only eluding the trap, but luring the Nizam to Bhopal, where Baji's force then surrounded him and laid siege. The Nizam was forced to sue for peace. Baji's terms included the formal cession of all of Malwa and Bundhelkhand; he was hereafter to receive the revenues of all lands between the rivers Narmada and Yamuna, as well as payment of half a million rupees in cash as indemnity for the war. It was the Nizam's most humiliating defeat, yet he was glad to escape with his life and army. Maratha power was now greater than any other single force in India's fast-fragmenting political jigsaw puzzle.

At this juncture, the Persians decided to wait no longer for Muhammad Shah's help. Nadir Shah advanced to take Kandahar and Kabul alone; then, at the end of 1738, he crossed the Indus where Alexander had centuries before him. Lahore fell, the Sikhs fled to the hills, and still the Persians advanced toward Delhi, which the emperor and his court left in a desperate attempt to make a stand at Karnal in the spring of 1739. The Mughals were easily defeated by Nadir Shah, whose army then sacked Delhi, slaughtering an estimated thirty thousand people and looting a billion rupees worth of gold and jewels—including the peacock throne, which was carried back to Persia. Shah Jahan's city was left a smoldering shell by May, when the invading hoard finally withdrew, sated. The once glorious Mughal Empire was a shambles.

While India's Great Mughal unity was thus destroyed by a combination of interregional religious wars, court incompetence, greedy factionalism, and traditional invasions in the northwest for plunder, European commercial enterprise quietly prospered on the fringes of the subcontinent. In the closing decades of the seventeenth century, the British strengthened their position on both west and east coasts, under the "just and stout" leadership of the "brothers" Sir John and

Sir Josia Child. Sir John was president of the company's factory at Surat and governor of Bombay from 1682 to 1690, and under his vigorous direction the company completed its shift in west coast headquarters from the highly vulnerable factory to the powerful island fortress. Sir Josia was head of the Leadenhall Street Council of twenty-four "committees" of company directors in England from 1674 to 1689; he was perhaps the first Englishman to envision the goal of John Company's business as "to establish such a Politie of civill and military power, and create and secure such a large Revenue as may be the foundation of a large, well-grounded, sure English Dominion in India for all time to come."[1] By securing permission from Alamgir to trade at a new base in Bengal in 1690, the British made substantial strides toward the realization of that vision. The company's factory was built on the river Hughli, a Ganga tributary to the Bay of Bengal, where a village shrine to the goddess Kali stood, from which *ghats* ("steps") descended to the water; hence its name, Calcutta, may have been derived from the words *Kali-ghat*. Job Charnock, a merchant of the Bengal Council, founded the city that was to become the British Empire's first capital in India, and by 1700 its population grew to over twelve hundred Englishmen. Fort William was erected at this spot, about a hundred miles upriver from the Bay of Bengal, a terrible site for what would become one of the world's most populous cities and ports—though no one imagined how fast it would grow, or how important a city of imperial power, wealth, and "dreadful night" it was soon to become. It was, however, neither Bombay nor Calcutta that served as the incubating ground where British merchants would first learn to mix their quiet commercial activity with vigorous political involvement and territorial acquisition on a large scale. The venue for that historic development was Madras. Nor did the British discover the key to Indian conquest alone; they learned the most from their most recent European rivals in India, the French.

The French East India Company (*Compagnie des Indes Orientales*) had been started at Jean-Baptiste Colbert's instigation with a capital of fifteen million *livres tournois* (about £600,000) in 1664. A decade after the company's birth, François Martin established its Indian headquarters at Pondicherry, some eighty-five miles south of

[1] Quoted in William Wilson Hunter, *A History of British India,* new impression (London: Longmans, Green, 1912), vol. 2, pp. 272–73.

Madras on the Coromandal coast. Emulating the British tripod pattern of settlement, the French established subsidiary factories at Surat and Chandarnagar on the river Hughli by the end of the century, and they were soon vigorously competing against English factors for Indian merchandise. By seizing the islands of Mauritius and Bourbon in the Indian Ocean after 1721, France was able to keep its own fleet ready for swift and effective offensive as well as defensive action against any Indian port. A decade later the French reached their peak of profit in India, reaping an average of 25 percent annually on their investment, while the much larger, more cumbersome British company's profits were down to under 10 percent, though English imports from India were now valued at over one million pounds per year. The earlier era of fellow-European cooperation and friendship in the "hostile Orient" was rapidly being replaced by the tensions of competitive trade, national rivalry, and personal jealousy. In 1741, when Pondicherry came under the presidency of Joseph François Dupleix (1697–1764), son of his company's director-general and a statesman of singular foresight, its population was approximately the same as that of Madras, close to fifty thousand. Five years later the European war over Maria Theresa's disputed claim to the Austrian succession lured England and France into a conflict that erupted in Southern India in the summer of 1746. The British captured several French ships, which provoked Dupleix to call for the fleet under Admiral Mahé de la Bourdonnais's command at Mauritius, swiftly turning the tide along the Coromandal coast in France's favor. Madras was easily captured by the French in September 1746, and among those taken prisoner at Fort St. George was young Robert Clive (1725–74), a "writer" in the company's civil service who had earlier been so bored with his counting house job that he had tried unsuccessfully to blow out his own brains with a pistol that misfired. Clive soon discovered that shooting others would as effectively relieve his tensions, and he learned from Dupleix that all of India was ripe for imperial conquest. The "game" Dupleix embarked upon has come to be called "nabobism," the English corruption of the Mughal title *nawab*.

Dupleix had not fully appreciated his own power until Anwar-ud-din, the *nawab* of the Carnatic, who had been appointed to his throne at Arcot by the Nizam, insisted that the French turn Madras over to him in October 1746. Dupleix's reluctance to surrender his prize induced the *nawab* to send an army of some ten thousand of his best cavalry to attack Fort St. George. This Indian force was defeated at St. Thomé by 230 Frenchmen supported by 700 of their trained sepoys, using cannon, muskets, and European methods of disciplined warfare with deadly efficacy. The otherwise minor engagement at St. Thomé marks the dawn of an era of revolutionary change in the balance of political power in South India. Following the rout of Anwar-ud-din's army, Dupleix had become *nawab* of the Carnatic in all but name. Had France been ruled by almost anyone other than Louis XV and Madame de Pompadour, neither of whom had the interest or inclination to concern themselves with Eastern politics, Dupleix might have become *de facto* emperor of India. Even with almost no support from home, he managed to emerge virtual *nizam* as well as *nawab,* after the death of Nizam-ul-Mulk in 1748.

How did he do it? First of all, he knew India, having lived for almost two decades in the subcontinent. Dupleix understood Indian attitudes and politics better than any other contemporary European. He was, moreover, sufficiently paranoid and power mad to flourish in the frenzied atmosphere of infighting and fragmentation that characterized eighteenth-century Indian politics. In 1749 he paid 700,000 rupees to ransom Chanda Sahib, son-in-law of the deceased *nawab* of the Carnatic, who claimed to be rightful heir to that throne, from a Maratha lock-up. Using Chanda Sahib as his puppet *nawab,* Dupleix was able to rally popular support for an army that drove Anwar-ud-din from Arcot and from life itself before that fateful year was over. Appreciating the ceremonial demands of any Indian monarch's role, Dupleix thrice refused the proferred throne of Arcot himself, choosing rather to remain the power behind that and other South Indian thrones without exposing himself to any of their formal responsibilities or dangers. While assisting Chanda Sahib to take the throne of Arcot, Dupleix also worked at elevating Muzaffar Jang, grandnephew of Nizam-ul-Mulk, to the grander throne of Hyderabad. The Nizam's son, Nasir Jang, who had rightfully inherited the throne, was assassinated in 1750, while Muzaffar was proclaimed *nizam* and advanced with Dupleix's French lieutenant, the Marquis de Bussy, to Hyderabad. Armed with three hundred French and fifteen hundred sepoy troops, Bussy was the prototypical European "political agent," who enthroned and dethroned many an Indian prince. When young Muzaffar was killed en route to his Deccan capital, Bussy quickly picked a new puppet, Salabat Jang, one of the former *nizam*'s sons, and placed him on the throne. The capital of the *nizam* was now under firm French control.

Had the European peace treaty of Aix-la-Chapelle not restored Madras to the English in 1749, Dupleix's position of primacy in the south could have been challenged by the Marathas alone. With the new lease on life granted to them from home, however, the British quickly learned to play the game Dupleix taught them. Anwar-ud-din's younger son, Muhammad Ali, had fled from Arcot to Trichinopoly in the southernmost tip of the peninsula, where his claim to be rightful *nawab* was supported by British arms. Chanda Sahib led a vast army against his rival, but instead of destroying Trichinopoly and its puny garrison, as he might easily have done, he laid siege to it and waited too long. While Trichinopoly was besieged, "Captain" Clive, who was especially commissioned for his daring new job, volunteered to lead an expedition of some two hundred Englishmen and three hundred sepoys out of Fort St. David, Madras's small satellite south of Pondicherry, on a forced march of more than a hundred miles north to Arcot in the blistering summer of 1751. Chanda Sahib had virtually denuded his capital of all troops for the Trichinopoly siege, and Clive's men marched unopposed through Arcot, watched by a million helpless spectators, to capture the *nawab*'s palace fortress. It was the most daring demonstration of military diversion in recent Indian history. Not only did it electrify the Carnatic with word of British audacity, but it cloaked Clive with the reputation of being a "heaven-sent" general and helped catapult him to the *"nawab-ship"* of Bengal. Chanda Sahib was obliged to withdraw over four thousand troops from Trichinopoly to return to Arcot, where Clive and his daily diminishing band of heroes held out for over fifty days and nights, until the tide of battle turned as Maratha troops, under Raghunath Bhonsle of Nagpur, rushed to the aid of Muhammad Ali and the British at Trichinopoly, leading to Chanda Sahib's capture and death in 1752. Now Clive became the Carnatic's *nawab*-maker, placing Muhammad Ali on the throne at Arcot. Though Dupleix fought to continue his struggle, the French at home lost confidence in their Napoleonic empire-builder, who was recalled by his directorate in 1754 for wasting too much of their investment on unprofitable ventures.

Europe's second war over the remnants of Charles VI's empire, the Seven Years' War (1756–63), served again to trigger Anglo-French conflict in India and to further bolster British fortunes, this time primarily in Bengal. After 1740, when Ali Vardi Khan became its *nawab,* Bengal attained virtual independence of Delhi. By encouraging European traders along the Hughli, Ali Vardi filled his treasury at Murshidabad. His major concern was not with any potential threat of French or British power, but the present danger posed by the Marathas. Baji Rao had died in 1740, but his son, Bālājī Rāo (1721–61), succeeded him as Poona's *peshwa* and for two decades led the Maratha Pentarchy in its continued expansion to the east as well as the north. Maratha expansion had, in fact, been so rapid during Baji Rao's *peshwa-ship* that Poona by now served more nearly as central bureaucratic headquarters of a loose-knit confederacy than as the capital of a kingdom. There were four extremely powerful Maratha generals, each of whom carved out his own dynastic domain: the Gaekwar in Baroda, Holkar at Indore, Sindia in Gwalior, and Bhonsle at Nagpur. These four paid nominal allegiance to the raja of Satara, whose *peshwa* was, however, the real head of what had thus actually become a pentarchy.

Raghunath Bhonsle was closest to Bengal and launched an expedition against Orissa (then still a subprovince of Bengal) in 1741, and by the following year Maratha horsemen were camped along the Hughli and even threatened Calcutta, around which the British then dug their famous Maratha ditch. When the *peshwa* demanded his share of the Orissan revenue Bhonsle had taken, however, he was ignored, and he therefore decided to respond to Ali Vardi Khan's plea for help against his own fellow Maharashtrians. The *peshwa*'s army moved, in fact, at imperial Mughal request to "assist" the *nawab* of Bengal against the plundering Maratha "invaders." Balaji Rao and Ali Vardi met near Plassey in Bengal, the *nawab* agreeing to pay over two million rupees and the annual *chauth* of Bengal to the *peshwa* for "ridding" his province of Bhonsle's horde. Balaji took the Mughal money and chased the weaker Marathas in Bhonsle's service back to Nagpur. Ali Vardi Khan was thus able to retain control of Bengal by pitting one Maratha army against another. When he died in April of 1756, however, Ali Vardi left no sons, and he had made the mistake of designating his youngest daughter's son, Mirza, known to history as Nawab Siraj-ud-daula, as heir to his throne. The twenty-year-old *nawab* was as impetuous as his grandfather had been cautious, and, failing to perceive the treachery lurking inside his own court or to probe the hidden mines that lay beyond the immediate goal of his power-hungry grasp, he was driven by impulse headlong toward his own destruction.

Siraj learned that the English were adding fortification to Fort William at Calcutta without his permission, and he decided to teach them a lesson, marching his elephants south along the Hughli in the summer of 1756. It took Siraj only eleven days to move his army of some fifty thousand the 160 miles from Murshidabad to Calcutta, coming upon the British garrison of little more than a thousand at Fort William. Calcutta's governor, Roger Drake, rushed for his boat as soon as he heard the first shots. The stampede of able-bodied men who abandoned Fort William that wretched June day forced many women and children to stay behind, for the boats were not large enough to carry everyone to safety down river. J. Z. Holwell, whose vivid imagination as the author of the "Black Hole Tragedy" was far greater than his generalship, commanded the garrison of 170 English soldiers left behind. There were two iron mortars in the fort, but most of the powder was damp, and the shot had lain so long in moist storage that it was almost all eaten by worms. Siraj attacked on Sunday, June 20, 1756; by midday Holwell raised his white flag. That night, according to his account, which would ignite generations of British schoolboys with passionate indignation and outrage against the "uncivilized natives" of India, 146 English prisoners, including one woman and a dozen wounded officers, were thrown into Fort William's lock-up, called the "Black Hole," an airless dungeon measuring fourteen by eighteen feet. When the door was opened at 6 A.M. the next morning, Holwell reported, only twenty-three of the prisoners emerged alive, the rest having died of suffocation or shock. A more careful study of the record by Professor Gupta[2] indicates that only sixty-four prisoners had been incarcerated, twenty-one of whom emerged alive, and that Siraj-ud-daula himself had neither ordered such torture nor been informed of it. Irreparable damage to Anglo-Indian relations was, nonetheless, done by the impulsive *nawab*'s underlings, and Eng-

---

[2] Brijen Kishore Gupta, *Siraj uddaullah and the East India Company, 1756–1757; Background to the Foundation of British Power in India* (Leiden: E. J. Brill, 1962).

lish retaliation would come swift and harsh, putting the torch of martyrdom to the tinder of greed and personal ambition.

Lieutenant Colonel Clive was dispatched from Madras in October with nine hundred Europeans and fifteen hundred sepoys under his command, carried in five transports and escorted by Admiral Watson's fleet with the same number of warships. Blown to Burma by tempestuous winds, the fleet did not reach Fulta, where Drake and his band of refugees found safety, until December, and Fort William was not recaptured until January of 1757. Taking advantage of the renewed war in Europe, Admiral Watson bombarded the French at Chandarnagar, while Clive attacked his rival's fortress on the Hughli, capturing it in March. With the French removed from Bengal, Clive now had ample opportunity to win the support he required from Hindu bankers in Calcutta, especially the Jagat Seth, India's Rothschild, whose fabled wealth bought and sold many a princely state in the latter part of the eighteenth century. Hindu bankers were by now as dissatisfied as the British were with the upstart *nawab*. Jagat Seth's candidate for the Murshidabad throne was Siraj's more malleable greatuncle, Mir Jafar, who felt that he should have succeeded his brother-in-law, Ali Vardi Khan. A threefold alliance, including Clive, was thus hatched and brought to its "revolutionary" fruition on the field of battle at Plassey, the "Mango Grove" between Calcutta and Murshidabad where Bengal's fate was decided on June 23, 1757. With only eight hundred European troops and little more than two thousand sepoys, Clive fought the *nawab*'s army of fifty thousand, winning the day thanks to Mir Jafar's treachery and that of his co-conspirators in the rebellion-ridden ranks of Siraj-ud-daula's divided force. Jagat Seth had lavished a small fortune on troops, who were paid more not to fight than they would have been paid for risking life and limb. Clive personally led Mir Jafar to the throne in Murshidabad a week later, and just a few days after that Siraj-ud-daula's body floated downriver.

Clive reaped an immediate personal fortune of £234,000 for his role in the coup and was made a *mansabdar* of six thousand with an annual salary of £30,000 in rent from the Twenty-four Parganas ("regions") of Bengal, some 880 square miles south of Calcutta, which became his private *jagir*. Overnight, at the age of thirty-two, he became one of England's wealthiest subjects, first of the reviled "nabobs," soon to return to London with bags of Indian jewels and gold that he used to buy up shares of company stock and rotten borough seats galore in Parliament. Before returning home, however, Clive secured the British position of primacy won by his astute exploitation of Indian pluralism and treachery. The faction-riddled court at Murshidabad was, of course, no worse than those at Delhi, Poona, or Hyderabad, but Clive had learned his lessons brilliantly from Dupleix and took full advantage of India's caste, class, and communal divisions within the Bengal base where fortune had placed him.

The war-bloated cadre of company servants in Bengal were now in a unique position of power without responsibility, and they fanned out on their own initiative to Bengal's village hinterland (*mofussil*), trading at enormous personal profit without paying a rupee in revenue or taxes of any sort. As *nawab,* Mir Jafar found himself presiding over a bankrupt kingdom, stripped of his specie by Clive, hopelessly indebted to Jagat Seth, distrusted by his closest relatives, and reviled by those still loyal to Siraj-ud-daula's memory. By 1760 he could endure it no longer and stepped down in favor of his eager son-in-law, Mir Kasim, who won the company's backing by promising to turn over revenue from several more provinces of East Bengal to Clive and his insatiable cohorts. Thus another half million pounds in silver poured annually from the soil of Bengal into the bottoms of British vessels, dramatically reversing the specie drain that had so long distressed Western European mercantilists. In 1760 Clive returned home and bought two hundred shares of company stock, at £500 each, thereby purchasing control over Leadenhall Street's directorate. He also bought enough seats in the Commons to ensure his company's fortune and secure his newfound empire from London rivals.

Delhi had never really recovered from the Persian invasion of 1739, though Muhammad Shah remained nominal emperor until his death in 1748. By that time, Nadir Shah had also died, and his Kabul lieutenant, Ahmad Shah Abdali (later changed to Durrani), proclaimed Afghanistan independent of Persia. To help sustain his newly consolidated kingdom, Ahmad Shah Abdali plundered Lahore and launched a series of raids into the Punjab, the last of which, in 1761, led to his conquest of Delhi itself. The Mughal crown now descended to a series of weak and impotent heads, each of whom served the greedy needs of court backers until their rivals managed to depose

them. Vazir Safdar Jung of Oudh and his rival, Imad-ul-Mulk of
Hyderabad, the Nizam's grandson, led armies that turned Delhi into
a field of Mughal civil war during 1753. The battle raged until Safdar
Jung and his followers were forced to retreat eastward to their stronghold in Oudh, which had by now become another independent kingdom under its own powerful *nawab,* Shuja-ud-Daula. Vazir Imad-ul-
Mulk could only rule Delhi with the support of the *peshwa*'s army,
thus rousing the wrath of orthodox Muslims at court, like the learned
scholar Shah Wali-ullah (1703–62), who appealed to the Afghans to
"save" Delhi from its "Hindu Raj." Ahmad Shah Abdali soon responded to this invitation and recaptured most of the valuable Punjab.
Amritsar remained the capital of Sikh power at this time; and though
the *khalsa* was divided into eleven *misls* ("equal units" of regional
rule) to give it the greatest mobility and local initiative in the face of
the ever-present threat of Mughal, Maratha, or Afghan attack, attempts were now made to regularize Sikh rule by introducing a standard revenue demand of one-fifth of each harvest in return for the
promise of *khalsa* "protection" (*rākhī*) for the Punjab peasantry.
During the winter of 1756–57, the Afghans launched their fourth invasion of India since 1748, and Abdali's plunderers turned Delhi into
a nightmare city of rape and carnage for more than a month of
Pathan occupation. The Sikhs harrassed the loot-laden Afghan force
as it marched back across the Punjab, but no Indian army—other
then the *peshwa*'s—was now strong enough to be pitted against Shah
Abdali's power.

The *peshwa* sent his younger brother, Raghunathṛao, at the
head of a mighty Maratha army, including Malharrao Holkar, to
Delhi in the wake of the Afghan orgy. Raghunathrao installed a new
imperial puppet on Delhi's throne, on the advice of Imad-ul-Mulk,
and then marched victoriously through the Punjab, driving Shah
Abdali's son and general from Lahore, advancing west to the Indus
without major opposition. Never had Poona's "empire" reached so
far or seemed so mighty as it did in 1758, when Delhi and Lahore
were both hardly more than provincial outposts of the *peshwa*'s domain. Young Raghunathrao, however, was impatient to return to his
Deccan home, and Holkar was eager to collect the Rajput revenues.
Dattaji Sindia alone of the Maratha generals was sent to reside in
Delhi, but he was hardly adept enough to handle the complex intrigue

at court. By mid-1759 Shah Abadali and his army returned to invade the Punjab, and by the year's end he had reached the environs
of Delhi, where Imad-ul-Mulk had panicked and murdered the emperor, Alamgir II, for fear of his favoring Shah Abdali. Then in
January of 1760, Sindia was killed while trying to hold off the Afghan
invaders, and his Maratha force fled in despair. Shah Abdali was now
offered the Mughal throne, but he had little interest in living out his
days under Delhi's hot sun. The Mughal prince, Mirza Abdulla, who
lived under the protection of the *nawab* of Oudh, Shuja-ud-Daula,
proclaimed himself emperor after his father's assassination, taking as
his royal name Shah Alam. His tragic reign would last forty-seven
years (1759–1806), but it was born in exile and proved a crown
more of thorns than of pleasures, especially after he was blinded by
an Afghan chief in 1788.

When news of Sindia's death reached Poona, the *peshwa* appointed his trusted cousin, Sadashivrao, to command what would be
the most formidable Maratha army ever sent from the Deccan to
Delhi. Thirty thousand crack Maratha horsemen, among them the
peshwa's own sons, left Maharashtra in March 1760, gathering Rajput
volunteers as they marched north, swelling to a total of some two
hundred thousand troops and noncombatant supporters before they
assembled for their fatal battle in January 1761. Shah Abdali remained in Delhi, awaiting the Maratha force and bolstering his own
army with Mughal and Rohilla Afghan troops, who eagerly joined
what was proclaimed to be a Muslim *jihad* against Hindu infidels.
Even Shuja-ud-Daula joined this alliance. The two armies met on the
blood-drenched field of Panipat, where India's fate had so often been
decided before. Each side dug in for what would be an agonizingly
prolonged defensive confrontation. The Marathas soon found themselves unable to purchase food or replenish their supplies from the
predominantly Muslim population of the Doab region around them.
By mid-January, Sadashivrao could hesitate no longer, for his men
and horses were starving to death. Forced to abandon their entrench-
ments, the Marathas attacked. Shah Abdali was by far the better general, and his army was much stronger and better positioned than the
peshwa's. Before the sun set on Panipat that mid-January day in
1761, close to seventy-five thousand Marathas were slaughtered; another thirty thousand were taken prisoner and later ransomed. Shah

Abdali now recognized Shah Alam as the Mughal emperor and appointed his Rohilla Afghan helper, Najib-ud-Dawla, as paymaster and governor of Delhi. Abdali's troops, sick of India and restive from lack of pay or plunder, forced their shah to return with them to Kabul. With Maratha power crushed and the Afghan armies withdrawing, no force remained in India that was strong enough to deter further British expansion. The nominal Mughal emperor himself stayed on at Lucknow, dependent on his *"vazir"* host, the *nawab* of Oudh, for sustenance. The *peshwa* was so shattered by news of the slaughter at Panipat that he died that June. Najib-ud-Dawla ruled Delhi timorously, venturing beyond the walls of his fort only to subdue the Jats west of the Yamuna, which was barely beyond the ever-narrowing suburban circle that had come to represent the last venue of a now defunct Mughal "empire."

For French as well as Maratha power, 1761 proved a year of dire defeat, a crucial turning point beyond which there would be no significant prospect of emerging as heir to Mughal rule. At the start of the Seven Years' War, Count de Lally had been sent from Versailles at the head of a French army that was ordered to drive the British from Madras. He failed, but he managed to seize Fort St. David and summoned Bussy from Hyderabad for assistance and advice. Clive sent Sir Eyre Coote from Bengal with a British army that defeated Lally decisively at Wandiwash on January 21, 1760. A year later Pondicherry fell, and though it was returned to the French by the Treaty of Paris in 1763, its fortifications were permanently dismantled. Thus emasculated, French enclaves remained on Indian soil without posing any serious challenge to British dominance again, except for one brief hour when Napoleon launched his abortive expedition toward the Red Sea in 1798.

The company's plunder of Bengal became so rapacious by 1762 that even Mir Kasim, willing puppet though he was, found his position as *nawab* intolerable. Driven to distraction by continued British greed, especially at Patna, Mir Kasim was soon engaged in battle with the merchant looters of his land. He was forced to flee Murshidabad in 1763 and was replaced by the tottering Mir Jafar, who served less than two years as puppet *nawab* before his death in February 1765. In a fit of frenzied frustration, Mir Kasim ordered every Englishman in Patna murdered, then raced to Lucknow, where he appealed for aid from Shuja-ud-Daula and his imperial guest, Shah Alam. It must have dawned on these once mighty Mughals that unless they joined forces, submerging their selfish ambitions to some higher cause of unity, there might be no empire left to defend. But it was already too late. The army raised by this Mughal alliance was soundly beaten by Major Hector Munro and his far smaller force at Buxar (Baksar), on the Ganges between Banaras and Patna, on October 22, 1764. It was a more significant victory for British arms and imperial aspirations than Plassey had been, for at Buxar the company's troops faced what was virtually the united remnant of Mughal power, not the faction-ridden ranks of an unpopular young *nawab*. In the past seven years, the company had amassed sufficient wealth and power from Bengal to shake the crumbling foundations of the Mughal Empire forcefully enough that the British could have reached for the mantle of Delhi's authority, had they so desired.

When Clive returned to govern Bengal in 1765, however, he wisely recognized that though his company may have had the power to march victoriously to Delhi and beyond, such an adventure would hardly have proved a profitable, or necessarily an enduring, conquest at so early a stage of Anglo-Indian relations. "If ideas of conquest were to be the rule of our conduct," Clive wrote his directors, "I foresee that we should, by necessity, be led from acquisition to acquisition, until we had the whole empire up in arms against us; and whilst we lay under the great disadvantage of fighting without a single ally, (for who could wish us well?) the natives, left without European allies, would find, in their own resources, means of carrying on war against us in a much more soldierly manner."[3] At Buxar, for example, he noted that Mir Kasim's troops fought harder than Indians had ever fought before. It was an important cautionary insight, proving how well Clive had learned the game of Dupleix, for under the cover of "native" *nawabs* and other such princely puppets, the company would long continue to prosper and expand its power base. By offering martial "protection" to Indian leaders, the British could keep them dependent without arousing much popular resentment or stirring widespread protest against foreign usurpation of powers. The one thing

[3] From Sir J. Malcolm's *Life of Clive*, vol. 2, p. 310; reprinted in Ramsay Muir, *The Making of British India 1766–1858* (Manchester: Manchester University Press, 1923), p. 82.

Clive did take, however, was the revenue of Bengal, Bihar, and Orissa, which was granted to the "honourable" company "from generation to generation, for ever and ever," by Shah Alam in his edict of August 12, 1765, proclaiming the company his *diwan* for those provinces. In return for the right to collect and keep the many millions of pounds' worth of revenue yielded by that vast region of the eastern Gangetic basin, as well as for full freedom to trade without charge throughout those hundreds of thousands of square miles of India's most populous area, the British agreed to pay the emperor £260,000 annually, to support him in his much diminished imperial style at Allahabad. John Company thus became an official "servant" of the Mughals, bolstering that fast collapsing facade as the most convenient screen against the glare of Indian consciousness.

THIRTEEN

JOHN COMPANY RAJ

(1765–93)

After 1765 John Company sought to establish its *rāj* (rule) over Bengal, Bihar, and Orissa on as sound and permanent a basis as possible. Governor Clive found himself, however, in the awkward position of having to warn junior agents in the company that they would no longer be permitted to do as he had done, but were expected to be honest as well as abstemious, taking no bribes and extorting no lavish gifts from natives within the company's domain. Having personally plundered the region, Clive now recognized the difficulties inherent in seeking to govern a land without sufficient resources. Not that Clive much concerned himself with governance, for such problems remained, at least nominally, under the *nawab* in Murshidabad, who retained formal responsibility for Bengal's criminal justice system, as well as its military administration (*nizamat*), though he had no revenue—all of which was collected by the company as *diwan*—with which to pay his police, soldiers, or legal officials. The company, of course, was supposed to have apportioned the land revenue equitably enough to assure effective administration of the triple province, but none of its servants found time, in their hectic private trading and company trading, to worry about the *nawab*'s unbalanced budget or the breakdown of law and order in remote villages. It was the worst of all possible dual governments: those with responsibility possessed no power, and those with power felt no responsibility.

Clive himself had reported in 1765 that since the restoration of Mir Jafar to the *nawab*-ship, "such a scene of anarchy, confusion, bribery, corruption, and extortion was never seen or heard of in any country but Bengal; nor such and so many fortunes acquired in so unjust and rapacious a manner."[1] Yet four years later (the first two

---

[1] From Sir J. Malcolm's *Life of Clive*, vol. 2, p. 379; quoted in Ramsay Muir, *The Making of British India 1756–1858* (Manchester: Manchester University Press, 1923), p. 76.

of which were Clive's second governorship), an Englishman wrote home from Bengal that "since the accession of the Company to the *Diwani* the condition of the people of this Country has been worse than it was before . . . this fine country, which flourished under the most despotic and arbitrary Government, is verging towards its Ruin."[2] By the end of 1769, when the monsoon rains failed, Bengal was left naked, stripped of its surplus wealth and grain. In the wake of British spoliation, famine struck and in 1770 alone took the lives of an estimated one-third of Bengal's peasantry. The company stored enough grain to feed its servants and soldiers, however, and merchant speculators made fortunes on the hunger and terror of less fortunate people, who bought handfuls of rice for treasures and were eventually driven to cannabalism. Company directors in London bemoaned the fact that Englishmen in India were "profiting by the universal distress of the miserable natives," yet the Bengal Council did nothing to help the starving, taking refuge behind its nominal lack of administrative responsibility.

What finally roused British parliamentary concern over the state of Bengal was not the plight of India's peasantry, but the company's professed inability to pay a promised annual tax of £400,000 to the treasury in 1767. As every member of Parliament could see, many individual servants of the company returned home from India with larger private fortunes than what they carried in their gunny bags. Clearly, something was wrong with the state of Bengal—still six months away by clipper ship from Greenwich, and on some voyages east, more than a year's wind-blown distance away. Remoteness added to the company servants' sense of total independence and general indifference to the wishes of those at home. The personal hardships and risks inherent in eastern exile intensified individual proclivities toward greed and sharpened the desire to get as much as possible as soon as possible, and then retire back home in comfort. Perhaps the emptiness of such motivation was best exposed by Clive's successful suicide after he finally had everything his ill-won fortune could buy. Yet as Burke was soon to declaim in the Commons, "new

[2] Richard Becher's letter from Bengal was written May 24, 1769, and laid before the Secret Committee of the Court of Directors (Secret Comm. Cons., Indian Office Records, Range A 9), reprinted in *ibid.*, pp. 92–95.

flights of birds of prey and passage" continued to leave Britain's shores for India, "animated with all the avarice of age and all the impetuosity of youth."

Parliamentary inquiry forced John Company to make some effort at reform and drove its directors to turn from the swashbuckling bullying of a Clive to the coldly methodical leadership of Warren Hastings (1732–1818), who was appointed governor of Fort William in 1772. Like Clive, Hastings had been reared in the company's service, but his was a personality more atuned to study Latin, Persian, and Urdu than to make forced marches or defend breached walls. His scholarly pursuits proved practical enough, however, by arming him with greater insight into and intimacy with the Indians, and much of his success may be attributed to the fact that he functioned more as a "native" tyrant, once he attained the power he so assiduously coveted, than as a foreign one. Nevertheless, he could be vindictive, cruel, arrogant, and violent when he was provoked, or when he sensed the value of such emotional displays in teaching others a lesson in power politics.

What Hastings did, first of all, was to redesign the revenue-collecting system of Bengal so that more money actually reached the coffers of the company rather than sticking to the fingers of its Indian agents and their English supervisors, who robbed peasant and company alike with equal impunity. Under his administration, all revenue flowed directly to Calcutta, instead of first being diverted to Murshidabad or Patna; in addition, clearer notice was given to peasants of their rights to property and their indebtedness to government, thereby eliminating the middleman *nawab* and his henchmen at court as continuing encumbrances upon Calcutta's Council and adding more formal and awesome authority to the company's annual demand upon each peasant's income. Though Bengal's *nawab* remained on his throne, Hastings cut his stipend in half and removed whatever powers had been left to him after 1773. Hastings also stopped payment of the annual tribute Clive had promised the Mughal emperor and cut many other such remunerations from the company's ledgers; these retrenchments saved over half a million pounds annually, while increasing the efficiency of revenue collection. Despite the aftermath of Bengal's famine, then, Hastings was able to restore a positive balance to the company's ledgers and revive the deflated spirits of the shareholders. Parliament remained, nonetheless, sufficiently concerned by reports of the company's inept management to take official action designed to ensure the more rational use of resources in the wake of the company's appeal for a loan of one million pounds to cover its "deficit" in Bengal.

Lord North's ministry granted the company a loan of one and a half million pounds in 1773 but at the same time passed a Regulating Act to limit the company's dividends to no more than 6 percent until that loan was repaid and to restrict the Court of Directors at home to terms of four years, obliging one-fourth of the twenty-four old men of Leadenhall Street to retire each year. The Regulating Act also brought the hitherto independent presidencies of Madras and Bombay under the official control of Bengal, whose governor was hereafter exalted to the title governor-general and whose council was augmented by four members, all of whom were named in the act. Three of them—General Clavering, Colonel Monson, and Philip Francis—were sent to India with no prior experience and voted as a ministry block against most of the measures Hastings proposed for the next few years, until Monson's death in 1776. As governor-general, Hastings had a deciding vote in the event of an equal division among himself and his councillors, only one of whom, Barwell, consistently agreed with his positions. Clavering was a crony of George III; like Monson, he knew nothing of Bengal and became the willing tool of Francis. Only thirty-three when he went east, Francis was brilliant, ambitious, and arrogant; he was convinced he would make a far better governor-general than Hastings and determined to do all in his power to prove just that. At the end of 1780, however, after being wounded by Hastings in a pistol duel, Francis abandoned Bengal. He continued his struggle against Hastings from England, where he won Burke to his side and was instrumental in provoking impeachment proceedings.

In dealing with Indian opposition, Hastings proved no less tenacious and formidable than he was with his own councillors. He was a shrewd adversary, but he was also loyal to his allies, as he proved in 1774 when he supported the *nawab-vazir* of Oudh, Shuja-ud-Daula, by providing English mercenaries to fight for him against the Rohilla Afghans, Oudh's northwestern neighbors. The use of Englishmen in Asia, as Hessians were being used by the English in America, roused moral resentment in London. Hastings was paid handsomely for his brigade of soldiers, however, and their victory in Oudh strengthened the company's foremost Indian buffer state and assured its loyal attachment to the British—at least, until Hastings' shrewd policy was abandoned and Oudh was annexed in 1856.

If the support of Oudh in its dependence on British arms was the first pillar of his foreign policy, Hastings's second highest priority was to undermine Maratha power. The stunning defeat at Panipat not only shattered Maratha hopes of imperial succession, but it also exposed Poona itself to violent attack from the south by the *nizam* in 1761. Within the Maratha heartland, moreover, a desperate struggle for power ensued after Balaji Rao's death, between his sixteen-year-old son and heir, Madhav Rao, and the new *peshwa*'s "guardian" uncle, Raghunathrao, who personally coveted the Maratha premiership. Raghunathrao's ambition overwhelmed his nephew, and a Maharashtrian civil war undermined the sources of Maratha strength, wasting Poona's wealth on fraternal fighting; yielding revenue-rich domain to the *nizam;* and losing troops and allies as well as irreplaceable resources and human confidence. After more than two years of such ruinous conflict, however, the Marathas recognized how self-defeating their tactics had been and united once again to march upon Aurangabad, bringing the *nizam* to surrender territory worth over eight million rupees in revenue by the Treaty of Aurangabad in September 1763. Madhav Rao's youthful vigor and resilience prevailed over his uncle's treachery, and the *peshwa* enjoyed eight more years of active rule from Poona, during which Maratha power gradually recovered some of its former dimension. Mahadji Sindia (1727–94) now emerged as the *peshwa*'s strongest ally in the Maratha Pentarchy, and Nana Phadnis (1742–1800) as his leading adviser in Poona. Mahadji and the *nana* would remain the strongest and shrewdest bulwarks of Maratha power until the end of the century.

In 1770 the Marathas were powerful enough to occupy Agra and Mathura, and the following year Mahadji Sindia captured Delhi and invited Shah Alam to return to his capital. With the Mughal emperor now under Maratha protection, Poona felt revitalized, though precious little money was available to restore substance as well as spirit to the *peshwa*. Moreover, Bhonsle and Holkar remained disgruntled leaders of opposition to the *peshwa* within the pentarchy, and Raghunathrao continued to conspire against his nephew. Before the end of 1772, Madhav Rao died of tuberculosis in Poona, reviving his uncle's aspirations and unleashing a new storm of palace intrigue and civil war. The new *peshwa,* Madhav Rao's younger brother, Narayan Rao, ruled for only nine months, assassinated in his own palace within view of his uncle at the end of August 1773. Raghunathrao now became *peshwa,* but his brutal plot against his own nephew roused such revulsion among the brahmans at court that he soon found it impossible to exercise effective leadership over the Maratha bureaucracy. Nana Phadnis organized Poona's opposition to the hated Raghunathrao, who was deposed by the raja of Satara in 1774 and finally forced to flee Poona by the "Council of Twelve Brothers" (Barbhais) led by Nana Phadnis.

At this juncture the Bombay Council of the British Company, eager to gain more bases in the west, took advantage of Raghunathrao's plight and used him as their puppet *peshwa* to validate the company's seizure of the island of Salsette and the port of Bassein. The Treaty of Surat, concluded between Raghunathrao and the British in March of 1775, promised no fewer than twenty-five hundred armed troops to the deposed *peshwa,* of which seven hundred would be Europeans, in return for sufficient gold and jewelry to support that force, plus the ceding of a number of islands around Bombay to the company "in perpetuity." The Bombay force proved incapable of restoring Raghunathrao to his throne, however, and had to surrender when the *nana*'s superior army threatened to overwhelm it on the outskirts of Poona in 1778. Had it not been for Warren Hastings's prescience earlier in dispatching a British army, under Colonel Upton, to cross India from Bengal and counter the influence of recently arrived French advisers in Poona, the Marathas would have remained clearly dominant over the Bombay Council and might well have won the support of other South Indian powers in a grand alliance against the British. When news of Burgoyne's surrender at Saratoga reached Calcutta, Francis and his friends were quick to urge Hastings to recall his forces and not risk any further British defeat at arms. But Hastings insisted that if, in fact, British arms had suffered so severe a setback in the Western world, then it was all "the more encumbent on those who are charged with the interest of Great Britain in the East to exert themselves for the retrieval of the national loss." His judgment was more than vindicated by the course of events, and thanks to the risks Hastings took with such bluff tenacity, Britain won a second empire in the East just as she was losing her first in the West.

Maratha power was not as yet defeated, but by the Treaty of Salbai concluded with Sindia in 1782, it was effectively neutralized for twenty years, giving the company the time it so desperately needed to muster strength for its next deadly drive against the only remaining Indian force that might challenge British primacy. Hastings also salvaged the Madras Council, which all but sank under the burden of corruption and treachery imposed by Benfield and Company, creditors to the slothful *nawab* of Arcot, who sought to repay his debts by seizing the domain of his neighbors, especially the rich land of Mysore, which had been conquered in 1762 by an energetic Muslim soldier of fortune named Haidar Ali Khan. The hostilities launched by the *nawab* at the goading of his creditors drove Haidar Ali to such fierce retaliation that his cavalry galloped within view of the very walls of Fort St. George in 1780 and might well have destroyed the British base in the Carnatic had it not been for Hastings's vigorous and timely assistance. Sir Eyre Coote was sent with an army from Calcutta to relieve Madras in November 1780, and at the same time Hastings dispatched political envoys to the *nizam* and the Maratha raja of Berar (Bhonsle), seeking to subvert Haidar Ali's efforts to weld a triple alliance of Indian powers against the foreign company. Haidar Ali was probably the most farsighted Indian monarch of his day, recognizing that Hindu-Muslim unity alone could successfully combat the ever-growing force of the British, who might still have been beaten in 1780, but who would never be stopped a decade or more later.

The drain of company resources incurred by the Maratha and Carnatic wars left the Calcutta Council so bankrupt that Hastings felt obliged to extort vast sums from his richest Indian tributaries, the raja of Banaras and the *begums* of Oudh, thereby exposing himself to subsequent parliamentary impeachment. As a loyal, lifelong servant of the company, Hastings saw nothing at all wrong with "squeezing" wealthy natives to support India's "pacification"; in fact, he was amazed to find so many Englishmen back home whose moral scruples and feelings about the sacrosanctity of property led them to think otherwise. "I enlarged and gave shape and consistency to the domin-

ion which you hold," Hastings subsequently protested to his directors.
"I preserved it; I sent forth its armies with an effectual but economical hand, through unknown and hostile regions, to the support of
your other possessions . . . I gave you all, and you have rewarded
me with confiscation, disgrace, and a life of impeachment."[3] Warren
Hastings left Bengal in 1785, an embittered, lonely man, longing for
the peerage that would never be offered to him, so blinded by the
arrogance of what had been a potentate's power that he could not
understand or countenance any of the criticism that was heaped upon
him during the decades of his forced retirement. He resigned the
governor-generalship when he learned of the passage of Pitt's India
Act of 1784, by which ultimate power over the administration of
Bengal would pass from the company's Court of Directors to the
crown's new Board of Control, which consisted of not less than three
nor more than six members of the British cabinet.

The act was a compromise formula devised by the younger Pitt
and his astute political "manager" for Scotland, the wily Henry
Dundas (1742–1811), to placate Fox and Burke, who would have
much rather had the British government take direct and immediate
charge of the company's property and affairs in India, yet were willing to live with this parliamentary half-loaf system instead. The
directors optimistically showered lavish gifts on the king, Lords, and
Commons so that they would be permitted by this act to enjoy their
continued, though diminished, lease on Bengal, Bombay, and Madras,
but Hastings read the fine print in Pitt's legislation to mean that his
days, at least, in high office were numbered. He chose to resign before being fired. The new act went much further than Lord North's
Regulating Act had done in asserting parliamentary supervision over
the company's affairs and domain, but it fell far short of the Act of
1858, when the British government would assert full and direct authority over India. Like much British legislation, Pitt's measure was
merely sufficient to permit the minimal requisite tinkering with a system that was never discarded, only gradually remodeled.

The new Board of Control could send "secret orders" to India
through a specially selected Secret Committee of the Court (consisting of no more than three men) on any matters "concerning the levy-

[3] From Sir G. W. Forrest, *Selections from the State Papers of* . . .
Hastings, vol. 1, p. 290; quoted in Muir, *op. cit.,* p. 152.

ing of war or making of peace, or negotiating with any of the native princes or states in India." The president of the new board, who for its first eighteen years was Henry Dundas, and the chairman of the committee could, in effect, and often did, make vital policy decisions without consulting anyone else in England or India. Under Pitt's India Act, the directors retained their formal patronage powers of appointment to all ranks of the company's services, civil, military, and judicial, including statutory powers to appoint the governor-general as well as the presidency governors of Bombay and Madras. The crown, however, on the advice of the president of the board, was empowered "to recall the present or any future Governor-General of Fort William at Bengal, or any other person" in the company's employ. That article sufficed to make Hastings resign. The directors readily accepted Lord Cornwallis (1738–1805), a general who had never been in the company's employ, as the best man to replace him.

Cornwallis went to India not only as governor-general of Bengal but also as commander-in-chief of its army. He hoped not so much to regain his military reputation, which had been tarnished in the American colonies—for Pitt's Act insisted that "to pursue schemes of conquest and extension of dominion in India, are measures repugnant to the wish, the honour and policy of this nation"—but rather, to bolster the confidence of English stockholders in India's solvency. There was something so reassuring to Englishmen of his day about Cornwallis's reputation for honesty, integrity, and God-fearing faith in property rights that his utter ignorance of India seems to have counted for nothing in the calculations of Pitt, Dundas, the king, and the directors, all of whom were overjoyed at the noble Marquess' willingness to "sacrifice" himself by accepting their invitation to further service. It was, in fact, Cornwallis, more than Hastings or Clive, who was to be the true architect of John Company Raj.

The new Raj that Cornwallis established was based on his firm belief that Britons were the best qualified people to govern anyone, and that the best British civil servants were those with sufficient virtue, integrity, and salary to resist temptations of private trade or peculation of any sort. He therefore not only fired those members of the Board of Trade and those district collectors who were found lining their own pockets with company produce and revenue, but he also convinced the Court of Directors that if they hoped to keep servants honest, they would have to pay them sensible salaries, not the pittance of £5 or £10 hitherto offered as annual wages to writers and factors in the company's employ. To pay those higher salaries, while covering the cost of Britain's commercial investment and dividends, the company required a steady and minimal assurance of annual revenue. Pitt's act called for a thorough investigation into claims to land made by "rajas, *zamindars,* and other native landholders" and the drafting of "permanent rules" for Bengal so that in the future, revenue collection might prove less chaotic and unreliable. The revenue councillor, John Shore, was deputed by Cornwallis to carry out that job, and after painstaking analysis, Shore recommended a ten-year settlement, at some £3.75 million a year, with Bengal's *zamindars,* the Mughal tax-collectors who lorded over Bengal's countryside. Initially, Cornwallis agreed with Shore's arguments that the settlement should be made for a decade, subject to possible upward evaluation thereafter, but by 1793 the Whig landowner in Cornwallis prevailed, and he recommended to Dundas that the settlement with Bengal's *zamindars* be made permanent. How else, after all, could Bengal's "gentry" feel enough stake in the future to assure their loyalties to the Company Raj? Was not the sacrosanctity of property at the root of Britain's prosperity and power? Why withhold such a blessing from Bengal? Dundas and Pitt agreed.

The Permanent *Zamindari* Settlement of 1793 radically transformed the character of Bengal's rural relationships, for traditionally *zamindars* only held "interests" in land "assigned" to them by imperial power for tax collection and the maintenance of law and order. Private ownership of land in traditional Indian society was virtually unknown. Between the raja or *padishah,* who ruled over and "protected" all his domain, and the vast "herd" (*ri'yat*) of peasant tillers of the soil, there were, naturally, various intermediaries, endowed with a variety of landed interests. Some of these intermediaries were local hereditary chiefs or petty rajas whose domains had been conquered but who accepted the overlordship of a higher monarch, thus retaining many of their traditional rights in the service of the new order. Others were valued generals or servants of the emperor, whose reward was the grant of a *zamindari* holding for life, an area to be administered and gleaned of its harvest surplus both for the support of the *zamindar* and his establishment and to contribute a "share" to

the imperial coffers. Some soldiers were rewarded even more generously with grants of *jagirs,* which meant control over the total revenue of a number of villages for life. Whatever the origin or precise
nature of the intermediary, however, his "interest" in the fruits of the
soil of any given region did not give him the right to alienate such
land, or to drive those who tilled the soil under his charge away from
the land in which they, too, had clearly and traditionally defined basic
interests. After Cornwallis's settlement, all of that changed, for British
common law, with its strict protection of private property, would
hereafter be marshalled in support of those who held deeds to land;
they could do whatever they pleased with that land, so long as they
paid the specified and permanently limited annual revenue to the
Company Raj. The awful implications and full meaning of the powers
associated with private ownership, especially the right to exclude
others from one's land, would only slowly become clear to Indians
living under the Company Raj, usually after prolonged and expensive
conflict in a court of law. The fabric of traditional rural interdependence was, however, irreparably torn apart by the settlement of 1793
and those that followed.

The British found the Cornwallis settlement attractive for two
reasons, the first being its permanent resolution of a problem that had
long plagued and preoccupied them; the second, its securing a class
of loyal Indian supporters for the British Raj, who remained faithful
throughout the remaining years of company rule. But Cornwallis's
hope that the original *zamindars* with whom his settlement was
reached would remain the actual landlords of Bengal proved illusory.
After the first year or two of poor rains or flooded crops, when
zamindars were unable to pay the British what they had contracted
to remit, their deeds were taken over by Calcutta bankers and moneylenders, who emerged as a new generation of absentee *zamindars.*
Many of the old Mughal aristocracy of Bengal were thus displaced by
Hindu families—the Roys, Sens, and Tagores—whose scions would,
in the next generation or two, become leaders of cultural syncretism
and Westernization in Calcutta society. By the outbreak of the War
of 1857–58, these families were so strongly attached to the British
Raj that their loyalty proved critical in averting the spread of "mutiny" to Bengal. Another positive result of the settlement was that,
since they were obliged to pay no more revenue than the amount

fixed in 1793, the *zamindars* were induced to bring uncleared and untilled soil within their domain into fruitful production—a fact that helps explain the rapid growth in Bengal's population, as well as the real value of its land throughout the nineteenth century.

As commander-in-chief of the Bengal Army, Cornwallis totally Europeanized its commissioned officer corps. Just as no civil servant in the company's employ earning more than £500 a year was to be Indian after 1790, so too would no sepoy rise through the ranks to attain commissioned status. The British-supremacist stamp so indelibly fixed upon the services by Cornwallis was to remain a distinctive feature of the Company Raj, and it continued during the following nine decades of direct crown rule to alienate India's population from their foreign masters.

Despite a solemn disclaimer in Pitt's act of any desire for further territorial acquisition, Cornwallis found himself lured into an imperialist war against Mysore in 1789, just in time to save the company's Bengal and Madras merchants from the necessity of declaring themselves bankrupt for lack of silver, which swiftly flowed east in abundance as soon as word of a fresh Carnatic war reached London. The company's merchants in Bombay joined with those of the other presidencies in competing for the immediate profits derived from Cornwallis's war against Tipu Sultan, who was forced to cede most of the Malabar coast to the British in 1792 so that he might be permitted a final six years of peaceful existence. The *peshwa* and the *nizam* nominally supported the British during this war, rejecting Tipu's impassioned appeal for a united South Indian alliance. Mahadji Sindia had, since 1785, been Emperor Shah Alam's *vakil* ("regent"), a position even higher than that of *vazir,* combining with the premiership the responsibilities of imperial paymaster (*mir bakshi*). Despite his titular power, however, Mahadji lacked sufficient wealth to pay his own troops and could not even secure a loan from his friends in Poona. He was finally obliged to open a new mint in Mathura to hammer out much-needed rupees. Mahadji Sindia wasted his energies and resources in fighting the Rajputs, while the Rohilla Afghan leader Ghulam Qadir attacked Delhi, captured the Red Fort in 1788, and extorted the rank of *mir bakshi* for himself before blinding the emperor. The Rohilla terror was driven out of Delhi near the end of 1788 and Ghulam himself was killed the next year by Mahadji's troops, but his prestige as well as the emperor's had suffered irreparable deflation. The British in Bengal could now well afford to ignore Mahadji's demands that they forward the promised revenue share to the emperor. He sought in vain to secure the aid of the Sikhs and Afghans in forging a North Indian alliance against the British, but regional greed and mutual mistrust were still the most powerful forces dictating Indian diplomacy. The British, by contrast, for all their interpresidency rivalry and squabbling, were paragons of united action and national loyalty. More than military tactics, mercantile enterprise, or technological advantage, it was this capacity of the British to sublimate individual desires, greed, and dislike of other Englishmen to the imperative of working together toward a national (or company) goal and obeying orders accordingly, that helps explain the paramount position the Company Raj came to hold over all of India.

Shortly before leaving India in 1793, Cornwallis enshrined his administrative system of government in Bengal within a Code of Forty-Eight Regulations. This Cornwallis Code, as it came to be called, laid the foundations for British rule throughout India, setting standards for the services, courts, and revenue collection that remained remarkably unaltered over time. The first of those regulations was the Permanent *Zamindari* Settlement, which remained in force in Bengal, Bihar, and Orissa until after the termination of British rule in 1947. In his judicial regulations, Cornwallis abolished the *faujdari* courts that had been presided over by Indian judges, establishing in their stead four provincial circuit courts, presided over by British judges. To expedite criminal justice, British magistrates were empowered summarily to try petty offenses where they found them. This merging of police and judicial authority in the hands of a single official, usually a young man recently arrived from England, often incapable of understanding the language in which criminal charges and countercharges were made, was changed after 1817, when the functions of magistrate and collector were united and justice was made a separate function. The collector, under Cornwallis's system, was invariably British and had control over the revenue, for which it was usually necessary for him to keep a guard; hence the addition of magistrate powers after some two decades of experience. The collector would, however, employ many Indian assistants (*āmils*), clerks, and servants of every variety. Internal duties and former local taxes were now all abolished by regulation, and John Company asserted its right to impose whatever taxes or charges it deemed necessary anywhere within its domain. The company's monopoly over the collection, sale, and importation of salt was asserted at this time, and salt became one of the most important sources of British Indian revenue, as well as the least popular, since the burden of the salt tax fell most heavily on the vast majority of impoverished peasants. The production and sale of opium was now also officially regulated, and it remained a company monopoly, as well as the most lucrative item of British India's trade with China. British home demand for Chinese tea and silk had grown at so remarkable a rate that were it not for the continuing expanded cultivation, production, and sale of Indian opium to China, the company would have been obliged to ship fortunes in silver to its factories at Canton and Shanghai. Thanks to Bengali opium, however, it found the one nonspecie product for which China had a perfectly elastic demand, despite official prohibition. Throughout the nineteenth century, British India's major contribution to China remained its opium, produced by government monopoly in Bengal.

FOURTEEN

THE NEW MUGHALS
(1793–1848)

Sir John Shore ruled without reigning from Calcutta for five years after Lord Cornwallis left him with the keys to the kingdom. Shore's successor, Richard Colley Wellesley (1761–1842), came to the governor-generalship not as a middle-class merchant, but as a New Mughal. He set a tone of aristocratic arrogance and high social style that was to become the model for generations of British proconsuls and servants of humbler rank. Wellesley reached India in 1798, the year Napoleon set sail for Egypt; it was a fateful year for Indian history, marking the final European struggle for power over India. Wellesley went east, determined to secure the Indian subcontinent for British rule. It was hardly a dream conceived in his own vigorous though still somewhat inexperienced mind, but rather the fruit of Henry Dundas's mature ambition, transmitted in lengthy personal briefings to his proconsul at the Board of Control before Wellesley embarked on his passage to India. Obsessed by Francophobia, Dundas selected Wellesley from his board to serve as Britain's martial arm in "cleansing" all remaining pockets of native power that had become contaminated by French influence. That meant first of all Tipu, and second, the Marathas, especially Mahadji Sindia.

Assisted by his younger brother, the Iron Duke of Wellington, then still Sir Arthur Wellesley, Lord Wellesley added more territory directly to British India and brought more princely real estate under the "protection" of the company's subsidiary alliance than did any other governor-general. Tipu fell before the full force of British arms within a year of the Wellesley brothers' arrival in India, killed when his capital, Seringapatam, was stormed on May 4, 1799. Almost half of Mysore was now annexed to the company's domain, linking Madras to the west coast; a portion of the northern region of the state was

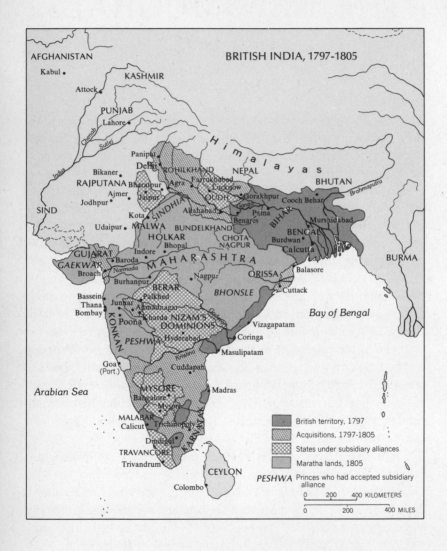

BRITISH INDIA, 1797-1805

AFGHANISTAN

Kabul •

Attock •

KASHMIR

PUNJAB

Lahore •

Chenab

Sutlej

Indus

_H i m a l a y a s_

Brahmaputra

NEPAL

BHUTAN

Cooch Behar

BURMA

Panipat

Delhi

ROHILKHAND

Farrukhabad

Agra

Lucknow

OUDH

Gorakhpur

Bikaner

RAJPUTANA

Bharatpur

Ajmer

Jaipur

Jodhpur

SIND

Kota

Udaipur

SINDHIA

Allahabad

Ganges

Patna

Benares

BIHAR

Murshidabad

MALWA

BUNDELKHAND

BENGAL

HOLKAR

Bhopal

CHOTA NAGPUR

Burdwan

Calcutta

Indore

GUJARAT

GAEKWAR

Baroda

Narmada

Broach

M A H A R A S H T R A

Nagpur

ORISSA

Balasore

Burhanpur

BERAR

BHONSLE

Cuttack

Bassein

Thana

Bombay

Junnar

Palkhed

Ahmadnagar

Kharda

NIZAM'S

DOMINIONS

Vizagapatam

Poona

Godavari

Coringa

KONKAN

PESHWA

Hyderabad

Masulipatam

Goa (Port.)

Krishna

Cuddapah

Bay of Bengal

Arabian Sea

MYSORE

Madras

Bangalore

Mysore

MALABAR

Calicut

Trichinopoly

KARNATAK

Dindigul

TRAVANCORE

Trivandrum

CEYLON

Colombo

British territory, 1797

Acquisitions, 1797-1805

States under subsidiary alliances

Maratha lands, 1805

PESHWA Princes who had accepted subsidiary alliance

0   200   400 KILOMETERS

0   200   400 MILES

turned over to the *nizam,* the company's latest subsidiary ally, as compensation for his support in the military operations of 1799. The other half of Mysore, its isolated center, was restored to the child maharaja, whose Hindu family had been deposed by Haidar Ali. The subsidiary alliance system, as perfected by Wellesley in his dealings with Mysore and the *nizam,* proved a most economical and convenient method for the rapid expansion of British power and effectively removed rebellion from the commonplace position it had acquired in the hitherto turbulent world of native states' politics. By reducing all Indian princes—one at a time—to virtual impotence, British power gradually brought an almost revolutionary state of peace and tranquillity to most of the subcontinent. Soon after Hyderabad's *nizam* was obliged to cede his cotton-rich region of Berar to the British to support Wellesley's regiment of praetorian guards at his palace gates, the *nawab-vazir* of Oudh was similarly brought within the ring of British control. In 1801, Wellesley ordered British troops to strip Oudh of its rich western Doab and Rohilkhand, which Warren Hastings had once rented British troops to secure for his staunch native ally. Lucknow was surrounded by British soldiers, and its ailing *nawab* was forced to disband his own army, assured that Oudh would in future receive all the protection it required from British-led soldiers, who were to be sumptuously sustained by revenues from the ceded half of Rama's once mighty state. The wealthy port-state of Surat was also now absorbed by fiat of the New Mughal, as was Tanjore in the south.

Wellesley's sole remaining competitor was the divided Maratha Pentarchy, which lost its most sagacious leader with the death of Nana Phadnis in 1800. The young *peshwa,* Baji Rao II (1775–1851), had only nominally ruled Poona since his accession in 1796, and after the Nana's death, the long-standing struggle between Sindia and Holkar erupted violently once again. Daulat Rao Sindia and his army initially dominated Poona as well as Delhi, but in 1801, Yeshwant Rao Holkar launched a frontal assault against the traditional center of Maharashtrian power, forcing the puppet *peshwa* to flee from Poona and seek refuge aboard a British warship that took him to Bassein, north of Bombay. At Bassein, on December 31, 1802, the displaced Baji Rao II signed a treaty with the Company Raj by which he became Wellesley's subsidiary ally, ceding the Maratha domain he had abandoned to British arms. This *de jure* surrender of Maratha power would be sealed in blood some sixteen years later. The Treaty of Bassein may, however, be said to herald a century of almost unchallenged rule over India's subcontinent by Britain's New Mughals.

As news of renewed and protracted warfare against the Marathas, with its additional expenditure and heavy casualties, reached London, however, Britain's home government lost confidence in Wellesley, who had treated the cabinet and company directors almost as imperiously as he treated his subsidiary allies. In 1805 the governor-general was recalled by His Majesty's Government, halted at the very threshold of imperial power over India, much the way Dupleix had been stopped by the French home authority. Though Wellesley was never actually impeached, he did not escape the censure of his Board of Directors, whose mercantile minds first trembled and then became enraged by the often arbitrary, violent, and unaccountable behavior of this most ambitious of British governor-generals, who was charged with having "goaded the whole country into a state of revolt." Yet while Wellesley himself was rewarded almost as negatively as Warren Hastings had been, he left enough chips off his New Mughal image, strategically situated in key service posts throughout India, to assure the continuity of Britain's conquest. Wellesley's leading protégés were John Malcolm (1769–1833), Mountstuart Elphinstone (1779–1859), Charles Metcalfe (1785–1846), and Thomas Munro (1761–1827). Malcolm, Wellesley's resident in Mysore and plenipotentiary extraordinary, "settled" central India, bringing fierce *pindari* raiders (unemployed Maratha soldiers turned freebooters) to rest content with a system of British rule that replaced looting with revenue collection and robber gangs with regiments of sepoys. He later succeeded Elphinstone as governor of Bombay (1827–30). Elphinstone was the first British *peshwa* of Poona, as shrewd a young Scot as ever was sent by Dundas to India, reared in the service of "Old Villainy" (as he fondly called Wellesley), a classical scholar, amateur historian, diplomat, and later autocrat of the Deccan. As British resident in Poona, Elphinstone sought to reduce Baji Rao II to the puppet he was expected to be, but when Baji proved recalcitrant and tried to revive Maratha pretensions to power, Elphinstone used the troops under his command to assert his full authority, and in June 1818 Maratha power was finally crushed. Baji Rao was pensioned off and shipped north to Bithur, a castle on the outskirts of Kanpur (Cawnpore), which would later become one of the centers of the 1857 "mutiny." Some of the former *peshwa*'s displaced courtiers clung to the purse strings of his pension, but most were forced to fend for themselves in Poona, losing at one stroke both income and status, stripped by the British Mughals of all but their pride and their brahmanic learning. The Chitpavin brahman community that had ruled Poona's bureaucratic roost, revered by many of their own people as gods on earth, were to provide leadership to India's National Congress before the end of the century, however.

. Elphinstone administered the newly conquered Maratha territories, soon to be integrated into the presidency of Bombay over which he presided as governor from 1819 to 1827, with impartiality and efficiency. He had the good sense to continue to use Indians in his administration wherever possible and even maintained government subsidies to devout brahmans and their temples (though at one-tenth the *peshwa*'s rate). The lands of the mightiest and most hostile Maratha *sardars* ("nobles") were confiscated for British use, but most *jagirdars* were permitted to hold their estates once they vowed their loyalty to the new Raj, and they soon became staunch allies of the Company Bahadur ("Great Company"). British success in retaining control over India, after wresting land from indigenous powers on the field of battle, was primarily due to the methodical, even-handed way in which newly conquered territories were "settled." Land revenue demands were no lower than those previously claimed by the Marathas, but peasants soon learned that once their share was paid to the British collector, they were free to live quietly for the rest of the year, unassailed by neighboring robbers demanding another quarter or third of their wealth. The petty pilfering and princely warfare that had become endemic to the Deccan, and in fact to most of India, during the latter half of the eighteenth century was virtually eliminated by the strong hand of the British Raj in the early part of the nineteenth century. Maratha brahmans would soon be able to tell their families in awed delight that a man could "carry gold at the top of his walking stick from Poona to Delhi without being molested by robbers."

Metcalfe was hardly sixteen when he reached Calcutta, sent east by his director father to serve under Wellesley in 1801. He joined the first class of Wellesley's college for company servants at Fort William and was soon packed off to Agra, assistant to the resident there, and thence to Delhi, where he came to know the environs of faded Mughal imperial might, over which he was one day to preside. In 1813, at the tender age of twenty-seven, he was made resident at Delhi, and though his power there did not yet extend inside the walls of the still-reigning Mughal emperor's Red Fort, Metcalfe administered justice throughout the teeming city, whose twilight glow was softened by the moderation of his enlightened measures. While the old Mughal remained in his harem, relishing the last petty perquisites of the power that Akbar once wielded, Metcalfe and his New Mughal collectors scoured the bazaars beyond the palace walls, raking in whatever wealth they could gather from the merchants and artisans of North India's greatest city, shrewd enough not to awaken unmanageable animosity by violent usurpations or excessive punishments. Like Malcolm and Elphinstone, Metcalfe was a pragmatist, testing each toehold cautiously as he climbed the perilous ladder of power, wisely leaving untouched whatever native institutions, officials, and procedures he found that were not strongly antipathetic to the new Raj. The last thing any of Old Villainy's disciples wanted to do was to "reform" India overnight or remodel it into an Eastern reflection of contemporary England. They were, in fact, continually amazed at the swift success of their daring ventures, and they daily expected the uprising or outcry from the population at large that—given effective leadership—could so easily have overwhelmed Britain's minuscule force in India. But for a single abortive mutiny at Vellore (near Madras) in 1806, and localized uprisings of peasants in North India led by orthodox Muslim *alims* and *mullas,* who zealously sought to arouse violent opposition to British consolidation, there was nothing but passive acquiescence.

Munro was a soldier, as tough and leathery a man as ever went from the Highland moors to South India's Madras, where he fought under Cornwallis against Tipu and subsequently settled most of the land taken from Mysore by Wellesley. He soon perceived that as long as he did not tamper with Hindu ritual or custom and kept his tax demands just a bit below what Tipu or the *nizam* had asked of the same plot of land, he would win the loyal allegiance of most of the peasants he met. Those who proved ungrateful were few enough to be crushed or silenced. As governor of Madras, Munro introduced the *Ryotwari* ("Peasant"), Settlement, which, in contrast to the *Zamindari* Settlement in Bengal, imposed the new Raj's demand for revenue directly upon the peasant tiller of the soil for a fixed period of time (from ten to thirty years), thus giving the British periodical opportunities to raise their revenue demands as land value increased and giving the peasants closer sense of identification with the otherwise remote foreign power that ruled them. Munro's Highlander sense of peasant propriety predisposed him to this form of settlement, which helped secure the loyalty of more peasant cultivators to the new Raj than had the permanent settlement in Bengal. Not all the *ryots* with whom the government dealt could continue to meet their annual revenue payments, however; many were forced to turn for assistance to local moneylenders, who took as collateral property deeds on which they later foreclosed. A class of absentee land-owning bankers thus emerged in Madras, much like the one that already flourished in Calcutta. The *Ryotwari* Settlement served, nonetheless, as an important stimulus to social change in South India, and it gave the district officer greater responsibility and more opportunity to deal directly with the people under his jurisdiction, going round the village circuit to hear grievances and ensure that the revenue collected was fair, adequate, and remitted in full to the government coffers.

With the New Mughals, other Englishmen flocked to India, not as officials or merchants in the service of company or crown, but as missionary reformers seeking the salvation of heathen souls. From its inception, the company perceived the threat to its profits if it permitted such "salvation-mongers" to settle freely in its factories and preach the gospel in neighboring bazaars. With Mughal antipathy to the Portuguese Jesuits as a cautionary precedent, the tough-minded merchants of Leadenhall Street were as loath to admit missionaries to their presidencies as they were to surrender their monopoly of trade. The first British missionaries to arrive in Bengal before the end of the eighteenth century, Baptists Joshua Marshman, William Carey, and William Ward, were considered so "subversive" a menace to tranquillity by Wellesley and his officers that they were banned from Calcutta and forced to settle in the nearby Danish enclave of Serampore, where they were welcomed and established their mission and their prolific press. This talented Serampore trio was to become instrumental in fostering Anglo-Orientalist education and advancing the study of Indo-Aryan languages, due to their assiduous labors in translating, printing, and teaching the Bible in several Indian languages. By 1801 even Wellesley had to admit their usefulness in instructing company servants in Bengali and Urdu at his Fort William College, and Carey was appointed to a professorship there. The missionary influence and its growing popularity in London had made itself felt in Calcutta. Shore and Grant had their evangelical impact on the company's directors, as well as upon Parliament. Joining the "saints" of William Wilburforce's Clapham Sect, they succeeded by the Charter Act of 1813 in removing the company's blanket ban on missionary enterprise, opening India to "licensed" missionaries as well as private traders. The spirit of Adam Smith's *laissez-faire* thus marched hand in hand with the gospel of evangelicalism through the narrow portals of company privilege, auguring many changes in the nature of Britain's new Indian Empire.

The struggle between British officials, whose primary concern was survival through stability, and British preachers or lay reformers, who were preoccupied with salvation through conversion or reform, was to remain a source of tension in the superstructure of the new Raj throughout the nineteenth century. It would, in fact, never fully be resolved until the end of British rule, for Britons were no more united on how to handle and rule their Indian Empire than Indians were in their responses to the consolidation of British rule. The simple motivating force of greed that had nurtured British factors through the hardships of the seventeenth century, mixed as it was with a lust for power that helped them survive the rigors of Mughal collapse and French antipathy in the eighteenth, was suddenly thrown into a state of confusion by such complex questions as international conscience and concern for "civilizing" India's "poor, benighted native souls." It was a problem to perplex the mind of a Jeremy Bentham and cudgel the brains of James Mill and his son John Stuart, both of whom drafted much of the company's correspondence in London while seeking to resolve utilitarian equations based on Bentham's calculus of pleasures and pains. The one Englishman who came closest to reconciling the polar impulses of pragmatic administration and idealistic principle was Thomas Babington Macaulay (1800–59), whose codifications of Anglo-Indian law remain classic monuments to his brilliance and diligence, but whose victory in the defense of English education for Indians accelerated demands for the demise of British rule by arming India's elite with the words in which to call for it.

The impact of English education, much like that of Christian missionary preaching, proved at best a mixed blessing to British rule. Elphinstone had known how dangerous a two-edged sword Western education would prove to be when he had the courage to introduce it in Bombay; indeed, he called it "our highroad back to Europe." The company was obliged to teach English to at least enough natives to facilitate and sustain the effective administration of the territories recently brought under its vast new imperial umbrella. The alternative of expecting young Britishers to learn enough Indian languages to carry on the daily chores of administrative collecting, spending, and punishing was simply too expensive and intellectually unlikely. The missionaries were dedicated linguists who made remarkable strides in their study of Bengali, Urdu, Marathi, Tamil, Telugu, and other Indian languages, but thousands of them would have been needed to teach every collector and his assistants fluency in all the regional languages they might require. Indians, of course, proved far more adept at mastering English than their masters proved at learning any Indian tongue, and when the Act of 1813 obliged the company to set aside £10,000 (100,000 rupees) for purposes of "education," it seemed a golden opportunity to teach the language of Clive, Hastings, and Cornwallis to the Roys, Basus, and Tagores of Bengal. Some British scholars of Indic culture, Orientalists in the tradition of Sir William Jones (1746–94), who founded the Asiatic Society of Bengal in Calcutta in 1784, felt that Indians would benefit more from learning about Indian civilization through the study of Sanskrit and Persian than they would by racking their brains to assimilate foreign concepts taught in an exotic tongue. Horace Hyman Wilson, the brilliant Sanskritist who produced the first Sanskrit-English dictionary while employed in the Calcutta mint, considered it "a visionary absurdity to think of making English the language of India,"[1] as did other learned contributors to the Asiatic Society's *Journal*. They felt that whatever

---

[1] H. H. Wilson to Ram Camul Sen, August 20, 1834, quoted in David Kopf, *British Orientalism and the Bengal Renaissance: The Dynamics of Indian Modernization, 1773–1834* (Berkeley and Los Angeles: University of California Press, 1969), p. 242.

sums the company was willing to set aside for education would best
be spent on strengthening knowledge of India's indigenous languages,
whose many pundits and patrons were impoverished and dethroned.
But the East India Company was not the Ford Foundation, and
Orientalist arguments would be easily outgunned by the combined
firepower of mercantile utility and Anglican pride once Macaulay
turned his attention to this issue in 1835.

For the decade following the Charter Act of 1813, however, the
united influence of British missionaries, free trading, and Orientalist
education helped stimulate a remarkably vital response from Bengal's
youthful intelligentsia. This cultural offshoot of Calcutta's early phase
of modernization has come to be known as the Hindu Renaissance,
but it may more accurately be viewed as the seed time for modern India's emergence, microcosmically reflected in the mirror of Calcutta's
bustling urban society. For just as Delhi and Agra housed the palaces
and tombs of old Mughal glory and decline, Calcutta became the
cradle of the New Mughals' rise to continental paramountcy.

"The many-sided, the smoky, the magnificent City of Dreadful
Night," as Kipling called it, Calcutta had a population of close to a
quarter of a million by 1820. It was capital of an empire, home of
princely palaces and prostitutes, a modern Babylon along the Hughli,
more than a century remote from those sleepy villages Job Charnock
had chosen as the sight of Fort William. Soon it would have no rivals
in all of Asia, and few urban centers the world over could approach
its extremes of wealth and poverty. Not surprisingly, it was in Calcutta that a new synthesis of Anglo-Indian culture began to emerge,
a synthesis that would subsequently fuel the fires of nationalism without ever losing its attachment to the British ideas and habits of life
that were to be indelibly stamped on the consciousness of India's
emerging elite.

Ram Mohun Roy (1772–1833), long revered as the father of
the Hindu Renaissance, best epitomizes the cultural synthesis that
evolved between the era of Warren Hastings and that of Lord Moira,
the Marquess of Hastings, who presided over Calcutta's Raj from
1813 to 1823. Born a Bengali brahman, reared first on the classics of
Sanskrit, Persian, and Arabic and later those of English, Latin, and
Greek, Roy lived much of his mature life in the heart of Calcutta but
died in England, the first and only "raja" emissary of Delhi's Mughal
emperor to the Court of St. James. There was symbolic poignance to

Roy's death in Bristol, for by the end of his learned life, Ram Mohun had in many ways become more English than Indian, alienated from the cultural environment in which he was born, yet unable to survive the alien climate to which he had been lured by historic fortune. Roy was employed in the Revenue Department of the East India Company as a young man, and he had risen to the rank of native assistant to the collector, the highest office to which an Indian could aspire, by 1814, when he retired to enjoy the most fruitful years of his life in Calcutta. While in the service, Roy acquired so fine a command of English that he was able to cross literary swords with British missionary friends, who assiduously sought to convert him to Christianity. To arm himself against their importunate attempts to "save his soul," Roy studied the ancient texts of Hinduism, especially the Upanishads, which gave him fresh appreciation of the poetry, power, and philosophic wisdom of his heritage. In 1815 he started meeting regularly with Bengali friends from the educated upper classes (the *bhadralok*), who had organized themselves into a "Friendly Association" (*Amitya Sabha*) to discuss questions of theology and philosophy. One regular member of the group was Dwarkanath Tagore (1794–1846), grandfather of the poet; he was one of the leading *zamindars* of Bengal and a singularly enterprising entrepreneur[2] in the vanguard of Indian banking and mercantile development. The Tagore family remained patrons of Roy's varied efforts to revitalize Hindu philosophic religious dogma in the light of Christian criticism. Roy's *Brahmo Samaj* ("Society of Brahma"), founded in Calcutta in 1828, was to have a seminal impact upon generations of young Bengalis, and though there were never more than a few hundred who personally participated in the activities of that society, they were, like the leaders of the Florentine Renaissance, beacons of change, auguring the rebirth of pride and faith in Hinduism among the young intelligentsia. By harking back to Upanishadic Vedanta for monistic inspiration, Calcutta's Brahmos armed themselves with more sophisticated ideological doctrines of unitarian thought than any of the Christian missionaries, who sought in vain to convert these brilliant "heathen."

The fact that a Roy could train himself to become an English-

---

[2] Blair B. Kling, *Partner in Empire: Dwarkanath Tagore and the Age of Enterprise in Eastern India* (Berkeley and Los Angeles: University of California Press, 1976).

man while remaining a Bengali brahman was the first portent of Indian independence, though it would take more than a century after Ram Mohun's death for the full fruition of that promise. A Scottish watchmaker and educator, David Hare, with the support of H. H. Wilson and other Calcutta Orientalists, helped the Hindu College (founded in 1816) provide the sons of this new Indian elite with the Western learning they coveted. Anglo-Indian poet Henry Louis Derozio (1809–31) was that college's most brilliant teacher, inspiring Bengal's youth to seek universal learning and creative expression. Though he was a Christian, Derozio worshipped India as "My Native Land" and wrote a poem with that title. In 1818 India's first public library was opened in Calcutta with what was then the world's best collection of Orientalia, totaling more than eleven thousand printed books and manuscripts. Publication of a number of newspapers and journals was started in Calcutta at this time, several by Roy himself, thus more widely communicating the new ideas and syncretic cultural concepts that blossomed in this urban soil.

The era of utilitarian liberalism initiated by Lord William Bentinck in Calcutta during his governor-generalship from 1828 to 1835 was both an affirmation of British cultural arrogance and a challenge to young India's capacity for cultural assimilation. By launching legal assaults against such violent Hindu customs as *sati* (the immolation of Hindu widows on the funeral pyres of their dead husbands) and *thugi* (ritual murder by strangling and highway robbery in the service of the Mother Goddess Kali) on the grounds of humanitarian principle, Bentinck risked the wrath and unpopularity of India's orthodox majority. It was, after all, one thing for a handful of missionaries to advocate radical social reforms, but quite another for the governor-general to enact them in council. Even Ram Mohun Roy, who agreed with Bentinck "in principle" that *sati* was uncivilized, could never bring himself openly to advocate its abolition by British law. That Act of Abolition was, however, passed in Calcutta in 1829, for "the good of mankind," as Lord Bentinck's minute put it. Mere legal proscription could not, however, abolish so deep-rooted a religious custom overnight, and *sati* deaths continued to be reported in remote regions of India well into the twentieth century. Bentinck's courageous act, nonetheless, established a precedent for government intervention into the most sacred areas of Indian ritual practice. The British Raj was no longer merely a company of plundering merchants or Machiavellian administrators whose interest in their Indian subjects extended only as far as the limits of profit or power.

Calcutta's administration actually proved a pale reflection of the shining image of British reform in this era of evangelical concern for salvation mixed with utilitarian interest in brightening the here and now. The Liberal Charter, as the Charter Act of 1833 came to be called, was passed the same year in which slavery was abolished in British Africa and the West Indies (though not in India until 1843); in the same year, the first British Factory Act prohibited the employ of children under nine years of age in mines and mills. Landless Indian peasants are, in fact, still "slaves" to the ploughs and Persian wheels they work all day, and childhood has proved no effective shield against exploitation in South Asian villages and towns. Still, the new act became a landmark of liberal change for India and remains one of the noblest British statements of equality of human opportunity. As Article 87 proclaimed: "No Native of the said Territories, nor any natural-born Subject of His Majesty resident therein, shall, by reason only of his religion, place of birth, descent, colour, or any of them be disabled from holding any Place, Office, or Employment under the said Company." The report of the committee of Parliament that studied the "Affairs of the East India Company" during that landmark year further elaborated: "On a large view of the state of Indian Legislation, and of the improvements of which it is susceptible, it is recognised as an indisputable principle, that the interests of the Native Subjects are to be consulted in preference to those of Europeans, whenever the two come in competition; and that therefore the Laws ought to be adapted rather to the feelings and habits of the Natives than to those of Europeans." Though these noble sentiments were more often quoted by Indian nationalists as evidence of the hypocrisy of British rule rather than proof of its magnanimity, they did set a standard toward which not only British, but independent Indian governments could ultimately aspire.

The Charter Act of 1833 also opened India to unrestricted British emigration and enterprise, mercantile as well as missionary, and, in the spirit of *laissez-faire* that now blew so freely over the British Isles, it abolished the company's monopoly over all trade (except opium and salt). As Britain's Midlands fast became the industrial workshop of the world, not only did the demand for Indian cloth diminish rapidly in London and Lancashire, but cheaper machine-made cottons from England became increasingly popular in Calcutta, Patna, Bombay, and Madras. The East India Company's cloth factories in Dacca were all abolished by 1820, and before 1833 British cloth swept Dacca muslims from Bengal's vast market wherever the two competed for customers, in village bazaars as well as cities and towns. Between 1813 and 1833, the decades that marked the collapse of Bengal's vast home-spun cotton industry, millions of Indian women and men were thrown out of work by machines half a world away. Bengali unemployment reached unprecedented levels as the industrial revolution rocked India's peasant economy, transforming what had hitherto been an interdependent but self-sufficient state of relative economic prosperity into a precariously dependent market of peasants, whose numbers would continue to swell during the remaining decades of the nineteenth century, increasing the economic pressures on India's arable land. This quiet revolution from economic self-sufficiency to foreign dependence was to prove a far more powerful factor than the English language in first binding India to Britain and later impelling her to seek independence. For Calcutta's Brahmo elite, however, British rule proved the springboard to economic fortune as well as cultural syncretism. Dwarkanath Tagore, for example, learned and earned enough assisting the British collector in the Twenty-Four Parganas region of Bengal to become the first Indian bank director in 1828 and the founder of Calcutta's Union Bank a year later. One of young India's first industrialists, Tagore cultivated jute and developed indigo mills, silk factories, and sugar processing plants. Dwarkanath joined Messrs. Carr and Prinsep in 1834 in what was the first Anglo-Indian Managing Agency, Carr, Tagore and Company, and founded the first Indian colliery a few years later at Runigunj. Like his dear friend Ram Mohun, however, Tagore died while abroad, at the peak of his fortune in London, a second symbol for India of the perils of assimilation.

The pace of change accelerated after Macaulay joined Bentinck in Calcutta as the first law member of the council in 1835. The new cabinet post in itself was testimony to Britain's long-range desire and commitment to govern India with justice as well as by force, and Macaulay, the Burke of his era, was personally devoted to the utilitarian dream of seeking to legislate "happiness" and to the even more astounding idea of one nation ruling another in "the best interests" of the latter. Macaulay was only thirty-six when he was entrusted by his peers in Britain's Parliament with the task of becoming "Parliament" for India. In his famous minute on education, Macaulay could unashamedly admit to having "no knowledge of either Sanskrit or Arabic," yet he argued "that all the historical information which has been collected from all the books written in the Sanskrit language is less valuable than what may be found in the most paltry abridgements used at preparatory schools in England. In every branch of physical or moral philosophy the relative position of the two nations is nearly the same." His rhetoric convinced Bentinck and the majority of his Calcutta Council, for in 1835 they decided that the company's annual allocation of funds for education would better be spent on teaching Western learning to young Indians in the English language, rather than Oriental learning in any native tongue. The system designed by Cornwallis was thus modified, as Macaulay argued in his minute, "to form a class who may be interpreters between us and the millions whom we govern; a class of persons, Indian in blood and colour, but English in taste, in opinions, in morals, and in intellect."

While Calcutta's select corps of Indians studied English and learned to memorize more of Milton, Chaucer, and Shakespeare than most Britishers did, an even smaller group of Englishmen systematically embarked upon the study of "Orientals" at the East India Company's College at Haileybury, their first stop on officialdom's passage to India. Picked by the directors and Board of Control from the ever-widening pool of younger sons of relatives, friends, and lesser gentry-at-large who could find no place in the Horse Guards, no Church "living," no suitable job in London, several hundred annual appointees to the Covenented Indian Services spent two years of indoctrination at Haileybury between 1809 and 1858 learning their first phrases in Persian and Hindusthani, taking their lessons in Indian history, which began of course, with the Black Hole and Plassey. Processed through Haileybury between the ages of fifteen and twenty-two, a motley crew of lads from Eton, Oxford, and Westminster imbibed their gruel of Indic learning, seasoned with a thick sauce of Anglo-Christian prejudice, Malthusian pseudoscience, and Benthamite optimism. Officer candidates for India's army were similarly trained at Addiscombe. The young civilians and cadets who sailed for India in the early decades of the nineteenth century were taught that they were born to lead "heathen natives" toward the wisdom they themselves epitomized. Many of Britain's brightest, dullest, or simply best-connected young men joined the growing stream of leaders who bravely shouldered their burdens of thankless labor in a sweltering subcontinent still but half "civilized."

After 1818, there was only one martial rival to British rule left in India: the young Sikh kingdom of the Punjab. Ranjit Singh (1781–1839) ruled the Land of the Five Rivers as maharaja from 1799 until his death forty years later, owing to his unique adroitness at diplomacy and his unrivalled skill in battle. In 1809 the river Sutlej was formally agreed upon as the boundary between Ranjit's domain to its west and that of the British to the east. His British neighbors diplomatically pacified, Ranjit directed his attention toward the settlement of his western and northern frontiers. He kept the divided Afghans at bay beyond the Indus with a stunning victory at Attock in 1813 and followed that blow several years later by wresting Kashmir from Afghan control in 1819. By 1820 Ranjit's kingdom thus embraced a quarter million square miles of South Asia's richest and strategically most significant domain. His standing "army of the pure" (*khalsa*) numbered nearly a hundred thousand Sikhs, a mighty force forged in the fires of Mughal oppression, tempered by constant conflict with Afghan Pathans, and welded into a weapon of righteous action by their faith in the scripture left to them by ten gurus.

Wisely assessing Sikh power as too formidable to assail, the British focused their fire upon the remote kingdom of Burma on the subcontinent's eastern wing, and in the First Anglo-Burmese War, 1824–26, they stripped King Thibaw's domain of its coastal provinces of Arakan and Tenasserim, adding the inland plum of Assam to Bengal and weaning the neighboring princely state of Manipur out from under Burma's royal umbrella. The underlying causes of that war were Manchester's search for new markets and Calcutta's concern for the security of its ever-widening circle of rural environs. By 1824 the company's three presidency armies contained no fewer than 170 sepoy regiments, in addition to 16 European regiments.[3] This military

---

[3] Nagendra Singh, *The Theory of Force and Organisation of Defence in Indian Constitutional History* (New York: Asia Publishing House, 1969), pp. 190 ff.

force of approximately two hundred thousand was the greatest army in Asia, and it seems to have been no less reluctant than other powerful armies throughout history have been to demonstrate its combat effectiveness at the slightest provocation, real or imagined.

Governor-General Amherst, Britain's "best barren choice"[4] to stand at the helm of Indian Empire from 1823 to 1828, was hardly statesman enough to resist mercantile or martial pressures from his friends back home or in Calcutta. With jingoist reports urging anti-Burmese action fed to him by young David Scott, Amherst's agent on the northeastern frontier, the governor-general was quick to launch an expedition into the Bay of Bengal on February 24, 1824. By February 1826 the Burmese king had surrendered his coastal provinces and all claims to Assam and Manipur. A commercial treaty was also negotiated, and the British resident, who moved into Ava, busied himself with opening Burma's market to British cottons and woollens. The results of such enterprising effort and martial sacrifice proved disappointing, however, and a quarter of a century later Britain decided to take another bite of the Burmese plum, which could only be devoured in its entirety by three separate gunboat parades up the Irrawaddy.

The First Burmese War was followed by the peaceful interlude of Bentinck's seven years in control of Calcutta's government, while the new Imperial Raj sought to recoup the fortune wasted in malarial jungles and to consolidate its power through internal reforms rather than overseas conquest. Perhaps Bentinck's proclivity for peace was strengthened by the ominous mutiny of a sepoy regiment on the outskirts of Calcutta, at Barrackpur, during the Burmese War, the first mutiny since Vellore, which had led to Bentinck's recall as governor of Madras in 1806. The forty-seven Bengal Native Infantry refused to board ships to cross the "dark waters" to Burma, a trip that would have defiled the high-caste Hindus who comprised most of the regiment. British-manned artillery opened fire on the troops, whose regiment was thus "erased" from the Indian Army, its terrible fate serving as a lesson to all other natives foolish enough to think their

[4] The worthy quote was Canning's, and I am indebted to Dr. Percival Spear for it, from his paper "Patterns of British Leadership in British India—Theme with Variations," presented to the Leadership in South Asia seminar at the School of Oriental and African Studies of the University of London, October 12, 1972.

faith could take precedence over the orders of a British commanding officer. Soon after Bentinck sailed home, when memories of Barrakpur had dimmed, older temptations conspired with more recent anxieties to lure British India once again toward frontier squabbling, this time on the northwest.

Nineteenth-century British fears of a Russian invasion of India seem so outlandish that we still speak of them as phobias, yet they were all too real to Anglo-Indians who could never quite believe that Napoleon's death had put an end to European dreams of invading the subcontinent. The most direct route for possible Russian invasion, of course, would be through Afganistan and over the rugged Khyber Pass, though if Persia were to help, an easier passage might be forged through lower lying Baluchi country into Sind. During the late 1820s and 1830s, British interest in Sind blossomed as fears of Russia mounted in London and Calcutta. After 1828, as Russian influence expanded in Persia to the detriment of British influence there, Whig fears of the "creeping bear" fed similar concerns long endemic to India, where the northwest frontier has always been viewed as a highroad for invading armies. The Muslim *amirs* of Sind had signed a treaty with the company in 1820, promising to exclude all "European or American settlements" from their predominantly desert domain, but a border battle was fought in Cutch in 1825, which led to a journey by Alexander Burnes up the Indus to Hyderabad two years later. The report of Burnes's voyage, which was published in 1831, added commercial and diplomatic incentive to the strategic advocacy of those who, like Colonel Sir De Lacy Evans, advocated the conquest of Sind by British arms before the Russian Cossacks could gallop there. Mercantile interest in Sind's major waterway was, indeed, so strong that in 1832 the British concluded a new treaty with the *amirs,* opening the Indus to navigation. Binding both nations "never to look with the eye of covetousness on the possessions of each other," that solemn treaty inaugurated what was to be the last decade of independence enjoyed by the *amirs* of Sind. Among other things, it also promised that "no armed vessels or boats" and "no person" bearing "military stores" of any description would be sent by the British into or through Sind along the Indus. Wiser and more honest were the Sindi peasants who prophesied, "Our land is lost now that the British have navigated our river."

Lord Ellenborough, thrice chairman of the Board of Control, felt "confident" in 1829 that the Company's Raj would "have to fight the Russians on the Indus." He would have liked to avert that threat, as would his successor, Sir John Hobhouse (Lord Broughton), to whom Russophobia was even more of an obsession. Hobhouse's fears were shared by George Eden, who was sent by him to rule over India as Lord Auckland in 1836. Auckland himself quickly fell under the pernicious influence of his young secretary, Sir William Macnaghten, who was determined to play the great game of realpolitik on the chessboard of Asia, whose principalities and kingdoms he viewed as pawns in a titanic struggle between Britain and Russia. Afghanistan by this time had come firmly under the rule of Dost Muhammad Khan of the Barakzai clan, who controlled not only Kabul but Kandahar. His ousted predecessor, Shah Shuja of the Durrani clan, had been forced to flee his native land in 1813, first to the "protection"[5] of Ranjit Singh in the Punjab, and a year later across the Sutlej to Ludhiana, where he received hospitality from the British Raj. Macnaghten—infected, it seemed, with the Great Mughal concept of Afghanistan as the westernmost province of Delhi—was eager to restore the shah to his Kabul throne as a British puppet, and shortly before the end of 1836, Auckland dispatched Captain Burnes on a "commercial mission" to Kabul's court. Dost Muhammad, however, was too stubborn and independent to take orders or arms from the British, while Shah Shuja was willing to promise whatever might be asked of him, insisting that "tens of thousands" of his people were only waiting for him to appear on the road to Kabul to rise in support of his claim to the throne. As for Ranjit Singh, that one-eyed lion of the Punjab was clearly the shrewdest statesman of all. He understood British diplomatic ambitions and commercial lust better than did most of the company's servants, and he was even better than Shah Shuja at making promises he had no intention of keeping. In June 1838 Macnaghten personally concluded his infamous Tripartite Treaty with Ranjit Singh and Shah Shuja, the final gesture in the diplomatic charade that served as prelude to the First Afghan War.

Auckland had hoped that Ranjit Singh's army would do most of his dirty work by invading Kabul directly and escorting Shah Shuja

[5] Ranjit spared Shah Shuja's life after the latter's wife turned over the Koh-i-noor diamond, which she had carried from Kabul.

to his throne. Ranjit not only refused to provide Sikh troops, however, but even denied the British permission to pass across his domain in what would have been the most direct line of march toward the Khyber. Macnaghten, of course, knew that his "treaty" had been concluded with a passive partner and a straw man, but he was sufficiently infatuated with the myth of British invincibility to believe that his Raj really needed no assistance against so small and backward a land as Afghanistan. He persuaded his governor-general to launch a mighty army up the Indus. It was a violation of all previous agreements that the British signed with Sind, but the *amirs* of Sind were much weaker than Ranjit Singh. Macnaghten was despatched as "political officer" with the force of some sixteen thousand of British India's finest troops under the command of Sir John Keane. They left at the end of 1838 and sailed up the Indus to Shikarpur, then marched over the Bolan Pass to Quetta and Kandahar, which they reached five months later. The welcoming army of loyal Afghans that Shah Shuja had promised Macnaghten never materialized. The Pathans were at best indifferent or sullen; at worst, they fiercely resisted the invasion of their homeland. The British army was, however, strong enough to march north against Ghazni and Kabul, taking the latter by storm in early August of 1839. Dost Muhammad fled to the Russian border, his son and brother surrendering to the now restored Durrani claimant. Auckland was made an earl and Macnaghten a baronet, but the war of Afghan attrition had only begun.

Guerrilla warfare, assassination, local uprisings, and looting became daily occurrences wherever British Indian troops were found in the land of the Afghans, in the bazaars of Kabul and Kandahar, along the road to the Khyber, in the palace itself. Rumors of a Russian invasion to restore Dost Muhammad became so common in Kabul that in 1839 Macnaghten urged Auckland to give him permission to launch a new expedition north to the Hindu Kush. Before the end of 1840, however, Dost Muhammad had surrendered himself to Macnaghten in Kabul and had been taken to India. Shah Shuja could inspire no mass support, however, and each month the cost of maintaining a full British Indian army of occupation in Afghanistan consumed whatever surplus wealth had been generated in both Indian and Afghan treasuries. In 1841 Macnaghten was appointed governor of Bombay and Burnes was named to succeed him in Kabul, but neither lived to take up their new burdens of high office. By December, when Macnaghten decided to negotiate the Raj's retreat, it was too late. Winter frost and Afghan fire cut down the sixteen thousand who abandoned Kabul and rushed in panic toward the Khyber Pass. One man, Dr. Brydon, reached Jallalabad, which was held by another British force, to report that he was "the army of the Indus." Shah Shuja remained behind, pleading with his countrymen for the chance to continue to rule them. He survived until April 1842, when he was assassinated.

Ellenborough, who sailed out to take Calcutta's helm as soon as word of Auckland's "folly" and his ensuing personal breakdown reached London, immediately organized India's armed forces for "some signal and decisive blow upon the Afghans" to prove to all natives that Britain's power of "inflicting punishment" remained unimpaired. Tiny Afghanistan had shown all of India that an army trained and officered by Britons could not only be defeated, but annihilated. Not since Clive's capture of Arcot had anything so revolutionary occurred within the subcontinent. Dost Muhammad was released in 1843 and sent home to pick up the pieces of his shattered land, retaining the reins of royal power in his firmly independent hands until his death twenty years later. The utterly futile war cost British India twenty thousand lives and more than fifteen million pounds sterling.

In the wake of the Afghan disaster, the British turned first to Sind, then to the Punjab, for compensatory conquest. Sind had served as the company's supply base throughout the Afghan War, during which the British seized control of Karachi and Sukkur and used them as their major depots, maintaining a resident with attendant troops in the central Sind city of Hyderabad. The example of Afghan independence and strength inspired the Baluchi *amirs* of Sind to seek to reclaim from Ellenborough at the war's end the freedom and self-control they had surrendered perforce to Auckland. They had no idea that the company's directors, as well as the Raj's Calcutta rulers, had long since decided that British dominance over Sind was a "necessary" prerequisite to the "permanent" navigation of the Indus, which British commercial and strategic interests "demanded"; nor could low-lying, riverine Sind, with its predominantly passive population of tenant farmers, hope to pose anything resembling the military resistance of Afghanistan's gun-bearing Pathans.

In September 1842, General Sir Charles Napier disembarked at

Karachi to take formal command of all British Indian troops in Sind and Baluchistan. Fashioned in the mould of Clive, Napier was determined to use Sind as a launching pad for his dreams of glory, and he bullied, goaded, and berated the *amirs* of Sind before taking to the field to destroy them. Napier's sadism was matched only by the mystic fervor of his religious zeal, for he believed himself to be a divinity incarnate. Such was the man whom Ellenborough trusted to decide how best Great Britain could "help mankind" by deposing the *amirs* of Sind. Leading his men on forced marches across the desert, Napier proceeded in January 1843 to blow up the unresisting Imam Garh fortress. The following month, he stalked his prey to Miani, attacking on the morning of February 17, 1843. Five thousand Sindis were butchered, and he lost 256 of his own troops, but Napier won Sind for Britain and himself, becoming the region's first British governor when it was formally annexed that June.

Before his death in 1839, Ranjit Singh had once remarked, as he viewed a map of India on which British-held territory was colored red, "All will one day become red!" With Sind firmly in hand, the British turned covetous eyes toward the last major area on the map of South Asia that was not yet red: the Punjab. By February 1844 Ellenborough openly referred to Anglo-Sikh relations as a state of "armed truce." Sikh chieftains had one another assassinated as their struggle for succession to Ranjit's throne continued into its fifth year, while the British quietly moved troops and artillery into positions of new-found strength along the Sutlej and northwest frontier. In 1838 there had been 2,500 British troops confronting the Sikh kingdom's Sutlej border; by 1843 that number rose to 14,000; and a year later it mushroomed to 32,000 troops with 68 field cannon, in addition to 100,000 reserves held ready for action at the great British base of Meerut, north of Delhi, just 250 miles east of the Sutlej. The garrison towns of Ferozepur and Ambala sprang up overnight as the Bengal Army's sepoy regiments marched and galloped into positions from which were soon to be launched some of the bloodiest battles in Indian history.

With two armies as large as the Sikh and the British Indian maneuvering for position along several disputed frontiers, there was no dearth of incidents to choose from as incitements of the First Sikh War. British diplomatic intrigue among various Sikh claimants to

Ranjit's power served, moreover, to inflame political rivalries and factional conflicts within the Punjab, so that by December 1845 the cold war of nerves exploded all along the Sutlej. In less than three months the British army conquered Lahore, and nearly twenty thousand Sikhs and sepoys lay dead following the battles of Mudki, Firozshah, and Sobraon. On March 9, 1846, a new treaty was signed, surrendering to the British all the fertile Sikh lands between the rivers Beas and Sutlej, plus strategic mountain country, including Kashmir. Moreover, the Sikhs were obliged to pay the cost of the war and to disband their armies, limiting them in future to a mere fraction of their fighting force. Kashmir was, in turn, given to Gulab Singh, a Dogra Hindu Rajput chief who treacherously deserted his Sikh allies at Sobraon, much the way Mir Jafar had abandoned his grandnephew at Plassey almost a century earlier, weaned away from traditional loyalties by British promises of wealth and power. The crown jewel of India's princely states, whose population was overwhelmingly Muslim, thus came under the autocratic control of the family of Hindu maharajas, who were to retain their nominal power over Kashmir for the ensuing century of British rule.

The Second Sikh War was simply a continuation, after a two-year interlude, of the first. The Sikh spirit of national independence was too powerful to be crushed by a single round of defeats, and British interests in expanding commercial markets, securing greater revenue, and extending their domain to the western border of its "manifest destiny" along the Indus sufficed to ring the bell for round two of the Anglo-Sikh struggle in April 1848. A new governor-general, James Andrew Broun Ramsay, Marquess of Dalhousie (1812–60), had just been sent by Peel's ministry to Calcutta after helping develop Britain's network of railroads as president of the Board of Trade. British India's most talented and energetic governor-general, Dalhousie, saw no reason to delay further the job of painting the entire Punjab red, and he instructed his commander-in-chief, Lord Gough, to do so. The Sikhs fought harder than was anticipated, and at Chilianwala in January 1849 they took so high a toll of British Indian dead (2,357) that Her Majesty's Government felt obliged to send Napier back out with orders to succeed Gough. Before Napier arrived, however, Gough recaptured much of his countrymen's lost faith by his decisive victory in the battle of Gujrat a month later. On March 30,

1849, the entire Punjab—a hundred thousand square miles of India's most fertile soil, destined to become the breadbasket of the British Empire and later the heartland of Pakistan—was taken directly under the Company's Raj. Dalhousie appointed a board of governors, the brothers John and Henry Lawrence and Charles Mansel, to finish the the work of pacifying, and administering the newly annexed domain. This talented troika has received such effusive praise from historians for its administrative labors of "settlement" (in the sense of revenue as well as law and order) that it is perhaps necessary to underscore the fact that their work was uninhibited by any concern for public opinion, whether emanating from the Punjab or Parliament. They functioned as despots, imbued with Platonic ideals of benevolence and restrained by Calvinist convictions and fears, yet despots nonetheless, armed with a staff of fifty-six of the brightest civil and military servants in the company's employ, and some sixteen thousand crack troops, who stood ready to shoot anyone so bold or so foolish as to oppose their decisions. The action-minded, rough-and-ready "Punjab school" of British administration thus set its brand upon India's new frontier, where men were men and things got done swiftly; where officials shot first and asked for evidence later; where higher officials always closed ranks behind subordinates, for that was how you survived in a seething sea of native sycophants, conspirators, and rebels. The myths that grew and flourished around these "titans"[6] inspired future generations of Punjab civilians and governors who, until after World War One, continued to act as though they would always be faced with the same arduous, impossible tasks as those so boldly overcome by the Lawrence brothers, Mansel, Nicholson, Edwardes, Lake, and Lumsden. These were the men who burnt themselves out with work in the districts, riding all day and writing their reports through the night, charting revenue maps, planning canals and digging wells, maintaining jails, building roads, bearing their White Man's Burden without a whisper of complaint or regret.

By 1850, with the Punjab pacified and locked into the Raj that spread from Bengal to the Indus, from Kashmir to Comorin, the British had all but consolidated their grip over the entire subcontinent.

[6] See Philip Woodruff (Mason), "The Titans of the Punjab," his chapter title in *The Founders of Modern India,* vol. 1 of *The Men Who Ruled India* (New York: St. Martin's Press, 1954), pp. 324–43.

What remained to be done, however, was to integrate the many areas of princely power and permanently alienated lands, bequeathed by deposed monarchs to a nobility that had grown effete yet paid no taxes, into the "regulated" system of British rule and revenue collection. There were also regions along both northern frontiers that continued to lure the imperial giant into military ventures. In Burma, these forays were destined to result in renewed victories in 1852 and 1885; in Afghanistan, they led to renewed frustration and tragic losses in 1878–80. With the frame of British conquest completed so early in his reign, Dalhousie could direct his extraordinary energies to the problems of unifying his empire and fostering its modernization—or what he and his cohorts would have called "civilizing the natives." Why else, after all, had so many of Britain's best and brightest young men ventured so far from home to live in so disease-ridden, hot, and dark-skinned a land?

FIFTEEN

UNIFICATION, MODERNIZATION, AND REVOLT

(1848–58)

The policy of the government of India, pursued under Dalhousie's direction from 1848 to 1856, was primarily one of internal unification and modernization. Though no one was quite so optimistic as to imagine that India could be turned into an Asian version of Britain in a matter of years or decades, there was, nonetheless, a spirit of dynamism and self-confidence, typified by the Punjab's regime, that seemed to radiate from the new governor-general's office, reaching out to the most remote backwater districts. The revenue surplus derived from the conquest and pacification of the Punjab was so substantial that, even from its first year, the total cost of the two army corps needed to control that turbulent region, in addition to all expenses of civil government, could be paid from it and still leave a "permanent surplus of fifty lacs (5 million rupees) per annum."[1] In addition, British irrigation technology was applied to the Punjab's fertile soil and soon augmented the region's yield so greatly that the directors of Leadenhall Street assured Dalhousie of their warmest support.

Dalhousie's more significant contribution to the government's coffers, however, came not as a result of costly martial conquest, but rather through the direct annexation of lands still owned by the company's princely "allies," who were stripped one by one of their privileged domains under the spurious legal doctrines of "lapse" and

[1] From the Court of Directors to Dalhousie, October 26, 1853, quoted in Ramsay Muir, *The Making of British India 1756–1858* (Manchester: Manchester University Press, 1923), p. 350.

"paramountcy." It was not, of course, easy for the British, having so recently and firmly implanted their faith in private property and legal contract on Indian soil, to turn a blind eye to both simply for the sake of more revenue; they could always find something "debauched" or "depraved" about the Indian princes, however, that would justify tearing up treaties and stealing their states. The doctrine of lapse was first used against the raja of Satara, direct heir to Shivaji. Indian custom had always treated royally adopted heirs with the same respect (or lack thereof) accorded to first-born natural sons in Britain, but in 1848 Dalhousie decided that "heirs and successors" in the treaty with Satara—and all similar treaties with princely states—applied only to "heirs natural." When the "worthless" raja died that year with only an adopted son as his "successor," the company seized direct control over Satara and its revenue. Sir John Hobhouse, president of the Board of Control, not only agreed with Dalhousie's novel and ingenius interpretation of the law, but clearly anticipated it, suggesting: "I have a very strong opinion that on the death of the present prince without a son, no adoption should be permitted, and this petty principality should be merged in the British Empire."[2]

After 1849, when the doctrine of lapse found its precedent in common law, the doctrine was pursued with increased assurance by the new, steam-rollering Raj. The small states of Jaitpur in central India and Sambalpur in Bengal were taken by lapse in 1849; in 1850 Baghat of the Punjab hill states fell; and in 1852 the Rajput state of Udaipur "lapsed," followed by the central Indian state of Jhansi in 1853. (The rani—who would, in fact, have ruled the latter had Dalhousie recognized her husband's adopted heir—proved four years later that she was braver, bolder, and better than most at fighting on horseback, and she has come to be known as the "Joan of India.") In 1854 the huge and once powerful state of Nagpur, home of over four million Marathas, was also plucked by lapse, its eighty thousand square miles painted red overnight, closing the company's ring fence solidly round Hyderabad to its south. Additional revenues of approximately four million pounds sterling were thus added to the company's

[2] Hobhouse to Dalhousie, quoted in Edward Thompson and G. T. Garratt, *Rise and Fulfilment of British Rule in India* (London: Macmillan, 1934; reprinted in photo-offset by Central Book Depot, Allahabad, 1962), p. 401.

annual resources by the stroke of a lawyer's quill rather than through
the more costly and messy method of war. British self-confidence inspired further financial acquisitions by introducing a somewhat different form of lapse, one that applied to pensioner's titles and their pensions, which had been awarded as compensation for lands long since
seized. The most famous of these deposed leaders of defeated native
powers, was the former *peshwa* of Poona, whose pension of £80,000
was declared to have "lapsed" when he died without any natural heir
in 1853. His adopted son, Nana Sahib, was soon after to become the
focal point of rebellion against the Raj in Cawnpore, on the outskirts
of which the former *peshwa*'s palace stood rotting in disrepair, a
swarming beehive of Maratha conspiracy. The title and pension of
the *nawab* of the Carnatic also lapsed that year, as did that of the raja
of Tanjore two years later. Dalhousie sought to dispense with the
Mughal emperor's title and yearly stipend as well, but on that point
he met with conservative resistance in London, where great royalty,
however weak its actual powers might become, found enough vigorous
support that the throne of old Bahadur Shah II was preserved for
what was soon to be a more wretched fate.

The process of territorial unification so forcefully pursued by
Dalhousie was but the first step in his master plan for the development of an Indian empire in emulation of modern Britain. As Dalhousie explicitly informed the directors in his summary minute to
them, on the eve of his departure, he had labored to harness to India's
bullock-cart civilization "three great engines of social improvement,
which the sagacity and science of recent times had previously given to
the Western nations—I mean Railways, uniform Postage, and the
Electric Telegraph."[3] Though the possibility of building a railway in
South India had been discussed as early as 1832, nothing was done
about actually starting one until 1850, when Dalhousie finally insisted on commencing construction of two experimental lines; one
from Howrah, opposite Calcutta on the Hughly's right bank, to the
coal fields at Raniganj 150 miles inland; the other from Bombay city
east to Thana, only 21 miles inland. The Great Indian Peninsular
Railway opened its Bombay line to public traffic, India's first railway
passenger service, on April 16, 1853. There had been apprehension

[3] Minute by Dalhousie, February 28, 1856, reprinted in Muir, *op.
cit.*, pp. 352–78.

that Hindus might object to traveling on such smoke-filled monsters, but the trains proved most popular from the start and inaugurated a new era of rapid economic development and social change. Dalhousie now proposed a broad general scheme for railway construction throughout India, designed to link the three presidency capitals of Bombay, Madras, and Calcutta, with a trunk line up the Ganges from Calcutta to Delhi and Lahore. He favored continuing private British development of this ambitious network, under the guaranteed-interest agreements negotiated by the board at 4 percent, and he quite presciently anticipated that his railroad plan, if adopted by the court, would "confer upon India the greatest boon she has ever received from the power of England."[4]

Commercially, the railroad was perceived as a way to facilitate distribution and sale of British-made imports to India's still "unopened" interior and to bring back from that interior, on the reverse run, such raw materials as coal and cotton. Manchester mill owners had begun to feel uncomfortably dependent upon the United States for raw cotton and hoped India might offer them a much cheaper, more secure source of that one commodity indispensable to their prosperity. The American Civil War was to prove their worst anxieties well founded, and the raw cotton of Nagpur and Berar, both regions of central India so firmly brought under the Raj's direct control by Dalhousie, provided some of the alternate supply required in Britain during that later decade. In 1854, however, with the outbreak of the Crimean War, Dundee's raw supply of jute, which had hitherto come from southern Russia, was completely cut off, and the newly opened Bengal railway helped transport raw jute as well as coal to Calcutta. Strategically, the railroad was a recognized source of rapid transport for troops from any central base to within easy range of potential trouble spots in the interior or on the frontier; moreover, once the net was completed, it allowed the British substantially to reduce their Indian garrisons (as they did again during World War One) without fear of rebellion. Politically and socioculturally, the railroads inaugurated a revolution, for from their first year of operation, Indians took to them in such numbers and with such eager interest and persistent

[4] *Dalhousie's Diary*, 1853, part i, May 8, 1853, quoted by M. N. Das, *Economic and Social Development of Modern India* (Calcutta: Mukhopadhyay, 1959), pp. 80–81.

delight that there have never since been sufficient seats in all the third-class carriages to accommodate half the passengers waiting to ride. During its first year of operation alone, some half million passengers travelled on the short Bombay line, while in Bengal there were twelve thousand third-class passengers crammed into the few available carriages every week. "The fondness for travelling by the rail has become almost a national passion among the inferior orders," noted the *Indian News* by mid-1855.[5]

India's first electric telegraph line was laid by the ingenius Dr. William O'Shaughnessey in Bengal in 1851, and after testing its effectiveness for almost a year, Dalhousie strongly supported O'Shaughnessey's recommendation that Calcutta, Agra, Lahore, Bombay, and Madras be linked by wire. The telegraph cable had already joined England to France, and there was talk in London of running a submarine cable from England to Bombay, before the directors of the East India Company actually approved funds for rigging some three thousand miles of electric wire across India. Dalhousie personally mapped the route of the first major Indian telegraph, from Calcutta to Banaras, Allahabad, Agra, Ambala, Lahore, and Peshawar. He rightly predicted that the most critical use to which such a line would be put was political, and indeed it may have saved the empire for Britain in 1857, when word of the "mutiny" was flashed 824 miles from Agra to Calcutta along that fragile imperial lifeline before it could be severed by rebel sepoys. Commercially, however, the telegraph proved second in value and importance only to the railroad, providing instant and continuing information and contact to entrepreneurs across the subcontinent, stimulating industry as well as trade. India's first long-distance telegraph message was flashed by O'Shaughnessey in Agra to Dalhousie in Calcutta on March 24, 1854, and two weeks later, when the great Ganges Canal, 525 miles long, was opened at Roorkee a thousand miles northwest of Calcutta, the governor-general received telegraph confirmation the same day. In one year a total of twenty-five hundred miles of telegraph line had been completed, linking India's major cities, including Calcutta and Bombay, sixteen hundred miles apart. Because of this revolution in communication technology, Dalhousie was able to send reinforcements to Turkey a month faster

[5] July 17, 1855, quoted in Das, *op. cit.*, p. 96.

than would have been possible a year earlier, thus helping England win the Crimean War.

The wire's potential value to India's postal service was early recognized, though not immediately taken advantage of. India's first Post Office Act had been passed in 1837, the year Sir Rowland Hill unveiled his revolutionary penny post idea in Great Britain. The powerfully unifying and commercially stimulating concept of carrying any letter any distance within a nation for a penny was not, however, introduced to British India until 1854. Again, it was Dalhousie who had the personal interest and initiative necessary to transform India's inefficient system of private, regional posts into a modern, national miniature of Britain's Post Office Department. The half-*anna* per letter postage revolutionized individual powers of communication throughout India, and the one-*anna* uniform rate for newspapers transformed their opinion-making potential from the local to the provincial, and even national, arena. Within three months the number of letters posted in Calcutta increased by 50 percent, and in June 1855 the postal and telegraph systems were officially linked, providing India with its first swift, efficient, and astonishingly cheap means of nationwide communication. Letters could hereafter be carried much further in India than anywhere in Britain for the equivalent of only three-quarters of a penny. Though most of India's population would long remain illiterate, the poorest village peasant could now afford to send a message to a distant relation, if he could find someone to write it for him and read the response. The penny post thus served not only to unite the subcontinent as nothing else had ever done, but it became a most important new stimulus to learning, literacy, literature, and sociocultural change of every imaginable kind. It was, indeed, more "miraculous" than the railroad or telegraph alone, for it reached beyond cities, towns, and military camps to the villages of rural India, stirring stagnant backwaters that were overgrown by traditional isolation and lifting clouds of dust with the new winds of thought, report, and possibility raised in its fast-moving wake.

Had Dalhousie left India in 1855, the combined impact of his policies of modernization, annexation of the Punjab and Lower Burma, and integration of the princely states might well have earned him the title of Father of Indian Nationalism. For, though what he did was clearly inspired by the company's need for more revenue and by selfish British desires to link India more effectively to imperial interests, the results, nonetheless, were internal unification and the creation of the technological base for a modern South Asian nation. Dalhousie never anticipated, nor would he have desired or condoned, the long-range political implications of the changes he set in motion— no more than he anticipated the violent aftermath of his reign, which was primarily a reaction to the events of 1856, and most of all to the annexation of Oudh. Even if all the British complaints about the "effete," "corrupt," "debauched," "capricious" king of Oudh and his equally "slothful, lazy, and stupid" courtiers were true, the annexation of Oudh in February 1856 was surely the company's and Dalhousie's worst blunder. Never before had the company blatantly ignored a treaty it had honored for more than half a century. The columns of troops that marched north from Cawnpore to take control of Lucknow were, moreover, ranked with men who had been born in the very land they were now invading. No shot was fired. King Wajid Ali came in his mourning robes to place his turban in the hands of Chief Commissioner Sir James Outram, pleading for his homeland and his legal rights, but when that proved futile, he raised no armed resistance within or outside his palace walls, leaving peacefully for Calcutta with his entourage on February 7, to appeal directly to Dalhousie. But the governor-general was on the eve of his departure for home, a sick and lonely man with no more interest in or power to comfort the king whose throne he had usurped than Outram had expressed. Oudh was one of the richest regions of India, the Gangetic plain's heartland, yet its revenue surplus was hardly a tenth of that of the neighboring Northwestern Provinces, and the company could not permit that potential revenue to be frittered away on courtiers and courtesans. Wajid Ali thought that in London he might find a more sympathetic audience, and justice, so he took his case and ministers there, only to confront the same expressionless wall of British faces he had seen in Lucknow and Calcutta. It is hardly surprising, then, that he returned to India to conspire with some of his countrymen, who were among the forty thousand brahmans and kshatriyas of Oudh who formed the backbone of the Sepoy Army of Bengal.

The annexation of Oudh undermined the Bengal Army's faith in the Raj it served, but 1856 brought other equally disturbing shocks to the sepoys, who had prided themselves on being "true to their salt."

Lord Charles John Canning (1812–62), the company's last governor-general and the crown's first viceroy, enacted several unpopular measures during his first year in Calcutta, most distressing of which to sepoys was the General Service Enlistment Act, requiring Indian soldiers to accept service "anywhere," whether in their home province or across "dark waters." The act was designed to assure a steady stream of troops for the growing garrison in Burma, where the Thirty-eighth Regiment had refused orders to march in 1852. Some soldiers said that British missionaries were conspiring with officials to send all brahmans and other high-caste Hindus overseas so that they would become permanently polluted, thus easier targets for conversion to Christianity. The Enlistment Act was but an added straw of discontent, yet it seemed to gain weight when Canning's government passed another act in 1856, permitting Hindu widows, those lowliest of outcasts, to remarry—as though the abolition of *sati* had not been intrusion enough into the household customs and traditions of India's majority community. Not since 1829 had so radical a social reform been legislated, though Calcutta's vernacular press was quick to remind its readers that in 1850 Dalhousie had carried through his council a Caste Disabilities Act, which permitted native converts to Christianity to inherit property. It was beginning to look like a concerted policy, a Christian conspiracy aimed at undermining the very foundations of Hindu orthodoxy, even as Dalhousie's annexations had uprooted so many princely pillars of independent Indian rule.

Early in 1857 the last straw was added, in the unlikely form of a new breechloading Enfield rifle. The British showed great ingenuity in its manufacture, but incredible stupidity in smearing its cartridges with animal fat and lard. To take full advantage of the Enfield rifle's swift firepower, sepoys were instructed by their British officers to bite the tip off the cartridge before inserting it into the open breech. The Enfield could thus be ready to fire in less than half the time required to prepare a muzzle loader, but for Muslims the prohibited taste of pig fat, as for Hindus the mere scent of sacred cow grease, counterbalanced any ordnance advantage the new weapons afforded. Muslims and Hindus alike were convinced that the cartridges were proof of an insidious missionary plot to defile them and force their conversion to Christianity, and the phobia swept through Bengal. The faith that sustained martial discipline, allowing just over 40,000 British troops to command 232,000 sepoys, defending and controlling a land of more than 200 million Indians, was irreparably shattered.

From Barrackpur to Meerut, regiment after reluctant sepoy regiment was lined up, with British-manned artillery aimed at them from high ground on their flanks, and each time the order to "load rifles!" was refused, another unit was erased from the Bengal Army's list, its troops ignominiously stripped of their insignia and left to walk home to their villages—mostly in Oudh—without pay, without the prospect of pension, without a tinge of the pride they had once felt in their uniforms. The sepoys were left with only bitter hatred for the Raj that had seduced, deceived, and finally rejected them. That hatred spread from Calcutta to Meerut and, mixed with older strains of resentment already prevalent at Delhi, Lucknow, and Cawnpore, it grew particularly strong in those centers of deposed Muslim and Maratha nobility. Rumors of water supplies polluted with dead pigs and of cow bones ground into sugar were added to tales of The Cartridge. Strange incidents soon were noted by loyal sepoys, who reported to British officers that "holy men" came to the gates of their cantonment with *chapattis* or a lotus, and that the wheat cake or flower would be passed solemnly from soldier to soldier until every sepoy in sight had touched the talisman, whose substance seemed mystically to unite them, before it was handed back to the silent *mulla* or *fakir,* who would then quickly depart. Was it a signal of some sort? A promise? A warning? No clear evidence of widespread conspiracy has emerged, though Wajid Ali's and Nana Sahib's men ventured far and wide in the first five months of 1857 and may have made contact with members of Bahadur Shah's court in London, Calcutta, or Delhi, as they did with Dost Muhammad's agents from Kabul.

The "mutiny," which was soon to become a full-scale Anglo-Indian war, was ignited at Meerut on Saturday, May 9, 1857. That morning, as the sun blazed down on the cantonment's parade ground, the Meerut Brigade stood at attention watching eighty-five of their sepoy comrades who had refused to load rifles being manacled and shackled in irons by blacksmiths before being led off to prison cells. Those hammer blows of British discipline proved to be Meerut's battle cry of rebellion. The next day all three of the sepoy regiments at the camp rose in revolt while their British officers were at church. They freed the prisoners, killed several officers who tried to stop them, and headed for the capital of the Great Mughal, only thirty miles to the south, shouting *"Chalo Delhi!"* ("Let's go to Delhi!"). The British troops at Meerut were not ordered to pursue the rebel force, and in Delhi itself there were but a handful of Englishmen, in charge of the munitions magazine. Indian soldiers inside Delhi welcomed the mutineers with open gates, and though the British officers blew up their powder magazine, the palace and its populace were "liberated" from British control on May 11, and the reluctant Bahadur Shah II was "restored" to imperial glory. The mutiny had become a revolt, the Great Mughal its rallying cry and symbol. Had the emperor himself been younger or more ambitious, he might have led his new army down the Gangetic plain to Cawnpore and Lucknow, or at least to Agra and perhaps to Lahore, to rally other sepoys and stimulate mass support for his cause. But the old man cared more for miniature painting than he did for the pursuit of power, and his children and courtiers could not subordinate personal ambitions to mobilize themselves or the army that had placed itself at their disposal. So the initial victory, a bold action that dealt a paralyzing blow to British power, was left to dissipate with the heat of May as the sepoys reveled in Delhi's urban pleasures. Nevertheless, as news of the revolt spread east and west, other sepoys took similar action, and the British, who had only a single regiment of English troops between Calcutta and Agra, did virtually nothing to stop or punish the mutineers. In Oudh there was, in fact, nothing less than a national uprising by the end of May. From the courtiers at Lucknow, to *taluqdars* (landed barons) across the plains, to village peasants, many of whom were related to Bengal sepoys, Oudh echoed with cries of revolt. Sir Henry Lawrence (1806–57), one of the company's wisest servants, who had too late been made chief commissioner at Lucknow, herded his flock of Europeans and Indians into the residency, which he had fortified and provisioned with enough ammunition and food to hold out until a relief column arrived that November. At Cawnpore, in early June, General Sir Hugh Wheeler tried in vain to protect some four hundred men, women, and children in the entrenched camp he had been less careful to fortify or provision. They held out for only eighteen days before surrendering to Nana Sahib, who promised them "safe conduct" down river to Allahabad. No sooner did that ill-fated force climb into their waiting boats, however, than treacherous fire rained down at them from the *nana*'s troops on both banks. Only four men escaped to tell the tale.

During the summer of 1857, the British lost control over North India's Gangetic heartland and of portions of the Punjab and Deccan as well, but in the latter regions only for brief interludes of localized rebellion. In Simla and Calcutta, panic was at first widespread among the European communities, but neither seat of the Raj's central government was ever seriously threatened or attacked, nor did any major units of the Bombay or Madras presidency armies rebel. Delhi, Lucknow, and Cawnpore were the only great centers of revolt, each with its court of restoration-minded rebels, two Muslim, one Hindu Maratha, none able or willing to join forces with the others in a unified national struggle. Oudh came closest to presenting a united front of opposition to British imperial rule, but Oudh was hardly India, and the region's former king remained captive in Calcutta. Delhi sought to assert Mughal leadership, but so many conflicting factions emerged around Bahadur Shah that no one proved effective. "Proclamations" were issued in the emperor's name, including one at Azimgarh, near Banaras, dated August 25, 1857, which began: "It is well known to all, that in this age the people of Hindoostan, both Hindoos and Mohammedans, are being ruined under the tyranny and oppression of the infidel and treacherous English. It is therefore the bounden duty of all the wealthy people of India . . . to stake their lives and property for the well being of the public. With the view of effecting this general good, several princes belonging to the royal family of Delhi, have dispersed themselves in the different parts of India."[6]

The impact of the military mutiny was magnified by a simultaneous series of rural rebellions, reflecting, as Professor Stokes has noted, not only "peasant resentment at . . . their loss of land control to 'new men' or urban money-lending castes," but also local grievances against "excessive, differential taxation."[7] Rajput and Jat peasants, who enjoyed strong clan structures, could disown their traditional rural leaders with relative ease. Elsewhere in North India, local

[6] "The Azamgarh Proclamation," reprinted in *1857 in India*, ed. Ainslie T. Embree (Boston: D. C. Heath and Company, 1963), pp. 1–3.
[7] Eric Stokes, "Traditional Resistance Movements and Afro-Asian Nationalism: The Context of the 1857 Mutiny Rebellion in India," *Past and Present*, 48 (August 1970): 110.

"rajas," like Devi Singh in Mathura and Kadam Singh near Meerut, emerged overnight to rally a generally reticent peasantry to rise up against authority. Some districts, however, remained loyal enough to collect revenue, which was transmitted in part to the British Raj, even during the revolt, by local landlords strong enough to keep their peasant vassals under control. While no nationalist leadership (in the modern sense) emerged, the revolt had many subnational Indian currents, the most powerful of which was the reassertion of tradition-bound monarchies. The greatest landed magnates of Oudh, like Man Singh and Rana Beni Madho, had virtually been feudal monarchs before the annexation; now they saw themselves stripped both of their traditional martial lordship and of substantial village revenues. In such regions, the "mutiny" is more accurately seen to have been a "postpacification" revolt. Once aflame with rebellion, the rugged badlands of central India proved most difficult to pacify, and Bundela *thakurs* and other Rajput and Maratha local leaders launched effective guerrilla sorties there well into 1858. But the traditional inability of Indian rajas and *nawabs* to subordinate personal ambitions and jealousies to national goals plagued rebel ranks from the Great Revolt's inception.

The British, on the other hand, never seriously doubted their military capacity to win back the ground they had lost; nor, except for some superannuated generals at Meerut and more than a few merchants and missionaries at Simla and Calcutta, did they ever lose heart or lose faith in their Raj. The only things that almost all of them lost, with the notable exceptions of Canning and John Lawrence (1811–79), were their tempers and their tolerance toward "nigger natives." As the first word of the murder of British men and women reached Lahore, Peshawar, Simla, and Calcutta, a terrible racial ferocity, unknown since the Black Hole tragedy, erupted and inspired British vengeance. Wanton attacks on passive villagers and unarmed Indians, even faithful domestic servants, became common practice in the wake of the mutiny. Virtually all the bridges so painstakingly erected between the British and Indian cultures were destroyed by fear and hatred. Men like John Nicholson, whose merciful behavior toward Sikh prisoners had won him their undying loyalty and high regard, went mad at the news of sepoys slaying English women and seriously proposed "flaying alive" any Indian guilty of such a crime, finding "simple" punishment such as "hanging" intolerable. Captured "mutineers" were, in fact, generally blown away from cannon to which they had been securely strapped. Entire villages were put to the torch for the "crime" of proximity to Cawnpore. Delhi was recaptured by British and Sikh troops from the Punjab. (The Sikhs, still smarting perhaps from the memory of how Bengal sepoys had defeated their armies, became Britain's staunchest allies in the War of 1857–58, after which they remained one of the major sources of military recruitment.) Nicholson, who led the attack into the breech opened by the blown Kashmir gate, was mortally wounded, but by September 20, 1857, Delhi was firmly under British control. Bahadur Shah was exiled to Burma, where he died in 1858, and his sons were murdered in cold blood by a Captain Hodson, who thus took upon himself "the total extinction of a dynasty, the most magnificent that the world had ever seen."[8]

The war continued to rage over northern and central India until late in 1858, with the *rani* of Jhansi and Nana Sahib's artillery expert, Tantia Topi, giving Britain's Generals Sir Colin Campbell and Sir Hugh Rose the most arduous run and combat. On July 8, 1858, when Lord Canning proclaimed peace, Nana Sahib was still uncaptured. The *rani,* however, died on her horse, and Tantia was caught and hung, as were most of the other "rebels." It was the fiercest, bloodiest war ever fought on Indian soil, the last desperate struggle of many an *ancien regime,* united by their fears and their hatred of the foreigner, whose Western Raj had become too powerful to destroy. It was far more than a mutiny, the name that British pride has always preferred, yet much less than a first war of independence, as some Indian nationalists like to call it. It proved to be the final convulsive death gasp not only of the Mughal Empire, independent Oudh, and the Peshwai, but of the Honourable Company's rule in India, which had lasted precisely one century after Plassey.

[8] John William Kaye, *A History of the Sepoy War in India, 1857–1858,* 4th ed. (London: W. H. Allen and Co., 1880), vol. 3, p. 646.

SIXTEEN

CROWN RULE—
A NEW ORDER
(1858–77)

The legacy of the War of 1857–58 was of fundamental and far-reaching import to British India. On August 2, 1858, the British Parliament passed the Government of India Act, transferring "all rights" that the company had hitherto enjoyed on Indian soil directly to the crown. Most members of Parliament had come to believe that it was the cumbersome machinery of a "dual government" at home (the Board of Control and Court of Proprietors) and the three rusty and outmoded presidency armies in India that made so tragic a war possible. Eloquent defenders of John Company's "innocence," including its brilliant secretary of correspondence, John Stuart Mill, sought in vain to convince their countrymen that it was less the company's antediluvian administration that had caused the conflict than it was Dalhousie's excess of zeal in hurrying India to keep pace with Britain's own modernization. But with so many Britons slaughtered in a war that cost England a full year's worth of Indian revenue (£36 million) to win, whatever its actual causes, some scapegoat had to be found. Moreover, the company had virtually ceased to exercise any but negative authority, since even its patronage powers had been stripped by the Charter Act of 1853, which introduced a competitive examination system—the gift of the Chinese to world bureaucracy—for appointments to the Indian Civil Service.

Under the Government of India Act, one of Her Majesty's secretaries of state was vested, through the cabinet, with full power and responsibility for the government and revenues of India, thus inheriting the duties of both court and board. The company was left with seven members of the now dissolved court, who were appointed to the new fifteen-member Council of India. The council, established at Whitehall, advised the secretary of state on all Indian issues, especially those concerning finance. Lord Canning remained as governor-general, but he assumed the title of viceroy as well when Queen Victoria's Proclamation of November 1, 1858, announced to India's "Princes, Chiefs and Peoples" the momentous changes in governance and policy. The first, and in many ways the most important, new policy introduced in the wake of the war was the rejection of Dalhousie's doctrine of lapse and the wooing of India's princes, who were now promised that "all treaties and engagements made with them" would be "scrupulously maintained."[1] This retrograde policy, which would leave more than 560 enclaves of autocratic princely rule dispersed throughout India during the ensuing ninety years of crown rule, reflected British fears that further annexations might only provoke another mutiny. Hereafter, the princes were faithfully assured that they might adopt any heir they wished as long as he vowed loyalty to the crown, whose queen became "empress of India" in 1877. Just as Cornwallis's generous settlement with the *zamindars* of Bengal was viewed as the major reason those Calcutta-based barons remained loyal and helped keep Bengal passive during the conflagration, so would this new policy be seen as a continuing source of aid and support to British rule throughout the forthcoming era of India's nationalist agitation. The princely states soon came to be viewed as comforting loyalist "breakwaters" in a turbulent sea of political troubles, whose rising tide might otherwise have washed British rule from the subcontinent long before 1947.

Similar fears concerning "native" sensitivities to social and religious changes of any sort inaugurated an era of socioreligious *laissez-faire* that put an end to further reform legislation in India for more than three decades. Couched in high-sounding phrases that made this policy of indifference to the plight of women, untouchables, and exploited children sound as noble as Victoria, the proclamation declared "it to be our royal will and pleasure that none be in anywise favoured, none molested or disquieted, by reason of their religious faith or observances, but that all shall alike enjoy the equal and impartial pro-

---

[1] Queen Victoria's proclamation is reprinted in C. H. Philips, *The Evolution of India and Pakistan, 1858 to 1947. Select Documents* (London: Oxford University Press, 1962), pp. 10–11.

tection of the law; and we do strictly charge and enjoin all those who
may be in authority under us that they abstain from all interference
with the religious belief or worship of any of our subjects on pain of
our highest displeasure." Missionaries were instructed to cut back
their attempts at conversion, and from that time on, most missionaries
in India preoccupied themselves with educational labors, especially in
the analysis, translation, and teaching of various Indian languages.
The Christian impulse and proselytizing tone that had inspired Sir
Charles Wood's Education Despatch of July 1854—which led to the
birth not only of India's three presidency universities in 1857 (a
singular tribute to British self-confidence), but also to many mis-
sionary-run vernacular schools funded by government grants-in-aid—
was subdued and curtailed. The scheme that Wood hatched had initially been inspired by his missionary friends, especially John Marshman, and was designed at least as much for the salvation of souls and
the improvement of manners and morals as it was to teach young Indians reading or mathematics. The war and its aftermath cast chilling
shadows over such purposefully planted seed, and with the government of India pressed to pay back the entire cost of British military
operations from its revenues, educational expenses were the first to be
cut.

The one institution in which dramatic positive reforms were
made after 1858 was the army. The company's presidency armies
were reorganized as a martially coordinated royal machine designed
to prevent any recurrence of rebellion. The ratio of Indian to British
troops was reduced to between two and three to one, total manpower
in 1863 being 140,000 Indian to 65,000 British soldiers. The British
were, moreover, given exclusive control over artillery and other
"scientific branches" of the service, so that in the event of any threat
of mutiny, they could immediately use such weapons to overpower it.
Indian regiments were recruited only from those natives whose loyalties had been tested in the war's flames, which eliminated Bengalis,
Biharis, and Marathas from the list of the eligible, but placed Sikhs,
Gurkhas, Pathans, and some Rajputs in the front ranks. The British
soon developed their spurious theories about "martial races" and
"nonmartial races," based for the most part upon their experience
with "loyal" and "disloyal" troops during the mutiny. Not only were
there "martial races," but also—and here Hinduism supported British

prejudice—"martial castes," least of which was the Bengali or Maharashtrian brahman, whose rebel leaders had fought so bitterly. Within each regiment, troops were to be mixed "promiscuously," and neighboring regiments were to be recruited from distant regions of India, ideally from peoples who spoke different languages or were of different castes and communities.

The land settlement of what is now the state of Uttar Pradesh (United Provinces), but was then called the Northwestern Provinces and Oudh, was the cause of much official controversy in the wake of the war. Canning had initially confiscated all *taluqdar* property, but he soon came to accept the arguments of his own advisers, as well as those of Her Majesty's Government, that it would perhaps be wiser to set up the *taluqdars* as aristocratic pillars of crown rule, returning all of their lands to them in perpetuity. The Cornwallis formula was thus applied to the neighboring provinces, and in October 1859 Canning conferred special property contracts (*sanads*) upon the assembled *taluqdars* at a glittering *durbar* (court) held in Lucknow. *Taluqdars* were thus added to princes and *zamindars* as staunch allies of British rule, and in order to keep their baronial status undivided, a primogeniture clause of inheritance was appended to their *sanads* before Canning left India in 1862. When most *taluqdars* subsequently fell so deeply into debt as to be threatened with possible loss of their estates, the government bailed them out with loans and elaborate legislative schemes designed to keep their bankrupt estates from collapsing. In 1861 the *taluqdars* of Oudh organized themselves into the British Indian Association, a loyalist, conservative group whose primary mission was to defend and protect the special privileges of its wealthy members and to support the Crown Raj it served whenever political questions aroused otherwise apathetic public opinion. The *taluqdars* proved particularly useful to British officials in serving as local magistrates over their regions, thereby resuming the powers of law and order that they had earlier held under the *nawab*. Bengali *zamindars* were similarly restored to their Mughal magisterial roles, as were a number of powerful landed gentry in the Northwestern Provinces and Sikh *sirdars* (nobles) in the Punjab. Dalhousie's egalitarian policy of treating all natives in much the same manner was now reversed with such shrewd effect that during the most turbulent eras of subsequent nationalist agitation, the British could always call upon their aristocratic lobby in every region of the subcontinent to defend the imperial cause with vigor and conviction. Even in the newly created Central Provinces (now Madhya Pradesh), formerly mostly Nagpur, a petty noble class of Mughal revenue collectors called *malguzars* was exalted to "permanent" (the 1862 settlement was initially made for thirty years) landlord and magisterial status as one of Canning's final acts.

Yet if, in the wooing of princes and lesser gentry, the postmutiny policy of crown rule may be viewed as a reactionary reversal of egalitarian settlements made in the latter years of Company Raj, Dalhousie's measures of technological modernization were continued with accelerated enthusiasm. The telegraph had clearly proved its worth throughout the war, its humming silver wires helping to "hang" many a "mutineer." The railroad and the improved metal-surfaced roads (like the Grand Trunk Road that linked Calcutta to Delhi before the war and was subsequently continued on to Peshawar) were advanced as swiftly as possible, for it was only with the extra insurance of knowing they could maintain rapid lines of communication that the British could so confidently reduce the overall size of their army in India. The postwar era proved, moreover, to be a period of unprecedented European capital investment and commercial agricultural development in British India. Victory convinced private enterpreneurs throughout Britain and Western Europe that the British Empire in India was destined to endure, capable of countering effectively any internal challenge.

Railroad construction advanced so rapidly that the 432 miles of track laid by 1859 had grown to more than 5,000 a decade later, and by the end of the century there were 25,000 miles of steel uniting India. All of the tracks, locomotives, cars, and railroad equipment were shipped directly from Britain, where they were produced by private Midlands steel companies whose market was, thanks to India, one of constant expansion and guaranteed profit. By 1869 there were no fewer than fifty coal mines operating at Raniganj, the fueling center of India's railways. The opening of the Suez Canal and the simultaneous shift from sail to steam transport on the sea route to India in 1869 temporarily depressed Raniganj sales, however, for by reducing the average voyage between Bristol and Bombay from three months to less than three weeks, it became cheaper to ship coal from Newcastle to West India than around Cape Comorin from Bengal. In 1870 a rail link between Calcutta and Bombay was officially opened through Allahabad. The broad-gauge Bengal-Nagpur Railway, which significantly diminished the mileage across central India, was not opened until 1888, however, and India could finally attain self-sufficiency in power, though not, as yet, in the industrial capacity to replace any of its imported machinery. By 1880 there were not only nine thousand miles of railroad, but twenty thousand miles of telegraph wire crisscrossing India. British industry invested some £150 million in the twenty-two years of India's postmutiny boom in public works production, much of which was lavished on constructing irrigation canals that increased agricultural yields on eight million acres of land in the United Provinces,[2] the Punjab, and Sind.

Perhaps the most pervasive impact of the war, however, was the psychological wall of racial distrust it raised between Britain's white and India's native "black" populations. The bitter legacy of countless atrocities committed in 1857 remained to poison the memories of Englishmen and Indians alike decades after the dust of battle settled. Tales of mutiny murders, of mutilated bodies stuffed into Cawnpore's well, of "loyal" sepoys and servants who slashed the throats of their sleeping officers or masters, fed the imaginations of generations of young Englishmen as they stepped forward to take up their White Man's Burdens: "And bid the sickness cease; / And when your goal is nearest— / The end of others sought— / Watch sloth and heathen Folly / Bring all your hope to nought." Kipling was the poet of their love-hate relationship with India and all the ambivalence it evoked in the British mind and heart. Obliged by the tempest of history to share the same insular cave, Britain's Prospero tolerated his native Caliban with all the loathing and contempt that dependence inspires in hearts keenly conscious of their own merit and worth.

The Indians, on the other hand, were haunted by similar memories of murder, rape, arson, and the brutal execution of their comrades. The wall that insulated white *sahib* society from the natives suddenly loomed impervious to any but a handful of princes and landed gentry, and even they complained of feeling mistrusted, eternally suspect, outsiders. New towns and suburbs, called civil lines and

---

[2] See Elizabeth Whitcombe, *Agrarian Conditions in Northern India,* vol. 1, *The United Provinces under British Rule 1860–1900* (Berkeley and Los Angeles: University of California Press, 1972).

camps, were now built for British officials and their wives, with grand "bungalows" on wide, tree-lined streets and spacious roads through which a regiment of troops could gallop swiftly, if needed, to put down any "trouble." The postmutiny separation of "races" brought a boom to the overseas market for British brides, now that fear of the treachery of Indian "housekeepers" made most young servants of the crown prefer the comfort and security of a British spouse to the availability of a native mistress. Closer contact with home, following the opening of the Suez Canal, further encouraged them to bring their families to India and to rear their children there, since it was now possible for crown servants to spend their holidays in London, via the Peninsular and Orient steamship, rather than on an up-the-country hunt or in one of India's cities. The British Club soon emerged as the insular nucleus of European society, and it was to retain its mid-Victorian mannerisms and peculiar morality on Indian soil long after Edwardian attitudes and fresh ideas liberated London from the narrow formalism and rigidity of such inhibiting mores. Distrust, frustration, and fear fed upon one another throughout the era of crown rule, as the liberal hopes and idealist dreams of Bentinck and Macaulay melted in the caldron of a war whose aftermath brought progress without civilization, confirming all of India's traditional disunities and converting the subcontinent into a land held hostage by the merciless martial force of the world's foremost power.

Though the personnel of the government of India at first remained unchanged with the transfer of power from company to crown, the administration at Calcutta was modernized somewhat in 1861, when the Indian Councils Act converted the Viceroy's Executive Council into a mini-cabinet, with each of its five members placed over a department—home, revenue, finance, military, and law. The commander-in-chief continued to attend council meetings, when he wished, as an extraordinary member. The viceroy himself always handled the Foreign Department, the bulk of whose business dealt with princely states and relations with neighboring Afghanistan, Burma, Persia, and Nepal. Since 1854 several "additional members" had been appointed by the governor-general to meet with his council for "legislative purposes," and by the Act of 1861 their number was increased to not fewer than six nor more than twelve, at least half of whom were to be "nonofficials." Here was the thin end of a wedge of

Indian participation in the highest council of the land, but until the next Councils Act in 1892, appointments were entirely limited to loyalist conservative gentry, hand-picked by the viceroy for their political docility and total dependability. They were the first token Indians of the Raj, nonetheless, and they could always be pointed to as evidence of British sincerity in seeking to sound "native opinion" on matters of legislative concern. Calcutta's annual Legislative Council meetings were "open" to a limited "public" audience as well, and became a faint symbol of Britain's continuing interest in educating its native dependency in the higher forms of its own political genius. As far as Provincial and district governments were concerned, however, administration in the postmutiny era became even more exclusively British, more despotic or paternalistic (depending on the character of individual officials) than it had previously been. No matter how well Indians spoke or wrote English, they were no longer trusted by British civil servants, who spent more time touring their districts, reporting on local conditions, and checking subordinates than had been normal administrative practice before the mutiny. The Punjab system was generally judged to be best, because the Punjab had not only held firm, but had provided troops to help suppress the revolt. John Lawrence, the dour first chief commissioner of that province, was, in fact, returned to India as viceroy (1864–69) shortly after Canning left, so that he might more firmly entrench the system he had directed for the Punjab throughout the entire empire.

The tough new paternal order of this era of early crown rule, combined with heavy investments in privately built public works and modern communications, linked India more tightly to British industrial needs and markets and accelerated the commercial agricultural revolution that had tenuously started in the prewar period. Except for indigo, opium, and cotton, there had been no widespread growth of commodities for export from India prior to the mid-nineteenth century. The influx of private European merchants, traders, and planters following the removal of the company's monopoly privileges in 1833, however, led to the development of new plantation industries in tea and coffee, which first blossomed in the wake of the war. Europeans moved in early to take control of much of the indigo industry in Bengal and Bihar, and their exploitation and cruel mistreatment of "coolie" labor provoked a "Blue Mutiny" in 1859–60, the first labor strike by Indians against British management. It was settled only after bloodshed impelled the government to appoint an Indigo Commission, which investigated and reported in support of the Indian peasants' claims and demands. Tea plantations were developed in earnest after passage of the Assam Clearance Act in 1854, which gave away up to three thousand acres of prime tea land to any European planter who promised to cultivate tea for export. Britain's taste for tea continued to grow, while the supply of Chinese tea began to be threatened by the turmoil of internal rebellion and anti-Western agitation in China. In 1850 there had been only one large tea plantation in British India, producing 200,000 pounds of tea annually, but by 1871 the number of estates was 295, with a total yield of some 6.25 million pounds. Tea cultivation spread across Assam to the mountain regions of the Punjab around Kangra; it also increased greatly in the southern Nilgiri hill country near Mysore, and after 1873 moved to within sight of Everest in Darjeeling, which became one of the affluent British colony's favorite summer "hill stations." The cultivation of coffee had been introduced to India in the seventeenth century, but the first European plantations were started in Mysore during the 1830s, and they multiplied tenfold in the two decades after the war, until disease blighted that industry after 1885. By that date, however, over a quarter million acres of South India were planted in coffee, producing more than thirty million pounds of beans a year.

Though cotton spinning and weaving had always been India's major home industry and Indian cotton had, at least since the dawn of the Christian era, clothed countless bodies in Africa and other parts of Asia, the development of European-style cotton factories in India was purely a post-1850 phenomenon. By 1861 there were eight mills in Bombay with almost two hundred thousand spindles and over twenty-five hundred looms. The Civil War in the United States then intervened to cut off Britain's major source of raw cotton, raising the price of raw Indian cotton so rapidly as to dampen further growth of cotton mills in India itself until the crash, which followed that boom five years later. India's premier industry recovered strength slowly, but by 1900 there were about two hundred mills (mostly in Bombay, though Ahmedabad had also begun to emerge as a great cotton center), with over five million spindles, producing close to five hundred million pounds of cotton yarn, about half of which was sold overseas.

The production of mill cloth continued, however, to meet less than 10 percent of India's home demand, most of which was satisfied by the British imports that had all but destroyed the rich varieties of Indian handicrafts and cottage industry. In the era of crown rule, Lancashire received even greater support for its mill products, which comprised between one-half and two-thirds of India's annual imports during the last quarter of the nineteenth century. One result of India's dependent relationship to Britain was the inequitable competition in cotton manufactures, with Britain's Government of India using its powers to assist Lancashire against Bombay by reducing or removing all import taxes and imposing heavy excise duties, rather than helping the infant native industry to grow under the shelter of an officially supportive tax structure. The most blatant example of such imperial favoritism occurred in 1879, when Viceroy Lytton actually overruled his entire council to accommodate Lancashire's lobby by removing all import duties on British-made cotton, despite India's desperate need for more revenue in a year of widespread famine and tragic loss of life throughout Maharashtra. Just as cotton remained the main textile industry of Bombay and much of Western India, jute and its manufacturing became the preoccupation of Calcutta and most of Eastern Bengal. The Crimean War gave impetus to the expansion of jute planting for export to Scotland, where Dundee mills would otherwise have been faced with disaster after their supply of Russian hemp was cut off. In 1854 the first steam-driven jute mill in India was started by a Scotsman at Serampore, and after a decade of little growth, Bengal's jute industry made such rapid strides that by 1882 there were twenty large mills lining the Hughli River, employing some twenty thousand workers.

By 1877 Britain's new order had taken such firm hold over the subcontinent it had almost lost two decades earlier that Prime Minister Disraeli was able to convince Queen Victoria of the wisdom of adding "empress of India" to her regalia. Lord Lytton eagerly proposed a grand *durbar* of princes to celebrate the occasion, reiterating what had been the keystone of crown policy since Victoria's proclamation, that "if we have with us the Princes, we shall have with us the people."[3] It must have been a comforting assumption, but the

[3] Lytton to the queen, May 4, 1876, quoted in S. Gopal, *British Policy in India, 1858–1905* (Cambridge: Cambridge University Press, 1965), p. 114.

pace of change set in motion by the harnessing of India's modernization to industrial Britain was too swift to allow so simplistic a doctrine
to hold true much longer. In fact it was during Lytton's viceroyalty,
from 1876 to 1880, that several new forces of enduring protest began
to make themselves felt.

SEVENTEEN

INDIAN NATIONALISM— THE FIRST MOVEMENT

(1885–1905)

Indian nationalism has always been a theme scored with religious, class, caste, and regional variations. The emergence of national consciousness among Indians during the nineteenth century was primarily the product of responses, both negative and positive, to the consolidation of British power. All Indians, whatever their religious, caste, or regional origins may have been, were immediately conscious of the "foreign" character of the white Christian *sahibs* who ruled their land, if they had any direct contact with these new rulers at all. Though many foreign invaders had conquered portions of India in the past, with the Aryans and Mughals extending their conquests over virtually the entire subcontinent, the British were the only power to control their Raj from an imperial base thousands of miles away. The influx of missionaries, the funding of English education, the opening of India to private trade, and the continuing process of British unification and modernization, served only to intensify Indian perceptions of their "native" differences, cultural, socioeconomic, and political, from the officials who ran the Company Raj. Some Indians, like Ram Mohun Roy, started newspapers and cultural societies to assert their "traditional" distinction through "modernist" institutions. Others, more outraged by British "treachery," like the Muslim *taluqdars* of Oudh, took up arms in revolt, as did so many sepoys, but the bitterness of war only added racial distance to the already gaping gulf between white *sahib* and native societies. Nonetheless, it was from British examples of political consolidation, technological integration, administrative unity, and the sublimation of personal interest and identity to the impersonal laws and "higher" needs of national purpose that Indians learned their positive lessons in modern national consciousness. The British Raj thus provided whip as well as carrot to India's nascent nationalism, whose roots went as deep as the Indus civilization and beyond, but whose modern emergence was part of the searing aftermath of the War of 1857–58.

Perhaps the clearest indicator of the ambivalent role played by British rule in the growth of India's first nationalist movement was that all of the major leaders of that movement had received some English education. Most of this elite cadre of young men studied law and were teachers, journalists, or authors of belles-lettres. At first, they were practically all high caste, middle-class intellectuals from Bengal, Bombay, and Madras, living primarily in Calcutta, Bombay City, Madras, and Poona. Since at least the 1830s, Calcutta set the pace of Indian change, and what was argued or thought about in the homes, clubs, inns, and courts of its intellectual "gentle people" (*bhadralok*) soon became food for thought, concern, and action elsewhere in India. Before 1876, however, Bengali public activity was confined to socioreligious reform and the lobbying of loyalist landholder associations, such as the British Indian Association, founded by *zamindars* in 1854 for the preservation of their privileged status. Talented young *bhadralok* took advantage of whatever educational opportunities the British presented to ally themselves to the Raj and seek its perquisites of money and position. The introduction of the Indian Civil Service (ICS) examination system after 1854 offered the promise of anything up to the position of chief commissioner to any boy born in British India, and though only one Indian so much as passed the coveted examination during the first fifteen years of that "open door" to officialdom, it still in theory assured all native sons of the soil an equal opportunity to rise to the top. When the brilliant young Bengali brahman, Surendranath Banerjea (1848–1926), ventured to London in 1869 after completing his B.A. in English literature at Calcutta and scored higher on the ICS examination than did most of his British competitors, it almost seemed as if the dream of equal opportunity would come true. For Banerjea was not only rich enough to have been able to pay his fare to London (no examinations were held in India until 1921), but he proved intelligent enough to answer questions in Latin, English literature, and moral philosophy, strong enough to keep his seat on a horse galloping over the requisite hurdles, and personable enough to shine in English as well as Bengali society. The British bureaucracy managed, however, to disqualify him for having "lied" about his age (Indians traditionally counted the nine months in the womb as part of life), though the Queen's Bench held in his favor upon appeal. For three years Banerjea was permitted to serve in the post he won at so high a cost, but was then summarily dismissed from the service for a minor infraction of rules. Once again Banerjea traveled to London to plead his case, but this time he lost and, as a man marked by the "disgrace" of official discharge, was not permitted to enter the bar. "I had suffered because I was an Indian," he wrote in painful reflection upon his plight. "The personal wrong done to me was an illustration of the helpless impotency of our people."[1] Surendranath, who subsequently won the sobriquet of "Surrender Not" for his leadership in the struggle against the first partition of Bengal, returned to Calcutta to work as a teacher-journalist, founding Bengal's first nationalist political organization, the Indian Association, in 1876, the first year of Lord Lytton's viceroyalty.

Several years earlier in 1870, another provincial political organization had been formed in Bombay, the Poona *Sarvajanik Sabha* ("All People's Association," hereafter called PSS), founded by the ingenius Chitpavin brahman, Mahadev Govind Ranade (1842–1901), who graduated with highest honors from the first class of Bombay University in 1862, went on to receive his law degree in 1866, and was appointed a subordinate judge in Poona in 1871. Despite his precocity and erudition and because of his Indian birth and public-minded activities, Ranade was kept waiting for almost a quarter of a century after his subordinate judicial appointment until he was finally elevated to Bombay's High Court. Even then he continued to be distrusted and treated as an outsider by Britain's small-minded governors, who could never forget that he was a member of the former *peshwa*'s brahmanic community and one of the guiding lights behind the formation of the Indian National Congress in 1885. Ranade sought to help his countrymen win their freedom by making full and effective use of all the political institutions and self-governing ideals

[1] Quoted in J. H. Broomfield, *Elite Conflict in a Plural Society: Twentieth-Century Bengal* (Berkeley and Los Angeles: University of California Press, 1968), p. 66.

embodied in British society and English literature and law. He always armed himself with facts and drafted his appeals for social, economic, and political reform with the brilliance and persuasive force of the great legal mind he was, inspiring generations of equally moderate disciples to take the same essentially emulative approach in presenting India's case for greater national independence and more equality of opportunity to British authorities.

Gopal Krishna Gokhale (1866–1915), Ranade's foremost disciple, sought with the same forbearance and sweet reasonableness of his mentor to win the hearts and minds of British governors, viceroys, secretaries of state, and other cabinet members on behalf of India's many appeals for political justice and economic equity. The Ranade-Gokhale school of nationalism also insisted upon the need for Indians to reform their own social and religious ideas and resolve internal conflicts as a prerequisite to political independence. They feared that unless such "modernization" occurred within Indian society, political freedom would mean little more than a return to India's eighteenth-century regional and religious wars and the continuing subjugation of women to men and of untouchables to the higher castes. Ranade himself was one of the leading Hindu reformers, and in 1869 he helped found the liberal *Prarthana Samaj* ("Prayer Society") in Bombay, western India's version of Calcutta's *Brahmo Samaj*. Ranade worked throughout his life to ameliorate the plight of Hindu widows through education and legal reform. Though his position on the bench prevented him from formally joining the Indian National Congress, he founded and was to remain the leading spirit within India's National Social Conference, started in 1887.

Poona also emerged as the home of India's first revolutionary cultural nationalist leadership, drawn from the same English-educated elite of Chitpavin brahman birth to which Ranade and Gokhale belonged. The dream of reviving Maratha power, of reestablishing the *svaraj* ("self-rule") for which Shivaji fought and struggled throughout his mature life, never left the hearts of many of Poona's penurious brahmans, one of whom was Vasudeo Balwant Phadke (1845–83), who learned just enough English to scratch out his livelihood as a petty clerk in the government's Military Accounts Department for fifteen years, then assumed the title of Minister to Shivaji II in 1879 and rode off into the hills of Maharashtra to raise an army against the crown in order to restore *svaraj* to his nation. He survived for al-

most four years, leading his British hunters in a wild but hopeless race
for freedom. Another reformer, Maharashtra's schoolteacher-poet
Vishnu Hari Chiplunkar (1850–82), was inspired in part by Phadke's
act of desperation, and he, too, decided to "kick off my chains,"
quitting his government job the year Phadke did but opening a private school in Poona in 1880, rather than galloping off to the hills.
Chiplunkar's courage and inspiring Marathi poetry and political essays attracted many young comrades, the most famous of whom was
Balwantrao Gangadhar Tilak (1856–1920), who came to be hailed
throughout India as the *Lokamanya* ("Revered by the People") and
was India's leading cultural revolutionary nationalist before Gandhi.

In Bombay City the greatest early nationalist leaders emerged
from the small but wealthy and elite mercantile community of Parsis.
These followers of the Zoroastrian faith had fled Persia in the seventh
century rather than convert to Islam. After centuries of obscurity and
penury, they established their community in Bombay, where they
flourished in mercantile collaboration with Portuguese and British
traders, emerging in the late eighteenth century as one of India's
wealthiest minorities. The "Grand Old Man" of Indian nationalism
was Parsi Dadabhai Naoroji (1825–1917), who established the first
Indian business firm in London and Liverpool in 1855. Author of Indian nationalism's major economic battle cry concerning the constant
"drain" of resources and wealth from India by England, Dadabhai
was also the first Indian to be elected to Britain's Parliament, and he
was to be chosen three times to serve as president of the Indian National Congress. Pherozeshah Mehta (1845–1915), the "Uncrowned
King of Bombay," also a Parsi, established his title to fame and national power through his dexterity at the bar, lured into the mainstream of public protest by Lord Lytton's Vernacular Press Act of
1878, which was passed in a matter of hours to silence opposition to
the government's unpopular Afghan War, but which served only to
strengthen the growing spirit of Indian protest and national awareness. Parsi leaders in Bombay, thanks to their combined mercantile
wealth and political power, emerged as what C. A. Bayly has called
"urban magnates" (*rais*), reflecting the pattern of urban control in
Allahabad from 1880 to 1920.[2] Leading Allahabad *rais,* however, like

[2] C. A. Bayly, "Local Control in Indian Towns—The case of Allahabad 1880–1920," *Modern Asian Studies* 5, 4 (1971): 289–311.

the great bankers Jagat Narayan and Ram Charan Das, were mostly Punjabi in origin, their ancestors having left the Karnal-Panipat region between 1760 and 1830 to follow the river trade east.

The Second Afghan War, like the first, was generated by a mixture of Russophobia and imperial megalomania. Canning's northwest frontier policy of "butcher and bolt" had sufficed to keep the Pathans relatively quiescent, and after Dost Muhammad died in 1863, Lawrence wisely refrained from seeking to name his successor, leaving the old *amir*'s sixteen sons to fight their own battles until 1868, when Sher Ali emerged victorious. Calcutta then recognized the new master of Kabul and subsidized him to help retain the power he had won. Sher Ali, however, sought more generous assistance, including arms and gold, which the government of India was most reluctant to export. Russia's advance into Turkistan and Samarkand sufficiently concerned both Disraeli and Salisbury by 1874 so that, when the Torys came to power, they pressed the government of India to act more aggressively with Kabul. Lytton was sent to Calcutta with the mission of bending Sher Ali to his will and making Afghanistan safe for British merchants. The *amir* was sufficiently Afghan, nonetheless, to refuse to admit Lytton's emissary to his hermit kingdom, provoking the viceroy to characterize Afghanistan as "an earthen pipkin between two metal pots." By 1878, when he felt ready to crush that pipkin, Lytton launched an invasion over the Khyber and Bolan passes. Sher Ali fled his capital, dying in exile the following year. The British Army occupied Kabul and Kandahar, but once in power, it faced the same unrelenting opposition and hatred that had plagued the first army of the Indus. Lytton's Macnaghten was Major Louis Cavagnari, a headstrong, ambitious, impetuous, "forward school" frontiersman who became political resident at Kabul and dominated his puppet *amir,* Yakub Khan, as he did his viceroy. Cavagnari sent Lytton glowing reports of the "success" of his policy, the "restoration of calm" that he witnessed, the growing "support" he was daily receiving from the average man in the bazaar, and the like. On September 2, 1879, he cabled his chief two words, "All well," and on September 3, Cavagnari, his staff, and his entire escort were assassinated inside their residency in Kabul. The British Army returned to retaliate, proving to Afghans that Englishmen could be as brutal and barbaric as Pathans. Stories of atrocities committed in Kabul sickened London,

as word of the endless financial drain imposed by this futile war
shook the self-assurance of Lytton, who realized too late that the
pipkin was less fragile than his pot. The Second Afghan War brought
Disraeli's government down in the spring of 1880, and Lytton was
replaced by Lord Ripon (1827–1909), a Liberal convert to Catholicism who was to become India's most popular viceroy during his brief
tenure in that high office from 1880 to 1884.

As Gladstone's political reflection in India, Ripon immediately
repealed Lytton's repressive Press Act and put an end to Simla's ad-
venturist forward policy on the northwest frontier. He sought next to
introduce some measure of local self-government, and by his resolution on that subject of May 18, 1882, he inaugurated what might have
been a new era of Anglo-Indian cooperation at all levels of administration were it not for the fact that his liberal goals were sabotaged by
a stiff-minded bureaucracy that failed to implement them effectively.
Ripon's idea was to create municipal and district government boards,
at least two-thirds of whose members would be elected local nonofficials, who best appreciated the needs and problems of their own "par-
ish," thus making good use of the "rapidly growing . . . intelligent
class of public spirited men who it is not only bad policy, but sheer
waste of power to fail to utilise."[3] The viceroy had hoped in this way
to restore Indian confidence, not only in British rule and its virtues,
but in their own capacity to guide and govern themselves. The district
officer and members of the official hierarchy were to serve only as
sounding boards and purveyors of information when required. In
practice, however, officialdom took charge of virtually all local boards
and managed to direct, bully, and control native members into agreeing to whatever the local ICS desired. Nevertheless, these local
boards, rural and urban, were to be breeding grounds of Indian nationalist leadership in the decades and century ahead.

It soon became painfully clear to more and more middle-class
Indians, however, that, no matter how well intentioned or powerful
individual Englishmen like Ripon might be, the system they served
was fundamentally unresponsive and hostile to many basic Indian
needs, aspirations, and desires; it was cold, imperious, paternal, and
foreign. The last of these lessons was more bitterly learned after

[3] Quoted from Ripon's resolution in Hugh Tinker, *The Foundations
of Local Self-Government in India, Pakistan, and Burma* (London:
Athlone Press, 1954), p. 44.

Ripon's new law member, Sir Courtney Ilbert, introduced his Bill on Criminal Jurisdiction in February 1883. The bill was innocuously designed to remove racial restrictions enacted in 1872 that made it impossible for Indians in the British Judicial Service to try cases involving Europeans, even for minor offenses. Ripon agreed with his law member that there was no place for such bias in any British legal system, and no member of the Executive Council in Calcutta anticipated any protest against a measure whose object was "simply the effectual and impartial administration of justice," as Sir Courtney put it when he presented his bill to the Legislative Council. Calcutta's nonofficial European community thought otherwise. The Ilbert bill raised so frenzied a storm of protest that it could only be passed in a thoroughly emasculated form, permitting Europeans tried in a court presided over by a native judge to demand a jury at least half European in complexion. British-born "Anglo-Indian" associations (in 1881 the British-born population in India was 100,000) were organized overnight in Calcutta to protest Ilbert's "horrendous measure," attacking the viceroy, running newspaper ads, and arguing in the plainest and boldest English that "nigger natives" were not "the peers or equals" of Englishmen. Never had British public opinion in India raised its voice to so united, outspoken, and vile a manner, crucifying Ripon personally, subjecting him to such unrelenting pressure that he was forced to capitulate. The government of India thus reversed itself on a matter of principle as well as practical administrative significance. This lesson in the tactics and impact of public pressure was a unique eye-opener to Bengal's intelligentsia. Shouting, ranting, signing petitions, and printing newspaper notices had achieved something that most of the Bengal Army, the Great Mughal, Burma, and the *peshwas* all failed to accomplish—defeating the government of India. If but a handful of determined Europeans could do so much, what could a nation full of equally determined Indians not accomplish—if only they could forget caste, class, and regional differences for once and join together in a united movement. If only, as Allan Octavian Hume (1829–1912), "Father" of the Indian National Congress, wrote to Calcutta University's graduates in 1883, if only fifty men could be found "with sufficient power of self-sacrifice, sufficient love for and pride in their country, sufficient genuine and unselfish heartfelt patriotism to take the initiative," India could be reborn.

In 1885 not fifty, but seventy-three men were found ready to serve as representatives from every province of British India in the first annual meeting of the Indian National Congress, convened at Bombay City on December 28, under the presidency of Womesh C. Bonnerjee, a Calcutta barrister. Ranade had been instrumental in inviting this historic body to western India, and only because of an eleventh-hour outbreak of cholera in Poona was the venue shifted to Bombay. Surendranath Banerjea did not attend the meeting, having earlier scheduled a national conference of his own Indian Association in Calcutta during the Christmas holiday season, a meeting that he doubtless felt should have sufficed to serve as a clarion call to the nation. The Bombay meeting, however, was more prestigious and broadly based, and the Congress became, in fact, the organizational vehicle for India's first great nationalist movement, embodying the dreams and aspirations of "New India."[4] Its membership from the beginning was overwhelmingly Hindu, however, and though a substantial representation of Parsis and Jains attended the first meeting, only two Muslims were present. Though eighty-three of the six hundred delegates attending the third annual session of Congress in Madras were Muslim, India's largest minority community never really came to feel "at home" inside Congress and would, therefore, turn to other political organizations of a communally separatist character, giving birth to a second and an inchoate third nationalist movement several decades after Congress was launched.

Most of the first Congress representatives were high-caste Hindus and Parsis, all of whom spoke English and had attended college, mostly to study law, though some were journalists, teachers, and businessmen. There were a handful of wealthy landowners and merchants, but no princes, zamindars, or other petty nobles. Several Englishmen attended and played key organizing and advisory roles, especially Hume, who had retired from the ICS in 1882 and was a confidant of Ripon, a political radical, and a mystic Theosophist. The liberal William Wedderburn (1838–1918) also appeared, and would later be elected twice as Congress president. The delegates met for three days. All proclaimed their loyalty to the crown, but everyone who spoke expressed some political grievance toward the government of India. As President Bonnerjee put it, what they desired was "that the basis of

[4] See Briton Martin, Jr., New India, 1885 (Berkeley and Los Angeles: University of California Press, 1969).

the Government should be widened and that the people should have their proper and legitimate share in it." These leaders of New India's ambitious middle class wanted the effective opening of the ICS to Indians and opportunities to serve on various government councils. They wanted less of India's wealth wasted on war, and more spent on internal development. (Ripon's mediocre successor, Dufferin, lost the confidence of Congress during his first year in office by embarking upon British India's third and final war against Burma, which completely swallowed that helpless kingdom, but at so high a financial cost that all measures of local reform sponsored by Ripon were abandoned.) Congress also called for the abolition of the secretary of state's council in Whitehall, which was viewed as a major obstacle to progressive legislation for India and a wasteful drain of Indian revenues. From its first meeting, Congress passed repeated resolutions calling for the reduction of military expenditure, which by 1885 accounted for some 40 percent of India's total annual revenue and would at times rise above 50 percent. Though often ignored by the viceroy and his government, such resolutions were the platform of Indian nationalism. They were initially framed in a spirit of loyal cooperation and emulation by the moderate early leadership of Congress, but they became increasingly revolutionary and independent after 1905 and turned totally uncooperative after 1920. Dufferin condemned Congress as a "microscopic minority," claiming that he better appreciated and was more selflessly concerned about the real needs of India's silent majority of 200 million illiterate peasants than these self-selected, self-serving advocates could possibly be. The viceroy's paternalistic pride in "understanding" what was in the "best interests" of his peasant "children" reflected the sentiments of a broad spectrum of British officialdom, including the best of the district officers. The leaders in Congress, on the other hand, believed themselves to be, as Dadabhai Naoroji said in his second presidential address to the Congress, "the true interpreters and mediators between the masses of our countrymen and our rulers."

In Poona, the PSS and the Deccan Education Society (DES) were the major regional organizations focusing on political or general public questions. Less effective in English than Gokhale was, Tilak turned more to his own language, Marathi, as the medium through which he aroused mass support for his causes. He edited the Marathi-language newspaper *Kesari* ("The Lion"), using its columns to reach the lower-middle-class populations of Deccan towns and cities, tens of thousands of whom never learned a word of English but took great pride in their knowledge of Marathi. A brilliant journalist, Tilak managed by innuendo to convey all his negative feelings about the British Raj without ever really writing anything disloyal, though he was convicted several times of "seditious conspiracy." Tilak used the unique popular appeal of Hinduism most astutely to weld his mass following into a cultural nationalist body. In 1891, when Sir Andrew Scoble moved his "Age of Consent" Bill—which was designed to raise the age of statutory rape, for intercourse with or without the consent of one's child bride, from ten to twelve years (the first significant British act of social reform since the mutiny)—Tilak opposed the measure, raising the battle cry of religious tradition in danger. He won populist adulation and substantial support from wealthy Hindu backers, whose orthodoxy found Tilak's advocacy of religious *laissez-faire* more comforting than the English, secular concerns for "women's rights" of liberals like Gokhale. In 1893 Tilak helped revive an old Maharashtrian festival to commemorate the birth of Shiva's elephant-headed son, Ganesh, one of Hinduism's most popular gods. For ten days of raucous annual celebration, villagers poured into neighboring Deccan towns and cities and were treated to a variety of songs, dances, sweets, speeches, and recitations of Hindu epic poetry and Puranic lore. The grass roots of Indian cultural nationalism were thus tapped for the first time, but even more than in Congress itself, here the "nation" was strictly Hindu. The annual Ganesh (or Ganapati) festivals were usually accompanied by violent conflicts between Hindus and Muslims, especially when paramilitary bands of young "Ganesh guards" in Poona, or Nasik, marched noisily past mosques inside which Muslims were trying to pray in silence.

In 1895 Tilak inaugurated a second annual Maharashtrian festival in honor of Shivaji's birth. "Hero worship," Tilak insisted, was at the root of "nationality, social order, and religion." Yet once again, by celebrating the courage of a hero whose revolutionary leadership was directed against Alamgir and the Muslim sultanate of Bijapur, Tilak's popular festivals helped alienate the Muslim quarter of India's population. By identifying himself with Shivaji, however, Tilak became the embodiment of Hindu Maratha dreams of revived national independence. He was soon to adopt Shivaji's demand for *svaraj* as his personal *mantra,* and he is best remembered for his passionate outcry, "Svaraj is my birthright, and I will have it!" The Shivaji festival of 1897 was held under a cloud of deepening gloom for Poona, as bubonic plague, which had been carried inland from Bombay, spread its shroud of "black death" across the city. Tilak lectured on the subject of Shivaji's murder of Afzal Khan, asking if that "act of the maharaja" was "good or bad." The answer to that question, Tilak argued, was to be found in the Gita, where Krishna preached that "we have a right even to kill our own guru and our kinsmen. No blame attaches to any person if he is doing deeds without being actuated by a desire to reap the fruit of his deeds . . . Do not circumscribe your vision like a frog in a well; get out of the Penal Code, enter into the lofty atmosphere of the Shrimat Bhagavad Gita and then consider the actions of great men." Several days later, one of Tilak's young Brahman disciples assassinated a British official who was leaving the governor's Jubilee Ball in honor of Empress Victoria. It was the first successful act of terrorism in India's nationalist movement. The cultural revival had given birth to violent revolutionary offspring.

In Bengal, the nationalist popularity and provincial leadership of Surendranath Banerjea was unique, and his newspaper, *The Bengalee,* could only be rivaled by the weekly *Amrita Bazar Patrika,* edited by the brothers Motilal and Shishir Ghosh. Motilal Ghosh (1856–1922), Tilak's closest friend in Bengal, was more orthodox than Banerjea, more of a cultural nationalist. Revolutionary nationalist terrorism did not blossom in Bengal prior to the first partition of 1905, however, though one of its leading activists, Arabinda Ghosh (1872–1950; later Shri Aurobindo) had as early as 1893 criticized the Bengal Congress for "dying of consumption." He protested that "in an era when democracy and similar big words slide so glibly from our tongues, a body like the Congress, which represents not the mass of the population, but a single and very limited class, could not honestly be called national."[5] Arabinda, who was to become "the high priest" of young Bengal's new literary national spirit, helped stimulate interest in the work of Bengal's most brilliant novelist, Bankim

[5] R. C. Majumdar, ed., *British Paramountcy and Indian Renaissance,* vol. 10 of *The History and Culture of the Indian People* (Bombay: Bharatiya Vidya Bhaven, 1965), part 2, p. 573.

Chandra Chatterji (1838–94), from whose novel *Anandamath* ("Abbey of Bliss") he borrowed the Bengali poem *Bande Mataram* ("Hail to Thee, Mother") for his literary journal. That same poem, set to music by Rabindranath Tagore (1861–1941), was to become Indian nationalism's first anthem. To Arabinda, devout mother goddess worshipper that he was, nationalism was a religion and the soil of India sacred land to be loved and defended, if need be, with the blood of her children. His ardent devotional Hinduism made him even more passionate than Tilak in writing about and working for political independence. In the aftermath of the 1905 partition he was to become a terrorist and to flee from Bengal to save his life, emigrating in 1910 to the French enclave of Pondicherry, where he founded his world famous religious *ashram* with the help of his French "mother" in 1914.

Madras was far less successful at developing the vigorous nationalist leadership for which both Bombay and Bengal were acclaimed, but followed a similar pattern of early public and political organization. The highly conservative Madras Native Association was originally formed as a branch of Bengal's British Indian Association in 1853 to lobby Parliament on behalf of wealthy landowners and Indian aspirants to the ICS. In 1878, however, the Association was reborn under the leadership of G. Subramaniya Aiyar and M. Viraraghava Chariar, two of Madras's first representatives to Congress and cofounders of their presidency's leading newspaper, the *Hindu*. G. S. Aiyar later followed in Tilak's journalistic footsteps by shifting from the English-language *Hindu* to a popular paper he edited in his own regional language, Tamil, named *Swadeshamitram* ("Friend of Our Own Nation"). These same precocious leaders of Madras's provincial public life also organized the Triplicane Literary Society, where the region's young intellectuals gathered to discuss Tamil poetry and politics. In the wake of the Ilbert bill catastrophe, the Madras Native Association was destroyed by factional conflict and became virtually extinct when Aiyar and Chariar quit to organize the more liberal and outspoken Madras *Mahajana Sabha* ("Great People's Society"), founded in 1884. The *Mahajana Sabha* received substantial monetary as well as personal support from leaders of the Madras brahman community, from which most of its English-educated elite were drawn, but its roots did not penetrate below that upper echelon to the mass of nonbrahman peasants or workers in the highly polarized Dravidian society of southern peninsular India. The *sabha* initially included some seventy-seven members, several of whom were Theosophists, whose society had headquarters at Adyar, a suburb of Madras. Under the auspices of the *sabha,* the first regional political conference was called in Madras at the end of 1884, and the second was convened just a few days before the Indian National Congress met in Bombay. The tone and resolutions passed by the Madras conference were similar in their moderation and loyalty and in the class interests they reflected to those of the early Congress sessions. After 1893, when Annie Besant (1847–1933) moved to Adyar to direct the Theosophical Society following the death of its founder, Madame Blavatsky, Madras politics gained greater notoriety. During World War One, Mrs. Besant was the first woman elected to serve as a Congress president.

The *Arya Samaj,* founded by Dayananda Saraswati in Bombay in 1875, would prove an important stimulus to the development of nationalist political activism in the Punjab after 1905, but it was initially created as a society for Hindu revival and Vedic proselytizing. It took several decades for this activist organization, based on Western models, to focus upon such a modernist problem as politics, however. Without either the financial resources or the local leadership required to keep it alive for more than the few days of its annual meetings, furthermore, even the Congress languished, yet it served as a historic lifeline for national aspirations and commitment throughout the era of increasingly centralized and paternalistic rule that characterized the Crown Raj in the final decade of the nineteenth century and the first five years of the twentieth. "It is the Congress of United India," declared Surendranath Banerjea in his presidential address at Poona in 1895, "of Hindus and Mahomedans, of Christians, of Parsees and Sikhs, of those who would reform their social customs and those who would not."

The most serious and potentially destructive challenge to India's nationalist movement came from Muslim leaders like Sir Sayyid Ahmad Khan (1817–98). Born to Delhi's Mughal nobility and reared in the British service, Sir Sayyid was determined to reverse the anti-Muslim sentiment roused in the British by the War of 1857–58, and he became India's first human bridge between Islamic tradition and Western thought. Though politically conservative, Sir Sayyid was religiously radical, proposing reforms in Indian Muslim thought and life that were designed to lure his community from its tents of Perso-Arabic mourning for the demise of Mughal glory into the market place of vigorous competition with Hindus, Parsis, and Christians for ICS positions and the privileges of Anglo-Indian power. Sir Sayyid visited England in 1870 and returned to India to found the Muhammadan Anglo-Oriental College at Aligarh (between Delhi and Agra) five years later, modeling his school on Cambridge. He invited a brilliant young Tory named Theodore Beck to serve as his first principal, and Aligarh quickly became and remained the premier center for the higher education of modernist Muslim leadership in India. Sir Sayyid never accepted the Congress as a truly national organization, insisting that "India is inhabited by different nationalities" and considering "the experiment which the Indian National Congress wants to make fraught with dangers and suffering for all the nationalities of India, especially for the Muslims." His college became the intellectual breeding ground for the Muslim League and Pakistan. Though Sir Sayyid was the most brilliant and effective Muslim modernist, he was hardly the only member of his faith to be assiduously cultivated and regularly consulted by the British during the latter part of the nineteenth century. India's Muslims, like its landlords and princes, were viewed as blocs of potential support for the new Crown Raj, especially after the seas of nationalism began to swell and rise. The Muslim community might, after all, serve as an effective breakwater to tidal waves of national opposition; not that Islamic sectarianism, combined with the often conflicting interests of social class and provincialism, ever made it possible to appeal to the Muslim quarter of India's population as a single entity. In key provinces, nonetheless, such as Uttar Pradesh, government officials labored well before the turn of the century to "control" their landlords and Muslims as "the twin props of British rule."[6]

[6] F. C. R. Robinson, "Consultation and Control: The United Provinces' government and its allies, 1860–1906," *Modern Asian Studies* 5, 4 (1971): 316.

EIGHTEEN

THE MACHINE SOLIDIFIES

(1885–1905)

Despite the crescendo of speeches and the mounting mass of resolutions passed annually by Congress, the bureaucratic machine by which the crown governed India secured its steel grip ever more solidly over the world's largest colonial domain from 1885 through 1905. It mattered little whether the viceroy was as insincere a Liberal as Lord Dufferin (1884–88), or as upstanding and righteous a Whig as Lord Lansdowne (1888–94), as insecure as Lord Elgin (1894–98), or as arrogant as Lord Curzon (1899–1905). The system had by now gathered momentum enough to carry on by itself, as it were, each district administered by its collector and magistrate-judge, who ruled their roost with that firmly judicious touch of control for which the ICS became known near and far, pig-sticking with one hand, surveying a canal or mapping a road with the other. If the natives grew restive, a brigadier and his troops were never too far from the nearest telegraph and post office. At provincial capitals there were the lieutenant governors or chief commissioners, older, hence presumably wiser, versions of the district officer in the *mofussil* ("hinterlands"). They too had generals handy, and their army of clerks, most of them *babus* ("boys"), who did paper work well enough, but really could not, one felt, be trusted. In Calcutta and Simla, the machine achieved the ultimate in complexity and self-perpetuation, for there each member of the council was backed by his own secretary and department, within which the clerks had clerks. As George Curzon so aptly put it, "Round and round like the diurnal revolutions of the earth" went file after file in the bureaucratic daily dance, "stately, solemn, sure and slow."

Lord Lansdowne was prepared to introduce the elective principle directly by adding members to his Legislative Council, though Her Majesty's Government was not quite ready to accept so radical a reform for India. The Indian electorate were to be members of municipal corporations, who were themselves elected only by local propertied ratepayers, and Lansdowne appreciated how conservative a body such an electorate was. He was willing to risk even immoderate criticism in his council chamber, where he rightly noted it would have none of the inflammatory impact of similar attacks launched from improvised platforms in local bazaars. Government would, moreover, retain its official majority, and it had more than enough votes to outgun any opposition. The Indian Councils Act of 1892 actually permitted the governor of India to initiate the elective principle indirectly in 1893, increasing the total additional membership of the central Legislative Council to not less than ten nor more than sixteen, and the councils of the governors of Bombay and Madras to not less than eight nor more than twenty. Though all of these members would still have to be "nominated" by the governor-general, Lansdowne encouraged the drafting of rules that gave limited Indian constituencies, such as municipalities, chambers of commerce, landowners' associations, and universities, permission to submit lists of elected representatives from whom he would then select the actual members of the council. Tilak was thus briefly (1895–97) given the opportunity of sitting on Bombay's council, and Gokhale not only succeeded him there, but was elected in 1901 to represent Bombay's elected members on the viceroy's council in Calcutta after Pherozeshah Mehta resigned that seat. Similarly, in Madras and Bengal, leaders of Congress moved after 1893 within earshot of their governors and governor-general.

The machine, however, did manage to remain virtually impervious to the criticisms and demands of "elected" nonofficials throughout this period. Such bureaucratic indifference to popular desires was perhaps most clearly demonstrated when officialdom closed ranks, not simply in defiance of the often repeated Congress demand that ICS examinations be held simultaneously in India and England, but also in the face of a House of Commons resolution passed to that effect in 1893 yet baldly rejected by the government of India. The official blocs of both the Supreme and the Provincial Legislative Councils invariably voted—as Banerjea, who had personal experience with both, reported—"in a solid phalanx." By 1892 some 25 percent of the government of India's total annual expenditure went to support the India Office, to pay British pensions, and for "charges" on the steadily mounting annual debt, the cumulative home charges that drained Indian wealth. Even under the most favorable financial circumstances, these charges were a heavy burden for impoverished India to pay. During the 1890s, however, the world slump in silver, caused by the discovery of new mines in America and by India's and Germany's demonetization, as well as by the Latin Union's abandonment of bimetallism, brought the Indian rupee to its lowest value in relation to the British pound, which was on the gold standard. The value of a rupee in the early nineteenth century, two shillings threepence, settled at about two shillings during the 1870s, but by 1892 it was down to one shilling and a penny. India was thus forced to pay 50 percent more by 1894 to discharge the same amount of debt owed in London two decades before. The economic depression that ensued was worse than any the subcontinent had known in its recent history, compounded by exchange fluctuations that unnerved most European investors and all but stopped commercial and industrial development by drying up the flow of capital to India. No sooner did the rupee begin to recover after 1895 than a series of monsoon failures followed that brought famine to virtually all of the Deccan; in addition, some ships from China practically closed the port of Bombay after bringing in rats with bubonic plague in 1896. This historic confluence of tragedies actually caused India's population, which had been increasing by over two million a year since the first census taken in 1872, to decline absolutely between 1895 and 1905, the only decade in which combined depression, famine, and plague took so devastating a toll of life. Though the government of India assured Parliament that "everything possible" was being done to use the railroads to ship grain from North India to the stricken south, such shipments were, in fact, hardly more than token measures of relief, invariably organized too late to save the starving, and in any case too little to be more than a salve for British consciences. The bleak economic reality was that grossly inflated home charges had consumed all of India's grain surplus during the years immediately preceding the monsoon failures of 1896–97 and 1899–1900. The commercial crop revolution had earlier weaned India's peasants from their time-honored custom of storing extra grain in each village as a natural defense against lean years of drought, and they were left at the mercy of a rapacious world market and monetary system, which reaped its whirlwind harvest as the century roared to its close.

While India's peasant population hovered on the knife edge of survival, government revenue demands continued to grow in order to feed the fires of wasteful frontier fighting, east and west. The military machine, whose forward-school momentum gathered fresh steam from the final conquest of Burma in 1886, pressed on to consolidate its power over the eastern border states of Assam and Manipur, where British tea planters could always be relied upon to call for more "protection." In 1890 the hitherto indpendent kingdom of Manipur was thus brought under British paramountcy, its capital invaded, its palace placed in a political agent's custody, its treasure plundered to pay for the "protective" guard of troops left at Imphal. On the northwest frontier, similar acts of preemptive protection were taken in 1892 against the *khan* of Kalat. Since Russia could always be cited in Parliament as posing a clear and present danger to Kashmir, more troops were sent to guard that protectorate. While in the area of Gilgit, British forces saw how beautiful a state Chitral was, and in 1893 a British envoy was sent to that remote region, more than a hundred miles northwest of the Indus, provoking tribal hostility that led to further conflict and almost constant confrontation along the frontier for the rest of the century. The military command at Simla grew so powerful in the wake of these continuing wars that a special new army corps was created for the northwest frontier and the Punjab in 1895. The Ninth Earl of Elgin, who served as viceroy from 1894 to 1898, tried but was unable to effectively countermand his military high command, who did much as they pleased along the frontier throughout his viceregal reign. Only when his generals recommended an advance to Kabul in 1897 did he finally stop them, abiding by the agreement that had fixed the boundary between British India and Afghanistan in 1896 along the Durand line.

George Nathaniel Curzon (1859–1925), Baron of Kedleston, was at once the most brilliant and the least popular viceroy, whose talented devotion to the job he was so well trained to perform proved by its colossal failure the inherent impossibility of sustaining British rule over India much longer. For not only was Curzon "a most unusual person," as his chums at Eton and Balliol knew, but he had consciously groomed himself for the post of viceroy for over a decade before taking up that burden, through world travel, study, and service in Whitehall as undersecretary of state both in the India Office and in the Foreign Office. Curzon was the best-informed, most hardworking British ruler India had known since Warren Hastings. He had visited Kabul and befriended the *amir,* Abdur Rahman, and had wandered over the Pamirs and through Chitral with Major Francis Younghusband (1863–1942) as guide. He did it all with a steel brace on his weak back, a brace he had learned to live with from the time he was nineteen, which may help explain his compulsive drive, stamina, and seemingly inexhaustible capacity for desk work. Curzon's governing passion and byword was "efficiency," and he sought by setting a personal example of efficient dedication to duty similarly to inspire the machine over which he was to retain nominal command for six years. He appointed a special commission to study administrative procedures, proposing reforms that helped somewhat to lubricate the gears, but he could not fundamentally redesign the machine, for it was at once too large and embodied too many vested interests. He did create a new province in 1901, however, the Northwest Frontier, extricating it from the control of the Punjab and bringing it directly under the eye of the viceroy. He also added to the central government a new Department for Commerce and Industry, with its own council member. He recommended important police reforms, but they were not ready for implementation until the eve of his departure in 1905. He removed the Indian Railway from the Department of Public Works, established in 1874, and placed it more soundly under its own board, which added almost six thousand miles of track to the rail net during Curzon's era. He appointed a director-general of archeology, who helped unearth India's ancient past. He hoped, of course, to do far more; he wanted the governors of Bombay and Madras to send their weekly reports to himself rather than to the secretary of state in London, to bring these often dilatory and recalcitrant presidencies more tightly under rein. He sought directly to administer every department at Simla, and every local government, from his own desk, where he sat working on the average of from ten to fourteen hours a day.

In foreign policy, Curzon viewed India as the spearpoint of British dominance over all of Asia. He knew the frontier too well, however, to make the mistakes of his predecessors in seeking rashly to subdue its fierce tribes, wasting British arms and treasure on a task doomed from its inception. What he did instead was to pay Pathan Afridis and other tribals to "police" their own territory, buying the support of local "levies" for an annual toll far less than the cost of keeping ten thousand British troops between Chitral and Peshawar, the situation he had found upon his arrival in January 1899. He also saw to it that the railroad was extended as far as possible, with track stockpiled at the northwest end of the line so that British troops could be moved, if necessary, into border fighting positions with minimum delay. With this northwest buffer zone established, Curzon turned his attention toward Persia, which he felt would be best handled by dividing it into spheres of influence for Britain and Russia. In 1902 the cabinet accepted Curzon's ideas as official policy, and Lansdowne, who was then foreign secretary, spelled out British "interests" in Seistan and the Gulf to the shah in London. It was not, however, until the conclusion of the Anglo-Russian Convention in 1907 that this policy was brought to its diplomatic fruition. As for Afghanistan, Curzon's policy was to continue to treat Abdur Rahman as the British dependent he had become, subsidizing him handsomely for the anti-Russian posture he maintained. When the *amir* died in 1901, his son Habibullah was allowed peacefully to succeed him, but the price of his popularity in Kabul was Habibullah's assertion of greater independence from Simla. He went so far as to correspond directly with Russia, and in 1902 reports reached India of preparations for a Russian mission to Afghanistan. Curzon's reaction was to assert in no equivocal terms that he would not remain idle in the face of Russian "interference" in India's "sphere," which sufficed to put an end to whatever mission may have been contemplated from Petersburg but increased the tension of Indo-Afghan relations. At the end of his first term as viceroy, Curzon took leave to return to London in November 1904, immediately after which his "acting" successor, Lord Ampthill, sent Sir Louis Dane to Kabul to seek a new agreement with the *amir*. Whitehall had become alarmed at the truculence of Curzon's tone in communicating with Habibullah and adopted a far more conciliatory approach to Afghanistan than would have been possible with Curzon still in the saddle at Simla. Despite Curzon's advice that Dane be "withdrawn" at the first assertion of Afghan independence, the mission was ordered to stay in Kabul more than three months, and a new treaty was successfully concluded, assuring Habibullah peace with subsidies, until his assassination in 1919.

Curzon added Tibet to Britain's sphere of influence, claiming that the Dalai Lama was seeking Russian military support, thereby justifying his unprovoked invasion of that peaceful hermit kingdom beyond the Himalayas. Curzon convinced Whitehall of the "necessity" of sending his friend, Colonel Younghusband, over the passes from Nepal with an "escort" of several hundred soldiers in 1903. The ensuing invasion of Tibet in 1904 proved singularly successful from Simla's martial point of view, for the Chumbi Valley was "taken" with virtually no resistance, and Lhasa itself was entered and "subdued" (no Russian troops or diplomats were ever found there) and a "convention" exacted by Younghusband from the terrified Tibetan administration granting British India the "right" to occupy the Chumbi for seventy-five years, during which an annual indemnity was to be paid to support the requisite troops. A British trade agent, moreover, was to be based permanently at Gyantze. It was all accomplished with the loss of hardly a British life, and the massacre of little more than several hundred Tibetan monks at Guru. Only the British cabinet's growing conviction that Curzon had "Russia on the brain" saved Tibet from total absorption, dealing Curzon's dream of Asian dominance a somewhat sobering blow.

India's military machine came under the command of Lord Kitchener of Khartoum and Broome (1850–1916) in 1902, and Simla soon proved too small a hill station to contain his ego as well as George Curzon's. The battle of the titans, which ended by forcing Curzon to quit India in August 1905, only a few months after he had returned from home leave to start his second term as viceroy, began as a struggle over the time-honored question of civil versus military control in the government of India. Kitchener found the fact that one of his junior officers served on the viceroy's Executive Council as military member while he was only an extraordinary member of that august body, not only exasperating, but intolerably humiliating. He saw no reason whatever for having his plans for martial reorganization delayed by so much as one hour in order to follow the usual practice of submitting such schemes to another general for perusal and recommendations before they reached the viceroy. The system, of course, had been established both as a check upon a rash commander's advice, and as a formal means of maintaining civil dominance over the army, since the military member was obliged to hold no field command during his council tenure. It was also assumed that the commander-in-chief might wish to spend more time inspecting his camps or some front than sitting in Simla, and that for the sake of administrative continuity, it would be best to have a separate martial official at the central government level. Curzon tried, tactfully, to explain all this to the hero of Khartoum, but he failed to convince Kitchener of its wisdom, necessity, or validity. The irony is that Curzon had personally recommended Kitchener for his Simla command in 1900. Never before nor since had Curzon met his rival in self-esteem. Using his personal popularity at home and his strong support in both the palace and the cabinet to greatest advantage, Kitchener complained and intrigued enough to win the backing of Prime Minister Balfour and India's secretary of state, St. John Brodrick, who recommended a compromise "solution" to the conflict, giving the commander-in-chief a regular council seat with more administrative power, while retaining a second general in the council, to be called military supply member. The latter would not, however, have power to veto any proposal submitted by the commander-in-chief. King Edward personally favored the compromise, and Curzon reluctantly acquiesced, but no sooner did he submit a candidate to fill the new post than Kitchener vetoed the appointment by cabling the Army Department at home that Curzon's candidate was "unsuitable." When Brodrick asked Curzon to suggest someone else who would be more acceptable to Kitchener, the viceroy resigned. He was asked to reconsider, but refused. What had started as a power struggle between Kitchener and Curzon thus ended as one between the viceroy and the secretary of state, and Brodrick asserted his full authority over the government of India and all major appointments to it. Never before had a viceroy as powerful as Curzon been overruled on so relatively minor a matter, and never again would a viceroy so brilliant or imperious be appointed by Whitehall.

Curzon's final year in India was also marked by the first eruptions of an earthquake of nationalist protest generated by the 1905 partition of Bengal. Once again the machine, in its mindless manner, acted as though nothing were more important than bureaucratic efficiency, as if no human consideration—certainly no native consideration—counted. Bengal was, in fact, too populous a province (about 85 million) for any British lieutenant governor to rule efficiently, but the "cure" proposed by British India's Home Department proved far worse than the delay, confusion, and corruption that had long plagued not only Bengal, but every other provincial government as well. The question, therefore, was why Bengal had been chosen for this experiment in administrative perfection, and why Curzon decided to partition that province into two new ones precisely along a line drawn down its midsection just east of Calcutta and the Hughly. For the Bengali-speaking Hindu majority of Calcutta's *bhadralok,* the answer seemed all too simple. The official scheme, undertaken without consulting or considering Indian opinion, was to create a new province of Eastern Bengal and Assam, in which there would be a Muslim majority of approximately six million. The remnant of West Bengal had a Hindu majority, but included so many Bihari- and Oriya-speaking neighbors of the Bengalis that it ended up with a non-Bengali-speaking majority. Hence at one fell stroke of the bureaucratic machine's mighty pen, the outspoken Calcutta-based Bengali-*babu* was to be thrown out of power in his mortally divided motherland. It sounded like *divide et impera* with a vengeance to Bengali ears, and that was precisely how Congress as a whole came to view the partition.

"A cruel wrong has been inflicted on our Bengalee brethren," said Gokhale in his presidential address at the Twenty-first Congress in 1905, "and the whole country has been stirred to its deepest depths in sorrow and resentment, as had never been the case before. The scheme of partition, concocted in the dark and carried out in the face of the fiercest opposition that any Government measure has encountered during the last half-a-century, will always stand as a complete illustration of the worst features of the present system of bureaucratic rule—its utter contempt for public opinion, its arrogant pretentions to superior wisdom, its reckless disregard of the most cherished feelings of the people, the mockery of an appeal to its sense of justice, its cool preference of Service interests to those of the governed." Whether, indeed, Curzon, who later claimed to have been so distracted by other problems as to have taken little note of Bengal, deliberately sought to

divide his Bengali opposition in order to undermine their effectiveness, or if his underlings were guilty of political chicanery, as there
seems some reason to believe of several of them, including H. H. Risley, Theodore Morison, and Sir Andrew Fraser, the historic fact remains that millions of Indians *believed* them to be nefariously inspired. Hence the impact of the partition was immediate and enduring.
Mass protest rallies filled every field in Calcutta. Petitions could not
be made long enough to hold all the signatures that were affixed to
them. Never were so many newspapers printed or sold in Bengal as
during the months immediately preceding and following the October
partition. Surendranath Banerjea came to be known as "Surrender
Not," and Chatterji's poem, *Bande Mataram,* which the gifted scion
of the great *zamindari* family, Rabindranath Tagore, had put to music, was sung throughout the land: "Hail to thee, Mother! Rich with
thy hurrying streams, Bright with thy orchard gleams, Cool with thy
winds of delight, Dark fields waving, Mother of might, Mother free.
Glory of moonlight dreams, Over thy branches and lordly streams—
Clad in thy blossoming trees, Mother, giver of ease, Laughing low and
sweet! Mother, I kiss thy feet, Speaker sweet and low! Mother to thee
I bow!" Bonfires of protest illumined Calcutta's sky as the ancient
Aryan ritual of sacrifice to Agni took modern political form, consuming British-made saris and other cloth imported from Lancashire.
The cry of *sva-deshi* ("of our own country," referring to Indian-made
goods) was raised as the positive side of the boycott movement against
British imports, which emerged in the wake of partition as a key plank
in the new Congress platform that galvanized the nation. Polite petitions and pleas turned into irate protests and angry demands. Mill
owners as well as *zamindars* joined the Hindu lawyers, schoolteachers,
oil pressers, and shopkeepers of Calcutta in a nationalist movement
that gathered more momentum from the final year of Curzon's era
than it had in the preceding two decades of Congress's existence.
Ironically enough, only five years earlier, Curzon had seen Congress
as "tottering to its fall" and confessed that one of his "great ambitions" was to "assist it to a peaceful demise."

NINETEEN

REVOLT, REPRESSION, AND REFORM

(1905–12)

The first partition of Bengal inaugurated a half decade of intense revolutionary nationalist activity, initially countered by severe official repression, and followed by an Act of Reform. Curzon's resignation augured the collapse of Balfour's home government before the end of 1905, followed by a Liberal sweep at the polls that brought to India's helm in Whitehall Gladstone's brilliant lieutenant in the struggle for Irish home rule, John Morley (1838–1923). India had by then already been saddled, however, with one of her most mediocre viceroys, Gilbert Elliot, Fourth Earl of Minto (1845–1914), and Campbell-Bannerman's cabinet considered it unwise to risk replacing Minto so soon after his arrival at Calcutta, just when seas of protest were rising throughout Bengal. That unfortunate decision left India in the hands of Kitchener and the machine, both of whom believed in only one remedy for political agitation of any kind. Ruthless repression spread like plague throughout Bengal, but it could not stifle the impassioned opposition to government.

The boycott of British goods, especially cotton piece goods, whose imports were valued at £22 million in 1904, proved so successful that by 1908 imports were down more than 25 percent. Marwari merchants in Calcutta complained that they could not "give away" British cloth, and some Bengalis protested that Bombay mill owners took advantage of the boycott movement to raise the price of their *svadeshi* cloth exorbitantly. Many merchants changed the labels of their British cottons to read "made in Germany" in order to revive the sales of imports. Mills in Bombay, Ahmedabad, and Nagpur enjoyed a half decade of industrial boom, as some six million spindles and over fifty thousand Indian power looms raced to keep up with the growing demand for *svadeshi* cloth. Though coarser and more expensive than Manchester imports, *svadeshi* became a badge of national devotion, a symbol of India's unity and political potency, and it was worn with pride by women as well as men. The *svadeshi* movement was not only a stimulus to Indian industrial development, however; it also helped revive the dying cottage industry of hand spinning and weaving. It served, moreover, to stimulate indigenous production of most other things that had hitherto been imported from England, including sugar, matches, glass objects, shoes, and metal goods. The Tata Iron and Steel Company was started in West Bengal in 1907 without government assistance of any sort. It blossomed during World War One into a major center of production, and by World War Two it was the British Empire's largest single steel complex. Bombay's great Parsi industrialist, Jamshed N. Tata (1839–1904), India's Henry Ford, had made his fortune in cotton mills, not only in Bombay, but in Nagpur as well, where his huge Empress Mill was started in 1877. Tata's sons, especially Ratan, who carried on his father's industrial trail-blazing tradition, became strong financial supporters of Congress from this period on.

As antipartition passions grew bolder and *svadeshi* sales boomed in the wake of boycott, the government attempted to crack the movement with wholesale criminal prosecutions against its most outspoken advocates and through instructions to educational institutions to prevent students from playing any active "role in politics" while they were receiving grants-in-aid on penalty of losing their official assistance. University students were "harassed, persecuted and oppressed,"[1] while those at lower levels were "flogged, fined, or expelled." Nationalist marchers were attacked by police wielding long, metal-tipped poles, called *lathis,* which took their toll of many an ardent nationalist youth and leader throughout the ensuing decades of protest and repression. Bipin Chandra Pal (1858–1932) emerged as the most popular leader of Bengal's radical youth, editing a new journal called *Bande Mataram,* which he started in August 1906. He was assisted by Arabinda Ghosh, first principal of Bengal's National College. Pal and Ghosh came to believe it was essential to extend the boycott of British

[1] R. C. Majumdar, *History of the Freedom Movement in India* (Calcutta: K. L. Mukhopadhyay, 1963), vol. 2, p. 45.

goods to British institutions, including schools, colleges, law courts, and government service, thus making it impossible for the British to continue functioning in India. Japan's victory over Russia in 1905 inspired them to hope for success in their struggle, and boycott was their major weapon. It had worked initially in diminishing the sale of British goods in Calcutta. Soon it spread beyond Bengal—as *svadeshi* had—winning Tilak's support in Poona and capturing the imagination of his coterie of young brahmans, as well as firing the heart of Lala Lajpat Rai (1865–1928), the leader of the Hindu revivalist *Arya Samaj,* which spearheaded the Punjab's nationalist awakening.

The New Party—the revolutionary wing of Congress—was born from the smouldering frustrations following partition, a party with its major nerve centers in Calcutta, Poona, and Lahore, under the leadership of Lal, Bal (Tilak's first name), and Pal, whose names were chanted in that simple rhyming formula by students all across the subcontinent. It was so swift and popular a happening that Congress's older leadership, committed to the model of British liberalism, were almost as shocked by what they saw and heard as was British officialdom. John Morley and the Liberal government had, after all, only just come to power. True enough, in Morley's first official statement on partition in Parliament, he labeled that outrage a "settled fact," yet his Indian friends believed that this was merely a "political" pronouncement. Congress President Gokhale, who always looked to Morley as "towards a master," had been invited to the India Office in 1906 for lengthy discussions with the new secretary of state about the need for reforms. He had made important contributions to the planning of Morley's liberal reform measures, and he was loath to have his painstaking work thrown into jeopardy by surrendering Congress leadership to New Party "extremists." When Tilak's name was proposed for president of the Calcutta Congress of 1906, Gokhale refused to agree, and Pherozeshah Mehta was even more adamant in casting his veto.

While Bengal's partition served at once to catapult Congress to a new plateau of national popularity and internal division, it also proved a catalyst for separatist Muslim political consciousness and demands. The creation of a new Muslim-majority province of Eastern Bengal and Assam turned Dacca from a backwater *mofussil* town into a provincial capital, removing the deltaic plain's Muslim population of what is now the nation of Bangladesh from the political as well as the economic domination of Calcutta's Hindu landowners, money-lenders, and lawyers. From the vantage point of many Indian Muslims, therefore, partition was hardly a cause for complaint but, rather, a political encouragement, and whether the British planned it that way or not, the fact remains that Aligarh's Muslims took heart and joined forces with Dacca Muslim leaders and those of the Deccan to organize themselves into a deputation to Viceroy Minto in the fall of 1906. That deputation was the genesis of the Muslim League. Khwaja Yusuf Shah of Amritsar had wired Mohsin-ul-Mulk (1837–1907), then secretary of Aligarh College, to "wake" from his political "sleep"[2] and organize a Muslim deputation to Simla. The principal of Aligarh at this time was an Englishman named W. A. J. Archbold, whose friendship with Minto's private secretary, J. R. Dunlop Smith, lubricated the machine so effectively that an audience with the viceroy could be scheduled in less than two months. Thirty-five distinguished Muslim leaders, headed by the young Aga Khan, arrived at the government's summer capital in the hills on October 1, 1906, and were ushered directly into the presence of the viceroy. Morley had alluded in Parliament only a few months earlier to "hopes" of initiating constitutional reforms for India, and the Muslim deputation was the first Indian "lobby" to approach Minto for "a fair share" for their "community" of any possible expanded representation on official councils. The deputation argued that Muslim representation "should be commensurate not merely with their numerical strength but also with their political importance and the value of the contribution which they make to the defence of the Empire." As viceroy, Minto officially assured the deputation, saying "I am entirely with you." He went on to promise "the Mahommedan community" that "their political rights and interests as a community will be safeguarded by any administrative re-organization with which I am concerned." Minto's "work of statesmanship" was, as one of his admiring officials gushed to him that evening, "nothing less than the pulling back of sixty-two million people from joining the ranks of the seditious opposition." His policy in support of communal Muslim interests bore fruit two months after the Simla meeting when the same group of Muslim leaders, augmented

[2] William Harcourt Butler to "Harry," September 23, 1906, Harcourt Butler Papers, India Office Library, London, Eur. Mss. F 116/24.

by thirty-five more delegates from every province of India and Burma, met at Dacca on December 30 to found the All-India Muslim League, whose first joint secretaries were Mohsin-ul-Mulk and Nawab Viqur-ul-Mulk, Mushtaq Hussain (1841–1917), another disciple of Sir Sayyid Ahmad Khan.

"Time and circumstances made it necessary for Mohammedans to unite in an association so as to make their voice heard above the din of other vociferous parties in India and across the wide seas to England," Viqur-ul-Mulk announced. The so-called *nawab* of Dacca, Salimullah Khan (d. 1915), one of Eastern Bengal's largest landowners, whose debt-ridden property had been saved for him by a last-minute loan specially arranged, thanks to Curzon's intervention, through the government of India, hosted this historic first meeting of the organization under whose aegis Pakistan was to be born some forty years later. The Aga Khan continued generously to support the League as well, and in 1908 he was rewarded for his munificence by the executive committee, who gave him the honorary title of permanent president. As had the early Congress, the league vowed "loyalty to the British government," but explained that its major purpose was to "protect and advance the political rights and interests of the Musalmans of India." Its initial membership was far more conservative than that of Congress had been, consisting of titled nobility, wealthy landowners, and only a few lawyers, educators, and journalists. Membership was strictly limited to Muslims, and no more than four hundred would for some time be permitted to join this upper-class club of India's foremost minority community. Following Congress's procedure, the League met annually during the Christmas holiday season. Karachi, the port of Sind that would become Pakistan's first capital, hosted the second session, under the presidency of Bombay's merchant prince, Sir Adamjee Peerbhoy. Until 1913 the league remained a cautious loyalist lobby for elite Muslim interests, pampered by British official attention, condemned by Congress, and ignored by most Indian Muslims as well as Hindus.

For Congress, 1907 was a year of traumatic division. The New Party's popularity continued to grow, particularly among energetic young people, who went through towns and villages carrying the latest editions of Tilak's and Pal's newspapers and Lajpat Rai's pamphlets, spreading the gospel of freedom through boycott across the subconti-

nent. The old party continued to repose its hopes, faith, and promises
of reform in John Morley, who was philosophically with them but
practically constrained to work with Minto and the machine, both of
whom sabotaged, through the classic bureaucratic technique of delay,
the liberal proposals that reached them from Whitehall. In May 1907
the government of India cracked down even harder on the "disturbances" that spread from Bengal to the Punjab, issuing an "emergency ordinance" to empower district officers to "suspend public
meetings" without warning or due process of any sort and to call in
"punitive police" to be "quartered" in areas that were designated by
officials as "disturbed." All British pretension of granting to natives
civil liberties much the same as those cherished by Englishmen at
home was now abandoned. The police and military moved into high
gear to clear the streets of crowds, locking up noisy agitators for "reasons of state," which never required further definition. Morley lamely
protested from his insular perch thousands of miles away, but Minto
assured him that "the strong hand carries more respect in India than
even the recognition of British justice." The editor and proprietor of
Lahore's most popular newspaper, *Punjabee,* were charged with "sedition," found guilty, and sentenced to two years and six months of
"rigorous imprisonment." The Punjab's government came at this time
under the iron heel of Sir Denzil Ibbetson (1847–1908), Curzon's
crony. Ibbetson's only solution for public protest was "strong medicine," and soon after he came to power in Lahore, he wired Minto for
"emergency" permission to arrest and deport without trial Ajit Singh
and Lala Lajpat Rai, the two most popular leaders of the *Arya Samaj*
and Congress in the Punjab. Minto wired immediate approval, and
Lajpat Rai, the schoolteacher whose popularity with younger members of Congress made him a candidate for the Congress presidency
that year, was hustled off to Mandalay Prison in Burma on May 6,
1907. Fearing a tide of popular reaction to such official oppression on
the eve of the fiftieth anniversary of the mutiny at Meerut, Minto put
the Punjab under total repression by resurrecting the Police Act of
1861. When Morley heard that news, he shouted to his private secretary, "No, I can't stand that; *I will not have that.*" But instead of ordering Minto to reverse himself, or resigning, Morley swallowed his
principles and proved by his impotence that even the most sympathetic British friend India ever had in the highest post of imperial

power could really do painfully little to moderate the terrifying impact of the machine.

By the time Congress was ready to meet in 1907, Lajpat Rai had been released from prison and was the nation's most popular figure. He therefore became the New Party's first candidate for president, yet to Mehta and Gokhale, he was as unacceptable as Tilak had been the year before. Dr. Rash Behari Ghose, a distinguished Bengali educator, was nominated to serve as president of the Congress at Surat, a stronghold of Mehta's liberal supporters and thus considered a "safe" venue for the session. "Every one must admit that we are passing through a sad and eventful period—a period of stress and storm—and if ever there was a time when we ought to close up our ranks and present a firm, serried and united front, that time is this," Dr. Ghose would have said, had he been permitted on December 26, 1907, to deliver the address he had prepared. Before he could reach the podium, however, a factional fight began inside the Congress tent, which ended only after police cleared it. The Surat session was over, and India's premier nationalist organization would remain a house divided for the next nine years.

Impassioned youth now turned to the cult of the bomb, seeking to win by terror that which had been denied them. In Bengal, Arabinda Ghosh's younger brother Barindra Kumar led a group of young terrorists, including Abinash Bhattacharya and Bhupendra Nath Datta, who drew inspiration less from contemporary Russian anarchist activities than from Bankim Chandra Chatterji's novels, dedicating their lives to the service of Kali and Durga, worshipping the bomb and pistol through which they hoped to liberate their motherland from the yoke of foreign tyranny. The terrorists started their own newspapers, the most popular of which was *Yugantar* ("New Era"), whose circulation was over ten thousand by 1908, and which openly preached revolutionary action against "White *Feringis*" (foreigners). In Maharashtra, Tilak toured the Deccan speaking and raising funds to foster national education, a nationalist platform designed to liberate Indian youth from the "shackles" of Western learning and "thralldom" to British schoolmasters. He also campaigned to extend the boycott of British goods to government-licensed liquor shops in Poona, and this was the start of a popular nationalist prohibition movement that caught on in many parts of India.

On April 30, 1908, the first bomb to take British lives was thrown by young Khudiram Bose in Bengal, killing two English women whose closed carriage resembled and was mistaken for that of Magistrate-Judge Douglas Kingsford. The judge was notorious throughout Bengal for the pleasure he took in sentencing so-called political offenders to flogging, a punishment Morley had moved to outlaw as soon as he gained control of the India Office, but the government of India dragged their heels on the matter for more than two years. British reactions to the bomb were predictably strident; one newspaper, the *Pioneer,* demanded the arrest of all "terrorists," warning that in the future, "ten of them would be shot for every life sacrificed." Tilak tried to argue in *Kesari* that the bomb was as much the by-product of "exasperation produced by the autocratic exercise of power" as it was of the New Party's "wild talk." The remedy he suggested was the immediate granting of *svaraj* to the people. Soon after that, he wrote of the bomb as "magic," calling it a "sacred formula" and an "amulet." Two weeks later he was arrested, charged with "sedition," and tried in Bombay. He was convicted by the seven Europeans on his jury, though found "not guilty" by its two Parsi members, and was sentenced to incarceration for six years in the same Mandalay prison where Lajpat Rai had sweltered a year earlier.

While Tilak roasted in his Burmese prison, Bipin Pal moved to London and half froze, half starved in a cold-water flat, and Lajpat Rai journeyed to the United States, living in New York on the meager royalties from his writings. The New Party was thus left to languish leaderless, its isolated young terrorists, whose violence was directed against "moderate" Indians as well as Englishmen, periodically achieving notoriety through acts of murder, arson, or suicide. British repression, unofficial as well as official, continued to grow more vindictive. One young English "corporal" in "a fit of excitement shot the first Native he met," Morley noted, expressing his concern over such matters to Minto and inquiring, "What happened to the corporal? . . . Was he put on his trial? Was he hanged? . . . If we are not strong enough to prevent Murder, then our pharisaic glorification of the stern justice of the British Raj is windy nonsense."[3] In Madras, Chidambaram Pillay was sentenced to "transportation for life" for "seditious

[3] Stanley Wolpert, *Morley and India, 1906–1910* (Berkeley and Los Angeles: University of California Press, 1967), p. 120.

speeches," and his judge opined that "there was no lawful occasion, so far as he could see, for any man in this country to make a political speech." "This explains Bombs," commented the secretary of state for India. Minto's government, however, only rushed to pass more legislative measures to add to its arsenal of repressive powers: an Explosive Substances Act and a Newspapers (Incitements to Offences) Act, the latter giving magistrates total discretion as to when a newspaper might be "seized" and permanently shut down. More violence was reported in the wake of each repressive measure. Minto and Morley personally became terrorist targets, but a bomb thrown at the viceroy failed to explode, and the closest anyone came to killing the secretary of state was when Madan Lal Dhingra assassinated Morley's political aide-de-camp, Lieutenant Colonel Sir William Curzon-Wyllie, in London on the night of July 1, 1909. Hundreds of Bengali terrorists were behind bars at the time, and even the most notorious of Maharashtra's young revolutionaries, the poet-author Ganesh Damodar (Veer) Savarkar, had just been transported for "life" to the Andaman Islands for "waging war against the King."

Though faced with so discouraging a cycle of revolution and repression, Morley nonetheless labored to complete the fashioning and the Parliamentary enactment of reforms. In 1907 he appointed the first Indian members to his council in Whitehall, one totally apolitical Hindu official from Calcutta's Board of Revenue, K. G. Gupta, and one Muslim, S. H. Bilgrami (1842–1926), who had been reared in the *nizam*'s employ and was actively associated with the Muslim League. Congress was hardly elated at either choice, and though Morley felt that "their colour is more important than their brains," neither distinguished himself during his tenure on the council. Once established, however, the precedent of Indian membership remained and was extended the following year to that "holy of holies" the governorgeneral's Executive Council, which met in Calcutta and Simla. For the latter post, an Indian barrister of great distinction was selected, Sir Satyendra P. Sinha (1864–1928), advocate-general of Bengal, who took his seat as law member of the Executive Council in Simla on April 18, 1909, "much against" the "will" of His Majesty King Edward VII, but at the insistence of Morley and Britain's Liberal cabinet. Bigots like Kitchener were, hereafter, forced to moderate their racial tirades in council, but Sinha found the strain of that position too much for him, and in 1910 his "native seat" was turned over to a Muslim successor, Sayyid Ali Imam, a barrister who had previously served on the Legislative Council. Sinha was elected president of Congress in 1915 and went on to become the first Indian parliamentary undersecretary of state in 1919; as Lord Sinha, he became the first Indian governor, of Bihar and Orissa, in 1920.

Morley's major scheme of reform was the Indian Councils Act of 1909, which increased the "additional membership" of the central Legislative Council to no fewer than sixty; in the provinces of Bombay, Bengal, and Madras, membership was increased to fifty, a majority of whom were to be nonofficials. An additional fifty members could also be appointed to the legislative councils of the United Provinces and of Eastern Bengal and Assam, and up to thirty members in all other provinces. By abolishing official majorities at the provincial level, Morley followed the advice of Gokhale and his brilliant Congress comrade from Bengal, Romesh Chandra Dutt (1848–1909), who was then lecturer in Indian history at University College, London, a past president of Congress (1899), and the future prime minister of the state of Baroda (1909). Minto and the machine, including members of Morley's own council in Whitehall, protested against abolishing official majorities at any level of government, for they not only distrusted natives, but they had little faith themselves in the system of parliamentary government and its basic ideals. Morley's act also directly introduced—for the first time—the elective principle for Indian Legislative Council membership. The electorate was initially a minuscule minority of upper-class Indians enfranchised by virtue of the municipal property taxes they paid or the amount of higher edution they had completed. Some of the seats were reserved for separate Muslim members, to be elected only by Muslim voters, in keeping with Minto's earlier communal promise, but in 1910, after the first elections were held, 135 Indian representatives, elected by their own constituencies, took seats as legislators on councils throughout British India. It was a historic landmark for Britain and India, a major step forward on the road of constitutional reforms that led to India's attainment of full dominion status in 1947. Elected members of the new councils were also empowered to engage officials of government in questions and debate about proposed legislation, and the annual budget debate, which had hitherto been limited to the last hectic day of the year's legislative session, was to be extended over many days so that nonofficial criticism could have some immediate impact. Indians would, moreover, be permitted to introduce legislation of their own. Gokhale took immediate advantage of the latter provision to formulate his Elementary Education Bill, which would have empowered municipalities and district boards to opt for "compulsory" elementary education for boys between the ages of six and ten. As a requisite corollary, legislation outlawing child labor to the age of ten would have to be passed. Even so timid a first step was, however, too radical for the government of British India to accept in 1911, and it proved, indeed, to be decades ahead of its time.

Morley's parliamentary undersecretary of state during his last year at the India Office was a young Liberal, Edwin Samuel Montagu (1879–1924), who would, as secretary of state during and after World War One, pilot the next great Act of Reform along the course chartered by his mentor. On November 7, 1910, John Morley left the India Office, having prevented Kitchener, as his final act of assistance to India, from achieving his burning ambition of succeeding Minto as viceroy, choosing the Liberal Lord Hardinge of Penshurst instead. As his first major official act, Hardinge proposed the reunification of Bengal, to which Whitehall agreed. That decision, one of the best-kept state secrets in British Indian history, was made public by King George V on December 12, 1911, at his coronation *durbar* in Delhi. Bengal was reunited, but a new province of Bihar and Orissa was carved out of its wings, and the central government's capital was to be moved from Calcutta to Delhi, where a new city (New Delhi) would be born. Designed by Sir Edward Luytens and Sir Herbert Baker as the architectural reflection of the British Crown Raj, New Delhi would remain the capital of independent India as well, symbolizing at once the transience and the continuity of British imperial rule.

TWENTY

THE IMPACT OF WORLD WAR ONE

(1914–19)

Council reforms and the reunification of Bengal gave heart to the moderate leadership of the old Congress and to most Bengali Hindus, but jolted the Muslim League from its moderate loyalist stand. "The policy of the Government is like a cannon which passed over the dead bodies of Muslims," cried Viqur-ul-Mulk upon learning of the annulment of Bengal's partition. The *nawab* of Dacca angrily protested that the "real reason" partition had been undone was that Hindus participated in a "revolutionary movement," while Muslims were lulled to rest content with mere British "promises." The remedy he recommended must have been clear to young Muslims, for in Dacca and elsewhere throughout Eastern Bengal their slogan became, "No bombs, no boons!" When Viceroy Hardinge ceremonially entered Delhi in 1912 to inaugurate the crown's new capital, he was almost assassinated by a bomb thrown into his elephant's *howdah* by an unknown and uncaught terrorist. The following year the Muslim League abandoned its docile pro-British platform and followed Congress's political example by adopting "self-government" within the empire as its new nationalist goal. Thanks to this bolder stand, the League now attracted young intellectuals to its ranks, including the uniquely gifted Bombay barrister Mohammad Ali Jinnah (1876–1948), who was already a "Muslim member" of the central Legislative Council and of Congress.

Born in Karachi to a family of middle-class mercantile Muslim Khojas, who had earlier migrated to Sind from Gujarat's Kathiawad peninsular, Jinnah was sent to London at the age of eighteen to study law. He was called to the bar at Lincoln's Inn and found himself politically "caught up in the excitement" of Dadabhai Naoroji's Liberal campaign for a Central Finsbury seat in the Commons. Jinnah became the Grand Old Man's political secretary in 1906, and was thus lured into Indian politics as a member of Congress rather than the League. He came under the strong personal influence of Gokhale as well as Dadabhai at that time, and once said his ambition in life was to become "the Muslim Gokhale." A Liberal Anglophile by temperament, Jinnah prospered at the Bombay bar and seemed indeed destined to follow Gokhale's footsteps when he was elected to the supreme Legislative Council in 1910. In 1913, before joining the League, he had accompanied Gokhale to Europe and was to return to London the year after at the head of a prestigious Congress deputation sent to lobby Secretary of State Lord Crewe for more Indian membership on his council. He was the most promising young barrister, the shrewdest parliamentarian, and, as Crewe reported to Viceroy Hardinge, "the best talker of the pack" among the Congress deputation, and he was heir-apparent to the mantle of Bombay's Naoroji-Mehta-Gokhale moderate leadership over India's nationalist movement. Were it not for the outbreak of World War One and the impact of that cataclysmic event on the history of India, as well as the rest of the world, Jinnah might indeed have become Gokhale's successor and India's prophet of Hindu-Muslim unity, rather than the father of Pakistan.

On the eve of the outbreak of the war, another of Gokhale's disciples, Mohandas Karamchand Gandhi (1869–1948), also arrived in London, following his twenty-year exile in South Africa, to meet with his "political guru" before returning home. Gandhi was born in the Kathiawad state of Porbandar, whose raja his father had served as *diwan* (revenue minister), though their caste was the third-ranking *modh bania* and the meaning of *gandhi* in Gujarati is "grocer." He was nineteen when he first left for London to study law, and he was called to the bar of the Inner Temple in 1891. Gandhi proved far less adept as a barrister than Jinnah, and after trying unsuccessfully to establish a practice in Gujarat and Bombay, he accepted an offer to work for a Memon Muslim firm in Durban, South Africa, in 1893. There were at the time more than forty thousand Indians in Natal, most of whom had originally been sent to Africa under the "new system of slavery,"[1] as indentured laborers after 1860. Natal sugar planters desperately needed cheap "coolie" labor from India, since the abo-

[1] Hugh Tinker, *A New System of Slavery: The Export of Indian Labour Overseas, 1830–1920* (London: Oxford University Press, 1974).

lition of slavery throughout the British Empire in the 1830s had liberated black slaves. The Crown Colony of Natal depended on the indenture system as the cheapest solution to their economic problem: luring thousands of landless Tamil laborers from South India to work for virtually nothing for a minimum of five years. After the five-year indenture expired, however, laborers were free to quit or to sign up for another five-year period, at the end of which return passage to India was paid to them, or they could use the monetary value of such fare (about £10) to purchase land in Natal. Practically all of the formerly indentured Indians opted to remain in Africa, buying land and soon enjoying what was, in comparison to the landless penury from which they had emerged, a certain degree of prosperity. The Indian population's rate of growth was so rapid that European settlers soon became alarmed, and by 1891 they were shocked to find themselves outnumbered by "coolies" and middle-class Indian Muslim merchants, who had come to Africa to cater to the Indian community's needs and desires. In 1893, when Natal was granted "responsible" government, its white rulers decided to do everything in their power to prevent further Indian influx and settlement and to make life so miserable and expensive for the Indians already entrenched in Natal so as to convince them to leave. This was when Gandhi arrived as legal counsel to Durban's leading Indian entrepreneur, Dada Abdulla.

The bigotry and racial arrogance of Natal's white community fired Gandhi's acutely sensitive spirit, and in the cauldron of racial prejudice, his consciousness of national identity as an Indian intensified. While leading Natal's Indian community in its struggle against such selectively oppressive measures as the Natal Legislature's attempt to disenfranchise all "Asiatics" and to impose a punitive head tax on Indian residents over the age of sixteen, Gandhi developed his activist technique of *satyagraha* ("holding fast to the truth"). This nonviolent method of noncooperation and civil disobedience tapped the deepest roots of India's cultural heritage. Mahatma ("Great Soul") Gandhi relied not only upon the Aryan cosmogonic force of *satya,* but also upon pre-Aryan yogic powers, including meditation, fasting, silence, and "nonviolence to any living thing," *ahimsa.* In the course of his South African traumas and struggle, Gandhi transformed himself from a middle-class British barrister into the prototypical Indian *sadhu* ("sage"), impoverished but serenely self-contained, sexually continent and spiritually confident of his cosmic powers.

When Britain declared war against Germany on August 4, 1914, India was immediately notified by Lord Hardinge that she, too, was at war. With vivid memories of the past decade of "unrest" and "disturbances," the higher echelons of the Indian government naturally felt considerable apprehension and concern about the impact of that dramatic announcement. The response of Indian nationalist politicians of all parties, however, as well as of princes, was so overwhelmingly loyalist and supportive of the crown in its hour of dire distress that both the government of India and Whitehall were delightfully surprised. Congress leaders of all factions naively anticipated that British victory would bring freedom to India. Practically all the rulers of native states wired "total support" in defense of the empire. Twenty-seven of the largest princely states immediately placed their Imperial Service Troops at the viceroy's "disposal." A hospital ship, *The Loyalty,* was fully provisioned by the princes and ready to sail with the first Indian Expeditionary Force, made up of some 16,000 British and 28,500 Indian troops of the Lahore and Meerut divisions and the Secunderabad Cavalry Brigade, which embarked from Karachi on August 24, reaching the port of Marseilles on September 26, 1914. "What an army!" reported the New York *World.* "Its 'native' contingent belongs to a civilisation that was old when Germany was a forest, and early Britons stained their naked bodies blue." They reached the western front just in time to face the full fury of the first German assault against Ypres. Thanks to Indian reinforcements, the "thin and straggling" line of Allied troops guarding Calais and the Channel from the massed might of Falkenhayn's force held firm. Before October ended, the blood of hundreds of Baluchi and Afridi Pathans mixed with that of British boys on European soil, and Khudabad, a sepoy of the Duke of Connaught's "own Baluchis," was the first Indian to win the Victoria Cross—posthumously. Two months later some seven thousand troops were dead, missing in action, or seriously wounded along the western front.

The council reforms that Jinnah and his Congress deputation had come to London to propose were buried by an unsympathetic House of Lords and quickly forgotten. Gandhi, who urged London's Indians to "think imperially," organized a Field Ambulance Training

Corps as his personal contribution to the war effort, but within a month of its inception, he launched a *satyagraha* campaign against its colonel, R. J. Baker, who treated the corps of volunteers as a regular military unit. Though the campaign failed, it did permit Gandhi to resign from the service he had joined enthusiastically, and it saved him from going to France during the worst period of combat, allowing him to return to Bombay instead before the year's end. Tilak had also come home in 1914 from enforced exile, and he cabled the king-emperor to assure his loyal support. Annie Besant was eager to mediate a rapprochement between Tilak and Gokhale, but though discussions were held in Poona, no Congress reunification was achieved until after the death of Gokhale and Pherozeshah Mehta in 1915. William Wedderburn now called upon Britain to make "a declaration by the highest authority of whole-hearted trust in the Indian people." India had stood as a "tower of strength" in support of empire, and it was high time to repay such loyalty with a real devolution of political power. No one but Kitchener, who was in command of the War Office, expected the war to last more than a few months. A British victory was depicted by Lord Hardinge to his Indian legislative councillors in Simla as "the triumph of right over might, of civilization over the military barbarism of Germany, of ordered freedom over military slavery, and of everything that men held dear." Indians felt, therefore, quite justified in expecting a generous measure of freedom as the quid pro quo of their cooperation.

Disillusion set in early. As soon as the first casualty figures were reported from the western front, it became apparent to India that the war would bring death in distant lands for many of her sons long before fulfilling any dreams of freedom. The reality of war made its first direct impact on India after September 10, 1914, when the German cruiser *Emden,* which had steamed into the Indian Ocean undetected, sank the first of the dozen British freighters it would send to the bottom of the Bay of Bengal before it was destroyed in November. That single cruiser all but paralyzed British shipping for two months, during which time the loss of German trade and the confiscation of ships and supplies for overseas expeditionary forces brought India's foreign trade to a precarious new low, sending prices of imports skyrocketing and bringing shortages in essential commodities, including railroad and telegraph spare parts. When the *Emden* bombarded

Madras on September 23, 1914, setting several oil-storage tanks ablaze, tens of thousands of Indians were no longer convinced of British victory and began to wonder whether a German invasion of India had not perhaps begun. The sudden cessation of German trade not only stripped Indian bazaars of many of their best and cheapest manufactured products, but it cut off entirely what had by 1914 become India's second largest overseas market for exports. In 1913–14 Germany alone had purchased some £17.5 million worth of Indian goods, almost half the value of all goods shipped from the subcontinent to Britain, and at least £3 million more than the total purchased by the United States. When the value of Indian produce shipped to Austria-Hungary was added to that sent to Germany, the total rose to over £24 million. Indian trade with the Central Powers had continued to grow at so rapid a rate during the first few months of fiscal 1914–15, moreover, that from April 1, 1914, to the outbreak of the war in August, some three hundred thousand tons of Central European produce had been brought by German ships to India, over ten times the total tonnage imported during that entire year from both the United States and Russia. The collapse of the export market depressed India's major commercial crop prices, especially those of short-staple cotton, jute, and oil seeds. The price of all manufactured goods, on the other hand, rose precipitously, and food grains sold at "famine prices." Food hoarding and shortages soon became prevalent and remained standard wartime conditions in India. England's demand for Punjabi wheat, furthermore, grew throughout the war years, and millions of tons of that precious grain were exported annually to help keep Britain and the Allied forces alive.

An even greater source of wartime disillusion and growing tension between India and Britain, however, was the Ottoman Empire's decision to join forces with the Central Powers. For India's sixty-million Muslim minority, at least, the Indo-British invasion of Mesopotamia and that entire tragic, mishandled "bastard war" proved a constant tug of loyalties, since Muslims the world over looked to the caliph as universal "leader" of their community. British fears of German interests in Persian Gulf oil, however, overcame any concerns about Muslim Indian loyalties, and by mid-October the Sixteenth Infantry Brigade of the Poona Division, with its "squeaking, kicking country-bred ponies" stowed between decks, left from Bombay in a sixty-five-ship convoy headed for Bahrein. By November 1, 1914, when Turkey joined the Central Powers, sixteen thousand Indian troops were at the mouth of the Shatt-al-Arab, ready to occupy Abadan and "secure" its pipeline and refineries. Curzon's policy of asserting control over the Persian Gulf was thus brought to fruition by a shot fired at Sarajevo. Basra was easily taken in December, and by October 1915 the British Indian Army had moved as far north as Kut-el-Amara, hardly a hundred miles from Baghdad. Baghdad thus seemed within reach of British arms, yet less than two weeks after General Townshend's doomed force of some twelve thousand Indians started toward that mirage in November 1915, they were stopped by Turkish arms at Ctesiphon, forced back to Kut, surrounded there, and all but destroyed before their surrender in April 1916. The "Mespot" disaster turned into a scandal of cabinet-shaking proportions for Britain when it was learned in Parliament that thousands of Indian troops died for want of medicine and mosquito netting as much as for lack of food, shoes, and bullets. The Royal Commission, whose report on the subject was published in June 1917, laid most of the blame upon the lack of coordination and inept handling of the government of India's Army and Military Supplies Departments, and Secretary of State Austen Chamberlain resigned as soon as such facts came to light. Before the end of 1914, another twenty-nine thousand Indian troops had been shipped to Egypt to guard the Suez Canal and secure Egypt against possible Turkish attack. Anglo-French greed and diplomacy resulted in the Sykes-Picot Agreement in early 1916, whereby the Ottoman Empire was divided between these two Allied powers, under whose rule nascent Arab and Jewish nations were to struggle for their subsequent emergence as states.

Since reunification of their motherland, the Bengalis had virtually ceased their acts of terrorism, but in the Punjab a wave of violent Sikh revolutionary activity began with the *Komagata Maru* tragedy early in the war and persisted beyond its conclusion. The Japanese steamship *Komagata Maru* had been chartered to carry some four hundred Sikhs from Shanghai to Vancouver before the war began, but Vancouver authorities refused to permit the Indians to disembark. Denied permission to enter several other Canadian ports as well, the ship was finally obliged to turn around and carry its luckless passengers back across the Pacific, unloading them after months of exhausting travel at Budge Budge, near Calcutta, on September 27, 1914. Outraged at what they perceived to be a British imperial policy of racial exclusion (though it was actually Canadian), the Sikhs started to march on Calcutta to make their grievances known to the government. They were stopped by a line of soldiers and police, who opened fire, killing 16 Sikhs, leaving 144 others "unaccounted for," and arresting 100 more, who were taken in irons to Calcutta. By the time a Viceroy's Committee of Inquiry could order the release of the prisoners and they could be returned by express train to the Punjab, they had, of course, become irreconcilable revolutionary nationalists. The *Ghadr* ("Mutiny") Party, whose leaders were mostly Sikhs, was founded in the western United States in 1913 and grew under the leadership of Har Dayal (b. 1884) into a well-funded force seeking to overthrow British imperial rule from outside India. The party's headquarters were in San Francisco, and Har Dayal himself went to Berlin in 1915, hoping to convince Germany to invade India through Afghanistan. Several similar revolutionary "conspiracies" never quite came to fruition, and though Germany's most famous spy during the war took as her Sanskrit pseudonym *Mata Hari* ("Mother Vishnu"), she was actually Dutch.

The divided Congress continued to seek some formula of reconciliation, for each of its faction's leaders realized the futility of trying to unite a nation under leaders who could not unite themselves. Jinnah was also anxious to bring the Muslim League and Congress together, and in Lucknow in 1916 that unification was accomplished, thanks in good measure to his initiative and Tilak's political pragmatism. Tilak was by now ready to accept the moderate Congress leadership's pledged formulation of their goal, which was to seek "by strictly constitutional means" the "attainment by India of self-government similar to that enjoyed by the self-governing members of the British Empire." The Lokamanya had, after all, lived through six years of prison exile; he had outlived his major rivals, Gokhale and Mehta, and he was hailed as a virtual deity by hundreds of thousands of Maharashtrians, who considered him a reincarnation of Shivaji, if not an *avatar* of Vishnu. He could afford to be magnanimous about future goals and promises. He was even willing to concede the Muslim League's insistence that the number of Muslim representatives on all provincial and central legislative councils be "weighed" in proportions greater than their overall population percentage would have dictated, which was one of the League's demands in return for associating itself with Congress's nationalist goal of self-government. Tilak wanted the British to leave, and the sooner, the better. Whatever means would hasten that departure were, as far as Tilak was concerned, worth trying. He had none of Gokhale's moral or political scruples, none of Gandhi's religious and ethical concerns about truth or nonviolence.

The Lucknow Pact, or Congress-League Scheme of Reforms, called for elected majorities on expanded provincial legislative councils, as well as on an enlarged Imperial Legislative Council. It demanded "as broad a franchise as possible," and specified province by province the percentage of seats that were to be reserved for Muslim members, from 50 percent in the Punjab to 15 percent in Madras. Half the members of the Executive Council were to be Indians elected by the elected members of the Imperial Legislative Council. The secretary of state's council "should be abolished" and India Office expenses charged to British, rather than Indian, taxpayers. In reference to the racially restrictive policies of both South Africa and Canada, the pact declared that "Indians should be placed on a footing of equality in respect of status and rights of citizenship with other subjects of His Majesty the King throughout the Empire." A final demand that gained significance as a result of the war was that military and naval service, "commissioned and non-commissioned ranks, should be thrown open to Indians, and adequate provision should be made for their selection, training and instruction in India."

Soon after Montagu replaced Chamberlain as secretary of state for India, he rose in Parliament on August 20, 1917, to announce that "the policy of His Majesty's Government, with which the Government of India are in complete accord, is that of the increasing association of Indians in every branch of the administration and the gradual development of self-governing institutions with a view to the progressive realisation of responsible government in India as an integral part of the British Empire." Montagu then embarked on a personal passage to India, becoming the first secretary of state to do so while in office. It was, in fact, his second visit, the first having been in 1913, when he was undersecretary of state. Montagu's deep personal affinity for and attachment to India was, he himself suspected, a product of his

Jewish birth, which made him refer to himself as "an Oriental." At any rate, he was determined to produce new reforms, which could become "the keystone of the future history of India." Lord Chelmsford—Lord Hardinge's successor as viceroy since 1916—was far less liberal than his predecessor, however, and Montagu soon began to fear that all the government of India and Parliament would allow would be "a niggling, miserly, grudging safeguard, fiddling with the existing order of things."[2] Montagu traveled throughout India, meeting all of the major political leaders as well as officials, and finding Jinnah most "impressive-looking, armed to the teeth with dialectics, and insistent upon the whole of his scheme." When Chelmsford tried to dispute Jinnah's position, he was "tied up into knots." Montagu concluded that it was "an outrage, that such a man [Jinnah] should have no chance of running the affairs of his own country." He was less impressed by Gandhi, who "dresses like a coolie, forswears all personal advancement, lives practically on the air, and is a pure visionary. He does not understand details of schemes." It was, however, the Mahatma's "saintly style" of politics, which Montagu did not understand, that would prove so uniquely effective in India.

While Montagu worked as assiduously as Morley had with moderate liberals like Jinnah and Sinha, trying to hammer out constitutional schemes that would only be sabotaged in their final formulation or actual implementation by a hostile, or at best unsympathetic, bureaucracy, Gandhi marched to the pace of his South African experience, continuing to test and evolve his *satyagraha* method on the far more fecund soil of his homeland. He established his base in Sabarmati, a suburb of Ahmedabad, the mill city capital of Gujarat, calling the self-sufficient rural community he founded there *Satyagraha Ashram*. Soon answering a "call" from an oppressed indigo peasant who invited him to visit Champaran in Bihar after the Lucknow Congress, Gandhi found fresh application for his technique of nonviolent resistance. Always traveling third class, dressed as the poorest peasant—or holiest *sadhu*—the Mahatma drew crowds and attention at every platform stop, reaching India's masses as no politician before him had ever done, embracing poverty and suffering in his own person, experiencing daily the plight of the "lowest of the low," becoming

[2] All of the Montagu quotes are from Edwin S. Montagu, *An Indian Diary,* ed. Venetia Montagu (London: William Heinemann Ltd., 1930).

their guru, not just another political leader. His was a potent charisma, for in India no other appeal had as much force as a religious one.

Gandhi's victory over the Indigo planters of Champaran in 1917 brought him popular acclaim as the champion of India's peasants. His next *satyagraha* campaign lured him back to Ahmedabad to help the exploited in a cloth factory owned by his friend and benefactor, Ambalal Sarabhai. The plague had finally reached Ahmedabad, and Sarabhai generously turned one of his homes over to its victims for a hospital, but he refused to meet the meager demands of his workers for wages high enough to help them cope with wartime inflation. Gandhi mediated a strike settlement, after undertaking his first public fast in India. As a result of the strike the Ahmedabad Textile Labor Association was born, and by 1925 it had become one of the most important trade unions in India, with some fourteen thousand members. No sooner was the Ahmedabad *satyagraha* won than Gandhi went to Kheda, also in Gujarat, to take command of the peasant protest against payment of enhanced land revenues, as widespread crop failures throughout that district increased the likelihood of famine during the ensuing year. The popular demand for suspension of revenue had been organized under the leadership of Vallabhbhai Patel (1875–1950), a powerful Kheda landlord and *sardar* ("noble") who was to become Congress's "boss" and India's first deputy prime minister. It was the first and only time Gandhi led a *satyagraha* for the nonpayment of revenue, though the campaign proved so effective in rallying peasant support that the British did make substantial reductions in their assessments throughout the district that year. Many of Gandhi's more radical followers later urged him to call upon all of India's peasantry to stop paying land revenue for a single year as the fastest way to immobilize British rule, but he recognized the potential danger that popularizing so revolutionary a technique of protest might pose to any nationalist successor government, and he always held it as a tactic of last resort.

The war's impact on India thus proved all-pervasive, transforming its economic balance, giving birth to massive new industries (including Tata's great Iron and Steel Mill at Jamshedpur in Bihar and Orissa), and stirring the waters of political change as they had never been roiled before. New aspirations were awakened, new pride; new constituencies arose, and along with them, a new consciousness of

India's value to the survival of Britain's empire and an impatience to be free of imperial constraints. For the first time in history, large numbers of Indians had gone to Western Europe, where they saw how French and English "peasants" lived in comparison to Indians and found themselves treated as social equals rather than mere natives or coolies. More than a million Indians, noncombatants and soldiers, were shipped overseas during the war. Most of them returned home with strange, dreadful, and wonderful tales of the world of Europe and Africa, their minds opened, their visions and aspirations for their children—if not for themselves—irreversibly altered. Annie Besant and Tilak both started separate home rule leagues during the war, popularizing the demand for *svaraj*. Gandhi evolved his practical "experiments" in "self-rule" with *satyagrahis* brave enough to face possible arrest or death in asserting their human dignity. While the Mahatma's leadership mobilized masses who had been touched by no leader before, other than Tilak, it also appealed mostly to Hindu cultural roots through its use of Hindu symbols, and thus tended to polarize India's pluralistic communal society. In the hope of broadening this narrow religious appeal, Gandhi would focus his attention after the war on support of the popular pan-Islamic *khilafat* movement, led by the brothers Shaukat and Muhammad Ali.

Immediately after he returned home in 1918, Montagu hammered out his "Report on Indian Constitutional Reform," which embodied the principles of "complete popular control"—"as far as possible"—at the local government level, and some immediate "measure of responsibility," with the aim of "complete responsibility as soon as conditions permit," at the provincial level. To implement the latter reform a new technique called dyarchy was devised, by which several provincial departments of government were "transferred" to ministers elected by elected legislative council representatives, while other departments were "reserved" to officialdom. The reserved subjects were the most well-funded, powerful branches of government, finance and law and order; nonetheless, dyarchy was a substantial step toward independence. No such transfer of executive authority would as yet be introduced at the center, but the Imperial Legislative Council would be "enlarged and made more representative."

Montagu's reform principles were, however, soon followed by harsh, repressive recommendations made by Justice Rowlatt's committee, which had been appointed by the government of India to investigate "seditious conspiracy" and the possible need for peacetime legislation along lines of the martial "law" Defense of India Act that had been passed as a wartime "emergency measure" in 1915. The Rowlatt Committee recommended extension of the Defense of India Act and introduction of a tougher Press Act as India's post-Armistice Day "rewards." The repressive Rowlatt Acts were driven through India's Imperial Legislative Council at high speed, over the universal opposition of elected Indian members, who found their passage more humiliating and exasperating than the first partition of Bengal had been. Jinnah and several of his colleagues resigned their council seats when the "Black Acts" were passed in March 1919. "The fundamental principles of justice have been uprooted and the constitutional rights of the people have been violated at a time when there is no real danger to the State," wrote Jinnah in his letter to the viceroy, "by an overfretful and incompetent bureaucracy which is neither responsible to the people nor in touch with real public opinion."[3]

Armistice thus brought to India not peace but the sword of continued repression, accompanied by the most catastrophic epidemic in recent history. Twelve million Indian lives were reaped by the 1918–19 influenza outbreak. Gandhi viewed this tragic aftermath of Armageddon as nothing less than cosmic proof of the failure of Western civilization and the incapacity of British rule. He condemned the Rowlatt Acts as "symptoms" of "a deep-seated disease in the governing body." He called upon all Indians to pledge themselves to "refuse civilly to obey" such "unjust, subversive" laws, declaring a nationwide work "suspension" day during the first week of April 1919, as a prelude to the launching of a national *satyagraha* campaign. At the Sikh temple city of Amritsar in the Punjab, the Mahatma's call was answered by anti-Rowlatt Acts meetings organized by Drs. Kitchlu and Satyapal, both of whom were arrested and "deported" without formal charge or indictment by Deputy Commissioner Miles Irving on April 10, 1919. Their "crime" was daring to participate in political action. The arrests of Amritsar's most popular leaders incensed their followers to march from the old city toward the commissioner's bun-

---

[3] Jinnah to Chelmsford, March 28, 1919, reprinted in Stanley Wolpert, *Jinnah of Pakistan* (New York: Oxford University Press, 1989), p. 62.

galow in the camp. They were halted by troops who opened fire on them, killing several petitioners and turning the crowd into a terror-crazed mob that went on to burn British banks and attack English men and women. Brigadier R. E. H. Dyer was called in to restore "order." No further "disturbances" occurred in Amritsar until April 13, when General Dyer banned public gatherings of any sort; upon learning that a meeting was to be held in Jallianwala Bagh that afternoon, Dyer marched his Gurkha and Baluchi rifles across the narrow entrance to that otherwise walled field and ordered them to open fire without a word of warning. It was a Sunday, and some ten thousand men, women, and children, mostly peasants from neighboring villages, had come to the *bagh* ("garden") to celebrate a Hindu festival. Dyer's troops fired for ten minutes, pouring 1,650 rounds of live ammunition into the unarmed mass of trapped humanity at point blank range. Some four hundred Indians were left dead, and twelve hundred wounded, when the brigadier and his force withdrew at sunset from the garden they had turned into a national graveyard.

When Sir Michael O'Dwyer, governor of the Punjab, learned of the "incident," he voiced full approval, and two days later, he imposed martial law over the entire Punjab. Viceroy Chelmsford also approved at first, but when he learned the details of the tragedy, which were brought to light by Lord Hunter's Parliamentary Commission of Inquiry, he characterized Dyer's action as "an error of judgment." The deranged brigadier defended what he had done as "the least amount of firing which would produce the necessary moral and widespread effect it was my duty to produce . . . from a military point of view, not only on those who were present, but more specially throughout the Punjab." The moral effect was indeed widespread. Within the year, all of India gradually learned of the Jallianwala Bagh Massacre and of the terror wrought by British officers in the wake of a war supposedly fought against "Prussian butchers" by "civilized allies." At Montagu's insistence, Dyer was relieved of his command, but he returned to England a hero to British Conservatives, who collected thousands of pounds sterling for him and gave him a jewelled sword inscribed to the "Saviour of the Punjab." During the succeeding months of martial law in the Punjab, official atrocities—including flogging Indians and forcing them to crawl—were perpetrated by government servants "in defense of the realm."

"The enormity of the measures taken by the Government of the
Punjab for quelling some local disturbances has, with a rude shock,
revealed to our minds the helplessness of our position as British subjects in India," wrote Rabindranath Tagore to the viceroy on May 30,
1919. "The universal agony of indignation roused in the hearts of the
people has been ignored by our rulers—possibly congratulating themselves for imparting what they imagine as salutary lessons. . . . The
time has come when badges of honour make our shame glaring in the
incongruous context of humiliation."[4] Tagore thereupon resigned
the knighthood, which had been conferred upon him after he won the
Nobel Prize for Literature in 1913.

Millions of less famous Indians turned at this time from loyal
supporters of the British Raj to nationalists who were no longer content to follow its orders or trust in the "fair play" of its officials. When
Motilal Nehru, the father of India's first prime minister and the grandfather of Indira Gandhi, presided over the Amritsar Congress in
December 1919, he told its eight thousand delegates and some thirty
thousand visitors that "if our lives and honour are to remain at the
mercy of an irresponsible executive and military, if the ordinary rights
of human beings are denied to us, then all talk of reform is a mockery." Just a few days earlier, the king had by royal proclamation
enacted the Montagu-Chelmsford reforms, which His Majesty hailed
as opening "a new era." But if war had raised too many hopes too
high, its aftermath crushed them too brutally. By 1919 India's era of
late-Victorian liberal cooperation and Edwardian politesse was for-
ever ended.

[4] Tagore's letter is reprinted in D. G. Tendulkar, *Mahatma* (New
Delhi: The Publications Division, Government of India, 1960), vol. 1,
p. 263.

TWENTY-ONE

TOWARD INDEPENDENCE

(1920–39)

The aftermath of World War One brought such widespread disillusionment to India that Congress abandoned its policy of cooperation with the British Raj to follow Gandhi's revolutionary call for nonviolent noncooperation. The Mahatma launched his first nationwide *satyagraha* campaign on August 1, 1920, the day Tilak died. Gandhi was now undisputed leader of the Congress, rising above such powerful leaders as Chittaranjan Das (1870–1925) of Bengal and Motilal Nehru (1861–1931) of Uttar Pradesh, thanks to the broad base of popular peasant and urban labor support he had won and the lucrative backing of wealthy Gujarati Banias and Jains, Bombay Parsis, and Calcutta Marwaris. Moreover, by championing the cause of the Ottoman caliph (*khalifah*), whose empire was to be dismembered by the victorious Allies, Gandhi won a mass Muslim following. He thus emerged as the first Indian nationalist whose appeal not only reached below the urban crust of India's English-educated elite, but also bridged the communal gap. Though the *khilafat* movement died in 1924 with Ataturk's abolition of his own caliphate, unleashing pan-Islamic frustrations inside India for what would thereafter be the worst decade of communal rioting, Gandhi used the movement's revolutionary popularity to help mount his first great *satyagraha*. During the two decades from 1920 to 1939, under Gandhi's leadership, Congress transformed itself from an elite, moderate club to a mass national party, representing all of India's major regional and interest groups, capable of mobilizing millions in its revolutionary noncooperation campaigns. These years were also a time of accelerating urbanization and economic depression as the collapse of farm prices forced

millions of marginal peasants to swell the ranks of the urban unemployed. The British Raj tried to shore up its crumbling base of power
with a number of constitutional proposals and reforms, and before
the outbreak of World War Two, it lured Congress back to its side for
a brief interlude of "responsible provincial rule," from 1937 to 1939.
That interlude, however, served only to intensify Hindu-Muslim conflicts, and it was to prove more of an asset to Muslim League separatism than to Congress nationalism.

As a true Mahatma, Gandhi sought to pit his yogic powers of
self-control, abstinence, suffering, and meditation against the awesome might of the world's greatest empire. As though testing the
Upanishadic formula *tat tvam asi,* he weighed his great *atman* in the
cosmic balance of time against the *Brahman* of the Raj, reminding
his followers that "the purer the suffering (*tapasya*), the greater is
the progress. Hence did the sacrifice of Jesus suffice to free a sorrow-
ing world. . . . If India wishes to see the Kingdom of God established on earth, instead of that of Satan which has enveloped Europe,
then I would urge her sons and daughters . . . to understand that we
must go through the suffering." Gandhi was "practical idealist"
enough to know that *satyagraha* would be India's only effective technique of waging a nationalist struggle against Great Britain's war-
strengthened Machine. In the wake of Jallianwala Bagh, moreover,
insistence on *ahimsa* ("nonviolence") as central to his method of
struggle was as much a shrewd tactic of survival as it was the saintly
core of his political religion. But for a minuscule band of terrorists,
India was, after all, totally disarmed. For Gandhi, of course, *ahimsa*
was not simply a negative force. Its positive aspect was love, and
through the power of love, he urged his *satyagrahis* to "convert" their
opponents, to win the hearts of those who preferred "brute force" to
"soul force" in the resolution of conflicts.

Until *ahimsa* could work its magic, however, the multiple boycott would be used to help bring the wheels of British government to
a halt. Not only imported British cottons and other manufactured
goods, but also British schools and colleges, courts of law, councils,
titles and honors, and possibly even revenue demands, would fall
within the catch-all boycott resolution of Congress's special session
at Calcutta in 1919. To fill the economic and institutional vacuums
left by so comprehensive a boycott movement, Indians would natu-

rally have to rely more and more upon themselves, their own handicrafts and *svadeshi* cottons, their own national schools, and public organizations, their own governing bodies. Gandhi's revolutionary resolutions won over the opposition of more conservative leaders like Jinnah and Das, who failed to muster half the number of votes Gandhi now commanded. With some five thousand delegates and fifteen thousand observers, the Calcutta Congress was the largest popular forum in recent Indian history, and it marked the dramatic shift in political constituencies that followed Gandhi's rise to leadership, for anyone who paid the token price of admission to the Congress tent was permitted to vote, and special *khilafat* trains had brought in thousands of Muslim workers to support whatever Gandhi proposed. The upper-middle-class elite represented by Jinnah and Das were hereafter outvoted and outshouted by a mass of peasants and workers, most of whom could not read or write any English, or even any Indian language. They rallied behind Gandhi not merely because of what they had heard of his "magic" powers, but because of the peasant-like simplicity of his unpretentious appearance and manner and the power of his revolutionary program. For the Hindus, Gandhi's embodiment of the virtues and ideals of Sanskritic culture sufficed to make him a Mahatma; for the Muslims, his ardent defense of their caliph was enough to make him their friend. And for young intellectuals like Jawaharlal Nehru (1889–1964), reared in luxury in Allahabad and educated in England (Harrow and Trinity College, Cambridge), Gandhi was the incarnation of *Bharat Mata* (Mother India), traditional guru and revolutionary hero in one unique person.

The regular annual meeting of Congress was held in Nagpur in December 1920. Gandhi and his program of noncooperation were so overwhelmingly supported by the fourteen thousand delegates who went to Nagpur that Jinnah sought in vain to caution against *satyagraha,* was shouted down, and quit the Congress in disgust. Gandhi drafted a new Congress Constitution at this time, whose first article became its credo: "the attainment of Svaraj by the people of India by all legitimate and peaceful means." The party machinery was thoroughly reorganized along the centralized democratic lines that Gandhi had found useful in mustering mass support for his campaigns in South Africa, with village Congress committees electing representatives to subdistrict committees, which would in turn send delegates to

district, thence to provincial, and finally to the All-India Central
Committee of Congress, whose 350 members were to elect a fifteen-
member Working Committee to function as the party's executive authority. This reorganization legitimized the new constituencies of
Gandhi's following and assured Congress continuing grass-roots
national support. The support of Congress for its British Committee
was now discontinued, thus finally cutting the silver cord of dependence on London. The revolutionary impact of Gandhi's Constitution
was to prove as significant as that of his program. The Mahatma
promised his followers *svaraj* in one year—by December 31, 1921.
Gandhi defined *svaraj* as neither more nor less than "that we can
maintain our separate existence without the presence of the English,"
or put another way, "to get rid of our helplessness." India was the
fabled tiger reared among goats, and this was her "roar" of recognition, reminding this nonviolent tiger of her natural identity.

Young men like Subhas Chandra Bose (1897–1945) of Bengal
were so excited by Gandhi's revolutionary call that they gave up possible careers in the coveted ICS to devote full time to the national
struggle. Bose himself, however, was soon disappointed by the vague-
ness of Gandhi's plans for winning *svaraj* within a year, and he shared
none of the Mahatma's religious aversion to violence. His first ambition had been military rather than civil service, but as a high-caste
Bengali (*Kayastha*), Bose did not qualify for British "martial caste"
service. (Nevertheless, during World War Two he would lead an army
of his own against the British Empire.) Returning to Calcutta in 1921,
Bose apprenticed himself to C. R. Das, and he was soon to share with
Jawaharlal Nehru the nationalist leadership of India's youth and peasant societies (*kisan sabhas*), as well as several labor organizations.
These new politicized organizations, started among university, college, and high school students, as well as peasants and workers,
broadened the base of Indian public consciousness and reflected the
accelerated pace of urbanization and modernization after World War
One. During the first half of 1920 alone, there were some two hundred strikes in India, affecting over a million workers. The All-India
Trade Union Congress was organized and held its first national meeting in Bombay in October 1920, under the presidency of Lala Lajpat
Rai. By 1929 there were more than one hundred trade unions across
the country, with almost a quarter of a million paying members. Stu-

dents began to become aware of peasant illiteracy and penury, and while boycotting English schools and classes, many well-to-do, city-bred young Indians ventured for the first time into the countryside, not only to learn more about village life and values, but to try to help villagers improve their lives or to organize *kisan sabhas.*

Congress was not alone in this movement of spreading political awareness. At this time India's Communist Party was born, through the initiative of the Comintern and its Bengali Indian agent, Mana-bendra Nath Roy (1886–1954). The Bolshevik revolution and Len-in's rise to power made their impact on India, and many an Indian found the formula of class struggle and the prospect of violent revo-lution more attractive than Gandhian methods of nonviolent change. In 1920 the Maharashtrian brahman Communist leader Shripat Amrit Dange (b. 1899) published his pamphlet on *Gandhi and Lenin,* in which he argued that the latter was much the greater leader of op-pressed mankind and the true revolutionary. Whether thanks to the unique form of revolution led by Gandhi, or the deep-rooted religi-osity of India's population, communism did not pose a significant challenge to India's National Congress, however. The *khilafat* move-ment, on the other hand, continued to preoccupy most young Muslims, including Maulana Abul Kalam Azad (1887–1958), a Delhi Alim who was to be president of the Congress throughout World War Two and independent India's minister of education. A *hijrat* ("flight") was organized by *khilafat* leaders, luring thousands of Indian Muslims to emigrate over the Khyber and Bolan passes to Afghanistan as the first lap of their trek to the Sublime Porte in Constantinople and on to Mecca, but Afghanistan was hardly receptive to such an "invasion" of its domain, and instead of welcoming their Muslim brethren with open arms, sent them back to India without sufficient food or funds. The abortive exodus ended in death or bitter communal conflict when these depressed and exhausted Muslims returned to their villages to find that their lands and houses were occupied by Hindu neighbors.

The noncooperation movement gathered momentum throughout 1921, and British efforts to steal its thunder by arranging a visit for the Prince of Wales in November backfired, the black flag reception and boycott of royalty ominously demonstrating the depth of Indian alienation from and hostility toward the Raj. Government repression came swift and fierce in the wake of the prince's visit, and by year's end some twenty thousand Indians were jailed; instead of crushing resistance, however, such tactics only incited greater disaffection. Gandhi remained out of jail, but he was rapidly losing control over the movement, whose millions of adherents were hardly as disciplined or devoted to nonviolence as their Mahatma. In Malabar, Muslim Moplahs declared a *jihad* (holy war) in August 1921, ostensibly in order to establish a new *khilafat* of their own, killing Europeans and wealthy Hindus wherever they found them and forcibly converting Hindu peasants and laborers to Islam. These bloody riots continued intermittently to plague the Malabar coast for the next few years. The Hindu *Mahasabha,* a communal party founded in 1914 at Hardwar by Pandit Madan Mohan Malaviya (1861–1946) as a counterfoil to Muslim separatist demands, became more active at this time and joined the *Arya Samaj* in its campaign of "reconversion" and "purification" (*shuddhi*) of Muslims, initially in the Punjab and later in the Gangetic Plain, Deccan, and other parts of India as well. By forcing Muslims to "wash away their pollution" with total immersion in a river or water tank, Hindu gangs provoked communal rioting, and the always precarious Hindu-Muslim unity, so essential to Gandhi's campaign and the Congress program, was further eroded.

When repression mounted, Gandhi consoled his followers by redefining *svaraj* as "abandonment of the fear of death." His imprecation to "Forget fear," which Nehru recalled as the most important message of the Mahatma, was not, however, sufficient to curb the passions and violent proclivities of many Indians, and as 1921 passed into history without the advent of *svaraj,* there was widespread disillusionment among those who had literally believed in Gandhi's promised deadline. "Mass civil disobedience is like an earthquake," Gandhi told Congress near the end of 1921. "Where the reign of mass civil disobedience begins, there the subsisting Government ceases to function. . . . The police stations, the court offices, etc., all shall cease to be the Government property and shall be taken charge of by the people." Soon after there were violent riots in Bombay, with Hindus and Muslims uniting primarily to attack Parsis, Christians, and Jews in that city. In January 1922 Gandhi was ready to lead the Bardoli subdistrict of Gujarat, whose entire population was only eighty-seven thousand, in a total mass civil disobedience campaign, including "nonpayment of taxes." Bardoli was thus chosen to be the first region of

India brought to the promised land of *svaraj*. Before launching his campaign, the Mahatma issued one final appeal to the new viceroy, Rufus Isaacs, Lord Reading (1860–1935; viceroy, 1921–26), to "set free all the non-co-operating prisoners" and declare "in clear terms a policy of absolute non-interference with all non-violent activities in the country." The government rejected his demands, and Gandhi stood poised in Bardoli to lead his most militant campaign. Then word of a mob immolation and murder of twenty-two Indian constables inside their police headquarters at Chauri Chaura in the United Provinces (U.P.) was cabled throughout India by the press. Gandhi immediately withdrew to fast and ponder that dreadful news, emerging several days later to report that "God . . . has warned me the third time that there is not as yet in India that non-violent and truthful atmosphere which alone can justify mass disobedience, which can be at all described as civil, which means gentle, truthful, humble, knowing, wilful yet loving, never criminal and hateful." The "general" then abandoned his army on the eve of battle. In their prison cells, thousands of young Indian patriots learned to their amazement and consternation of Gandhi's startling decision. The Mahatma now turned from political agitation to what he called a "constructive program," which emphasized hand spinning and weaving of cotton cloth (*khadi*) and social welfare work (sanitation and education) in the villages. He hoped by such methods to prepare India's masses for the self-control required of true *satyagrahis*. Gandhi had been terrified by the prospect of finding India "liberated" overnight, without either the British army or police to assist Congress in maintaining order and preventing mass murder. Less than a month later, he was arrested at his *ashram* and charged with "promoting disaffection towards the Government established by law" in India. Gandhi pleaded guilty and was sentenced to six years imprisonment. After two years, however, he was released for an appendicitis operation, but he did not resume active political agitation until 1929.

While Gandhi languished in jail, the leadership of Congress passed to C. R. Das and Motilal Nehru, both of whom favored taking advantage of the new Government of India Act (1921). That act had transformed the Central Legislative Council into a bicameral parliament, made up of an Imperial Legislative Assembly with 140 members, 100 of them elected, and a Council of State with 60 members,

40 of them elected. Some five million land-owning, taxpaying, educated Indians were now enfranchised to elect representatives to expanded provincial councils under this act, while one million could vote for candidates for the assembly, and seventeen thousand for members of the more exclusive Council of State. The viceroy's Executive Council would now have at least three Indian members, and provinces would all have several responsible ministers to preside over such "transferred" departments as education, public health, public works, and agriculture. At least 70 percent of the members of all expanded provincial legislative councils would, moreover, be elected nonofficials. By standing for election to the councils and fighting from within their chambers to paralyze the machinery of government, Das and Nehru argued, Congress could advance more rapidly toward its goal of *svaraj*. In 1923 these *swarajists* won a majority of India's political activists to their policy of "noncooperation from *within* councils." Stalwart cohorts of the Mahatma, however, led by Chakravarti Rajagopalachari (1879–1972) of Madras, called C.R., and Rajendra Prasad (1884–1963) of Bihar, insisted upon a "no changes" continuation of Gandhi's policy, touring the subcontinent to encourage hand spinning and other aspects of Gandhi's constructive program.

Under the intelligent lead of Lord Reading, India's only Jewish viceroy, the government pursued a policy of trying to win middle-class Indian support by granting many of the liberal reforms Congress had demanded from its inception. In 1923 simultaneous examinations for the ICS were held for the first time in New Delhi and London, and Indians were admitted to military officer training. Indian fiscal autonomy was also established that year, with the creation of a Tariff Board in New Delhi, which within two years would entirely abolish the cotton excise. To compensate for the loss of cotton revenue, however, Reading's government doubled the salt tax, which fell most heavily upon those least able to bear any added financial burden, the poorest peasants and landless laborers who needed salt to replenish its daily drain from working under the hot sun. This increased salt tax was to become the rallying cry to revive Gandhian *satyagraha* in 1930. The *Swarajists* also protested the "cruel burden" of the tax when they contested council seats during the 1923–24 elections, winning a plurality in the central Assembly and sufficient power in Bengal and Bombay to prevent the government from establishing ministries in either of those key provinces. They also took control of India's major municipalities: C. R. Das became mayor of Calcutta, and Jawaharlal Nehru of Allahabad. After the death of Das in 1925, however, the party lost its most vigorous leadership, and Subhas Bose, his guru. Motilal Nehru was unable effectively to direct and coordinate internal obstruction of the government by himself, and many *Swarajists* became, in fact, cooperators with officialdom, while others turned back to the Mahatma for guidance.

Following the collapse of the *khilafat* movement in 1924, millions of Indian Muslims were left without a cause, and, frustrated by their now "leaderless" state, they turned against their Hindu neighbors. In several provinces of India, local "caliphs" led violent bands on *jihads* to "save" Islam. The orthodox Hindu *Mahasabha* and *Arya Samaj*ists responded in kind, while the British maintained a posture of noninvolvement, as communal rioting spread across the subcontinent. The causes of intensified Hindu-Muslim conflict in this era were economic as well as religious. In the Punjab and Bengal, Muslims were mostly the poorer peasants or landless laborers; the Hindus were generally the larger landowners and moneylenders. The same was true in Gujarat and Maharashtra. Muslim League despair at the shattered state of the Muslim minority, and Indian nationalist politics as a whole, led Jinnah, as president of the league in 1924, to call for a restoration of the spirit of the Lucknow Pact. Gandhi, who had by now recovered fully from his operation, spoke of Hindu-Muslim unity as "the breath of our life," deploring all communal conflict and violence. But the Lucknow Pact was as much a part of the past as was World War One, from whose era it had emerged. Gandhi's major energies, time, and concern continued to be directed toward work with villagers, and his program remained immediately practical and manual in orientation, rather than focusing upon formulas for the apportionment of power in councils he had no desire to join. He and Jinnah remained as incompatible and irreconcilable as they had discovered themselves to be at Nagpur in 1920.

In 1925, at the suggestion of Gandhi, Congress raised its annual dues from four *annas* to two thousand yards of handspun yarn per month. It was a toll that neither of the Nehrus could have paid but for the loophole that any member could have his dues "spun for him by another." The Mahatma believed that daily spinning would bring

India's leaders into closer touch with peasant life and enhance the dignity of labor in the minds of Indian intellectuals who had never done a day's work in their lives. The "music" of the spinning wheel (*charka*), which became Congress's symbol, would, he often remarked, become India's "song of freedom" when it was heard throughout the land. Touring the country, Gandhi came into daily contact with untouchables (whom he called *harijans,* "children of God"), as well as other peasants, and he began from this period to address himself to their plight, awakening upper-caste Hindu concern and awareness of India's most ancient and most blatant form of discrimination.

Dr. Bhimrao Ramji Ambedkar (1891–1956), the brilliant untouchable leader of the Mahars of Maharashtra, had just returned from abroad (he studied law in London and received his doctorate from Columbia University) to help form the Depressed Classes Institute in the Deccan. Thanks to Ambedkar's personal example, organizing talents, and untiring labors, untouchables began to develop a modernist spirit of political consciousness. In the South Indian state of Travancore, Brahmanic orthodoxy was most harsh in its ostracism of untouchables, obliging members of "outcaste" communities to carry bells, as though they were cattle, to warn high-caste Hindus of their approach. (A brahman was believed to be "polluted" if he permitted even the shadow of a pariah to fall upon him.) A *satyagraha* campaign to allow untouchables to walk the road that passed in front of the temple at Vykom had been in progress for more than a year when Gandhi arrived there in March 1925. The maharaja of Travancore sided with his brahman subjects and set up police barricades to keep watch over the public road around the temple. The Vykom struggle, Gandhi insisted, was for nothing less than "to rid Hinduism of its greatest blot," for in labeling some fifty million Indians "untouchables," upper-class Hindus condemned them to subhuman living conditions. It took months of intense negotiation for Gandhi to convince Travancore's government merely to open the roads on three sides of one temple to everyone, but the moral weight of his presence and the national publicity this problem now received eventually had wide impact. Within the decade, Hindu temples opened their doors as well as the roads outside their walls to the "children of God," several of whom Gandhi brought into his own *ashram,* for as long as "the curse of untouchability" continued to "disfigure Hinduism," the Mahatma declared, there could be no true *svaraj*.

The Conservative victory in Britain in 1925 brought Stanley Baldwin to Downing Street and Edward Wood, Sir Charles Wood's grandson, to India as Viceroy Lord Irwin from 1926 to 1931. Irwin was a devout Christian, and he shared the Mahatma's sense of messianic mission, though from the opposite side of the imperial battleground. No sooner did the new viceroy arrive in India than communal riots rocked Calcutta, virtually paralyzing that great city for six weeks in April and May. More than a hundred deaths and over a thousand injuries were inflicted before the violence subsided. Calcutta's leading Muslim politicians included Sir Abdur Rahim, former minister of justice and jails on the governor's Executive Council, and his son-in-law, Huseyn Shaheed Suhrawardy (1893–1963), deputy mayor of Calcutta. Suhrawardy later became the prime minister of Pakistan, and he was to be the Muslim League's premier in Bengal during the "Great Killing" in Calcutta just two decades later. One of Bengal's most popular and enduring political leaders was the lawyer-teacher Abul Kasem Fazlul Huq (1873–1962), who had served both as president of the league and as general secretary of his provincial congress in 1918. Huq organized the *Krishak Praja* ("Workers and Peasants") Party in 1929 and was Bengal's first indigenous premier, 1937–43. Political polarization and Hindu-Muslim conflict thus reached new extremes at much the same time.

In 1927 Britain's home government appointed a seven-man Indian Statutory Commission, led by Sir John Simon, to formulate what would be the "next stage" of Indian constitutional reforms. The fact that no Hindu, Muslim, Sikh, or Parsi was asked to serve as a member of that Parliamentary Commission raised political blood pressures throughout India, so that when Simon and his team, which included the young Labor M.P. Clement Attlee, arrived in India early in 1928, they were greeted with black flags and shouts of "Simon, go back!" Congress resolved to boycott the commission as "the only self-respecting course for India" and led mass demonstrations opposing Simon and calling for immediate *svaraj*. Wherever Simon went—Bombay, Delhi, Calcutta, Lucknow, Lahore, Patna, or Madras—the same chaotic scenes of mass demonstrations and armies of police arrayed against shouting mobs were repeated. Congress now took the initia-

tive and called an All-Parties Conference, under the presidency of
Motilal Nehru, to draft a Constitution acceptable to Indians.

The Nehru committee proposed a "Commonwealth of India,"
with all powers to be derived "from the people" and with all the free-
doms enjoyed by other dominions of the empire, within which it was
to remain. A hybrid of the British Parliament and the U.S. Congress
was recommended, without special provision for any minority membership. The Muslim demand that no less than one-third of the seats
on the central legislative body be reserved for Muslims was rejected.
The possible redistribution of provincial boundaries "on a linguistic
basis" was suggested, however, in keeping with earlier Congress proposals. In seeking to resolve their different political demands among
themselves, Congress and the League only moved further apart, and
by December 1928 Jinnah warned the All-Parties Conference that,
for Indian Muslims, the only alternatives to greater constitutional
"security" were "revolution and civil war." The Hindu *Mahasabha*
challenged Jinnah's credentials to "speak for the Muslims," to which
the man who was to become the founder of Pakistan responded: "Do
you want or do you not want the Muslim India to go along with you?
. . . If you do not settle this question today, we shall have to settle
it tomorrow." Nehru and the majority insisted that there was, in fact,
no "Muslim problem." When Jinnah walked out of the All-Parties
Conference in Calcutta, he went directly to Delhi to join the All-India
Muslim Conference convened by the Aga Khan. This Muslim conference reaffirmed its unanimous commitment to separate Muslim electorates, resolving "in no way" to consent to joint electorates under
any future constitutional formula. The Hindu-Muslim unity that had
resulted from the Lucknow Pact was never to be revived.

The first All-India Workers and Peasants Party Conference was
also held in Calcutta during December 1928, at which time the Communist Party of India (CPI) was revitalized. Communist strength
inside the All-India Trade Union Council had been amply demonstrated, and the party's new central executive hoped to consolidate
its gains on all fronts, using the growing unrest and unemployment in
India as springboards for gaining greater political power. Before the
party could put its plans into operation, however, the government
arrested thirty-one "conspirators," who were taken to Meerut on
March 20, 1929, and charged with "conspiracy to deprive the King

Emperor of his sovereignty." That Meerut conspiracy trial-without-jury dragged on for more than four years, throughout which period the accused were all kept behind bars, only to be released for lack of evidence at the end of 1933. The highly publicized Meerut case made martyrs of its "conspirators," whose cause was defended by Jawaharlal Nehru, and whose release was followed by a sudden growth of Communist Party membership and influence in the subcontinent. Nehru and most young Indian intellectuals viewed the trial as "one phase of the offensive" launched by government "against the Labour movement." He and Bose organized a Socialist Independence for India League in 1928, demanding "complete independence" from the British Empire, pitting themselves against Gandhi and Motilal and the majority of older Congress leaders, who were content with *svaraj* within the empire as their national goal. By mid-1929, however, Gandhi was politically astute enough to realize that, though he had the votes to be Congress president if he wished that year, Jawaharlal's popularity with youth, labor, and other left-wing groups was such that his election to lead the nationalist movement would help capture those important and powerful currents of contemporary society. Subhas Bose was equally active in presiding over student and youth conferences from Bengal to the Punjab during 1929, but Bose and Gandhi never saw eye to eye, and Jawaharlal was, of course, Motilal's son, a not insignificant consideration in explaining his rise to national leadership at the age of forty, as Congress's youngest president.

More important, however, than Jawaharlal's (the name means "Red Jewel") ancestry was his personal charisma, a complex product of intellectual brilliance, physical attractiveness and stamina, human sensitivity, and charm. He had the natural aristocrat's disinterest in power and its petty perquisites, which made him at once the hero of India's youth and the trusted spokesman of the older generations, who were less repelled by his radicalism than they were attracted and reassured by his urbanity. He was almost too perfect to be true, a melange of India's legendary heroes, the reincarnation of an Ashoka and an Akbar, as brave as he was handsome, as self-sacrificing of wealth as he was rich, surely destined to lead his nation's struggle for freedom. Vallabhbhai Patel had more powerful provincial Congress support than Nehru, but Jawaharlal was Gandhi's choice. The interlude of *Swarajist* Party leadership was over, for working through the councils had proved as futile in hastening the end of British rule as the Mahatma had predicted.

The repressive powers of British officialdom remained as formidable as ever. Indian poverty continued to spread, and India's resources continued to be drained away to pay for British soldiers and home charges. Just when Tory power seemed most inflexibly entrenched in London, however, British elections brought Labor Prime Minister Ramsay MacDonald's coalition ministry to power, and Irwin was called home for instructions from his new secretary of state, William Wedgewood Benn, and the cabinet. The viceroy returned to India in October 1929 to report that His Majesty's Government recognized the need for "seeking the greatest possible measure of agreement for the final proposals" on constitutional reforms before submitting them to Parliament, indicating that Indian opinion would play a much greater role in the process that had been inaugurated by Simon's commission. The Montagu declaration of August 1917 was now reiterated, and its "natural issue" was explicitly stated—for the first time officially—to be "the attainment of Dominion Status." The Working Committee of Congress welcomed the declaration and the promise, but called upon government to demonstrate its sincerity by specific acts of conciliation and clemency, including amnesty for all political prisoners. Irwin was hardly ready for such radical action, and Indian frustration led to further terrorist explosions. A mile from New Delhi the viceroy's train was struck by a bomb, which narrowly missed his carriage.

Congress met at Lahore in December 1929 under the presidency of Jawaharlal Nehru, raising its battle cry of *purna svaraj* ("complete independence") as India's new political demand. "Independence for us means complete freedom from the British domination and British imperialism," proclaimed Jawaharlal. As the decade of the 1920s drew to its end, India's tricolor was unfurled in Lahore amid deafening roars of *"Inquilab zindabad!"* ("Long live the revolution!"). January 26, 1930, was proclaimed Independence Day and was celebrated as such throughout India, with millions of nationalists reciting the pledge: "We believe that it is the inalienable right of the Indian people, as of any other people, to have freedom and to enjoy the fruits of their toil and have the necessities of life, so that they may have full opportunities of growth. We believe also that if any government deprives a people of these rights and oppresses them, the people have a further right to alter it or to abolish it. The British Government in India has not only deprived the Indian people of their freedom, but has based itself on the exploitation of the masses, and has ruined India economically, politically, culturally, and spiritually." Motilal Nehru resigned his assembly seat and called upon all Indian members to do the same. The nation stood poised on the brink of a second great *satyagraha.*

The salt tax was the issue Gandhi chose around which to mobilize mass support. Not only was salt the one commodity every Indian required, but its production remained an official, heavily taxed monopoly, and it was illegal for anyone else to manufacture or sell salt. If a peasant living near the sea dared to pick up and use natural salt, he could be arrested. Declaring that he would not return to his *ashram* till the salt tax was repealed, the Mahatma led a band of seventy-eight *satyagrahis* from Sabarmati on March 12, 1930. It was the start of a 240-mile trek to Dandi on the coast. Staff in hand, the sixty-one-year-old Mahatma seemed to many an onlooker a latter-day Moses leading the exodus of his people. The pilgrims gathered support from crowds as they walked, and each day the march was reported by newspapers and wire services across the land and around the world, inspiring millions of empathetic watchers to wait anxiously to see whether the government would arrest or otherwise try to stop the bald, half-naked little man, whose *ashramites* called him *Bapu* ("Father").

On April 6 Gandhi reached the sea. As he walked into the surf to pick up a lump of natural salt, Sarojini Naidu, India's poetess-*satyagrahi* who was to become the nation's first woman governor, shouted "Hail, law breaker!" There were no police, no government officials in sight or sound of that historic scene. Gandhi issued a public statement confessing his "breach of the salt law" and urging Indians everywhere to "manufacture salt" or "help themselves" to the natural salt found in creeks or pits or at the seashore. It was a formal declaration of war against the tax, and against the government that collected it. Across the subcontinent countless peasants took their Mahatma's advice and broke the law to obtain the daily salt they required. Arrests came swiftly and in numbers. Jawaharlal Nehru was sentenced to six months on April 14, and his father became acting president of the Congress in his stead. Gandhi's wife, Kasturbai, led women *satyagrahis* to picket liquor shops, and in this campaign women began to appear in great numbers on nationalist battle lines. Gandhi was arrested on May 4 and remained in prison until January of the following year. News of the Mahatma's arrest stimulated protests and strikes throughout the land, including violent outbursts in Calcutta, Delhi, and Peshawar. India's prisons were packed with tens of thousands of patriots. In June 1930 the entire Working Committee of Congress was arrested.

Sir Tej Bahadur Sapru (1875–1949) and Dr. Mukund Ramrao Jayakar (1873–1959), the leaders of India's National Liberal Federation, which had broken from Congress before the end of the war, went as emissaries from Lord Irwin to Gandhi, seeking to bring Congress to a Round Table Conference on Reforms. Sapru, like the Nehrus, was a Kashmiri brahman, who became a distinguished Allahabad barrister before turning to politics and had served as law member on Lord Reading's Executive Council. Jayakar was a middle-class Bombay Pathare Prabhu, who also became a barrister and a leading educator in Poona, serving near the end of his life as vice-chancellor of the university. The Nehrus were transported from their U.P. prison to Gandhi's cell in August. Motilal and Jawaharlal were released soon after, though Gandhi remained in prison a bit longer. Government convened its first Round Table Conference on Reforms in London in November 1930, a glittering assembly of Indian princes, *zamindars, taluqdars,* industrialists, liberals, and Muslims and other minority community leaders, but without representatives of Congress, it was like trying to stage *Hamlet* without the Prince of Denmark. Meanwhile, Britain's Tory opposition, led by Winston Churchill, almost launched a noncooperation movement of its own against the very concept of the Round Table Conference, insisting that Simon's commission and its report sufficed for the "next step" in Indian constitutional reforms. The first London conference ended with little accomplished. Before a second conference was convened, Gandhi was released from jail.

For the Muslim League, as well as Congress, 1930 proved to be a year of revitalization. The Punjab's great poet-philosopher, Muhammad Iqbal (1877–1938), presided over the league's annual meeting at Allahabad, proposing "the formation of a consolidated North-

West Indian Muslim state" as "the final destiny of the Muslims, at least of North-West India." Iqbal specified that this Muslim state should embrace the four provinces of Punjab, Sind, Baluchistan, and the Northwest Frontier, thus delineating what was virtually to become the post-1971 boundary of Pakistan. Significantly enough, this first major Muslim demand for a separate Islamic state in India did not include Bengal. Iqbal justified his demand with much the same argument Jinnah would use a decade later at Lahore: "We are seventy million and far more homogeneous than any other people in India. Indeed the Muslims of India are the only Indian people who can fitly be described as a nation in the modern sense of the word." At this same time in Cambridge, a number of Muslim Indian students, led by Choudhry Rahmat Ali, published a pamphlet called *Now or Never,* in which they used the name *Pakistan* for the Islamic "fatherland" in India. An Urdu word, derived from the Persian, meaning "Land of the Pure" (*Pak* is "ritually pure"), Pakistan was also explained as an acrostic: *P* for Punjab, *A* for Afghania (the Northwest Frontier Province), *K* for Kashmir, *S* for Sind, and *Tan* for the last part of Baluchistan. Here, too, whether by design or by oversight, not even symbolic justification could be found for including Bengal in this newly proposed nation.

Irwin hoped that, by releasing Gandhi on January 25, 1931, he would be able to reach an agreement with him that might heal the wounds so sorely afflicting British India. Talks between Irwin and the Mahatma began that winter at the viceregal palace in Delhi. Churchill found it "nauseating" to see pictures of the "half-naked fakir" striding to "parley on equal terms" with the king-emperor's viceroy. Gandhi had a fine sense of humor, however, and when asked by reporters if he planned to visit Buckingham Palace dressed in his usual attire, he quipped, "Some go in plus fours, I will go in minus fours." On March 5, 1931, a Gandhi-Irwin pact was announced by the viceroy. Civil disobedience was "discontinued," bringing a halt to all "organised defiance of the provisions of any law" and to all attempts "to influence civil and military servants or village officials against Government or to persuade them to resign their posts." The government, in turn, recognized and approved of the *svadeshi* movement, though not of the boycott of British goods, which was to be discontinued along with other aspects of *satyagraha*. Gandhi also agreed to

attend the next Round Table Conference in London. Subhas Bose and
other leaders of India's radical youth read the pact as nothing but
capitulation to government, hardly less than a sellout of nationalist
demands in return for vague British promises. They were specially
troubled by Gandhi's voluntary withdrawal of the demand for official
inquiries into numerous allegations of police atrocities. During the
past year, between sixty and ninety thousand men, women, and children had been jailed or booked by police for political activities. Between April and July 1930 alone, the government had admitted
killing over one hundred civilians in civil conflict and wounding some
four hundred more. Many Indians read the Gandhi-Irwin Pact as a
Congress retreat from *purna svaraj,* but Gandhi insisted it was not so
intended. He conceded the possibility of India's choosing to remain
within the Commonwealth "on terms of absolute equality"; precisely
what happened after 1947.

Irwin left India in April 1931, succeeded by the older and more
rigid Conservative Lord Willingdon (1866–1941), who had earlier
been governor of Bombay and Madras and who was to reign as viceroy until 1936. Britain's first Labor government coalition was now
displaced by depression politics with a Tory-dominated coalition
cabinet, and the Second Round Table Conference in London proved
little more than an exhausting experience for Gandhi, who made the
mistake of trying to represent Congress as its sole delegate. He had
hoped thus to symbolize the unity of India's nationalist movement,
insisting he went "with God as my guide," but proved instead the
frailty of human flesh and the weakness of his own judgment by attempting so herculean a task alone. From London Gandhi's voice was
heard for the first time in the United States, broadcast coast-to-coast
for half an hour in September 1931. "I feel in the innermost recess
of my heart," the Mahatma told America, "that the world is sick unto
death of blood-spilling. It is seeking a way out, and I flatter myself
with the belief that perhaps it will be the privilege of the ancient land
of India to show that way out to the hungering world." Gandhi's ardent public appeals for sympathy and support for India's poverty-
stricken masses would have been more warmly received were it not
for the growing ranks of unemployed Englishmen, especially in the
depressed mill towns of Lancashire, that darkened Britain's national
horizon at this time. For many a Labor M.P., the choice between his

own constituency's needs and Indian nationalist aspirations seemed no less clear than it had once appeared to the East India Company's directors.

Iqbal, the Aga Khan, and Jinnah represented the Muslim League at the London Conference; Master Tara Singh the Sikhs; and Dr. Ambedkar the untouchables (the "depressed classes"). All of them demanded separate electorates for their communities. Gandhi insisted, however, that the untouchables were Hindus and claimed to represent them "in my own person." He viewed British attention to Ambedkar's separatist political demand as an even more insidious example of "divide and rule" than the grant of separate Muslim electorates had been, and he would soon undertake a fast-unto-death against any such grant. On January 4, 1932, one week after Gandhi returned to India, he was arrested and taken back to jail. Willingdon's government decided to try a tougher line, outlawing Congress, rounding up all members of the Working Committee and the Provincial Committees, and imposing Ordinance Raj over most of India's provinces. Within four months eighty thousand Indians were again behind bars. Boycott spread across the land. Many peasants refused to pay taxes. *Kisan* and youth organizations were officially banned but became very popular. Terrorist activity increased, with women joining their brothers and husbands in the terrorist struggle. The U.P. and Northwest Frontier were armed camps, and British troops at Meerut and Peshawar were kept at combat alert.

In mid-August 1932 Ramsay MacDonald announced his Communal Award, a voting scheme to be incorporated in India's new Constitution. The separate Muslim electorate formula had been expanded to include Sikhs, Indian Christians, Anglo-Indians, Europeans, and depressed classes. A month later Gandhi launched his fast-unto-death, arguing that as "a man of religion," he had "no other course left open" to him. Gandhi had used the "fiery weapon" of fasting before, but his friends, as well as his opponents, wondered why he chose this particular issue for so terrifying a tactic, since it seemed directed more against the untouchables than the British government. Gandhi considered this issue central to the very survival of Hinduism, however, and for that reason he was willing to concede more reserved seats to Ambedkar for his party than the British had done. All he asked was that the depressed classes should not think of themselves as members

of any religious community but Hinduism. Ambedkar criticized
Gandhi and denounced Hinduism for treating untouchables "as lep-
ers, not fit for ordinary intercourse." By dramatically focusing caste
Hindu attention on the intolerable plight of the untouchables, however, Gandhi did convince his followers to open thousands of temples
to *harijans*. He also sensitized world public opinion to the inhumanity
of untouchability. As with his salt march, Gandhi's religio-political
instinct proved its universal revolutionary genius. "I believe that if
untouchability is really rooted out," he said, "it will not only purge
Hinduism of a terrible blot but its repercussion will be worldwide. My
fight against untouchability is a fight against the impure in humanity."
After almost a week of frenzied negotiations, Gandhi and Ambedkar
reached an agreement in the Yeravda jail. By that historic pact, depressed classes would be assured 147 seats on provincial councils,
rather than the 71 awarded to them by the prime minister's scheme;
they would also hold 18 percent of the Central Assembly seats, as long
as they ran for election by the general electorate rather than under
any separate rubric. The new pact impelled the Hindu Leaders Conference to resolve "that henceforth, amongst Hindus, no one shall be
regarded as an untouchable by reason of his birth, and those who
have been so regarded hitherto, will have the same right as other
Hindus in regard to the use of public wells, public schools, public
roads and other public institutions." It would take another two decades, however, for untouchability to be abolished by law, and many
more years than that before its insidious social ostracism and economic disabilities would be removed from the heart of Indian society.
Britain's cabinet accepted the Yeravda Pact as an amendment to the
prime minister's Communal Award, permitting Gandhi to end his fast
just a week after it started. The following week was celebrated
throughout India as Untouchability Abolition Week.

While the Mahatma's fast highlighted the misery of untouchables, life for the average Indian peasant at this time was hardly much
rosier. Between the census of 1921 and one taken a decade later, India's population had grown from some 305 to 338 million, but food-
grain production remained virtually static and world food prices fell
precipitously. The Calcutta index of cereal prices dropped from 153 in
1920 to 68 in 1932, and the sugar index fell from 407 in 1920 to
146 in 1932. In real terms, then, the average peasant dropped closer

to the bottom margin of survival, and many millions were obliged either to abandon their debt-encumbered land and become landless laborers at subsistence wages, or to move to the nearest city, where they found neither work nor food. Expenditure for expanded irrigation and other such public works was devoted mostly to raising more commercial crops for export, especially jute, cotton, and rubber, whose production increased by almost 50 percent after the end of World War One. Though food prices tumbled swiftly, the index of manufactured goods dropped less rapidly, at least until the impact of the Wall Street collapse in 1929. The boom in the Indian cotton industry, which in the year of the first *satyagraha* had sent the price index of cotton manufactures up to 325 (using July 1914 as the base of 100) was deflated to 119 by 1932, and Lancashire never fully recovered from the inroads on its Indian market made by *svadeshi* cloth. The combination of nationalist agitation and widespread industrial unrest and strikes, which reached their peak in 1928–29, diminished the flow of capital to India and brought economic development to a standstill. The pace of urbanization increased as the pressure of population and underemployment drove millions of Indians from some seven hundred thousand villages to neighboring towns and cities, which claimed over 11 percent of India's population in 1931. The largest cities attracted the highest percentage of migrants; in 1931 India's fifteen largest cities contained more than seven million people, and a decade later almost ten million. Though there were still no reliable statistics on urban unemployment, the visible army of beggars and street-dwellers who filled Calcutta, Bombay, Delhi, Madras, Lucknow, and Banaras, continued to spread like many other endemic urban diseases throughout the twentieth century. One visit to an Indian urban slum would suffice to explain Gandhi's wholesale rejection of "modern civilization" and industrialization in favor of village communal society, where some balance of human, animal, and vegetative life might at least be sustained in harmony, without either affluence or the total abandonment of human dignity.

Lord Willingdon's government combined harsh repression with encouragement of Gandhi's anti-untouchability agitation, sanctioning the introduction of Ranga Iyer's Untouchability Abolition Bill and Dr. Subbaroyan's Temple Entry Bill in Delhi's Assembly in early 1933, thereby undercutting nationalist civil resistance. In May 1933

Gandhi went on a three-week "self-purification" fast to reduce brahman opposition to those bills. The government released the Mahatma from jail before he started his fast, and he, in turn, suspended his second *satyagraha* for six weeks to allow his countrymen to devote themselves wholeheartedly to "the service of Harijans." Subhas Bose and other left-wing leaders of Congress were outraged. They called for the suspension of Gandhi, insisting "the Mahatma as a political leader has failed," that Congress required fresh leadership and "radical reorganization."

While *satyagraha* petered out, the machinery of the Raj continued its "reforms" with a Third Round Table Conference, held in November 1932. The publication of the government's White Paper in 1933, however, evoked almost universal criticism in India, since so much of its scheme bore but slight resemblance to conclusions reached at any of the Round Table Conferences. Nonetheless, through the mill of Parliamentary Joint Committee and the ritual of three readings in both Houses, following talismanic consent from the crown, the White Paper was transmuted into the Government of India Act of 1935, India's last British-made Constitution. Acknowledging the "chorus of disapproval" raised by his reforms, Secretary of State Sir Samuel Hoare countered by arguing: "No one in India has produced a workable alternative." In the Commons, the bill was vigorously attacked by Winston Churchill for granting all Indians too much power and responsibility, and by Clement Attlee for giving the princes too much but Congress too little. The act was to create a "Federation of India," which would have consisted of eleven totally responsible governors' provinces (the provinces of British India); the princely states; and a small number of ICS-run, centrally administered areas called chief commissioners' provinces. This elaborate federation never actually came into existence, since the princes would not agree on questions of protocol or other weighty matters, and Congress was hardly willing to share national power with a chamber full of despotic puppets who had inherited thrones held by the grace of the British crown.

Provincial autonomy, however—the second major principle embodied in the act—was inaugurated on April 1, 1937, following nationwide elections. Burma became a totally separate colony at this time. Within British India, Orissa was established administratively as a governor's province separate from Bihar, as was Sind from Bombay.

The Northwest Frontier Province was also advanced to this status of full responsibility. In six governors' provinces (Bengal, Bombay, Madras, U.P., Bihar, and Assam), there were to be two legislative chambers, a council and an assembly; in the other five provinces there was to be a unicameral assembly. All legislators were to be elected, the franchise expanded to include some thirty-five million propertied Indians, six million of whom were women, and 10 percent untouchables. In keeping with Ramsay MacDonald's Communal Award, as amended by the Yeravda Pact, separate electorates were retained. The governor, however, reserved so many special "certifying" and other "safeguard" powers that Congress initially argued that the act was no forward step at all, but merely a deception, proffering the shadow while retaining the substance of power. The Marquess of Linlithgow, who became viceroy in 1936, had been a member of the Parliamentary Select Committee on these reforms and was anxious to see them implemented, assuring the Congress Working Committee that the reserve powers were not meant to subvert the act's principle of full provincial responsibility. Nehru condemned the act as a "new charter of slavery" in his presidential address to Congress in 1936, but agreed the following year to direct the Congress ministries that took over in seven of the governors' provinces (Bihar, Bombay, Central Provinces, Madras, Northwest Frontier, Orissa, and U.P.) after the "notable victory at the polls" won by Congress early in 1937. In 1938 an eighth Congress ministry came to power in Assam. Though less than half of the 1,585 provincial legislative seats contested throughout India were open to the general electorate, Congress candidates won all of them, plus 59 more from among the separate electorate contests, to emerge with a total of 716 legislative members. The Muslim League, India's second most powerful party, won only 109 seats, not even a majority of the 482 separate Muslim seats reserved nationwide. The Congress victory was most impressive, its candidates sweeping 70 percent of the total popular vote. With seemingly good reason, then, Nehru could comment early in 1937 that there were "only two parties" in India, the government and the Congress. Jinnah, however, insisted there was a third party—Muslim India—and devoted the next ten years of his life to proving the political reality of his claim.

Since 1935, when Jinnah returned from self-imposed exile in London at the behest of his Oxford-educated friend, Liaquat Ali Khan

(1896–1951), he had worked at "reviving the dormant Muslim League." With virtually no financial and little popular support, the League had languished, despite its high point of aspiration articulated in Iqbal's stirring address in 1930 at Allahabad. Jinnah's dynamic leadership helped inspire renewed confidence, however, as he toured India selecting Muslim members for a central parliamentary board designed to offer some political cohesion to the many parties competing for the claim to represent India's eighty million Muslims. In the limited time at his disposal, and given his personal predilection for parlor-room politics rather than mass rallies, Jinnah could hardly do more than construct a skeleton for the League, which did not even find enough candidates to contest all the Muslim seats in the 1937 elections. The League's manifesto and platform, moreover, were mostly tepid restatements of Congress policies that were phrased in more passionate, soul-stirring terms by Nehru's brilliant, Socialist-oriented leadership, which focused on British imperial exploitation of Indian poverty as the key issue of its campaign. Despite the League's poor electoral performance, a careful analysis of the votes reveals that Muslim India was, in fact, hardly attracted to Congress candidates, even though they had to be Muslims to qualify for the "separate" seats. The only overwhelmingly Muslim province that went strongly for Congress was the Northwest Frontier, where the most popular political leader was one of the Mahatma's disciples, a giant Pathan, the "Frontier Gandhi," Khan Abdul Ghaffar Khan (1890–1988). The Frontier had suffered the harshest British repression during the second *satyagraha* campaign, feeling the blast of British bombs in Peshawar as part of Lord Willingdon's no-nonsense policy, and Abdul Ghaffar Khan, like Gandhi, spent many years behind prison bars.

In the wake of the 1937 elections, Jinnah realized that if the League was to become anything more than a middle-class Muslim pressure group, it would have to either attract enough mass support to win control of some governors' provinces, at least those of the Punjab and Sind, where Muslims were in a majority, or join Congress in coalition ministries throughout the country that would give its members of the legislative assemblies some real power over provincial administrations. Congress rejected the latter alternative, insisting that cabinets had to be united. The ensuing two-year interlude of Congress provincial responsibility was most important historically not because of what Congress accomplished in office, but rather for what it failed to do to assuage Muslim criticism and growing concern. This era thus became the seed time for the emergence of Pakistan. Congress ministers had suffered such personal hardship and waited so long for their first taste of political power that it was hardly surprising to find many of them taking immediate advantage of their new offices, appointing relatives or caste brethren to jobs at their disposal, awarding contracts to firms owned by inlaws, neighbors, or those willing to offer the greatest "consideration," and taking the time to answer some constituents' complaints but not others. In a word, they were human; or, what seems doomed to be worse, politically human. They rewarded friends, curried favor among the richest and most powerful in their communities, ignored minorities, and often proved vindictive to enemies.

For Congress, however, such universal human weaknesses proved particularly disastrous, since they could be pointed out by Jinnah and the League to Muslims everywhere as "proof positive" of the bitter fruit born of a Hindu Raj. The League's battle cry became "Islam in danger!" The brickbats of Muslim grievances could be hurled with devastating effectiveness from every League platform and newspaper headline against the glass houses of Congress ministries, who had no League counterpart at which to throw back. Muslim congressmen like Maulana Azad, one of Nehru's right-hand men in the high command that supervised the provincial ministries, could insist as loud or long as they wished that charges of "injustice" to Muslims were "absolutely false," just as the British had earlier denied Congress's charges of "divide and rule" and "economic drain." The Muslims could always find sufficient evidence—even if it were only that Congress made *Bande Mataram* the anthem sung by Muslim as well as Hindu schoolchildren in public schools, or that the Congress flag with Gandhi's *charka* in the middle flew from all public buildings—to "prove" Hindu bias. Once that seed of distrust was planted in the mind, much as with the mutiny's greased cartridges, nothing could dislodge it. Jinnah could now build his League into a political weapon powerful enough to tear the subcontinent apart. "On the very threshold of what little power and responsibility is given," Jinnah declaimed in his presidential address to the League in 1937, "the majority community have clearly shown their hand that Hindustan is for the Hindus."

Nehru and Azad continued to insist that there was no Muslim

problem, claiming that a hundred thousand nationalist Muslims were members of Congress and that Congress represented all of India. Meanwhile, Jinnah brought the vagrant flocks of diverse provincial Muslim parties into one political fold under his League, winning over Fazlul Huq's radical *Praja* Party and control of Bengal in October 1937 at the same time as he captured the reactionary ministry of the baronial landlord Sir Sikander Hyat Khan, whose Unionist Party ruled the Punjab. Nehru later attacked Jinnah's "astonishing course" as "aggressively antinationalist and narrow-minded," calling it "a negative program of hatred and violence, reminiscent of Nazi methods." By the spring of 1938, when the League met in Calcutta, Jinnah was able to report that "within less than six months we have succeeded in organizing Musalmans all over India as they never were at any time during the last century and a half." Nehru was now ready to meet Jinnah "whenever he cares to meet me," but the correspondence that passed between these two leaders in 1938 reflected only the increasing distance that divided them and would soon divide British India. Nehru's mind was "obsessed" at this time with "the critical international situation and the terrible sense of impending catastrophe that hangs over the world"; Jinnah's with how best to build his League and make good his claim for it of "complete equality with the Congress." Subhas Bose was elected Congress president in 1938. Jinnah met briefly with Gandhi, Nehru, and Bose that year, but these talks broke down over the League's demand that it be recognized by Congress as the sole party of India's Muslims, a claim that would have negated Congress's insistence that it represented the entire nation. Following the breakdown of discussions, at the League session in Patna in December 1938, Jinnah reported that Congress had "killed every hope of Hindu-Muslim settlement in the right royal fashion of Fascism."

British officials watched the process of factional disintegration and communal strife with a mixture of sadness and smug smiles as each report of nepotism, corruption, rape, and rioting hit the headlines or came in red-tape-bound "confidential" files across their desks. Paternalistic pride continued to convince elder British servants of the crown that they were, after all, best suited to rule India's peasant mass, Muslim as well as Hindu, though younger ICS officials, like Maurice Zinkin, who reached India in 1938, "mostly accepted the inevitability of self-government at some time" during their careers. Throughout this period of provincial responsibility, British stock rose in all parts of India as more and more people looked to the Raj and its steel frame of officialdom as the final arbiter of India's fratricidal disputes. Congress chief ministers, as well as the high command, also looked to their British official "advisers" and "servants" for the practical guidance and daily professional help without which the Machine could not have functioned. Whether Bombay's political leader was an Englishman or Indian mattered much less to the daily routine of that great city's life than did the ceaseless flow of bureaucratic paper, confidence in government credit, fear of its coercive force, and faith in its ultimate justice. British officials had always known that, of course, and Indian ministers now learned it, frustrated to find how little power they actually had once they were finally given what they had always insisted they wanted; ashamed to discover how easily tempted they could be; and relieved to learn that someone was still in the office next door, ready to answer questions they had never dreamed of during the years of political apprenticeship they had spent behind prison bars. It was a sobering awakening and a terribly deflating one; for men like Nehru, it was the source of brooding depression. Men who had grown old in the struggle for freedom were suddenly forced to admit—if only to themselves—that freedom alone would not suffice to solve India's complex problems.

At the end of 1938 Congress itself was torn asunder by the struggle waged by Subhas Chandra Bose at Tripura to retain its presidency despite Gandhi's dissatisfaction with his leadership and the old guard's united support of the Mahatma. Gandhi tried first to convince Nehru to take up the Congress burden once again, but Jawaharlal refused. Then Patel, acting on Gandhi's instructions, urged Bose to step aside in favor of Congress historian Pattabhi Sitaramayya, but Bose knew he had the support of student, worker, and peasant delegates of the All-India Congress Committee, and he decided to fight to retain the position he had held throughout the year. It was the first contested presidential election in Congress history, and Bose won by 1,580 votes to 1,375. The significance of that revolutionary victory was not lost upon Gandhi, whose position as moral leader of the party was challenged by the fiery *Netaji* ("Leader," a title Bose took in emulation of the *Führer*). Nor was Bose's rebellion and democratic victory accepted by the Mahatma or his loyal disciples on the Working Committee, twelve of whom resigned in protest against their own president in February 1939. Bose thus found himself presiding over a party that could not function for lack of cooperation within its own highest management ranks. In desperation he appealed to Gandhi for assistance, but the Mahatma ignored him. Nehru tried to mediate, but Patel, Pandit G. B. Pant (chief minister of U.P.), and Rajendra Prasad (Bihar's "boss," who was to become India's first president) wanted nothing further to do with the rebellious Subhas Chandra, who had broken party discipline and was being punished with ostracism by his "caste." Bose's health broke down, yet he tried to lead Congress alone on a stepped-up campaign against the Raj. Gandhi's opposition remained adamant, however, and Subhas was finally driven to resign in May, replaced as president by Prasad, who was "elected" by the Working Committee, to which he belonged. Bose and his brother Sarat Chandra now abandoned Congress to form their own Forward Bloc Party in Bengal, devoted to the revolutionary struggle for freedom. Before the end of 1939, however, mightier forces of world politics would intervene once again to alter the course of Indian history.

TWENTY-TWO

THE IMPACT OF WORLD WAR TWO

(1939–46)

World War One had served as a catalyst to unify India's nationalist movement, but World War Two shattered all hope of Congress-League reconciliation. On September 3, 1939, when Lord Linlithgow informed the people of India that they were at war with Germany, echoing Great Britain's decision, Congress felt humiliated and betrayed. Bitter memories of the aftermath of World War One remained vivid in the minds of Gandhi and his colleagues. Though Nehru and other Indian socialists sympathized personally with Britain's humanitarian struggle against Nazi and Fascist dictatorships, they were unwilling to support a government that had not consulted them prior to its declaration of war, nor would they consent to fight without the grant of *purna svaraj*.

"A free democratic India will gladly associate herself with other free nations for mutual defence against aggression and for economic co-operation," the Working Committee resolved, "[but] co-operation must be between equals and by mutual consent." Britain was not ready to address itself to the problem of Indian independence in the fall of 1939, however, and so in October Congress's high command ordered all of its provincial ministries to resign. Jinnah could barely believe his good fortune. The last Congress ministry abandoned provincial office by mid-November, and on December 2, Jinnah proclaimed Friday, December 22, 1939, Muslim India's Day of Deliverance. To mark their "relief" that "the Congress regime" had finally "ceased to function," the League passed resolutions and offered prayers, thanking God for "deliverance from tyranny, oppression and injustice during the last two and a half years." It was the death knell to Congress dreams of inheriting control over a united Indian nation.

Relieved of the task of attacking Congress ministries, Jinnah could focus his attention on the challenging problem of formulating his party's constitutional goal. In March 1940 that goal was to emerge from the Lahore session of the League as Pakistan—in all but name alone.

"The problem in India is not of an intercommunal character but manifestly of an international one," Jinnah told his Lahore audience of an estimated hundred thousand Muslims, "and it must be treated as such. So long as this basic and fundamental truth is not realised, any constitution that may be built will result in disaster and will prove destructive and harmful not only to the Musalmans but to the British and Hindus also. If the British Government is really in earnest and sincere to secure peace and happiness of the people of the subcontinent, the only course open to us all is to allow the major nations separate homelands by dividing India into autonomous national states."

There were at least six different proposals as to how that division of the subcontinent might best be accomplished, one recommending no less than three Muslim nations—Pakistan, Bengal, and "Usmanistan" in the South, to be carved from Hyderabad, whose *nizam*'s family name was Usman. The formula finally agreed upon at Lahore was hammered out as a resolution on the night of March 24, and Fazlul Huq of Bengal played so prominent a role in its drafting that there is ample reason to believe it was originally designed to create two independent ("autonomous and sovereign") Muslim states, Pakistan in the Northwest, and Bangladesh in the East, rather than one. Neither name was, however, actually specified, and ambiguity in the resolution's drafting permitted the press, whose morning headlines dubbed it the "Pakistan Resolution," to convince India's Muslims that a single Pakistan was what they really wanted. The Lahore Resolution stated "that no constitutional plan would be workable in this country or acceptable to the Muslims unless it is designed on the following basic principles, viz., that geographically contiguous units are demarcated into regions which should be so constituted with such territorial readjustments as may be necessary that the areas in which Muslims are numerically in a majority, as in the north-western and eastern zones of India, should be grouped to constitute 'independent

States' in which the constituent units shall be autonomous and sovereign." The League's Working Committee was, moreover, authorized in the resolution's final paragraph to "frame a scheme of constitution in accordance with these basic principles, providing for the assumption finally by the respective regions of all powers such as defense, external affairs, communications, customs and such other matters as may be necessary." The province of East Pakistan was never granted those sovereign powers, though almost a quarter of a century after Pakistan was born, Bangladesh fought for and won them with India's help.

Gandhi was "deeply hurt" by the Lahore Resolution, denouncing it as a call for the "vivisection of India." He appealed to Muslims' "good sense" and "self-interest" to desist from "the obvious suicide" that partition would mean. In his presidential address at Lahore, Jinnah had insisted that Hindus and Muslims never had and never could evolve "a common nationality," and he labeled the "misconception of one Indian nation" the root cause of most of India's troubles. "The Hindus and Muslims belong to two different religious philosophies, social customs, and literatures. They neither intermarry nor interdine together and, indeed, they belong to two different civilisations," declared the Father of Pakistan. Gandhi claimed, however, that the "two-nation" theory was not true. "The vast majority of Muslims of India are converts to Islam or are descendants of converts. They did not become a separate nation as soon as they became converts," argued the Mahatma. His argument was strongest when he focused upon Bengali Muslims, whose language, dress, appearance, food, and social life made them virtually indistinguishable from Bengali Hindus. To bolster its single-nation argument, Congress elected Maulana Azad its president from 1940 to 1945. The passionate dialogue started by Jinnah and Gandhi continued unabated throughout the next seven years, taken up by millions of lesser voices; volumes of historical "evidence" were produced in support of both positions, and learned "leading authorities" (living as well as dead) were cited to "prove" each argument in direct contradiction or refutation of the opposite side. Intellectually, India thus became a land divided by advocates of the one-nation and two-nation theories long before the subcontinent's partition in 1947.

Throughout the war, Jinnah proved a shrewder political leader than either Gandhi or Nehru, courting British support without actually committing the League to anything more than "benevolent neutrality." His brilliance as a barrister was put to good use on the political trail he blazed with unflagging zeal. "Where is the nation which is denationalised?" Jinnah asked in response to Gandhi's cry of vivisection. "Where is the central national Government whose authority is being violated? India is held by the British power and that is the hand that holds and gives the impression of a United India and a unitary Government. Indian nation and central government do not exist." It was the sort of argument British officials most appreciated, especially when their government was laboring to support a war whose personnel and materiel demands imposed enough strain without Congress complaints to worry about. Soon after Churchill's War Cabinet replaced "Munich" Chamberlain's discredited government in May 1940, Linlithgow invited Jinnah to Simla to confer on questions relating to India and the war. Jinnah submitted a list of tentative proposals to the viceroy that July, just after the fall of France, and they became the outline for Linlithgow's "August Offer" of 1940. To "more closely associate Indian public opinion with the conduct of the war," the viceroy offered to invite a "number of representative Indians" to join his Executive Council and proposed the creation of a War Advisory Council, to include princes and "other interests in the national life of India as a whole." Jinnah further recommended, and Linlithgow promised, that the government should adopt no future constitutional scheme without "the previous approval of Muslim India." "It goes without saying," agreed the viceroy, "that they [His Majesty's Government] could not contemplate transfer of their present responsibilities for the peace and welfare of India to any system of government whose authority is directly denied by large and powerful elements in India's national life." The important role played by Muslims in the Indian Army at home and abroad greatly strengthened Jinnah's hand. Secretary of State L. S. Amery now spoke in the Commons, bemoaning the "failure" of Congress leaders to represent all of India. "If only they had succeeded," Amery said, "if the Congress could, in fact, speak for all main elements in India's national life then, however advanced their demands, our problem would have been in many respects far easier."

That September Congress met and authorized Gandhi to launch

satyagraha, if he decided it was a necessary "next step." On October 17, 1940, the Mahatma initiated his individual *satyagraha* campaign by sending his foremost disciple, Vinoba Bhave, "ceremonially" to proclaim his resolve to resist the war effort nonviolently. Vinoba, subsequently revered as India's "Walking Saint," was arrested and sentenced to three months in prison. Nehru notified the police of his intention to follow in Vinoba's footsteps, and ten days later he, too, was arrested, receiving the much harsher sentence of four years. "It is 'do or die,' " Gandhi proclaimed. "There is no turning back. Our case is invulnerable. There is no giving in." By mid-1941 some twenty thousand *satyagrahis* had been convicted, and fourteen thousand were back in jail. Nor was Indian opposition to the Raj confined to nonviolent protest: in Bengal the terrorist movement revived following the arrest of Subhas Bose in 1940 and the banning of his Forward Bloc Party. Bose himself escaped house arrest on the eve of his trial in 1941 and journeyed in disguise across northern India to Afghanistan. He flew from Kabul to Moscow and, during the pre-June Russo-German springtime of friendship, from Moscow to Berlin. In Germany he was welcomed by Hitler and given high-powered radio facilities to beam daily broadcasts to India in several languages urging his countrymen to rise in revolt against British tyranny. Many of Bose's followers did as their Netaji urged, and throughout the war trains were derailed, British soldiers assassinated, and bombs exploded in public buildings, theaters, and bazaars as the violent wing of India's nationalist movement erupted.

As Hitler's armies seized control of the Balkans and marched across Russia, with Rommel poised within range of the Suez Canal, Franklin Roosevelt met with Winston Churchill aboard the cruiser *Augusta* in August 1941 to commit the United States and Great Britain jointly to peace aims that came to be known as the Atlantic Charter. Though the United States had not as yet entered the war, this proclamation affirmed "the right of all peoples to choose the form of government under which they will live" and expressed the "wish to see sovereign rights and self-government restored to those who have been forcibly deprived of them." Indians were at once heartened by this new charter of freedom, but they were not permitted to enjoy it long, for on September 9 Churchill told the Commons that the Atlantic Charter did not apply to "India, Burma and other parts of the

British Empire." The government of India was, nonetheless, anxious to win popular backing for its war effort and on December 3, 1941, released all Congress *satyagrahis* from prison, sensing, no doubt, that Japan was on the verge of entering the war. The December 7 attack on Pearl Harbor, which was followed by a succession of lightening Japanese strikes against Shanghai, Hong Kong, the Philippines, Malaya, Indochina, and Thailand, left the world dizzy and Allied forces reeling on every front. On February 15, 1942, the "unconquerable" island of Singapore, whose big guns pointed out to sea and were too solidly mounted in concrete to swing around, was taken from Malaya's southern tip by a Japanese army that caught Britain's command with their gin and tonics half down. Colonel Hunt surrendered his force of sixty thousand Indian troops without firing a shot, and this garrison of prisoners of war became the backbone of Subhas Bose's Indian National Army (INA).

With Japan at India's eastern door and China virtually swallowed, the importance of India as an Allied base was such that even Churchill conceded that His Majesty's Government could wait no longer in seeking to reach some constitutional settlement. On March 11, 1942, Churchill announced his dispatch of Sir Stafford Cripps, leader of the House of Commons and a member of the War Cabinet, on a mission to New Delhi. A Labour Party minister who was a friend of both Nehru and Gandhi and personally sympathetic to Indian aspirations for freedom, Sir Stafford raised nationalist hopes and expectations throughout India. He arrived in New Delhi on March 22 and remained there for three ill-fated weeks, proposing full dominion status for an Indian union "upon the cessation of the hostilities." Any provinces or Indian states that preferred not to belong to that union would have the right to "opt out," and British rule would remain essentially as it was throughout the duration of the war. "Why did you come if this is what you have to offer?" Gandhi inquired of Cripps. "I would advise you to take the first plane home." The Mahatma labelled Cripps's offer "a post-dated cheque on a bank that was failing." The Working Committee followed suit and rejected the offer, whose "novel principle of non-accession for a Province" was seen as a "severe blow to the conception of Indian unity and an apple of discord likely to generate growing trouble." The negative reaction from Congress sent Sir Stafford flying home in early April. Jinnah and the

Muslim League were prepared to accept the offer, since it essentially embodied their Pakistan demand, but Congress rejection left them no political option but to do likewise in order to compete most effectively for mass support.

In the wake of Cripps's failure, Gandhi launched his last great *satyagraha* campaign, the "Quit India" movement. "I am convinced that the time has come," stated the Mahatma on May 3, 1942, "for the British and the Indians to be reconciled to complete separation from each other." There was, he insisted, no "joint common interest" left to unite such distant and different nations. When asked by reporters to whom the British should relinquish their authority, Gandhi answered, "Leave India to God. If that is too much, then leave her to anarchy." Congress met in August to sanction the start of the new campaign, which took for its slogan, "Quit India," the shout that greeted every Englishman, woman, and child in the subcontinent for the remainder of the war. Jinnah and the League "deplored" what it called Congress's decision "to launch an 'open rebellion.' " The British treated it as such, arresting some sixty thousand Congress supporters by the end of 1942. Violence and repression rose to their peak at this time of terror, fear, and hatred. Gandhi could no more control his "troops" than the British could keep their police and soldiers from opening fire on what they considered "mobs of traitors." More than a thousand deaths and over three thousand serious injuries were directly attributed by official count to the riots that followed the Quit India *satyagraha*.

Despite widespread opposition to British rule and to the war, most Indians remained either peacefully apathetic or actively supportive of the Raj in its wartime effort to mobilize India's resources, which bolstered the economy, beefed up the army, and sent prices and profits into inflationary orbits that brought fortunes overnight to many merchant and industrialist families and penury to millions on pensions or fixed incomes. The population had grown by more than fifty million in the previous decade, with the rise more rapid in North India than in the South, the Punjab and Bengal experiencing the most rapid rates of growth. There was ample land and grain within the Punjab to absorb the increase in population during that decade, thanks to new canal irrigation of large tracts of fertile land, but Bengal, with its more densely populated area, could hardly support more mouths to feed. The average family holding in Bengal was about one acre. With wartime strains imposed upon internal transport, moreover, and the loss of Burma's rice surplus, the inflationary market for rice, heavily demanded by India's army (which grew from its prewar strength of 175,000 to almost two million men by 1943), led grain merchants and moneylenders in Calcutta to buy up peasant stocks that should have been kept in villages for food and seed. The resulting famine of 1943, a famine of wartime dislocation and official ineptitude, was the worst in recent Indian history, taking a toll of from one to three million lives, or ten to thirty times the total number of military deaths suffered by India during the war.

Industry grew more rapidly during World War Two than in any comparable previous era. The output of electric power increased by 45 percent from 1938–39 to 1943–44. Tata's iron and steel complex at Jamshedpur was the largest single producer of steel in the British Empire, turning out more than 1.5 million tons a year during the war. The production of cotton textiles, paper, and cement reached new peaks as demand soared not only at home but to help supply troops sent to the Middle East and North Africa. Industries that had never before been ventured in India, including bicycle, locomotive, and automotive factories, were started at this time. Bombay's chemical, pharmaceutical, and light engineering plants enjoyed an unprecedented boom, and Gujarat, Orissa, Bengal, Bihar, the Punjab, and U.P. all developed pockets of modern industry, whose standard wartime level of operation was slightly above "maximum capacity." Still, India remained industrially underdeveloped, and the wartime strain imposed upon its railway net was so great that many trains broke down from overloading, instead of from terrorist sabotage.

In June 1943 Field Marshal Archibald Wavell (1883–1953) was appointed viceroy, and he retained his command over India until the eve of its independence. The appointment of a soldier as viceroy was seen in India as proof positive of British determination to do nothing further about reforms during the war. Wavell himself, however, was far less "wooden" than he appeared, and he was more interested in politics than his inarticulateness and lack of social charm indicated to those who did not know him well. The American and Australian press welcomed the appointment as reflecting Britain's "determination to wage the eastern war vigorously," and Wavell was invited to attend

War Cabinet meetings during the summer of 1943 before sailing for India. The mounting tragedy of Bengal's famine seemed only further proof to most Indians of the destruction brought to their land by British rule and the impossibility of continuing to live under the incubus of foreign occupation. Earlier in the year Gandhi had fasted for three weeks to dramatize his protest against British rule, but the war diverted world concern, and the Mahatma's fast received little outside attention.

In the spring of 1943, Subhas Bose went on an "epic" ninety-day submarine voyage from Hamburg to Singapore, around the Cape of Good Hope. There were so few Indians residing in prewar Europe that the Free Indian Legion that Bose commanded in Berlin mustered fewer than a hundred soldiers, but sixty thousand troops awaited their Netaji in Southeast Asia. Tojo turned over all his Indian POWs to Bose's command, and in October 1943 Bose announced the creation of a Provisional Government of Azad ("Free") India, of which he became head of state, prime minister, minister of war, and minister of foreign affairs. Some two million Indians were living in Southeast Asia when the Japanese seized control of that region, and these emigrees were the first "citizens" of that government, founded under the "protection" of Japan and headquartered on the "liberated" Andaman Islands. Bose declared war on the United States and Great Britain the day after his government was established. In January 1944 he moved his provisional capital to Rangoon and started his Indian National Army on their march north to the battle cry of the Meerut mutineers: *"Chalo Delhi!"* On March 21, 1944, Subhas Bose and advanced units of the INA crossed the borders of India, entering Manipur, and by May they had advanced to the outskirts of that state's capital, Imphal. That was the closest Bose came to Bengal, where millions of his devoted followers awaited his army's "liberation." The British garrison at Imphal and its air arm withstood Bose's much larger force long enough for the monsoon rains to defer all possibility of warfare in that jungle region for the three months the British so desperately needed to strengthen their eastern wing. Bose had promised his men freedom in exchange for their blood, but the tide of battle turned against them after the 1944 rains, and in May 1945 the INA surrendered in Rangoon. Bose escaped on the last Japanese plane to leave Saigon, but he died in Formosa after a crash landing there in August. By that time, however, his death had been falsely reported so many times that a myth soon emerged in Bengal that Netaji Subhas Chandra was alive—raising another army in China or Tibet or the Soviet Union—and would return with it to "liberate" India.

Noncooperation brought nothing but detention and despair to Indian nationalist ranks. The Raj had proved more resilient than any of the Congress high command imagined it would be a year earlier. Without accusing the Mahatma of wishing "deliberately to aid the Japanese," Wavell did write in 1944 to say that he believed Gandhi and the Congress were hoping "to take advantage of our supposed military straits to gain political advantage." He now urged Gandhi to cooperate instead and help his government solve India's pressing economic problems, as well as to "make steady and substantial progress" toward self-government. Early in May Gandhi was released from detention, and that September he met with Jinnah in Bombay to talk about the Hindu-Muslim stalemate. *Qa'id-i-Azam* ("Great Leader") Jinnah, however, insisted that agreement on the creation of Pakistan was an absolute prerequisite to independence from British rule, but Gandhi refused even to discuss Pakistan until self-rule was granted. "I was meeting with the prophet of Pakistan looking for his Ali," Gandhi remarked after the talks broke down on the eighteenth day. Jinnah, who was left ill, weak, and depressed by the talks, paraphrased Gandhi's own caustic comment concerning Cripps, wondering aloud, "Why did Gandhi come to see me if he had nothing better to offer?"

The collapse of the Gandhi-Jinnah talks left Wavell to seek some official means of resolving India's most difficult dispute. The cabinet was almost ready to make a fresh attempt at "an Indian settlement," dealing out an offer similar to the one made by Cripps—from Britain's new position of military strength rather than its earlier weakness. The war years had reversed the balance of payments between India and Britain for the first time, and by 1945 Britain owed India millions of pounds sterling for the goods and services "borrowed" to help win the war. The combination of economic debt and terrorist activity, which made the daily life of Englishmen in India an unpleasant one at best, and mortally dangerous at worst, inspired Wavell and Amery to work on strategies of constitutional settlement that would make the closing months of the war—or, if Japan proved as powerful as many feared she was, its closing years—more bearable. Leading Indians in the ICS like V. P. Menon and liberal nonofficials like Sapru briefed Wavell extensively, but the viceroy himself rightly recognized by the end of 1944 that his government was "preponderantly Indian." He also optimistically felt that the communal problem was not insoluble. Wavell flew home in the spring of 1945 to consult with the cabinet and returned to Delhi with a new scheme, which he announced on June 14. The major proposed change was to be in the composition of his Executive Council, which would be chosen to give "a balanced representation of the main communities," including "equal proportions of Moslems and Caste Hindus." Only the viceroy and his commander-in-chief would remain British. Gandhi, Nehru, and Azad refused, however, to accept the formula of "communal parity" with the League, since it not only inflated the League's political position but meant that Congress could no longer appoint any Muslim members to executive power. The term "Caste Hindus," moreover, at least implicitly denied Congress responsibility for or authority over untouchables.

The Simla Conference, which was convened to consider the proposal, began on June 25, 1945, attended by twenty-one Indian leaders, including Gandhi, Jinnah, Nehru, and Azad. Azad was still Congress president, but Jinnah condemned him as a "show-case Muslim," refusing to "recognize" his political status. Jinnah insisted that all Muslim members of the conference be appointed by the League. Wavell tried in vain to alter Jinnah's stand on Azad, and on July 14 he announced that his conference had failed. Within two weeks of the collapse of the Simla Conference, Churchill's government was voted out of power and replaced by Attlee's. Lord Pethick-Lawrence, a vegetarian and friend or Gandhi, succeeded Amery as secretary of state for India. The surrender of Japan after the atomic bombs were dropped in early August also brought a radical change in Indian nationalist fortunes. It soon became clear that Britain's home government had no desire to retain control over its Indian empire for much longer. Nehru was jubilant at Attlee's victory, having always felt ideologically close to Britain's Party. Sir Stafford Cripps suggested holding new elections throughout India to allow the people to choose their own representatives to a constituent assembly. The call for general elections elicited a favorable response from Indian politicians, and on August 21, 1945, the viceroy announced that elections would be held during "the cold weather." The elections of 1945–46 gave Jinnah the mandate he called for, with the League capturing all thirty of the seats reserved in the Central Assembly for Muslims. As expected, Congress won 90 percent of the general electorate seats. Fifteen other seats were divided among Sikhs, Europeans, and independents. The February 1946 provincial elections confirmed this bipolar political picture, the League winning 439 out of 494 Muslim seats. Although the League had no absolute majority in any provincial assembly, Jinnah was able to form League-led coalition ministries in Bengal and Sind, but not in the Punjab, which remained under Unionist control. Congress formed ministries in all eight of the remaining provinces, including Assam and the Northwest Frontier, both of which were claimed by the League for Pakistan.

The captured officers of Bose's INA went on trial in Delhi's Red Fort that winter, and much to the chagrin of the government, those "rebels" quickly emerged as national heroes, defended by barristers of no less distinction than Jawaharlal Nehru, Bhulabhai Desai, and Sir Tej Bahadur Sapru. The trial became a eulogy for Bose, a tribute to India's martial courage, a symbol of her sense of total independence, a matter more of pride than of justice. It helped weaken the last ties of loyalty to the Raj, for here were Indian soldiers who had actually fired at Englishmen at Imphal and stood up to claim that they did so for "love of India," from "patriotic motives." All those convicted of "treason" were granted suspended sentences in recognition of their newly won national adulation. Other Indian soldiers, sailors, and air officers now began to question their own positions and attitudes. Were they, in fact, loyal subjects of the crown, or patriotic sons of *Bharat Mata?* How could they remain true to two such distant and different rules? Was it not their own stupidity and weakness that held India in thralldom to Britain? When did obedience to orders cease to be the highest martial virtue and become the lowest imperial vice? These rumblings among the ranks sounded more ominous than any heard since the Mutiny of 1857–58. At Dum Dum airbase several members of the Royal Indian Air Force refused direct orders, and on February 18, 1946, the Royal Indian Navy mutinied in Bombay harbor. The strong man of the Congress high command, Vallabhbhai Patel, went personally to negotiate a "surrender" of the ships, prom-

ising assistance from Congress for sailors who had "legitimate grievances." The day after Bombay's mutiny broke out, Lord Pethick-
Lawrence announced in the Lords, and Prime Minister Attlee told
the Commons, that His Majesty's Government had decided to send
a three-man cabinet mission to New Delhi to seek to resolve India's
constitutional problems.

The three "wise men" were Pethick-Lawrence himself; Stafford
Cripps, now president of the Board of Trade; and A. V. Alexander,
first lord of the admiralty. They put down at Palam Airport on March
24, 1946, and remained in India for more than three months, seeking
by every diplomatic device known to British statesmanship to break
the Congress-League deadlock before it broke up the subcontinent.
They were men of extraordinary patience, skill, and good will. They
carried the imprimatur of Britain's cabinet and crown in their collective decision-making power, and they were to be Great Britain's
last desperate effort to find a formula that would have allowed His
Majesty's Government to transfer what power it still held over India
to a single union. Their failure was not simply historic testimony to
the complexity of the problem they tackled, but one of the great
tragedies of India's history, for its aftermath brought the traumas of
partition and almost continuous South Asian war.

By the end of April the cabinet mission proposed a three-tier
(union, group, and province) federal structure to both Congress and
the League, inviting each party to send four of its representatives to
a round table conference in Simla in early May to seek resolution of
their remaining disagreements. The Second Simla Conference proved
hardly more successful than the first had been. Faced with continued
Congress-League deadlock, the cabinet mission and Wavell proposed
what they felt was the "best arrangement possible" for establishing
a new Constitution. On May 16, the mission conceded that "if there
is to be internal peace in India it must be secured by measures that
will assure to the Muslims a control in all matters vital to their culture, religion, and economic and other interests," but it rejected the
League's demand for a Pakistan consisting of all six provinces, since
substantial portions of those provinces contained non-Muslim minorities. The mission also rejected the alternative of partitioning Bengal
and the Punjab, convinced that any such "radical partition" would be
"contrary to the wishes and interests" of a very large proportion of

the people living in both provinces and would, in the Punjab, leave Sikhs "on both sides of the boundary." They also alluded to "weighty" administrative, economic, and military factors that made them decide against partition, as well as the question of how then to deal with princely states. Finally, they argued that the League's demand would create a geographically divided and highly vulnerable Pakistan, whose both halves "in war and peace would be dependent on the good will of Hindustan."

The cabinet mission's solution, therefore, was to propose a union embracing British India and the states, with central powers limited to foreign affairs, defense, communication, and "the powers necessary to raise the finances required" to support all three union functions. The union would be governed by an executive and legislature, and any "question raising a major communal issue" would have to be decided by a majority of "each of the two major communities" in both. All other subjects and residual powers were to be vested in provinces that would be "free to form groups," to be called A, B, and C. Group B was to embrace the Punjab, Sind, Northwest Frontier, and Baluchistan; Group C would contain Bengal and Assam; Group A, everything else. Recently elected provincial assemblies were to elect representatives to a forthcoming Constituent Assembly in roughly the ratio of one to one million inhabitants on a proportional basis for India's major communities: "general," Muslim, and Sikh. The general category would apply to all persons who were neither Muslims nor Sikhs. Congress would have thus received a 167-to-20 majority of general representatives in Group A; the Muslims would have a majority of 22 to 9 general and 4 Sikh representatives in Group B and a bare majority of 36 to 34 general representatives in Group C. Until the new Constitution was actually completed, the day-to-day tasks of administration would be carried out by an interim government, whose entire cabinet would consist of "Indian leaders having the full confidence of the people," and after the Constitution was in operation, any province could vote to reconsider its terms.

It was a remarkable scheme, the last rational hope for a non-violent transfer of power from British to Indian authority. "We and our Government and countrymen hoped that it would be possible for the Indian people themselves to agree upon the method of framing the new constitution under which they will live," concluded the cabinet mission in unveiling its plan, adding that since "this has not been possible," they had devised what they "trust will enable you to attain your independence in the shortest time and with the least danger of internal disturbance and conflict." The only alternative, they feared, would be "a grave danger of violence, chaos, and even civil war." Gandhi's immediate reaction to the proposal was to state: "Whatever the wrong done to India by British rule, if the statement of the Mission is genuine, as I believe it is, it is in discharge of an obligation they have declared the British owed to India, namely, to get off India's back. It contains the seed to convert this land of sorrow into one without sorrow and suffering." The Congress Working Committee, however, was less enthusiastic, insisting that their objectives remained "Independence for India" and "a strong, though limited, central authority." The League met a few days later, resolving that since the "basis of Pakistan" was "inherent" in the mission's plan, it would cooperate with the Constitution-making machinery proposed and hoped that such a process would ultimately result in the establishment of a "complete sovereign Pakistan." Azad, who took paternal pride in the mission's plan, convinced the All-India Congress Committee to vote its approval on July 6, but was then replaced by Nehru as Congress president. A few days later Jawaharlal held a press conference at which he was undiplomatically blunt in insisting that Congress remained "absolutely free" and "uncommitted" to any details of the plan. As for the matter of grouping, Nehru suggested there was a "big probability" that "there will be no grouping," since Group A might decide against it, the Northwest Frontier might choose to opt out of Group B, and Assam was opposed to joining Bengal. Concerning the central powers of the union executive, Nehru argued that it would require "some over-all power to intervene in grave crisis breakdown" and that such central power "inevitably grows."

Jinnah read Nehru's comments as "a complete repudiation of the basic form upon which the long-term scheme rests and all of its fundamentals." It was to the Qa'id-i-Azam one more example of Congress's treachery and of "its petty-fogging and higgling attitude." He probably knew by this time that the chronic "bronchitis" that left him so exhausted and easily bedridden was more than a transitory illness. He had precious little time left in which to achieve the burning ambition of his life, and no more time to waste on bargaining with *banias, pandits,* and *sardars.* He felt betrayed now not only by Congress, but by the cabinet mission as well, who had flown home to leave him to "deal" with people he had never trusted. Even Lord Wavell seemed to have sold out to Congress pressure, appointing six rather than five Congress ministers to his interim Executive Council, one to "represent" the untouchables, while the League was told to rest content with what was almost parity. Jinnah convened his council of the League in Bombay on July 27, 1946, denouncing the "bad faith" of government and Congress alike and calling upon his Working Committee to prepare Muslim India for "direct action."

"Never have we in the whole history of the League done anything except by constitutional methods and by constitutionalism," declared the Qa'id-i-Azam. "But now . . . we bid goodbye to constitutional methods." The League proclaimed August 16, 1946, Direct Action Day. "We have exhausted all reason," said Jinnah. "There is no tribunal to which we can go. The only tribunal is the Muslim Nation." In the annals of Calcutta, Direct Action Day was to be written in blood as the "Great Killing," launching a week of "unbridled savagery" during which Premier H. S. Suhrawardy's police were ordered to enjoy a special holiday, while the city's underworld took charge of the streets. Within seventy-two hours more than five thousand lay dead, at least twenty thousand were seriously injured, and a hundred thousand residents of Calcutta city alone were left homeless. It was only the beginning of the Civil War of Succession, which turned the final year of Britain's Crown Raj into an orgy of communal violence, terror, and slaughter. From Calcutta it spread to Dacca, from East Bengal to Bihar, and thence to Bombay, Ahmedabad, Lahore. Where Muslims predominated, more Hindus were murdered, and where Hindus were the majority, more Muslim bodies could be counted among the mounds of dead that blocked traffic in India's major cities during the winter of 1946–47. Wavell invited Nehru to form an interim government, which took office on September 2, 1946, despite the mortal stabbing the night before of one of its non-League Muslim members, Sir Shafatt Ahmed Khan.

The inauguration of the new Congress government was greeted with black flags flown from the homes of millions of Indian Muslims. By mid-October Jinnah relented, following a personal appeal from the viceroy, to the extent of permitting Liaquat Ali Khan and four of his colleagues to join the new cabinet. As minister of finance, however,

Liaquat Ali was able to "dictate policy" and obstruct the effective functioning of every department by arresting funds, proving by his noncooperative tactics (dictated by Jinnah) that a unified Indian government was simply impossible. Britain then made one last-ditch attempt to resurrect the cabinet mission's plan by sending Wavell, Nehru, and Jinnah to London for talks with Pethick-Lawrence, Cripps, and Attlee in December, but three days of unyielding argument sufficed to make Nehru fly home and inform an audience in Benaras that "we have now altogether stopped looking towards London." The League refused, however, to participate in a Constituent Assembly convened by Nehru on December 9, and by the dawn of 1947 all progress toward affecting any sort of peaceful transfer of power ceased.

Prime Minister Attlee announced in the Commons on February 20, 1947, that His Majesty's Government had resolved to transfer its power by no later than June 1948 to "responsible Indian hands," by which was meant a government "capable of maintaining peace." The British were by now eager to divest themselves of the burden of India, finding enough problems at home to preoccupy all of their administrative talents and time. Lord Louis Mountbatten (1900–79), Queen Victoria's great grandson and the dashing wartime commander of all Allied forces in Southeast Asia, had agreed to replace Wavell as viceroy, but only for a limited time. Mountbatten's royal charm was considered irresistible the world over, and many hoped he could successfully divide the Indian pie long before June 1948. He succeeded so well at his job that he was to be not only Britain's last viceroy, but the Dominion of India's first governor-general. Mountbatten and his staff flew into Delhi on March 22, 1947, with all the promise of a monsoon's first downpour after months of blistering drought. He met at once with Nehru and Liaquat Ali, seeking to break the deadlock between them that had paralyzed the interim government. Wavell warned Mountbatten that the real difficulty was Liaquat's proposal of "heavy taxes on all large incomes," which put Congress into the "invidious position of being called upon to protect its big business supporters." Nehru, however, insisted that the League was "determined to sabotage any economic planning," and he found himself thoroughly captivated by Mountbatten's personal diplomacy, soon falling under the spell not only of the viceroy but of his charming wife, Edwina, as well. Liaquat Ali was less easily charmed, for he

and Jinnah feared that Mountbatten had been sent to India to placate
Congress and knew that Wavell had been "sacked" because of strong
protests voiced against him to Attlee and Pethick-Lawrence by Nehru
and Gandhi, both of whom considered him pro-Muslim. Gandhi imp-
ishly suggested in his first meeting with Mountbatten that it might be
best for him to "dismiss" the Nehru cabinet and call upon Jinnah to
appoint an administration of his own. Mountbatten refused to enter-
tain the Mahatma's idea as a serious suggestion, but Gandhi insisted
he was "absolutely sincere." It was precisely the sort of outrageous
advice that made so many of Gandhi's closest colleagues in Congress
consider their Bapu "a bit senile."

While Mountbatten talked in Delhi, the Punjab burst into flames
of communal rioting and destruction. Though 56 percent of its some
thirty million inhabitants were Muslim, the Punjab had been administered by a precarious Unionist coalition of Hindus, Sikhs, and non-
League Muslims under Khizr Hayat Khan until his resignation in early
March. The League was then asked to form its own ministry, under the
Khan of Mamdot. Master Tara Singh, bearded leader of the militant
Sikhs, called for direct action by his *khalsa* against the League at this
time, igniting the powder keg of repressed violence that set the Punjab
ablaze with his cry of *"Pakistan Murdabad"* ("Death to Pakistan").
Tara Singh and his followers were demanding a Sikh nation of their
own, Sikhistan, and by demonstrating their willingness to die in defense of their homeland, they sought to prove the validity of their
claim. Amritsar and Lahore became centers of carnage, while all of
the major cities of the Punjab and much of its peasant countryside
witnessed murders, arson, looting, and street fights throughout the
last four fateful months of British rule. In the Northwest Frontier
Province the League launched a civil disobedience campaign of its
own, led by Dr. Khan Sahib, Abdul Ghaffar Khan's brother, against
the Congress ministry.

Meanwhile, Gandhi was busy walking through the Noakhali and
Tiperah districts of Eastern Bengal on a pilgrimage of love, seeking to
abate the communal strife that left thousands dead in these districts
alone. He sought, as always, to communicate directly with the masses,
taking time to teach peasant illiterates something about the great
changes that were occurring in this historic era, hoping to "convert"
all of mankind to his religious philosophy of *sarvodaya* (literally, "the
uplift of all"), and wishing he could live to be 125 so that he might

have sufficient time to see his dream of revolutionary rebirth come to fruition. Before he could complete his tour of Bengal, however, the Mahatma was called off to Bihar, where he had heard that Hindus were murdering Muslims, to offer what aid he could to that suffering minority. "To me," Gandhi told his friends in Bihar, "the sins of the Noakhali Muslims and the Bihar Hindus are of the same magnitude and are equally condemnable." At this time Gandhi began to recite prayers from the Koran at his public meetings, even when he spoke inside Hindu temples, and some fanatical young Hindu communalists considered him a traitor to his own faith, calling him "Muhammad Gandhi" and "Jinnah's slave."

By mid-April Mountbatten had come to think of Pakistan as "inevitable," though his closest advisers warned him against creating two Pakistans, fearing that the one in Bengal would remain a rural slum. Jinnah was apprehensive about winning "a moth-eaten Pakistan," but he soon resigned himself to the partition of both Bengal and the Punjab. On April 20 Nehru publicly declared that the League could have Pakistan if it insisted, but only "on the condition that they do not take away other parts of India which do not wish to join Pakistan," a position to which the League could hardly object on principle. Suhrawardy, as premier of Bengal, now proposed "a sovereign, independent and undivided Bengal," but he was icily ignored by both Congress and the League, each of which found his proposal equally unacceptable. Nehru was adamantly opposed to India's Balkanization, and Jinnah determined not to surrender the majority of Pakistan's population on the eve of his nation's birth. Continuing civil strife and the deteriorating morale not only of British officials and merchants, but also of India's armed forces and police predisposed both major parties to a formula finally evolved by V. P. Menon in consultation with Vallabhbhai Patel. British power was to be transferred swiftly to an Indian Dominion, from which certain areas might drop out if the majority of their population voted to do so. Privately, Nehru was reconciled to partition, quipping that he hoped that by "cutting off the head we will get rid of the headache." Gandhi alone refused to accept the vivisection of his motherland, but no one in power paid much attention to the Mahatma anymore.

On July 15, 1947, the House of Commons proclaimed that in precisely one month "two independent Dominions" would be established in India, "to be known respectively as India and Pakistan." A boundary commission was appointed under Sir Cyril Radcliffe, a legal expert, whose special claim to the impartiality required for chairing the body that partitioned the subcontinent was that he had never set foot in India before; nor would he ever return to that land, where his new borders were soon to be soaked with the blood of desperately fleeing refugees. Eight Indian high-court judges, four from Bengal, four from the Punjab, half chosen by Congress, half by the League, assisted Sir Cyril in drawing his tortuous lines, but all the most bitterly contested decisions were his to make alone, on small maps, often outdated, in rooms remote from the Punjab's dusty reality and Bengal's rain-drenched soil.

The countless practical consequences of partition, which had hitherto been neglected or purposely ignored by those who refused to believe the possibility of Jinnah's demand, now flooded the desks of every official in Delhi, Lahore, and Calcutta, whose jobs were to divide the assets and liabilities of British India, as well as its territory, into bundles of 82.5 percent for India and 17.5 percent for Pakistan. The civil service and armed forces had to be dismantled overnight, as were the railways, the police, and the revenue services, as were trucks, reams of paper, pens and pencils, rupees and pounds. In one month the accumulated possessions of centuries of imperial hoarding, building, and creating were torn apart, severed as though by caesarian section to permit two new nations to be born. The terror, the anguish of millions of Hindus, Muslims, and Sikhs who feared that they would awaken next morning to find themselves trapped in a nation fundamentally hostile to their faith and were thus driven to take what pitiful possessions they could carry and abandon their fields and homes, can barely be imagined. An estimated ten million people changed lands that summer of 1947, and approximately one million of them never reached their promised nation alive. A totally inadequate boundary force of fifty thousand troops, most of them infected with communal fever, had been assembled in the Punjab to help make the transition a peaceful one. For the most part they stayed in their barracks, cleaning their weapons and boots, while trainloads of Sikh refugees moving east were slaughtered by Muslims in Pakistan and Muslims headed west were butchered by Sikhs and Hindus in India. The stream became a flood, the flood a holocaust of pain, looting, rape, and murder. Gandhi called independence with partition a "wooden loaf" for Congress, one which poisoned those who ate it, yet would have left those who rejected it to starve. He no longer wanted to live 125 years.

Mountbatten convened a conference of India's princes on July 25, explaining that by August 15, when British paramountcy would cease, they should have acceded either to India or Pakistan, as geographic proximity dictated. Sardar Patel negotiated accession settlements with most of the princes, promising them much the same support they had received from the British crown, in return for their surrender to the Dominion of India the powers of defense, external affairs, and communications. Jinnah convinced the few Muslim princes within Pakistan's domain to do likewise. The two largest princely states, Kashmir and Hyderabad, and one small state, Junagadh, posed special problems, however, and were to present the new dominions with their most thorny immediate causes of conflict. Mountbatten had hoped to continue as governor-general of both dominions, at least to help carry them over the crises of their first traumatic months of life, but Jinnah did not have enough confidence in his impartiality and decided to become Pakistan's governor-general himself. Nehru, on the other hand, was delighted to continue to serve—at least nominally—under Lord Mountbatten following India's independence. Jinnah flew from New Delhi to Karachi on August 7, symbolizing by his departure from India's domain the partition that would officially come into effect a week later. "The rot began with the alien government," wrote Gandhi that day, adding: "We, the inheritors, have not taken the trouble to rectify the errors of the past." He did not wish to remain in Delhi for the celebration of India's independence and returned to Bengal to take up his labor of love as India's one-man "border patrol."

In New Delhi's crowded Constituent Assembly, as the midnight hour of August 14, 1947, approached, Jawaharlal Nehru, India's first prime minister, rose to declaim: "Long years ago, we made a tryst with destiny, and now the time comes when we shall redeem our pledge, not wholly or in full measure, but very substantially. At the stroke of the midnight hour, when the world sleeps, India will wake to life and freedom." He then called upon his fellow members of India's independent government to "take the pledge of dedication to the service of India and her people and to the still larger cause of humanity." The tricolor was unfurled atop Delhi's Red Fort, and in the darkness outside millions cheered the dawn of full dominion status for their nation, though as Nehru cautioned, "The past clings on to us still."

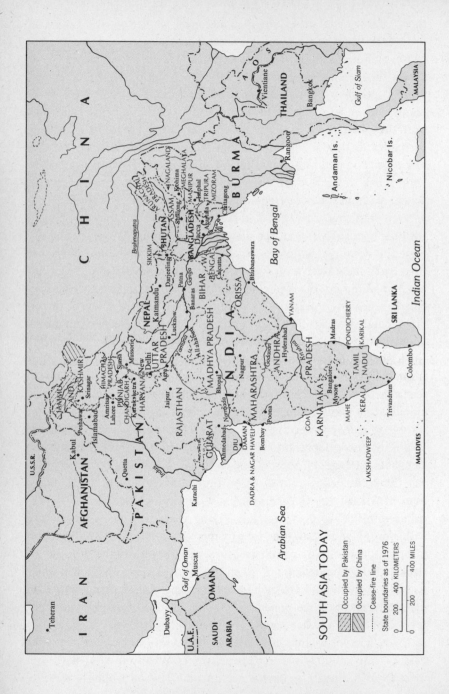

SOUTH ASIA TODAY

Occupied by Pakistan
Occupied by China
Cease-fire line

State boundaries as of 1976

0   200   400 KILOMETERS
0   200   400 MILES

TWENTY-THREE

THE NEHRU ERA
(1947–64)

India's first seventeen years of independence were dominated by the goals and dynamic leadership of its gifted prime minister, Jawaharlal Nehru (1889–1964). Panditji, as he was popularly called, seemed to be everywhere, at every major ceremonial inauguration, conference, congress, and assembly. He was India's royal figure and its matinee idol, that unique combination of brilliance, good looks, and brooding isolation who endeared himself to admirers of both sexes and all ages throughout the land. The fact that he achieved much less than he hoped to, hardly enjoyed the power that he seemed to, and inspired much more than he accomplished by no means diminishes his impact on Indian history or the significance of his legacy. He was the architect of India's foreign policy and its Five-Year Plans, the captain of its ship of state, as benevolent a despot as Ashoka or Akbar had been, in Fabian Socialist mental clothing.

For the first two and a half years of India's dominion status, Nehru was obliged to share his spotlight of central power with Sardar Vallabhbhai Patel, the Congress Party's blunt-spoken strong man, for whom the position of deputy prime minister was specially created, and who also held the home and states portfolios in Nehru's cabinet. The *sardar* had none of Nehru's cosmopolitan cultivation, erudition, or charm. He was a lawyer, sufficiently trained in India to pass the district pleader's examination and shrewd enough to establish a good criminal defense practice in his home district of Kheda in Gujarat, where he carved out a solid base of continuing political support. He later went to England, became a barrister of the Middle Temple, and returned to practice in Ahmedabad. Fourteen years older than Nehru, Patel was considered by most senior members of the Working Committee a more appropriate successor to Gandhi, but he lacked Panditji's broad-gauged appeal, especially among India's Socialist-minded youth. He was obliged, therefore, to remain in Nehru's shadow at the center of India's political spotlight, yet it was a position singularly congenial to Patel's temperament and technique of political maneuvering, for he could enjoy maximum power, especially over patronage, with minimal publicity. If Nehru was the front man of his nation, Patel was its backbone. Much like Gandhi, he epitomized peasant India's durability and native shrewdness.

As minister of states the *sardar,* assisted by his able secretary, V. P. Menon, convinced all but two—Hyderabad and Junagadh—of the 550 princely states within India's borders, plus Kashmir, the largest of all the 570 states in the subcontinent, to accede to the new dominion by August 15, 1947. Though the *nawab* of Junagadh and the *nizam* of Hyderabad were Muslim, the overwhelming majority of their subjects were Hindus, and it seemed only proper, therefore, in the spirit of democracy, to consider the preferences of the people rather than merely the despotic desires of their monarchs in deciding to which dominion these states would accede. Geographic position was also a most important determinant, at least in the case of Hyderabad, which was totally surrounded by India. The Nizam's reluctance to accede was thus viewed as little more than an autocratic whim, and India accepted a one-year standstill agreement from Hyderabad, giving the Nizam until August 1948 to make up his mind. On September 13, 1948, Indian troops invaded the state in what Delhi called a "police action," code-named Operation Polo. Two divisions of the Indian Army's Southern Command crushed Hyderabad resistance in four days. Junagadh was much smaller than Hyderabad, yet its case was similar, though being on the coast of the Kathiawar peninsula some three hundred miles by sea from Karachi, its *nawab* claimed that he had every right to accede to Pakistan, which he did on August 15, 1947. The government of India feared that the accession would have "undesirable effects on law and order in Kathiawar as a whole" and hence imposed an economic blockade and subsequently armed a "liberation army" of Hindu emigrees to take control of the state.

Kashmir posed a different sort of problem, since three-fourths of its four million people were Muslims, but its maharaja, Hari Singh, was a Hindu Dogra Rajput. With some eighty-five thousand square miles of beautiful and highly strategic real estate dangling over North India as well as West Pakistan, the state of Jammu and Kashmir was a most tempting plum for both dominions, so tempting, in fact, that it would lure them to war more than once over its sparkling Himalayan vale. The Dogra Rajput clan of Jammu ruled Kashmir through their able Kashmiri brahman civil servants, the Pandits, whose most famous family was the Nehru-Kaul clan. In 1932 the Muslim mystic, Sheikh Muhammad Abdullah (b. 1905), the "Lion of Kashmir," founded his All-Jammu and Kashmir Muslim Conference, demanding a Muslim share in the state's administration. Seven years later the sheikh was bolstered in his incipient nationalist movement by Hindus and Sikhs who joined his party and changed its name to the All-Jammu and Kashmir National Conference. Not before the end of World War Two, however, did the maharaja grudgingly transfer a small portion of his autocratic power, jailing Sheikh Abdullah for demanding more. As the countdown on Britain's transfer of power neared its conclusion, Hari Singh signed a standstill agreement with Pakistan, unable to bring himself to accede to either dominion and hoping that his mountainous state might be permitted to remain independent, a Switzerland of Asia. The communal complexion of most of his population, as well as the all-weather road link to Pakistan and the fact that Kashmir controlled the headwaters of the Indus and other Punjab rivers, certainly seemed good reasons to predispose Kashmir toward accession to Pakistan, but Hari Singh no doubt feared that his prospects of retaining power might be much weaker in a Muslim dominion than with India. While he temporized during late August and early September, the province of Poonch, in the southwest corner of Kashmir, flared in revolt, Muslim peasants rising against the oppression of Dogra Rajput landowners. The Poonch revolt was supported by neighboring Pakistani Muslims, who crossed the borders of the state in numbers to aid co-religionists in their agrarian struggle for freedom. Hari Singh viewed the revolt as a Pakistani plot to depose him. The maharaja turned to New Delhi for support and released Sheikh Abdullah from jail to fly there for talks with Nehru, who had supported his National Conference during its previous decade of struggle. On October 21, 1947, the sheikh informed Delhi's press that Kashmiris wanted "freedom before accession." Pakistan's British army lorries were, however, already packed with thousands of Pathans, armed to the teeth, barreling east over the Indus at Attock, where Alexander had crossed with his army, down the Jhelum river valley to the undefended Baramula Road, the highway to Srinagar.

On October 26, Hari Singh formally acceded to India and appealed for military support to defend Srinagar. Mountbatten insisted, in view of the predominantly Muslim population of Kashmir, that India's acceptance "be conditional on the will of the people being ascertained by a plebiscite after the raiders had been driven out of the State" and order was restored. Nehru and his cabinet ministers agreed. The airlift of the First Sikh Battalion from Delhi to Srinagar was launched the following morning; the Vale's capital was saved and the tribal forces turned west within a few days. Jinnah tried to hurl Pakistan's Army into battle at this time, but his British commander-in-chief, General Gracey, explained that he could not order his troops into combat against a sister dominion's forces without prior approval from his supreme commander, Field Marshal Auchinleck. Auchinleck flew to Lahore on October 28 to inform Governor-General Jinnah that if he insisted upon sending Pakistan's Regular Army into Kashmir, which had "legally acceded" to India, it would "automatically and immediately" require the withdrawal of every British officer serving with the Pakistan Army.

The fighting in Kashmir raged on until the state's de facto partition was effected on a line of battle stabilized just east of Uri and Poonch. The number of tribal raiders was about thirty thousand before the year's end, and these men became the regular army of the government of Azad ("Free") Kashmir, which held less than a quarter of the western portion of the state, excluding the Vale. Azad Kashmir, with its capital at Muzaffarabad, subsequently acceded to Pakistan. No plebiscite was ever held in Kashmir as a whole. Nehru insisted as a prerequisite to plebiscite that Pakistan "vacate its aggression" by withdrawing all its "invaders" from Kashmir, and Jinnah proposed simultaneous withdrawal of "the forces of Indian Dominion and the tribesmen." Sheikh Abdullah was flown back to Srinagar as premier. The cease-fire arranged by the United Nations would not come into effect in Kashmir until January 1, 1949.

As the undeclared Indo-Pakistan war over Kashmir continued, Hindu refugees in their tattered millions poured into Delhi and Cal-

cutta, squatting in open fields, on train platforms, in streets and alley-
ways, falling when they could move no more, and setting up "homes"
wherever they dropped. North of Delhi, a refugee camp was established at Panipat, named Kurukshetra, after the field of epic battle
whose ancient cries of pain and anguish were echoed once again that
winter. The Mahatma, fast losing patience with the Congress elite for
their self-serving attitudes amidst the growing misery of so many impoverished refugees and sickened by Hindu persecution of Muslims,
resolved that "my battle has to be fought and won in Delhi itself."

On January 11, 1948, Gandhi suggested that it would be better
"to dissolve" Congress than allow it to continue functioning as it was,
full of "decay and decline," a place of "corruption" and "power politics." The next day he announced his decision to fast "by way of protest" against the persecution of Muslims in India's capital, in an effort
to restore "heart friendship" among Indians of every religion. "Death
for me would be a glorious deliverance rather than that I should be a
helpless witness of the destruction of India, Hinduism, Sikhism and
Islam," the Mahatma told his prayer meeting audience that afternoon
at Birla House, where he lived. "Let my fast quicken conscience, not
deaden it. Just contemplate the rot that has set in in beloved India."
It was the last of his fasts. He ended it in less than a week, following
messages of sorrow and prayer, including one from Sardar Patel
promising to pay Pakistan forty million pounds sterling in cash assets, hitherto withheld by India. The entire cabinet gathered at the
Mahatma's bedside to confirm that promise. Nehru announced that
the loss of Gandhi's life would be "the loss of India's soul." There
were others, however, who chanted, "Let Gandhi die!" Fanatical
brahman members of the paramilitary Hindu communal *Rashtriya
Svayamsevak Sangh* ("National Volunteer Association"; hereafter
RSS), an offspring of the Hindu *Mahasabha*, plotted to assassinate the
"old man," who had "outlived his time." A Poona Chitpavan, Naturam
V. Godse, intellectual disciple of Savarkar, fired the fatal shots at
Gandhi as he walked to his prayer meeting platform in the garden of
Birla House just before sundown on January 30, 1948.

"The light has gone out of our lives and there is darkness every-
where," Nehru announced in his nationwide broadcast from New
Delhi that evening. The shock of the Mahatma's assassination helped
India's government repress the reactionary forces of Hindu commu-

nalism that had taken control of Delhi and many other great cities
and gave the secular-minded Nehru the leverage of popular indignation he needed to wrest control over his nation and its police from
the Sardar, whose position was now mortally weakened. The RSS was
outlawed; Godse, convicted of murder, was hanged; and antibrahman
riots swept Poona, Nagpur, and Bombay, the major centers of *Mahasabha* Hindu support. In Delhi India's Muslim minority rested more
easy. Through his martyrdom, Gandhi came closer to achieving his
goal of Hindu-Muslim unity than he ever did in his lifetime.

India's Constituent Assembly elected Dr. Ambedkar to chair its
drafting committee, which produced a Constitution for India's "Sovereign Democratic Republic and Union of States" in 1949. That Constitution became law on the twentieth anniversary of Independence
Day, January 26, 1950, when the Republic of India, whose sovereignty was derived from its people, was born. India's professed constitutional goals were "to secure to all its citizens . . . JUSTICE
social, economic and political; LIBERTY of thought, expression, belief, faith and worship; EQUALITY of status and opportunity; and to
promote among them all FRATERNITY assuring the dignity of the
individual and the unity of the Nation." These noble principles reflected Nehru's personal ideals, but they gained special power and
poignance from the fact that they were drafted by a committee led by
one who had been born an untouchable, yet lived to become his republic's minister of law. In stressing the secular and egalitarian ideals
of independent India, Nehru often inveighed against India's "caste
and priest-ridden society." The franchise was now universal for all
adult citizens, making India technically the world's largest democracy,
with 173 million people eligible to vote in 1951 for nearly four thousand representatives, though 80 percent of that electorate was illiterate and impoverished.

The president and vice-president had powers in keeping with
their roles as ceremonial heads of state, republican rather than royal,
except for one "emergency" clause, which had been borrowed from
the Constitution of the Weimar Republic and which allowed India's
head of state to "suspend" the Constitution entirely in the interests of
"security," for up to six months at a time. Dr. Rajendra Prasad was
India's first president, and the distinguished Sanskritist and Vedantic Philosopher, Dr. Sarvepalli Radhakrishnan, was the first vice-

president, succeeding Prasad as the nation's second president in 1962.
The prime minister (Nehru), as in the British model upon which India's system was primarily based, was the actual leader of the republic, head of the *Lok Sabha* ("House of the People"), the all-powerful
lower house of a bicameral Parliament. With Nehru personally traveling over thirty thousand miles in the 1950–51 winter campaign,
during which he appealed, as standard-bearer for his party, to an
estimated thirty million voters, Congress won a sweeping victory of
362 out of 489 seats in the *Lok Sabha,* though its popular vote was
less than half (45 percent) of the total 100 million cast.

In keeping with the historic tradition established by the British
Government of India Acts, India's republic retained its federal structure as a union of states, the center sharing its powers provincially.
The former provinces of British India were hereafter called "Part A"
states, of which there were at first only nine, with Bengal reverting to
its partitioned title, West Bengal. The former princely states were
amalgamated and integrated into the union as "Part B" states, also
totaling nine, Hyderabad, Jammu and Kashmir, Mysore, and Travan-
core-Cochin remaining essentially as they had been before independence and some 275 smaller principalities merged into five new administrative units: Madhya Bharat ("Central India"), Patiala and
East Punjab States Union (PEPSU), Rajasthan ("King's Land"),
Saurashtra ("Sea Country"), and Vindhya Pradesh ("Mountain Province"). Finally, there were sixty-one smaller and more primitive areas
that were reorganized into ten centrally administered "Part C" states.
State legislatures elected representatives to the central upper house,
the *Rajya Sabha* ("Council of State"), with relatively equal territorial
representation, as in the U.S. Senate, on which it was modeled. *Lok
Sabha* representatives were elected by the population, with no more
than one member for from every 500,000 to 750,000 constituents.

The central (union) government was charged with total responsibility for defense, atomic energy (which revealed remarkable foresight, since the subject was irrelevant to India before 1970), foreign
affairs, railways, airways, shipping, posts and telegraphs, currency,
and a total of almost a hundred lesser items (including "preventive
detention," which remained part of "Free" India's legal arsenal). The
states were responsible for police, the administration of justice, public health and sanitation, education, agriculture, forests, fisheries, lo-

cal government, and some sixty other functions. There was also a
"concurrent list" of forty-seven shared federal powers, including economic and social planning, trade and commerce, commercial and industrial monopolies, combines and trusts, and trade unions, making
it possible for the center to assert its will in any significant state
question. Untouchability was abolished and its practice in any form
forbidden.

"We are fortunate to witness the emergence of the Republic of
India and our successors may well envy us this day," Nehru told his
nation on its first Republic Day; "but fortune is a hostage which has
to be zealously guarded by our own good work and which has a ten-
dency to slip away if we slacken in our efforts or if we look in wrong
directions." Nehru and Congress had long recognized poverty as the
most critical and urgent problem confronting India's populace. Now
that the British were gone, they could no longer defer tackling that
national tragedy simply by shifting the burden of its blame to foreign
shoulders. As an intellectual Socialist, Nehru had long advocated centralized economic planning, and as early as 1931, he had moved a
resolution passed by Congress calling for the nationalization of key
industries. In 1948 Nehru chaired the Economic Program Committee
of his party, and in March 1950 he headed the National Planning
Commission, which was created to plan for the "most effective and
balanced utilization of the country's resources." In April 1951, however, when India's First Five-Year Plan was inaugurated, it was a
modest proposal calling for a modest 11 percent growth in national
income by 1956, an overall increase in the value of annual production of goods and services from about eighteen to twenty billion dollars. The continued staggering growth of India's population, estimated
at 360 million in 1950, more than counterbalanced any actual increase in food production during the first five-year period, especially
because of the loss of most of the Punjab's wheat basket and the addition of many millions of nonproductive refugees to feed. The average "free" Indian thus remained one of the world's most poorly nour-
ished human beings, lacking protein as well as vitamins in his daily
diet of gruel, victim to all the diseases of malnutrition, especially tuberculosis, which claimed more than half a million lives annually.
Indian life expectancy at this time averaged only thirty-two years, approximately half that enjoyed in most economically developed na-

tions. Still, India's population grew by from five to six million a year.

Of the total of $3.7 billion of public funds expended during the First Five-Year Plan, more than $3 billion went simply to restoring India's prewar consumer-goods production capability, repairing communications, and trying to improve agricultural yields. Industry and mining received only 7.6 percent of the total capital outlay, and power 11 percent. In 1951 India had to purchase some four million tons of wheat from the United States, and she was to remain a food-deficit nation for another two decades. Monsoon failures in 1952 and 1953 added to India's food deficits, as famine once again stalked the land and imports of food grains doubled and tripled. By the end of the first plan, however, total food-grain production had increased from 52 million tons to over 65 million.

Several major campaigns were launched during Nehru's Era to tackle the problems of peasant "backwardness" and rural poverty. Vinoba Bhave, Gandhi's leading disciple, sought to counter the appeal of India's Communist Party in the Telugu-speaking regions of Hyderabad and other impoverished parts of the Deccan in 1948 by asking wealthy landowners to "adopt" him as their fifth son and grant him a "gift of land" (bhoodan), which he could then turn over to their landless laborers. Many landlords responded with gifts, not exactly of a fifth of their property, but some portion of land, possibly the least arable or productive, yet something. Within a decade Vinobaji, as he was called, had collected more than a million acres, and his bhoodan movement won national acclaim and international fame. The frail whisp of a sadhu then sought to expand his idea by requesting "gifts of a village" (Gramdan) and of "human life" (jivan-dan). He did all of his traveling on foot, covering the length and breadth of India several times on his constant "pilgrimage" (yatri) in pursuit of sarvodaya, his guru's Socialist legacy. Like Gandhi, Vinoba attracted followers from many distant lands. He walked with his band of devotees from village to village and was welcomed as a visiting saint. Several famous leaders of India's nationalist movement offered gifts of their "lives," including one of the founding fathers of the Congress Socialist Party (later the Praja—"People's"—Socialist Party), Jaya Prakash (JP) Narayan (1902–79), who devoted himself to rural revitalization. JP called for the "reconstruction of Indian polity" as an "organic democracy," strongly supporting another idea that emerged from the Planning Commission's hope to accelerate the pace of rural development—*Panchayati Raj* (government by "village councils of five," or *panchayats*).

The central government launched its first Community Development and Rural Extension Program with assistance from the Ford Foundation on October 2, 1952, the anniversary of Mahatma Gandhi's birth. The initial expenditure was modest, but the idea was revolutionary, calling for the training of workers to help villagers "transform" their socioeconomic lives through self-help. The village-level worker, something of a pre-Peace Corps Indian jack-of-all-trades, would be sent to help a group of five villages in various practical ways, including reclamation of waste or unused land; augmentation of the water supply through minor irrigation or well-digging techniques; introduction of improved seed, livestock, and fertilizers; eradication of mosquitoes and other pests; and building of elementary schools and the like. Each project area would consist of one hundred villages, and three such areas were called a block. By the end of the first year, there were fifty-five blocks, each with a population of roughly two hundred thousand peasants, cultivating about a hundred and fifty thousand acres of land. The Ford Foundation funded the training of all project executive officers and their staffs. It was a most ambitious program, and it worked better on paper than in rural reality, where much effort and expenditure were wasted in pouring concrete for urinals that were never used, digging new wells in locations whose drainage soon led to pollution, or trying to teach old peasants new tricks they were reluctant, usually with good reason, to learn. Nor was the Ford Foundation much more successful than Gandhi had been in recruiting the thousands of brilliant and selflessly dedicated young people needed to translate their fine blueprint into effective action, though money was wasted on many workers who interpreted the program's goal of self-help too narrowly. Poverty, rural conservatism, illiteracy, and superstition, as well as the combination of increasing population pressure (ten million a year by 1959), limited water resources, and lack of chemical fertilizers, electricity, and power, all conspired to make Indian rural uplift one of the world's most difficult problems.

Panchayati Raj was launched in 1959 in India's second largest and most backward northern state, Rajasthan, whose 750-mile desert border with Pakistan to the west added strategic importance to the political significance of its choice as a testing ground for New Delhi's experiment in rural reconstruction. The idea was to strengthen the roots of economic planning by involving villagers in the actual process of nation building and development. The ancient *panchayat* was supposedly a tribal town council of the leading Aryan householders of each village, "elected" to administer their common affairs. The *panchayats* known to history were, in fact, groups of conservative elders, usually led by the village headman, whose hereditary powers were based on substantial landholdings that were inherited, won in battle, or granted tax free by higher authority for special service rendered to the Raj. These "natural leaders" of the village were, therefore, not necessarily the most enlightened members of any local community, nor were they generally interested in change. They did, of course, live among their neighbors and could thus reflect local interests, at least of their own upper-class caste. Members of each *panchayat* elected one of their group to the second-tier *Panchayati Samiti* ("Association") level, which corresponded to the Community Development Block. The *Samiti,* consisting of about fifty members, was chaired by the block executive officer and would in turn elect one of its members to meet with the *Zila Parishad* ("District Party"), the third-tier level in this pyramidal hierarchy of centralized democracy. There was supposed to be two-way communication among all three levels. From Rajasthan *Panchayati Raj* spread to most other states in the union, and by 1964 there were some three hundred *Zila Parishads* and almost five thousand *Panchayati Samitis* in India.

The Second Five-Year Plan (1956–61) was more ambitious than the first, spending almost three times as much money, raising total food-grain production to approximately 80 million tons, though by 1961 there were 440 million Indian mouths to feed. The government of India had decided in 1959 to endorse family planning, but it still did so without substantial commitment, and hence no real inroads were made in the illiterate ocean of village society. Industry and mining received some 20 percent of the second plan's funding, thanks to which the production of iron ore increased from 4 million tons in 1956 to almost 11 million tons in 1961, while coal jumped from 38 million to 54 million and India's power capacity doubled.

The production of finished steel fell short of anticipated growth, with only 2.4 million tons in 1961. Cotton textiles remained India's leading industry, with over five billion yards produced annually after 1956.

Congress adopted as its official party goal "the establishment of a socialistic pattern of society" in December 1954, and Nehru and others in his cabinet on the Planning Commission called for the "socialization of the vacuum" in Indian industry, escalating populist rhetoric as the date of new elections approached. But despite the rhetoric, private enterprise remained firmly entrenched, secure, and increasingly profitable throughout the Nehru era. Several attempts by India's government to tap the inordinate profits and rapidly accumulating wealth of its minuscule minority of business Mughals, the *banias,* Marwaris, Parsis, and Jains (many of whom, like the Birlas, Tatas, Sarabhais, and Bajajs, were the most prominent supporters of Congress and its candidates), proved profitless for all but the lawyers, who found loopholes in every legislative act large enough for hundreds of millions of rupees to escape through. Left-wing political opposition, Socialist as well as Communist, compelled Congress to promise more and more radical economic reforms, but no left-wing coalition seriously posed an electoral threat to Nehru's government, since he was so brilliantly effective a socialist on the political stage, stealing the thunder of Communist and Socialist Party candidates.

Right-wing political opposition to Congress was also neutralized, since its two major political parties—the *Bharatiya Jan Sangh* ("Indian People's Party"), founded in 1951, and *Swatantra* ("Freedom"), founded in 1959—had no real issues round which to rally mass support. The potential power of the *Jan Sangh's* orthodox Hindu revivalist appeal was great, for India, after all, was predominantly a Hindu state, for all its secular pretensions. As for the *Swatantra* Party, its major platform was *laissez-faire,* and its appeal for more "freedom" hardly attracted the peasant masses, a hundred million of whom were landless, homeless, or both.

Though agrarian backwardness, inequity of land distribution, and poverty continued to plague India's countryside, the Third Five-Year Plan (1961–66) catapulted India to the rank of the world's seventh most industrially advanced nation. New steel factories were built at Bhilai, Durgapur, Rourkela, and Bokharo, producing over a total of seven million tons of finished steel by 1966. Since the rest of India's economy was still so underdeveloped, however, some of that steel was exported to Japan. The production of iron was now over 25 million tons and coal over 74 million. Generated electric power was 43 billion kilowatt-hours, more than double the 1961 capacity. India produced eleven million tons of cement, half a million tons of chemical fertilizers, two and a half million electric fans, one and a half million bicycles, sixty thousand diesel engines, and more than ninety million tons of food grains. British, American, Russian, and West German capital and technical assistance helped Indian industry gain momentum in many key areas. The international Aid-to-India Club, a consortium of six nations (United States, Great Britain, Canada, France, West Germany, and Japan, with the United States as the major contributor) provided more than $5 billion to India's third plan. Industrialization during the Nehru era was thus in great measure an international enterprise, the economic return from Nehru's policy of "nonalignment," thanks to which India reaped benefits of technical and capital assistance from East as well as West, while the world remained sorely divided by Cold War differences.

As his own foreign minister, Nehru set the course of India's foreign policy, which, like that of every other nation, was dictated first of all by national self-interest; hence, the war over Kashmir, the Hyderabad "police action," and the "liberation" of Goa from Portuguese rule in 1961. The last of these actions was the culmination of years of fruitless negotiation with Portugal, who insisted that Goa, Diu, and Daman were "integral" parts of the Portuguese nation, despite their remoteness. France had been more accommodating, turning over Pondicherry, the last of its South Asian enclaves, to the Indian Republic in 1954. Portugal proved the most stubborn of all the Western European powers. A Goan Liberation Committee was established in Bombay in 1954, and several *satyagraha* campaigns were launched in the ensuing years, costing a number of Indian lives before Nehru finally allowed the Indian Army to move into Goa in December 1961 and take control. In general, however, Nehru worked to identify India's foreign policy with principles that were integral to India's history: anticolonialism and antiracism were preeminent among them, born in reaction to British rule. India's proclaimed interest in peace and predilection for the peacemaker's role may be viewed as an offshoot of Gandhian policies, which in turn had deeper roots in Buddhist, Jain, and ancient Hindu ideals.

Nehru's foreign policy was also aimed at securing leadership for India among a "third force" of "nonaligned" Afro-Asian nations. Though he convinced India to remain within the British Commonwealth, Nehru looked toward his own continent and Africa for natural allies. Even before India won dominion status, Nehru had convened an Asian Relations Conference in Delhi in the spring of 1947. It was attended by representatives from twenty-eight Afro-Asian countries. In 1949 he hosted a second conference of fifteen Afro-Asian powers to condemn the Dutch "police action" in Indonesia. Asia was in "ferment," Nehru insisted, and there was "such a thing as Asian sentiment." At the Afro-Asian Conference in Bandung, Indonesia, in 1955, Nehru found that he was not alone in hoping to stand forth as leader of the twenty-nine nations represented there, for China's Chou En-lai posed a formidable challenge to him. The key to Nehru's policy of Asian unity was Sino-Indian friendship, and in April 1954 India concluded a treaty of trade with China over Tibet, in which the "five principles" (*panch shila*) of that key were formulated: mutual respect for each other's territorial integrity and sovereignty; nonaggression; noninterference in each other's internal affairs; equality and mutual benefit; and peaceful coexistence. It was the dawn of a new era of Sino-Indian accord, and its slogan was *Hindu-Chin bhai bhai* ("Indians and Chinese are brothers").

The border between India and China had been drawn by Sir Arthur Henry McMahon in 1914 on the principle of the "highest watershed" dividing the two Asian giants, but it was never ratified by China, which continued to produce maps claiming over fifty-two thousand square miles of "Indian territory" as its sovereign domain. After the Chinese "liberation" of Tibet in 1950, Sino-Indian tensions mounted, and it was hoped that the *Panch Shila* Treaty would dispel Delhi's fears of conflict on the northern border. Tempers flared again in the winter of 1955, when Chinese troops crossed into India's Garhwal district in U.P., but the troops were withdrawn following Delhi's protests. In 1957 the Chinese constructed a road linking their "new province," Sinkiang, with Tibet across the Aksai Chin region of Ladakh, a region once claimed by Tibet but subsequently conquered and held by Kashmir. The area was so remote and India's claim to it so tenuous that it took Indian patrols over two years to discover the Chinese road. When the confrontation finally occurred, in October 1959, a number of Indian soldiers were killed by Chinese guards. The Dalai Lama's rebellion in Lhasa earlier that year had been repressed by Peking, and the young Lama fled with thousands of his monks to India for asylum. There had also been fighting between Chinese and Indian troops in the Northeast Frontier Agency (NEFA) outpost of Longju, another disputed zone. Nehru invited Chou to Delhi for talks in 1960, but the 2,640-mile border dividing the two nations was so difficult to survey and the maps relied upon by each leader so different that nothing could be resolved. The "cartographic war" continued to simmer until September 1962, when Nehru ordered his troops to take back "all our territories" that were at the time being held by Chinese "aggressors." Krishna Menon, who had been India's minister of defense since 1956, must have assured his old friend that such an order could be carried out, but the Indian Army proved no match for the Chinese. In October and November 1962 Chinese troops advanced over India's northeast frontier with such ease and in such numbers that, were it not for their unilateral decision on November 21 to withdraw as unexpectedly as they had invaded, they might easily have marched into Gauhati, the capital of Assam. For Nehru, as well as India, it was a rude awakening. Krishna Menon resigned. India's policy of nonalignment was scrapped in favor of American air and missile support. A new elite arm was added to India's military establishment, the specially trained Border Security Force, kept directly under the prime minister's personal command.

This was an era of social as well as political and economic change. As a so-called secular republic, India would have to transform itself in many ways, the most difficult yet fundamental of which was initiated by the constitutional abolition of untouchability. Until the Untouchability (Offences) Act was passed in 1955, however, providing specific penalties for continued high-caste Hindu discrimination against this submerged minority of more than sixty million citizens, there was no legal remedy to enforce the blanket constitutional abolition. To help compensate *harijans* and assist them in overcoming the handicaps inflicted by birth alone, the Indian government established special "ex-untouchable" quotas for all competitive central government services as well as seats on the state legislative assemblies and

Lok Sabha. They also reserved a percentage of university grants for this underprivileged "community." Ironically, such favored treatment and special opportunity for advancement to "escape" one's tragic past required ex-untouchables first of all to identify themselves as such. With Jawaharlal as its standard-bearer, Congress retained a secure majority in the *Lok Sabha* through three general elections, the last of which, in 1962, saw 216 million Indians going to the polls to return 361 Congress members to a *Lok Sabha* of 494 seats. Congress dominated state assemblies as well, winning 1,800 out of the total of 2,930 seats, a gauge of Nehru's unflagging popularity and the opposition parties' inability to join forces against Congress.

The struggle for women's liberation in India, dating back at least to the abolition of *sati* in 1829, has also been a long and painfully slow process. The subordinate and subservient position of women under Hindu as well as Islamic traditions and law conspired to inflict many hardships and disabilities on Indian women, depriving them of virtually all human rights. Married in infancy or childhood to men they had never seen before, they became the legal chattel of their husbands, powerless to divorce men they hated or despised, sinking to the oblivion of "living death" as polluted shadows if they were widowed, inheriting no real property, often much abused and maligned, illiterate and subordinate. Thanks to Gandhi, women entered India's nationalist movement in numbers. Their greatest leaders, like Sarojini Naidu and Nehru's only daughter, Indira, remained in politics throughout their lives, but they were "third-sex" exceptions to a rule of village female servitude that bordered on slavery. Suicide rates among Indian women are perhaps the highest in the world.

Independence significantly changed the legal status of Indian women, first of all by elevating them to the rights and privileges of citizens, which included the franchise. Women were offered the opportunity of rising to any position in the nation, including prime minister. In 1949 a Hindu Marriage Validating Act was passed to remove intercaste barriers to marriage, in emulation of Gandhi's constructive advice that more upper-caste Hindus should marry *shudras* and *harijans*. In 1955, the Hindu Marriage Act finally gave Hindu women the right of divorce; in addition, it raised the minimum age of marriage to eighteen for males and fifteen for females. Seeking to encourage monogamy among Hindus, that act permitted any wife to sue for divorce if a "co-wife or co-wives" were sharing her husband (though a man who took more wives could not legally divorce an earlier spouse, whom he was obliged to continue to support). The Hindu Succession Act, passed in 1956, gave female children equal claims with male siblings to inherit paternal property. That same year, an Adoption and Maintenance Act permitted women to adopt children, female as well as male, assuring any adopted girl the same legal rights as an adopted boy. It was hardly less than a legal revolution, providing the overwhelming majority of India's women (these acts did not change the status of Muslim women) with full equality before the law. It would, however, be some time before most of India's Hindu women either understood their new rights or were sufficiently self-assured and courageous to take advantage of them. By 1956–57 almost 40 percent of the 92 million women qualified to vote cast valid ballots, helping elect 27 women to the *Lok Sabha* and 105 women to state assemblies.

The effective pace of Indian reform of every sort was geared in great measure to independent India's educational system. Since the establishment of the British Indian Educational Service in 1864, that system had developed under British tutelage, with imported professional standards of English education that served to train an elite of Indian teachers and scholars for fully a century before Nehru's death. National education and Gandhi's *wardha* scheme, which stressed the need for more vocational and manual training in India's curricula, significantly amended the British system, yet it would take time for pedagogic innovation effectively to transform India's highly conservative and elite system into a truly national and democratic one. Gokhale's dream of free and compulsory elementary education for all Indian children up to the age of fourteen was written into India's Constitution as a desired "directive principle." The initial date set for achieving that national goal was 1960, but by that time hardly 60 percent of all Indian children between the ages of six and eleven were attending school for any portion of the year, and only 20 percent of those between eleven and fourteen. The Third Five-Year Plan's target was, therefore, to educate 75 percent of all Indians in the age group of from six to eleven, estimated at about 50 million, by 1966. Even this more modest goal proved impossible to achieve, however, for a third of a million new teachers would have been required and a

50-percent increase in the number of schools. Still, in the decade of the 1950s, educational progress seemed impressive, with almost 50 million Indian students of all ages attending close to half a million schools by 1959–60. The census of 1961 revealed growth in literacy to an average of 23.7 percent nationwide, though only 12.8 percent of India's women were as yet literate.

Literacy applied not only to English, but to India's many indigenous languages as well, especially Hindi written in the *Devanagari* script, which was designated the "official language of the Union" by India's constitution. Ever since the Nagpur Congress of 1920, Hindi had shared the platform with English as India's "national" language, though its use was initially confined to the U. P. Gandhi, whose mother tongue was Gujarati, chose Hindi for the national place of honor since it was spoken by a plurality of India's population and was closely related to all other modern Indo-Aryan languages of the North. At Nagpur the idea of redrawing India's internal provincial boundaries along regional linguistic lines rather than the more arbitrary historical lines of British administrative division was first proposed, and Congress had then organized itself into twenty-one linguistic provincial committees. It was, therefore, only natural to expect the Indian Republic to take appropriate action to implement such internal reorganization, but the terrible aftermath of partition served to dampen the ardor of those who had hitherto been staunch advocates of any internal change in provincial boundaries. Linguistic "separatist" demands could not, however, be silenced, and the first vigorous agitation for a "linguistic province" emerged in the Telugu-speaking region of northern Madras, which wanted a state of its own, to be called Andhra, after the ancient Deccan Empire. Nehru was less supportive of the linguistic provinces' movement than Gandhi had been, and he succumbed to its popular pressure only after Potti Sriramalu, the saintly father of the Andhra movement, fasted to death in December 1952, leading to the creation of Andhra on October 1, 1953. Nehru was now faced with a plethora of demands from many regions of India for linguistic states. He appointed a States Reorganization Commission (SRC) to study the problem, and he was thus able to defer further action until after its report in October 1955. The SRC recommendations were not implemented for another year, at which time India was reapportioned into fourteen states and six

centrally administered territories. All of the states were now supposed
to embrace populations speaking the same language. Thus, Kerala
was the home of Malayalam speakers; Karnataka (Mysore) the
home of Kanarese-speakers; and Madras, from which so much had
been stripped, was almost entirely Tamil in population—hence the
subsequent change to its name to Tamil Nadu ("Land of the Tamils"). Linguistic States Reorganization was designed to undermine
Dravidian separatist demands for a "Dravidistan" in Southern India
by offering provincial politicians new power bases. The *Dravida
Kazhagam* ("Dravidian Federation"; DK), founded in Madras in
1914 under the charismatic *Periyar* (Tamil *Mahatma*) E. V. Ramaswami Naicker (1880–1974), had long rallied non-Brahman *Harijans* and oppressed *Shudra* Dravidians to its "Dravidnad" platform,
by appealing to their "racial" pride as indigenes whose roots in South
Indian soil presumably antedated the Aryan invasions and subsequent Hindu conquest. The "two-race" theory thus to emerge in
Tamil Nadu led to vocal, often violent, separatist "National" demands that included burning India's flag as well as constitution, and
"South India's" rejection of Hindi as a "national language." C. N.
Annadurai, the Periyar's brightest disciple, who lost faith in his *guru,*
founded a more militant and popular *Dravida Munnetra Kazhagam*
("Dravidian Progressive Federation"; DMK) in 1949, and built a
broad-based following by his advocacy of radical land-reforms and
revolutionary economic change as well as a wide range of cultural
Dravidian demands. DMK control over Tamil Nadu remained secure,
and Delhi's central government was most careful and usually generous to a fault in dealing with these Southern "Princes" of regional
power, whose separatist potential was so dangerous.

The state of Bombay remained unaltered by the SRC Report,
despite the fact that the Deccan portion of its population overwhelmingly spoke Marathi, while the region north of Bombay City, including all of the Kathiawar peninsula, spoke predominantly Gujarati.
Within Bombay two new political parties suddenly rose to prominence—the *Samyukta Maharashtra Samiti* ("United Maharashtra
Party") and the *Maha Gujarat Parishad* ("Great Gujarat Party")—
sweeping Congress out of provincial office and proving that the power
of ancient linguistic regional loyalties was stronger than younger national bonds. Nehru and Congress resisted for several years, but the

demands for provincial partition only grew more strident, more violent, leading to riots in Bombay City itself, forcing Delhi to capitulate by announcing division of Bombay into the states of Gujarat and Maharashtra effective May 1, 1960. One other new state was born through the fires of regional revolt during Nehru's era, the remote and forested homeland of India's former head-hunting Nagas, who fought for and won their claim to independence from Assam on December 1, 1963, when Nagaland became India's sixteenth state.

Just as the Chinese invasion destroyed Nehru's faith in Asian unity, the spread of linguistic provincial agitation undermined his confidence in his ability to continue to hold the "union" of India's giant pluralistic federation together. In 1958, at the age of sixty-nine, Panditji sought to relieve himself of his burdens of office, but the storm of protest that greeted his announcement helped convince him not to retire. He alone seemed capable of bridging the growing ideological differences among right, left, and center Congress factions. On May 27, 1964, Prime Minister Nehru died, still in office. He was cremated the following day at *Shanti Ghat* ("The Steps of Peace") on the bank of the river Yamuna in Delhi, near *Raj Ghat* ("The Royal Steps"), where the Mahatma had earlier been cremated. A decade earlier, Nehru had written in his last will and Testament:

> I do not want any religious ceremonies performed for me after my death. I do not believe in any such ceremonies. . . . My desire to have a handful of my ashes thrown into the Ganga at Allahabad has no religious significance, so far as I am concerned. . . . I have been attached to the Ganga and Yamuna Rivers in Allahabad ever since my childhood. . . . The Ganga, especially, is the river of India . . . a symbol of India's age-long culture and civilization, ever-changing, ever-flowing, and yet ever the same. . . .

"Panditji amar rahe" ("Panditji has become immortal"), cried the crowd of millions who gathered to watch worshipfully as the sandalwood flames rose from the pyre of India's first prime minister, whose popularity and power exceeded those of any Mughal emperor or British viceroy.

TWENTY-FOUR

FROM COLLECTIVE LEADERSHIP TO INDIRA RAJ

(1964–77)

The immediate aftermath of Nehru's Era was an interlude of Congress Collective Leadership under Prime Minister Lal Bahadur Shastri (1904–66). Shastri ("Teacher"), famed for his party loyalty and moderation, surprisingly emerged the "unanimous" choice as premier of his Congress colleagues in both houses of parliament on June 9, 1964. The party's strongman "boss" at this time, Tamil-born President Kumaraswami Kamaraj Nadar (1903–76), "managed" Shastri's election as deftly as he would that of his successor. Kamaraj ("King of Love") had worked closely with Nehru in the final years of Panditji's Era, and in 1963 they unveiled the famed "Kamaraj Plan," supposedly devised for revitalization of Congress, by calling upon its most powerful central cabinet ministers as well as state chief ministers to resign from office to do grass roots work for the party back in the villages. The Plan was, in fact, a modified version of Mahatma Gandhi's earlier advice to the entire cabinet to resign; yet timed as it was so late in Nehru's career, cynics saw it simply as a clever way of removing powerful leaders like Finance Minister Morarji Desai (1896–1995) from New Delhi's political center stage in order to give Nehru's daughter and only child greater opportunity to shine. Indira Gandhi (her deceased husband Firoze was no relation of the Mahatma), shy and self-effacing, had been living in her father's home as his official hostess, nurse, and closest companion. She had started her political climb as leader of the Youth Congress and her party's left-wing "ginger group" and had travelled the world over with Nehru. She so diplomatically denied considering herself "qualified" to succeed her father that many observers believed perhaps she was. Yet Nehru's death came too soon for Indira, who had as yet no elective or cabinet experience, to move from the shadows of her deep mourning to the glaring heat of a contest for his mantle. Mrs. Gandhi rested content with the proffered cabinet office of minister for information and broadcasting in Shastri's government, a modest position that enhanced her public image, permitted her to travel widely, and taught her valuable things about media, dissemination of information, and thought control in modern society. Morarji Desai, denied Kamaraj's support and divested of his perquisites of ministerial power, struggled in vain to muster Congress backing for what many considered his "rightful" claim to Nehru's high office; he finally agreed to accept Shastri's compromise candidacy.

The "Syndicate," as the Shastri-Kamaraj board of collective leaders, who now ruled India, was called, included powerful chief ministers who were brought to Delhi to join the cabinet, such as Neelam Sanjiva Reddy (b. 1913) of Andhra, ex-Congress president and chief minister of Andhra till 1964, Shastri's minister for steel and mines, Speaker of Lok Sabha (1967–69), and president of India after 1977. Other popular provincial heads, like Yeshwant B. Chavan (b. 1913) of Maharashtra, who had served Nehru as defense minister in the wake of the Chinese invasion of November 1962, remained at that important post. Chavan was later to be shifted by Indira to home, finance, and external affairs. Swaran Singh, bearded Sikh of the Punjab, was to become Shastri's foreign minister after proving himself a tough negotiator with Pakistan's Bhutto over Kashmir. Shastri's premiership was brief and strife-torn. India's second prime minister was to be the nation's foremost victim of its 1965 "wars" with Pakistan.

For Pakistan, the first decade and a half of freedom had proved most turbulent and politically unstable. Early bereft of its "Great Leader" (*Quaid-i-Azam*) Mohammad Ali Jinnah, who died on September 1, 1948, Pakistan floundered on shoals of provincial conflict and ideological disagreement that all but tore that troubled dominion apart before consensus could be reached on a constitution. Pakistan's first prime minister, Liaquat Ali Khan (1905–51), Jinnah's leading lieutenant in reorganizing the Muslim League who proved more adept at negotiating with Nehru than mollifying his own people, fell victim to assassination in the army headquarters city, Rawalpindi, on October 16, 1951. Orthodox dissatisfaction with Liaquat's secular modernism may have conspired with Afghan-Pathan separatist sentiment in the North-West Frontier Province as well as marital demands for the "liberation" of Kashmir's coveted Vale from Indian "occupation" to remove his hand from Pakistan's helm. In the aftermath of Liaquat's murder, Pakistan fell under the control, first, of a series of pedestrian civil bureaucrats reared in British service traditions and, after 1958, under the steel frame of martial "law." None of the bureaucrats or politicians who struggled so desperately for power in Karachi found a formula to resolve either of Pakistan's perennial, most troublesome constitutional problems: how to provide East Bengal's majority population with their "fair" share of representative power without totally alienating Western Punjabi and Sindi national leadership, or how to administer a "modern" republic without losing support of Pakistan's orthodox Islamic leadership or their mass following. The political tensions and frustrations arising from repeated failures to resolve those two problems so weakened Pakistan's central government that it fell easy victim to martial coup in 1958. General Muhammad Ayub Khan (1907–74), a Sandhurst-trained Pathan, who led the coup, moved swiftly from commander in chief of Pakistan's army to prime minister, self-appointed president of his nation as well as its first field-marshal. In the Cold War of the 1950s Pakistan joined John Foster Dulles' Western Bloc "chain" of military alliances forged to "contain Communism." As a member both of CENTO (Central Treaty Organization) and SEATO (South East Asia Treaty Organization), Pakistan received lavish military and economic development assistance from Washington. Though his army remained much smaller than India's, Ayub deluded himself into believing that its modern Patton tanks and F-86 sabre jets manned by "martial race" Muslim warriors would more than compensate for mere numbers in the field. Before unleashing his new weapons, however, Ayub launched his young protegee, Berkeley-and-Oxford educated Zulfikar Ali Bhutto (1927–79), wealthy scion of an aristocratic landowning Sindi family, on a round of cabinet-level "talks" to try to negotiate a diplomatic settlement with India over Kashmir. Nehru's adamant refusal to permit any statewide "plebiscite" until Pakistan

"totally vacated its aggression" from Azad Kashmir had brought repeated UN efforts to resolve the Kashmir dispute to a standstill. Ayub
and Bhutto established cordial relations with China in the aftermath
of the latter's invasion of India, adding to the already strained relations between India and Pakistan on the eve of Nehru's death. U.S.
arms to India at this time helped New Delhi to muster and train ten
new "mountain" divisions, almost doubling the size of its standing
army.

Indo-Pak sparring started early in 1965 in the *Rann* ("Salt
Marsh") of Kutch, a desolate wasteland mostly submerged by monsoon rains every year, that forms 200 miles of the southern portion
of the Indo-West Pakistan border. Claiming that Pakistan "illegally
patrolled" the Rann north of the 24th parallel, Indian border guards
manned outposts along a line they insisted belonged within India.
Pakistani troops fired upon and cleared those outposts in April, and
before the end of that month, rolled through the Rann with heavy
Patton tanks and 100-pound artillery pieces. India charged Pakistan
with using the Rann of Kutch as a staging ground to test its new U.S.
weapons against India's martial "will power." The Pattons performed
well, easily penetrating ten miles into Indian territory after which
Ayub proposed a cease-fire, agreed upon by June 30 when a joint
commission was set up under UN auspices to demarcate the border.
Troops were withdrawn from the Rann just before monsoon rains
flooded it that summer. Pakistan seemed the winner of round one,
though India had not committed its major forces to the Rann. In
August-September the center of conflict and scene of confrontation
shifted north to Kashmir and the Punjab.

The number of "cease-fire" violations reported by the UN's
peace-keeping mission in Kashmir during the first half of 1965 increased "alarmingly," totalling over 2,000. India charged Pakistan
with training guerrilla units to infiltrate Kashmir and "soften up" the
state for invasion before winter set in. UN observers reported armed
men in civilian clothes crossing the cease-fire line from Azad Kashmir
into the Indian-held portion of Jammu and Kashmir. Radio Pakistan
announced reports of a "spontaneous War of Liberation against Indian Imperialism" that had suddenly broken out in Kashmir. On
August 14, Bhutto officially denied Pakistani involvement in the "uprising against tyranny" within Kashmir. The following day, Shastri

spoke from the ramparts of Delhi's Red Fort to commemorate the eighteenth anniversary of India's independence, charging that "Pakistan has invaded Kashmir" and promising that "force will be met with force and aggression against us will never be allowed to succeed." On August 30, India insisted it was "necessary" to send its own troops across the cease-fire line near Uri to "clear up the Pakistani raiders" around the Uri-Poonch "bulge." In response to that Indian move, regular units of the Pakistan Army crossed the line in the Chhamb sector, heading toward Jammu, on September 1, 1965. The Second Indo-Pak War was in full swing. Shortly before dawn on September 6, India launched a three-pronged tank drive aimed at Lahore. With an army of some nine hundred thousand troops, India had almost four times the martial manpower of Pakistan and, thanks to recent U.S. support against China, considerably more air power and armor. The one hundred and fifty thousand Indian troops in Kashmir easily retained India's position in the Vale as well as Jammu and advanced to seize the high ground at Uri. Indian tanks rolled to within three miles of Lahore, destroying over 450 Pakistani tanks. By September 23, when a UN cease-fire was agreed upon, both nations had run low on ammunition because of a U.S. embargo imposed at the war's start. India conquered less than 500 square miles of Pakistan, while Pakistan claimed to "control" more than 1,500 square miles of India (mostly Rajasthan desert), 340 of which were in Indian Kashmir. A close analysis of the three-week war, however, clearly revealed India's martial victory, whose impact was masked from the glare of world vision by India's restraint of her forces at the outskirts of Lahore, which then lay virtually defenseless.

For India, as for Pakistan, war, however brief, helped divert popular attention from growing internal conflicts and problems, uniting the nation as nothing but the nationalist movement had done before. India's negative position on a Kashmir plebiscite hardened as a result of her martial successes. Shastri called upon the Security Council to declare Pakistan "the aggressor," while Ayub demanded "justice."

In a diplomatic bid for greater influence in South Asia, Soviet Premier Aleksei Kosygin invited Shastri and Ayub to Tashkent for a summit meeting to discuss mutual problems. The Tashkent Summit of January 4–10, 1966, was Prime Minister Shastri's last conference.

India's premier suggested to Ayub at the opening of their historic meeting that "Our objective at this meeting should be not recrimination over the past, but a new look towards the future." He recommended that positive steps be taken to improve Indo-Pakistan trade and to ensure greater future economic cooperation rather than continuing armed conflict. "Heavy responsibility lies on our shoulders. The subcontinent has a population of 600 million. . . . Instead of fighting each other, let us start fighting poverty, disease and ignorance." A week of intensive negotiations ensued, ending with an agreement signed by Shastri and Ayub on January 10, declaring their "firm resolve to restore normal and peaceful relations between their countries and to promote understanding and friendly relations between their peoples." They reaffirmed their "obligation under the Charter not to have recourse to force and to settle their disputes through peaceful means," agreeing to withdraw their armed forces by not later than February 25, 1966, to positions they had held prior to August 5, 1965. They pledged that in the future, relations between India and Pakistan "shall be based on the principle of non-interference in the internal affairs of each other." They also promised to "discourage any propaganda directed against the other country," to restore normal diplomatic relations, and "to consider measures towards the restoration of economic and trade relations, communications, as well as cultural exchanges," all of which had been disrupted by partition. Finally, Shastri and Ayub agreed to repatriate prisoners, to "continue discussion of questions relating to the problems of refugees," and to "create conditions which will prevent exodus of people." It was the most promising agreement yet reached between India and Pakistan, and it seemed to mark the dawn of a new era of South Asian peace, friendship, and cooperation. Before dawn arrived, however, Lal Bahadur Shastri was dead. The fatal heart attack he suffered within hours of signing the agreement in Tashkent seemed to symbolize the futility of the struggle he had waged against the tragic legacy of communalism.

Indian attention now focused upon the problem of who would succeed Shastri. The bitter struggle for succession that ensued between Indira Gandhi and Morarji Desai, respective champions of the Congress Left and Right, could not be smoothed over and hidden from public light even by the deft political maneuvering of Kamaraj, who tried to convince Morarji to defer to Nehru's daughter. At sixty-nine, Desai felt he was too old to risk waiting for another chance at the job he considered himself best qualified to fill. Mrs. Gandhi was only forty-eight, after all, and not only a woman, but a widow, with relatively little administrative experience. Her natural national constituency was Socialist youth and women, an unlikely candidate, perhaps, to do battle against so formidable a conservative captain of industrial India's rock-ribbed leadership as Morarji Desai, yet her seeming weakness gave her strength. Kamaraj knew all too well that with Morarji as prime minister, his days of patronage and behind-the-scenes power were over, but Indira appeared to be much more humble and malleable. On January 15 Kamaraj convened a Delhi meeting of eight chief ministers, and after "an hour of deliberations" announced their unanimous backing for Mrs. Gandhi. By nightfall she was virtually assured 347 votes out of the Congress Party's 551 Members of Parliament. The balloting in both houses four days later took almost four hours to complete, but when the votes were counted Indira Gandhi had 355 and Morarji Desai, 169.

Prime Minister Gandhi immediately promised to continue the policies established by her father and to honor the pledges made at Tashkent by Shastri. "We should encourage the Tashkent spirit," she told her nation. "We should have peace at home and abroad, if possible." She had good reason to modify her pious hope, for several days later Mizo tribal rebels in the northeast were met by Indian army gunfire as the worst uprising since the birth of Nagaland imposed fresh strains of provincial divisiveness on India's pluralistic nation. Monsoon failure in 1965 had, moreover, raised the specter of famine, which loomed so large on the eve of Indira's election that the advantages of having a cosmopolitan prime minister who could effectively appeal for food grains and other desperately needed financial support abroad helped Congress make the choice it did. In her Independence Day message on January 26, Prime Minister Gandhi announced the promise of "large enough quantities of food grains" from the United States to "bridge" India's impending famine gap, adding, "We are grateful to the United States for her sympathetic understanding and prompt help." The Democratic administration in Washington had consistently followed a strongly supportive policy of foreign aid to India, especially in the face of famine. With much of Central India already in the grips of famine, and with the 1966 grain harvest expected to drop to seventy-six million tons (twelve million below the high reached in 1965), the Johnson administration's guarantee of continuing monthly shipments of U.S. wheat to India through the end of 1966 was crucial to survival itself for millions of India's peasantry. Mrs. Gandhi flew to Washington in March and received personal assurance not only of some twelve million tons of wheat, but also of $435 million in U.S. loans and credits for the first year of what was to be India's Fourth Five-Year Plan. "India, like the United States, is wedded to the democratic ideal," the prime minister assured Washington's National Press Club at a luncheon meeting in her honor. She added, however, the following important clarification as to what she personally and her party's collective leadership meant by that ideal: "Today democracy inescapably implies social welfare, equality of opportunity, reasonable living standards and dignity of individual. Man does not live by bread alone. But equally he needs bread to enjoy liberty."

Assured of substantial and continuing American aid, Mrs. Gandhi returned home to inaugurate the first year of India's fourth plan in April, announcing it would "lay the groundwork for a breakthrough in agriculture." Though India's national income was only $31.4 billion and its population now over 475 million, there were "several signs of change" that the prime minister noted as economically hopeful. In the previous year two million kilowatt-hours of power had been added to the national grid. New high-yield seeds were being tested in several regions of the land, and agricultural cooperatives in Gujarat were now processing milk, sugar, and other commodities far more energetically and economically than ever before in India's history. Most of Madras's eighteen thousand villages had been electrified, and sixty-eight million Indian children were in school. The growth in technical and managerial skills acquired by thousands of Indians during the first three five-year plans was impossible to quantify, yet "no less important" in adding to "the quality and pace of social transformation."

In June 1966, however, when Mrs. Gandhi announced the devaluation of India's rupee from 4.76 to 7.50 to the U.S. dollar, leaders of India's opposition parties all interpreted the act as an official confession of the failure of Indian economic planning and the prime min-

ister's "capitulation" to American "pressure." Government prestige
fell as precipitously as the deflated rupee, and hardly half a year later,
Congress would be obliged to fight its fourth nationwide election
campaign. Mrs. Gandhi insisted that devaluation was required to increase the sale of Indian exports to earn more foreign exchange. "No
nation has achieved economic growth or any kind of independence,
political or economic, without going through hell and fire," she informed India's press. "We have had a pretty soft life all these years.
We cannot hope to have a breakthrough unless we are willing to go
through that experience." It was a courageous, though unpopular,
stand, and the new prime minister fell to new lows in national confidence. Mrs. Gandhi regained some of her personal popularity at
home and restored her Socialist image of independence from Washington through her increasingly outspoken remarks in opposition to
greater U.S. participation in the Vietnam War. She visited Moscow
in July and called peace the "special responsibility" of "nonaligned
countries," working with Nasser and Tito to find a "just solution" to
the "bitter and bloody" war, a solution that would satisfy "the legitimate rights and hopes of the people of Vietnam." Her bid for the
international mantle of her father's Third World leadership, however, was frustrated by the Ayub-Chou alliance, and the "spirit" of
Tashkent grew ever more elusive. Internally, Mrs. Gandhi was faced
with a mounting tempo of strikes and food riots, as well as linguistic
and religious separatist agitation in the Punjab, where militant Sikhs
continued to demand a Punjabi *suba* ("state").

With some 33 percent of its otherwise-Hindu population Sikhs,
whose scripture was written in the *Gurumukhi* ("Language of the
Gurus") script and who spoke Punjabi rather than Hindi, the Punjab
remained a region of unresolved bilingual as well as communal conflict. Sikh demands for statehood were based on the predominantly
Sikh, Punjabi-speaking northwestern half of the Punjab, which was
more economically advanced than Haryana, the overwhelmingly
Hindu and Hindi-speaking southeastern region of the 47,000-square-
mile state. The beautiful new capital of the Punjab, Chandigarh (the
"City of Silver" designed by Le Corbusier), was also claimed by Sikh
separatists, who since the dawn of independent India had demanded
statehood, which at times meant national, though usually only provincial, autonomy. A militant revivalist Sikh party, the *Akali Dal,* fought

first under the charismatic direction of its intermittently fasting Master Tara Singh and subsequently won its demand under the leadership of Sant Fateh Singh. In December 1960, when the Master was in jail, the *Sant* ("Saint") entered Punjabi politics by fasting more than three weeks for the Punjabi *suba*. The *Arya Samaj* and *Jan Sangh* organized "save Hindi" marches and demonstrations in orthodox Hindu districts of Haryana. The Indo-Pak War of 1965 began just in time to postpone a fast-unto-death threatened by the Sant, who agreed after the cease-fire to abandon his fast while a cabinet committee reviewed the party's demand. The new states of Haryana and the Punjab thus emerged out of the former Punjab on November 1, 1966, but it was not until after Darshan Singh Pheruman's fast-unto-death in October 1969, however, followed by Sant Fateh Singh's threat to immolate himself, that Chandigarh was promised to the Sikh-majority state of the Punjab as its own capital, early in 1970.

The fourth general election, in February 1967, proved as disastrously deflating to Mrs. Gandhi's government majority as most political analysts predicted it would be. In the *Lok Sabha* the Congress majority fell from over 200 seats to a bare working majority of 20, capturing 279 seats in a house of 515 and winning only 40 percent of the popular vote. At the state level the full scope of the debacle became clearer, for there Congress lost a majority of the total seats, capturing only 1,661 out of 3,453. Non-Congress ministries were established by opposition coalitions in West Bengal, Bihar, Madras, Orissa, Kerala, and the Punjab by the end of March 1967. In April the Congress coalition ministry in Uttar Pradesh fell, and in Rajasthan the Congress-led ministry was so unstable that president's rule had to be instituted. The damage was so widespread, the popular loss of confidence in Congress leadership so deep and pervasive, that Kamaraj sought desperately to rebuild the party over whose virtual demise he had presided by persuading Mrs. Gandhi to invite Morarji Desai back to Delhi as her deputy prime minister, a post created by her father for Sardar Patel. Morarji had also wanted to be home minister, the other portfolio held by Patel, but Indira refused to give him control over India's police, offering him finance instead. She recognized that, though the election of 1967 had battered her party, it was really the right wing of Congress, at least its older and more conservative leadership, that had suffered the most stunning defeat. She thus accepted

Kamaraj's plan for rapprochement with Morarji reluctantly, as a compromise of her own political instinct for the sake of supposed party unity. The unlikely marriage of political expediency proved futile, however, and Congress, incapable of reforming itself as Mahatma Gandhi had long since urged it to do, of sublimating personal ambition, greed, and pleasure to national goals and public service, all but lost its monopoly of power in 1967, just twenty years after independence. The manifestos of every opposition party, right, left, and center, echoed the same sobering themes: "the country is today in shambles in all spheres of national life" (*Swatantra*); "The Congress Party . . . has lost the confidence of the people . . . has betrayed an unawareness of the nation's basic unity and integrity, ignored people's sentiments and has been callously indifferent to the common man's welfare" (*Bharatiya Jan Sangh*); "For twenty years, Congress Governments have given great disappointment to the people" (*Samyukta* Socialist Party).

The ousted old guard of Congress sought to regain control over the party's machinery, but Mrs. Gandhi resisted them, and took an increasingly leftist position. The U.S. decision that April to resume the Pentagon's flow of arms and "spare parts" to Pakistan, shipments that had been discontinued with the outbreak of war in 1965, helped confirm Mrs. Gandhi's determination to recapture the confidence of her nation by moving its government's policies more rapidly toward the realization of the "Socialist pattern of society" that her father and Congress had committed themselves to more than a decade ago. As a new South Asian arms race began, the spirit of Tashkent was buried. India now refused to sign the Treaty on Nuclear Nonproliferation proposed to the world at Geneva, insisting, as Mrs. Gandhi said, that "we must be satisfied that we have what I might call a credible guarantee for our security." Since 1964 India had been extracting explosive plutonium at its atomic plant at Trombay.

Determined to prove its claim to secular statehood, especially on the eve of growing tensions with Pakistan, India elected its Muslim vice-president, Dr. Zakir Husain (1897–1969), as third president of the republic on May 6, 1967. The Hyderabad-born Husain had earned his Ph.D. from Berlin University and had been vice-chancellor of Aligarh University for eight years following India's independence. India's new vice-president, Varahagiri Venkata Giri (b. 1894), a for-

mer labor organizer, had served as general-secretary of the All-India
Railwaymen's Federation, minister of labor and industry for Madras,
and minister of labor in New Delhi. Congress' days of unity, however, were numbered. The unwieldy coalition of radicals, liberals,
and conservatives that had served as the vehicle for bringing India to
its goal of nationhood could not sustain the internal tensions of basic
political differences that finally split the party asunder in 1969. Mid-
term elections early that year revealed further loss of Congress support in U.P., West Bengal, the Punjab, and Bihar, and steady growth
in more radical opposition parties and independents with the courage
to attack outmoded Congress policies and unkept promises. Mrs.
Gandhi vigorously defended her government's policy of keeping all
new heavy industries under public direction. On May 3 President
Husain died, and Vice-President Giri took over as acting president
pending special presidential elections.

Once again the rift between Mrs. Gandhi and the old guard
widened, for while she preferred Giri, he received additional support
only from Fakhruddin Ali Ahmed, the one Muslim on the powerful
Central Parliamentary Board of Congress, which chose *Lok Sabha*
Speaker Sanjiva Reddy as its official candidate. Instead of silently
bowing to the will of her party bosses, Prime Minister Gandhi urged
Giri to resign his acting office in order to campaign for the presidency,
letting it be known that he was her candidate. At the Bangalore meeting of Congress that July, Indira unveiled her proposals for nation-
alizing India's banks, pursuing a vigorous policy of land reform, and
placing ceilings on personal income and private property as well as
on corporate profits. On July 16, Prime Minister Gandhi personally
took control of the Finance Ministry, ousting Morarji Desai from
that office, as a result of which he immediately resigned his deputy
premiership. Three days later Indira issued an ordinance nationaliz-
ing fourteen of India's major banks, including the Central Bank and
the Bank of India, explaining that "our sole concern has been to accelerate development and thus make a significant impact on the problems of poverty and unemployment." One month later the electoral
victory of V. V. Giri confirmed that Prime Minister Indira Gandhi
had emerged stronger and far more popular than the party machine
that had put her in power just three and a half years earlier. The interlude of collective leadership was over; so was the era of united
Congress Party rule; that of Indira Raj had begun.

Prime Minister Gandhi's adroit outmaneuvering of her party's old guard led to her "expulsion" from the Congress for "indiscipline" on November 12, 1969, following which Indira rallied a majority of the Congress members of Parliament to her "New Congress" banner. The Working Committee split almost evenly between Morarji Desai's Organization (O) and Indira's Requisition (R) parties, which held separate sessions in Delhi that fateful month, but Indira retained the allegiance and loyal support of over two hundred *Lok Sabha* members of Congress, while Morarji mustered only sixty-five, enough to give his party parliamentary opposition status, but not enough to bring down the government. Lacking a *Lok Sabha* majority of her Congress (R) loyalists, Indira now led a national left-wing coalition that included both Communist parties, as well as the regional DMK and *Akali Dal* parties, all of whom supported her basic policy of more rapid economic development through increased governmental controls. Morarji's Congress (O) Party had the support of *Swatantra* and *Jan Sangh* members, but his right-wing coalition remained much smaller than the prime minister's force.

As India's government moved further to the left to allow Mrs. Gandhi to retain her hold over an increasingly impatient populace by seeking to fulfill long-deferred nationalist promises, Richard Nixon was completing his first year as president of the United States, and Field Marshal Ayub Khan had stepped aside to turn over Pakistan's sorely divided land to the martial grip of his Pathan protégé, General Aga Muhammad Yahya Khan (b. 1917). The war in Vietnam continued to escalate, and Pakistan's military command, lavishly supported by Washington, was poised on the brink of turning East Bengal into its own Vietnam. After twenty-one years of independence, the gulf that divided Pakistan's east from its west wing had only widened, the cultural differences deepening with time. East Pakistan remained economically underdeveloped and impoverished, sustaining over 55 percent of Pakistan's 120 million people on 15 percent of its total land area, earning most of Pakistan's hard-currency foreign exchange with its exported jute, yet reaping a smaller portion of national industrial development funds than did West Pakistan. In many ways East Pakistan was a "colony" of the West, administered mostly by bureaucrats from the Punjab who spoke and thought in Urdu rather than Bengali and kept under martial control by an army and air force run by Sandhurst- and Pentagon-trained Pathan, Baluchi, and Punjabi generals and colonels. Jinnah's dream of national integration dissolved in the acid of Ayub's regime, which crushed popular opposition by its strategy of alternating arrests and intimidation.

Mader-i-Millat ("Mother of the Nation") Fatima Jinnah (1893–1967), the Quaid-i-Azam's sister, sought in 1964 to depose the martial dictator, against whom she ran for "president" as the candidate of Pakistan's Combined Opposition Parties. Ayub's "constitution," however, had limited the franchise to eighty thousand so-called Basic Democrats, who were chosen—generally from among his supporters, often members of military families—to "represent" Pakistan's eighty million adults, and hence assured the field marshal of victory. The failure of Fatima's spirited campaign, despite the charisma of her name and the unanimous backing of opposition parties from both East and West, demonstrated the futility of Pakistan's democratic aspirations and the impotence of its popular opposition. Nonetheless, Sheikh Mujibur Rahman (1920–75), Bengali leader of the popular *Awami* League, proposed a militant Six-Point Program to his party in March 1966, calling for virtual autonomy for East Pakistan and demanding full provincial powers over taxation and revenue collection, control over foreign exchange earnings, separate militia or paramilitary force, separate but freely convertible currency, and a parliamentary type of central, federal government run by a responsible legislature directly elected by the people on the basis of universal adult franchise. Mujib's Six-Point Program was to become the platform of Bangladesh. In West Pakistan, Bhutto also strove to rally opposition to the regime from which he had resigned after Tashkent, launching his Pakistan People's Party in November 1967, based on a platform of "Islamic Socialism" at home and "freedom" for Kashmir, with Sino-Pak cooperation abroad. In 1968 Ayub's regime charged Mujib and eight conspirators with treason, and that same year Bhutto was arrested for "inciting the masses, particularly students, to violate the law and create disorder by resorting to violence." Instead of silencing opposition, however, such repression only stirred greater turmoil and popular discontent, driving Ayub to abandon his "Basic Democracy" and turn his tattered authority over to Commander-in-Chief Yahya Khan's bayonet-and-tank administration. West Pakistani reinforcements were flown into Dacca a few days before martial law clamped the lid on Bengali political protest in March 1969. Nixon sent his personal "congratulations and greetings" to Yahya, and the

World Bank's Aid-to-Pakistan Consortium assured him $484 million in further assistance for the forthcoming fiscal year.

With U.S. support and advice, Yahya announced in October 1969 that his government planned to hold general elections throughout Pakistan—the first since independence—in October 1970, after which he promised to "turn over power" to "elected representatives of the people." All political prisoners were soon released, but in March 1970, when Yahya presented his Legal Framework Order to the nation, explaining his ground rules for Pakistan's future constitution, he ominously insisted that "no power on earth" could separate East from West Pakistan, since they were "two limbs of the same body." At this time, however, he unveiled Pakistan's ambitious Fourth Five-Year Plan, which actually promised the East more public expenditure than the West. It almost seemed as though Pakistan had learned the lessons of its past mistakes.

India had by now made a spectacular breakthrough in agricultural yields, its so-called Green Revolution, thanks to the introduction of new U.S.-developed, high-yield Mexican wheat (Sonora 64) and Taiwan and Philippine rice (Taichung Native I and Tainan III and IR 8), which raised India's food-grain production to almost 100 million tons in 1968–69. Despite Indian population growth of some 13 million annually, with an estimated population of 530 million in 1969, average daily consumption of calories was over 2,100, and life expectancy had risen to 51 years. The slow-starting rural revolution had finally begun to yield rich harvests. The tempo of industrial production was also picking up, registering over 7 percent growth in 1969, which helped diminish unemployment slightly and gave Indira's government the economic good news it needed to consolidate her base of popular support. The revised Fourth Five-Year Plan presented by Mrs. Gandhi to Parliament in March 1970 proposed heavier expenditure on agriculture, calling upon the states as well as the central government to devote greater resources to strengthening the gains made in food production. In June 1970 Prime Minister Gandhi took personal control of her Home Ministry, retaining the atomic energy and planning portfolios as well, and thus assuming more direct responsibility for central administration than even her father had done. The era of Indira Raj was one of consolidation of central power, and of India's assertion of greater independence from the West.

In July 1970 floodwaters ravaged much of East Pakistan, leaving an estimated ten million homeless and starting a new wave of refugees moving west to seek food and shelter in India's Bengal. Internally, West Bengal had long been a center of rising discontent because of population pressures and unemployment, but in 1970 its deteriorating political climate took a sharp turn for the worse with the appearance of an extremely violent splinter group of Maoist Communists called Naxalites, after the Naxalbari district in northern Bengal where the movement was born. Rejecting elections and government alike, the Naxalites used murder as their major political tactic, assassinating public leaders and officials and turning Calcutta into a city that dreaded nightfall, as well as one of "Dreadful Night." The influx of East Bengali Hindu refugees—some hundred thousand of them in 1970—added to the congestion, violence, and pollution of Calcutta and its environs, and president's rule was imposed over West Bengal.

Indira's Congress proposed a bill in 1970 to abolish former maharaja privy purses and privileges, which had been vouchsafed by the government to those deposed princes since independence. The controversial bill, the Twenty-fourth Amendment to India's Constitution, received its requisite two-thirds majority in the *Lok Sabha* that September, despite outspoken opposition. When the *Rajya Sabha* failed to muster the requisite two-thirds majority, however, the princes were "derecognized" by simple presidential order, and Indira proved once again that she was more powerful than her united opposition. The Supreme Court struck down the presidential order against the princes on December 15, 1970, and several chief ministers refused to cooperate with Indira's proposed land reforms and other measures designed to distribute India's land and resources more equitably. "We are concerned not merely with remaining in power but with using that power to ensure better life to the vast majority of our people and satisfy their aspirations for just social order," Mrs. Gandhi announced as she called for new elections on December 27, 1970.

That same month, on December 7, Mujibur Rahman's *Awami* League swept the polls throughout East Pakistan, winning 160 of the 162 seats allotted to the East in Pakistan's National Assembly. It was Pakistan's first nationwide popular election, and no clearer mandate could have been won by any party or its leader than Mujib mustered for his Six-Point Program of Bengali autonomy. Bhutto's Pakistan

People's Party won a majority of the votes in the West, electing eighty-one members to the assembly. But Yahya was not ready to step down and turn over Pakistan's central power to a Bengali-run regime, nor was Bhutto willing to accept Mujib as prime minister. The newly elected National Assembly was supposed to meet on March 1, 1971, but Bhutto demanded that it be postponed, and Yahya agreed. Early in March Mujib called upon the people of East Pakistan to go on strike. Yahya imposed an immediate curfew, and a few days later he promised to convene the assembly on March 25.

On March 1, 1971, India's fifth general election began, and after ten days of polling the results were a stunning victory for Indira's ruling Congress, which won a clear majority of 350 out of 515 *Lok Sabha* seats, reducing the Congress opposition to a mere sixteen members. "The country has given a clear verdict in favor of the policies and programs of my party," Indira announced and specified that those policies were to "reduce economic disparities" and unemployment. "The elections have proved how strong the democratic roots in the country are and how discerning our people are," she noted. Her opponents had run on the slogan of *"Indira Hatao!"* ("Eliminate Indira"), while she went to the people with the slogan *"Garibi Hatao!"* ("Eliminate Poverty"). Slum clearance and the rehabilitation of slum-dwellers, the building of low-cost housing across the land, and the settlement of landless laborers on property they could call their own, were projects Indira's Raj planned for the immediate future.

On March 25, 1971, week-long talks in Dacca among Mujib, Bhutto, and Yahya ended abruptly as Bhutto and Yahya flew home to the West, leaving Lieutenant General Tikka Khan and his force of some sixty thousand crack West Pakistani troops in the East, which now proclaimed itself Bangladesh. Unfurling flags of national independence, the people of East Pakistan resolved to wait no longer for the freedom they demanded under the lead of their "Nation Unifer" (*Bangabandhu*), Sheikh Mujibur Rahman. Shortly before midnight Tikka Khan's troops and tanks left their barracks and moved across Dacca to open fire at student dormitories on the university campus and on crowded Hindu bazaars where thousands were asleep. At 1:00 A.M. Mujib was arrested and escorted by tanks from his home. The message he left for his "dear brothers and sisters of free Bangla Desh," however, was relayed throughout the countryside, to every

mango grove and blooming field of rice: "You are citizens of a free
country. . . . Today the West Pakistan's military force is engaged in
a genocide in Bangla Desh by killing Lakhs [hundreds of thousands]
of civilians. In their bid to rob the people of Bangla Desh of their
freedom, they have unleashed unparalleled barbarity on the golden
Bengal . . . Certain is our victory. Allah is with us. The world public opinion is with us. *JAI BANGLA* ['Victory to Bengal']!" Bangladesh was born.

On March 31 India appealed to the United Nations, insisting
that "the scale of human suffering is such that it ceases to be a matter
of domestic concern of Pakistan alone." Within a month nearly one
million terrified Bengali civilians fled from East Pakistan, crossing
India's borders at the rate of some sixty thousand a day by the end
of April. By December 1971 that flood of terror would drive almost
ten million desperate, starving people from their homes in war-torn
Bangladesh to neighboring India. Thousands of young Bangladeshis
joined guerrilla bands of "Liberation Forces" (*Mukti Bahini*), which
received arms and support from Indian troops across the border.

Nixon's White House said nothing and did nothing to halt its
lethal flow of arms to West Pakistan, which claimed that its problems
in the East were caused by an Indian "plot to break up" its nation.
As American eye-witness reports of the blood bath in Dacca, Chittagong and other Bangladesh cities reached the United States, Senators
Frank Church of Idaho and Edward Kennedy of Massachusetts
joined former Indian Ambassadors Chester Bowles and John K.
Galbraith to lead what soon became a wave of national protest
against West Pakistan's military atrocities and the continuing support
for Yahya's dictatorship from the White House and Pentagon. Church
reminded his colleagues in Congress that the "military largesse" of
some $2 billion in arms given by the U.S. to Pakistan since 1954 was
supposed to have been used "as a shield to protect the Pakistanis"
from "Communist aggression," not to kill and terrorize their "own"
population. But the Nixon administration continued to do business as
usual with Yahya's government. Nixon and Secretary of State Henry
Kissinger were at this time employing Yahya Khan as their Asian
middleman in negotiating a top-secret settlement with China. On July
8, Kissinger flew into Rawalpindi en route to Paris, and after talks
with Yahya his departure was announced to be "unexpectedly de-

layed," supposedly because of "an intestinal affliction." That was when he flew to Peking for his first meeting with Chou to arrange the details of Nixon's China visit.

On August 9, 1971, Mrs. Gandhi signed a twenty-year Treaty of Peace, Friendship, and Cooperation with the Soviet Union, stipulating that if either party were "subjected to an attack or a threat thereof," the two would "immediately enter mutual consultations in order to remove such threat and to take appropriate effective measures to ensure peace and the security of their countries." It was the "great power" support she required on the eve of India's third war with Pakistan. The roots of Indo-American friendship and cooperation had been poisoned by Nixonian diplomacy, and Nehru's policy of nonalignment, which had, in fact, always depended heavily on Anglo-American assistance and good will, was scrapped by his daughter in favor of a new Indo-Russian alliance.

By September it was costing India approximately $200 million a month to feed its eight million refugees; the entire 1965 war with Pakistan had cost her only $70 million. When the monsoon rains stopped in October, a revitalized, Indian-trained *Mukti Bahini* of some thirty thousand Bengalis moved into action throughout Bangladesh, disrupting communications, gathering intelligence, terrorizing Pakistani officials and soldiers, "softening up" the countryside for the final blow. In desperation, Yahya promised to convene the National Assembly on December 27 and to rewrite the Constitution. On November 23, Pakistan's ambassador to the United States called a press conference to report that troops of three Indian divisions, supported by armor and air cover, had "made an all out attack on Jessore, Rangpur, Chittagong Hill Tracts and Sylhet." Asked if war was "imminent" between India and Pakistan, Ambassador Raza replied: "The war is on. It is not only imminent, but it is on."

On the morning of December 3, Kissinger convened his Special Action Group in the Situation Room of the White House, reporting that he was "getting hell every half-hour from the President" for "not being tough enough on India." Nixon wanted "to tilt in favor of Pakistan" and had just ordered Kissinger to cut off all letters of credit to India, totaling some $99 million at the time, as well as another $72 million worth of Public Law 480 credit. That evening Pakistani planes attacked twelve Indian airfields in the West, including Amritsar, Agra, and Srinagar. The next day Indian forces struck hard in Pakistan as well as Bangladesh. With total air superiority and popular support, India advanced quickly in the East, while fighting a "holding action" in the West. The green, red, and gold banner of Bangladesh was unfurled in every town and village as the Indian Army closed its ring of steel round Dacca. On December 9 Jessore was liberated amid cries of *"Jai Bangla!"* Six days later, India's chief-of-staff, General Sam Manekshaw, personally accepted Pakistani General Niazi's offer to surrender, despite the latter's earlier vow to "fight to the last man." The Instrument of Surrender of "All Pakistan Armed Forces in Bangla Desh" was signed on December 15, 1971. India declared a unilateral cease-fire and Prime Minister Gandhi told a wildly cheering Parliament that "Dacca is now the free capital of a free country."

The success of Indian arms and diplomacy in the liberation of Bangladesh was the crowning triumph of Indira's Raj, irrevocably shifting the balance of South Asian power to India. Pakistan emerged from this third undeclared war with less than half its population, its army and economy on the brink of collapse, its myth of Muslim unity destroyed, its spirit sorely deflated, its dream of the conquest of Kashmir as dead as its hope of recapturing Bangladesh. On December 20 Yahya resigned, turning over the reigns of his remaining power to Bhutto, who bravely promised his countrymen to "pick up the pieces, very small pieces," and "make a new Pakistan—a prosperous and progressive Pakistan." For Bangladesh freedom was almost as painful, for so many had died and so much rebuilding remained to be done, yet from those ashes the world's eighth largest nation had been born, the People's Republic of Bangladesh, a secular state. On December 21, Bhutto released Mujib from jail, and two weeks later he was permitted to leave Pakistan, returning to his native Dacca on January 10, 1972. Mujib now took over as prime minister of his nation, with the former vice-chancellor of Dacca University, Abu Sayeed Choudhury, becoming its president. The millions of refugees began to return on December 22, and by March 25, 1972, Bangladesh's first "anniversary," nearly all of the 9,774,140 who had fled to India were back in their homeland.

In the wake of the war, India reorganized its northeastern region to try to ensure more effective control over its border zones with China and Burma. On January 21, 1972, three new full-fledged states were added to India's union: Meghalaya ("Abode of Clouds"), which was carved from the Khasi, Jaintia, and Garo Hills of Assam's Shillong plateau; Tripura (whose capital is Agartala), which had been the Union Territories of Tripura; and Manipur, with its capital at Imphal. The former Northeast Frontier Agency was renamed Arunachal Pradesh ("Dawn's Province"), but remained a union territory under Delhi's direct administrative control till 1979. The former Mizo Hills district of Assam, bordering on Burma and the center of Mizo uprisings, was kept under direct Delhi control as the new Union Territory of Mizoram. The sparsely populated (under fifteen million in 1972) state of Assam was thus stripped of some 65 percent of its territory by Indira's Raj. In 1975 the former "Protectorate" of Sikkim, gateway to Tibet, was granted Indian statehood.

Elections were held for assembly seats in sixteen of India's twenty-one states (Himachal Pradesh—"Snow's Province"—with its capital at Simla, had attained statehood in January 1972). Nearly two hundred million Indians were entitled to cast their votes for almost twelve thousand candidates, who contested close to three thousand seats across the land. As anticipated, Indira's ruling Congress Party made a clean sweep of the polls. The old guard Syndicate was routed as effectively as Pakistan's army had been, failing even to retain control of its once solid bastion of power, the Gujarat district of Surat. Indira could now move forward with her central government's policy of socialization and more rapid development. In August 1971 the Twenty-fourth Amendment empowered Parliament to change the Constitution's "Fundamental Rights," a decision that was challenged by a South Indian religious trust concerning its property rights. In October 1971 the disputed amendment was brought before the Supreme Court, which issued its judgment the following April, essentially supporting Indira's government. In December of 1971 the Twenty-fifth and Twenty-sixth Amendments were passed, substituting officially determined "amount" for "compensation" of nationalized property, and once again abolishing princely "purses" and finally "derecognizing" their still exalted status. Indira's critics charged her with "capitulating" to a Communist coalition and pressures from Moscow, with scuttling "democratic principles" for political expedience. She countered by insisting that no "selfish minority" of "rich capitalists" could continue to hold India hostage to its "monopoly interests."

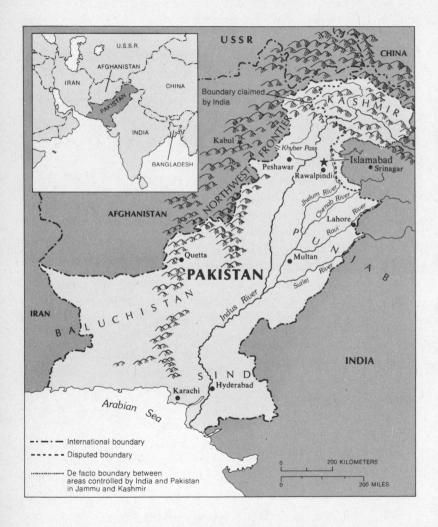

International boundary

Disputed boundary

De facto boundary between areas controlled by India and Pakistan in Jammu and Kashmir

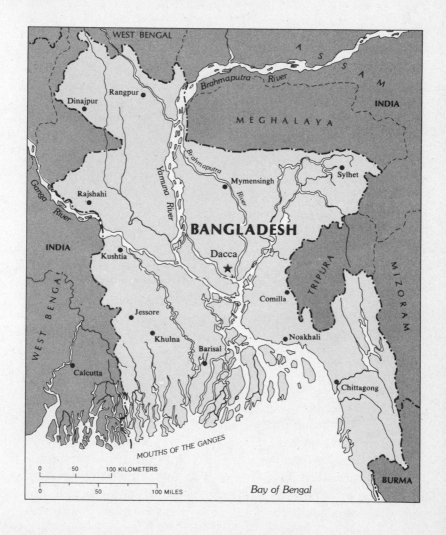

To fulfill her promise of abolishing poverty, Indira counted on more radical land reforms; the nationalization of more industries, including coal mines, which were brought under government control in 1973; and central-government-controlled wholesale food-grain markets (wheat was taken over by Delhi in 1973, but had to be returned to state control in 1974). But India's most difficult and age-old problem was not to be solved in a year or two or even ten, no matter how popular or powerful the Raj that attacked it was or how radical the measures of reform. The very government officials empowered to enforce the "tough" new laws were those same landed interests who would suffer most from strict enforcement. Similarly, harsh taxes on industrial profits and urban income or wealth continued to be unscrupulously evaded by leading urban backers of the Congress Party. Nor would nepotism, one of India's oldest traditions, be eliminated any more swiftly than poverty. Few Indians were any more startled to learn that Indira's son, Sanjay, had been appointed top manager of India's new automobile manufacturing industry, which took five years to produce its first car, than they were to recall that their prime minister was Nehru's daughter. If one did not help one's relatives, after all, who should one assist? Nor would India's problems of communal jealousy, distrust, ignorance, and hatred, which erupted periodically in Gujarat, Uttar Pradesh, the Punjab, Bihar, and Maharashtra, disappear overnight. Nor the equally dangerous conflicts of regional antipathy that pitted Tamil against Hindi, Bihari against Bengali, Gujarati against Marathi. The *karma* of India's complex pluralistic history would continue to bear fruit for long centuries to come.

Nevertheless, in 1974 India boldly launched its Fifth Five-Year Plan, with the "removal of poverty" and "attainment of economic self-reliance" as twin goals. It was hoped that food-grain production could be increased from the 114 million tons reaped in 1973–74 to some 140 million tons by 1978–79. With India's population close to 600 million in 1975, even as dramatic a growth in food production as that would not ensure self-sufficiency, but Indira's Raj was investing more heavily in birth control and family planning schemes, which received about $400 million during the fourth plan period and were scheduled to get close to $700 million during the fifth plan. With another $10 billion to be invested in industry and approximately the same amount in electric power, the fifth plan was designed to generate from between 8 and 10 percent more industrial production annually, which was to be linked to providing greater immediate employment on an extensive scale where jobs were most urgently required. Inflation, especially in oil prices during the 1973–74 "energy crisis," when the cost of Indian oil almost quadrupled, eroded the optimistic hopes of Indian planners. Improved relations with the United States in the wake of the Bangladesh War, however, and increasing trade with all the Americas helped India's economy weather its inflationary spiral. At the end of 1973 the United States relinquished claims to over $2 billion in P.L. 480 funds that New Delhi still owed Washington for previous aid, thus helping rectify the 1971 White House "tilt" toward Pakistan.

On May 18, 1974, India carried out its first successful underground nuclear explosion in the desert of Rajasthan. Mrs. Gandhi insisted that there was a great difference between a "nuclear country," which India had become, and a "nuclear weapons country," which she had "no intention" of becoming. "We have been taunted that a poor nation cannot afford this luxury," Indira argued in the wake of the explosion, but she recalled what her father had remarked a quarter century earlier, that "as far as one can see, the main source of power has to be atomic energy." In 1965 when China first exploded a nuclear device, India was swift to condemn that act as a "retrograde step," calling upon all nations to stop such explosions. China was now on notice that it no longer enjoyed a monopoly of nuclear power in Asia, and Pakistan had, of course, been close enough to virtually feel India's dramatic entry into the world's nuclear club. Delhi now refused to sign the United Nation's Test Ban Resolution.

Yet even as atomic power shook the sands of Rajasthan, another kind of explosion rocked Bihar and Gujarat, where tens of thousands of students and workers marched to protest against runaway inflation and Congress corruption. In the state of Bihar, where four-fifths of the population of sixty million were illiterate, over two-thirds wretchedly impoverished, and more than half landless, JP Narayan led a broad-based coalition against Congress in what he called nothing less than "a total revolution." In Gujarat, where two years of drought conspired with an inflation of 30 percent per annum to wipe out the savings and patience of middle-class office workers, schoolteachers, and professionals as well as students, Morarji Desai went on a fast-unto-death to bring down the corrupt Chimanbhai Patel Congress ministry, and after president's rule was proclaimed in February 1974, he undertook a second fast demanding new elections. The opposition's frustration and disgust with Congress finally led to the formation of a grand alliance of non-Congress parties, the *Janata Morcha* ("People's Front"), under the joint leadership of JP and Morarji. By 1975 the alliance included such radical left-wing groups as the Maoist Communists and Left-Socialists; the conservative *Swatantra* and communal *Rashtriya Loktantrik Dal* ("National People's Party"), a *Jan Sangh* splinter group; and the rightwing *Anand Marg* ("Path of Eternal Bliss"), the party founded by Tantric cultist Prabhat Ranjan ("Baba") Sarkar in 1955. In mid-1974 a national railroad strike threatened to paralyze India's economy, but Indira's government crushed it ruthlessly, and the following January, Union Railway Minister L. N. Mishra was blown up with his carriage at Samastipur. Shortly after that, Indira's hand-picked chief justice, A. N. Ray, was almost assassinated in New Delhi.

"India has not ever been an easy country to understand," Prime Minister Gandhi reflected shortly after her government's nuclear blast. "Perhaps it is too deep, contradictory and diverse, and few people in the contemporary world have the time or inclination to look beyond the obvious, especially because in our country we have the greatest scope for free expression of opinion and all differences are constantly being debated." Not for much longer, however. On June 12, 1975, Judge Jag Mohan Lal Sinha of Allahabad's High Court found Prime Minister Gandhi guilty of two counts (out of fifty-two charges brought against her) of campaign malpractice in connection with her race against Raj Narain for a seat in Parliament four years earlier. The conviction carried a mandatory penalty of barring Mrs. Gandhi from running for or holding any elective office for a period of six years. The very next day, moreover, the results of a hotly contested Gujarat election announced the victory of Morarji's coalition opposition. Both decisions were direct blows against the prime minister, who had personally testified in the High Court case on her own behalf and had vigorously campaigned throughout Gujarat on behalf of her party. The *Janata Morcha* immediately organized a mass sitdown protest outside the "President's House" (*Rashtrapati Bhavan*) in New Delhi, calling upon President Ahmad to fire the discredited prime minister. Politicians across the country, including some members of her own party, and many leading newspapers urged Mrs. Gandhi to "step down" from her high office, at least until the Supreme Court had time to hear the appeal of her conviction.

The clamor of protest mounted through June 25, when a huge opposition rally was held in Delhi, chaired by Morarji Desai and addressed by JP, who urged members of the police as well as the army to join a nationwide *satyagraha* campaign against corruption and not to obey "illegal" orders issued by any "disqualified" head of a discredited government. The rally was punctuated by loud and prolonged cheers. At 4 A.M. the next morning, Indira's elite force of Central Reserve Police started arresting all opposition political leaders (except those of the pro-Moscow CPI), including the seventy-two-year-old JP and the seventy-nine-year-old Morarji. At 6 A.M. she summoned her cabinet into session, and an hour later, as dawn broke on June 26, 1975, the president of India proclaimed a state of national emergency, "suspending" all civil rights, including *habeas corpus,* clamping a lid on the press, placing armored units on special alert, and grounding air flights over Delhi. It was all done with martial speed and precision. Not only Delhi, but Calcutta, Bombay, and other major cities where "subversive violence" was feared were completely "secured," and not a shot was fired as thousands of "subversive" politicians, students, journalists, and lawyers were hustled off to jail. Campus dormitories in many states were surrounded, and "trouble-making" faculty as well as students were arrested: fifteen hundred in Bihar, several thousand in West Bengal, more than a thousand in Maharashtra, several hundred in Delhi. By August, foreign newspapers reported that at least ten thousand political prisoners had been jailed, though opposition estimates ranged as high as fifty thousand and the government admitted arresting a "few thousand." "Those who are in should be out," shouted intrepid Piloo Mody (b. 1926), one of Janata's arrested heroes. "And those who are out should be in!"

"The decision to have emergency was not one that could be taken lightly or easily," Mrs. Gandhi reported to the *Rajya Sabha* on July 22, 1975, "but there comes a time in the life of the nation when hard decisions have to be taken. When there is an atmosphere of violence and of indiscipline and one can visibly see the nation going down, then the time has come to stop this process." There had indeed been more strikes, sit-ins, and public protests against skyrocketing prices during the past years than in any comparable period of recent Indian history, for there was also more tax evasion, corruption, and starvation. With less than two thousand Indian citizens admitting to more than $1,300 a year of taxable income and more than four hundred million living on the knife edge of starvation (spending less than twenty cents a day), there was considerable reason to be discontent. And with at least one-third of the more than sixteen million annual college graduates of India unable to find jobs of any sort, it was hardly surprising to find educated youth in the vanguard of the opposition. "Some members of the opposition have rightly blamed us saying why did we not take some action earlier," Indira confessed. "I do admit that had I taken action earlier it would have been less drastic action."

Less than a week after proclaiming the emergency, Indira unveiled her Twenty-Point Program of economic reforms over Radio India. Point one was to "bring down prices," for she rightly recognized that inflation had become her worst enemy. Points two through seven called for radical land reforms, including the "liquidation of rural indebtedness" and the "abolition of bounded labour," a populist appeal aimed directly at the hundreds of millions of deeply indebted peasant families and landless serfs at the bottom of India's agrarian pyramid. (These were not, however, the first land reforms proposed by Congress.) Points eight through ten reiterated goals of development, especially in the "handloom sector" and "supply of people's cloth." Point eleven was "socialization of urban lands"; points twelve and thirteen focused on "tax evasion" and called for "confiscation of smuggler's property"; point fourteen was to "prevent misuse of licenses"; and points fifteen and twenty both called for greater attention to workers in their employment and training as well as their "association" with industry. There was a point for "national permits for road transport," another for income-tax relief to "the middle class," and finally two points promising "controlled prices" to students for their hostel food as well as for books and stationery. In her Twenty Points Indira thus managed to promise something to most of the important segments of Indian society, especially those who had been so vocal in their opposition to her party, the intellectual young elite. The combination of stick and carrot proved quite effective. Within days bureaucrats who had never reached their offices before half-past ten in the morning were hard at work by nine. Black market hoarding and price gouging stopped overnight. Smugglers and tax evaders quickly realized that "Madam" was determined to enforce her program, even if to do so required arresting her former supporters and friends. Within a month the prices of rice and barley fell 5 percent, and they kept falling. Almost miraculously, it seemed to many Indians, unconscious of the irony of their proud remark, the "trains were running on time." Congress President Dev Kant Barooah (b. 1914) of Assam announced that "India is Indira, and Indira is India." Posters with the prime minister's picture sprouted everywhere, proclaiming, "She stood between Chaos and Order."

On July 4, twenty-six political organizations, representing extreme right- and left-wing opposition forces to Congress (R) power, Indira's Requisition Party, were banned under the emergency, and strict censorship was enforced on foreign as well as domestic press reports. Before the end of July, Mrs. Gandhi reconvened Parliament to consider her proposal to extend the emergency "indefinitely." The debate was surprisingly brief. In the *Lok Sabha* a majority of 336 voted to support the measure, and in the *Rajya Sabha* the majority was 134. Only 59 members of the lower house and 33 members of the upper house opposed, shouting "Shame!" as they walked out to boycott the remainder of the session. There was considerable speculation about what may have "detained" the 123 members of *Lok Sabha* and 76 members of *Rajya Sabha* who failed to attend that historic meeting. Parliament next approved the government's proposed Twenty-eighth Amendment to the Constitution, which barred India's courts of law from hearing any challenges to the emergency, and the Twenty-ninth Amendment, which retroactively exonerated the prime minister from any legal charges pending against her, as well as from all possible future charges of criminal actions while she was in high office. The president, vice-president, and speaker of *Lok Sabha* were also declared immune by this amendment to the rule of law. Finally, the amendment itself was declared immune to Supreme Court review. It passed unanimously by a vote of 336 to 0. Within twenty-four hours of Parliament's rubber-stamping of the emergency, all twenty state legislatures with Congress (R) ministries (only Tamil Nadu and

Gujarat remained under non-Indira party rule) met and voted unanimously to endorse the new amendments. "Unity is a good thing," Jawaharlal Nehru had noted in his autobiography, "but unity in subjection is hardly a thing to be proud of. The very strength of a despotic government may become a greater burden for a people; and a police force, no doubt useful in many ways, can be, and has been often enough, turned against the very people it is supposed to protect."[1]

"Like freedom, democracy too does not mean that everybody is free to follow his individual path," Indira informed her people on the twenty-eighth anniversary of India's independence. Her voice was beamed from the ramparts of Delhi's Red Fort to millions of villagers more than a thousand miles away by a new NASA "Satellite Instructional Television Experiment" (SITE), allowing her to reach more of India's populace than any Mughal emperor had ever dreamed of doing. "We have no intention of enforcing one-party rule," she promised. "Hard work, clear vision, iron will, and strictest discipline" was the new ethic that began now to blossom on billboards and signs throughout India. "The 'only magic' to eradicate Poverty," Indians were assured, "is hard work." The Ministry of Information and Broadcasting worked overtime at developing slogans and pithy messages to help a new and more unified India achieve its Twenty-Point Program. "Work more, talk less," they said, and "Rumour mongering is the worst enemy."

Six months before Indira declared her emergency, Mujibur Rahman had convened Bangladesh's Parliament (*Jatiyo Sangsad*) in Dacca also to "amend" the Constitution, abolishing representative government in favor of president's rule. "Severe economic harships, corruption, profiteering, hoarding, smuggling," and other problems had made it "necessary," Mujib explained, to trade in his mantle of prime minister for the more comprehensive cloak of president. Parliamentary debate was a "luxury" poor Bangladesh could no longer afford. In many ways, of course, the same was true of India, and it had long before been accepted by Pakistan. Democracy was such a "wasteful" process, slow and expensive, far less "efficient" than martial rule. The latter, however, entailed greater risks, for the autocrat usually required a palace guard of personal police, who generally

[1] Jawaharlal Nehru, *Toward Freedom* (New York: John Day, 1941), p. 278.

roused the enmity of the army. On August 14, 1975, Mujib and his family were all shot dead in a Dacca coup led by units of Bangladesh's Regular Army. Mujib had doubly diminished in stature by widespread charges of personal corruption and atheism. Several more coups kept Dacca in a state of turmoil throughout the rest of 1975. Major General Zia-ur Rahman emerged as chief martial law administrator over Bangladesh. Zia was pro-Chinese and Western rather than Indo-Soviet. In 1978 he felt sufficiently confident to put his popularity to elective test, and he emerged quite easily as president of his Islamic Republic. Zia was also chairman of his government's Bangladesh Nationalist Party (BNP). A new *Jatiyo Sangsad* was convened in 1979, and Dacca's relations with Delhi cooled, for since the *"Jai Bangla"* days of 1971 many Indo-Bangladesh disputes had surfaced over Ganga water sharing, Bay of Bengal oil resources, currency exchange, and illegal border crossings. On May 30, 1981, Zia was assassinated in Chittagong by Major General Manzur Ahmed, who was caught and killed two days later. Vice-president Abdus Sattar (b. 1905), acting president of Bangladesh, urged his tragedy-plagued people to "maintain peace and discipline."

Indira called her "Emergency" Raj Disciplined Democracy, stressing the importance of discipline as one of India's ancient attributes, the quality that gave yogis their "magic powers." She herself was a regular practitioner of yoga, and she clearly inherited her father's singular stamina and capacity for hard work. Of her two sons, the younger, Sanjay (1947–80), emerged during the emergency as leader of the Youth Congress, the second most powerful figure in India. His statements and actions were reported almost as extensively as those of his mother in the carefully controlled and censored press.

The 1975 monsoon brought a bumper crop of over 114 million tons of food grains, ending three years of near-famine conditions and helping Indira's Raj to keep essential commodity prices down to 1971 levels, a victory over inflation that was almost unique in the world. The emergency had eliminated strikes as well as increasing the productivity of those who worked in offices, raising industrial production by over 6 percent in 1975 and 10 percent in 1976–77. The climate of order and the "work is worship" ethic stimulated capital investment as it had never been stimulated before, to the rate of 30 percent during the first year of emergency. Despite growing Soviet influence in New Delhi and the ever-expanding presence of Russians and of USSR bookstalls throughout India, Marwari, Parsi, Gujarati, Jain, and other Indian capitalists obviously felt no threat to their fortunes or the future of their business enterprises from Indira's Raj. It was a uniquely Indian syncretism of socialism and capitalism, of state-supported free enterprise. The USSR's twenty-year promise of peace and friendship assured Indira of the heavy military support she might need, possibly even to keep her own army in line. Sanjay's Western-style auto manufacturing and his leadership of India's youthful business elite proved most reassuring to the banking and industrial sections of the nation as it moved toward a new era of central unification under autocratic rule of an Indian variety. In February 1976 the DMK ministry in Tamil Nadu was toppled by president's rule at orders from the prime minister on the grounds of "corruption." The Tamil Nadu government had been Indira's most outspoken critic, and the DMK was thought to have been on the brink of declaring Tamil Nadu an independent nation, following the precedent set by Bangladesh. Within a month of the smoothly enforced central government takeover, however, most DMK leaders had either "resigned" from their party or were under arrest. In Tamil Nadu, Congress (O) and (R) leaders quickly agreed to forget their differences and reunite, welcoming Indira back to Madras to bless the reconciliation. In Gujarat the following month, the *Janata Morcha* ministry collapsed on a vote of confidence after several of its members had been lured away by Indira's Congress. All vestige of visible or vocal opposition was thus eliminated before the emergency had completed its first year. Parliamentary elections that had been scheduled for early 1976 were "postponed" for another year.

Parliamentary government, democratic elections, freedom of speech and of the press, and the rule of law were, after all, part of the gloss of Western modernization introduced in the last phase of British rule to a civilization that found them as strangely exotic and foreign as their Anglo-Saxon Liberal and Radical advocates; while the "emergency" powers assumed by Indira to save her Raj seemed destined to outlast those imported institutions and ideals that were enshrined in India's Constitution as fundamental rights. Imbued with greater might than that enjoyed by Mauryan, Guptan, Mughal, or British imperial monarchs, Prime Minister Gandhi could well afford to insist that hers remained a truly "democratic" Raj, and a "constitutional" one. As long as prices stayed under control, investment continued high, and the monsoon was reasonably kind, it could also be a very popular Raj. For India's patient, tolerant, long-suffering populace had been inured by millennia of cultural continuity to accept authority as part of the divine order (*dharma*), which was law and religion combined. The wise *rishis* of old had known that the more things changed, the more they stayed the same; or, at least, the more they returned with cyclical certainty to what they had initially been. Name and form might look or sound new, but the *maya*-world was ingenious in its deceptions. Essential reality persisted beneath the veil; somehow, whatever was added would always be *Indianized*.

"I had always believed that Mrs. Gandhi had no faith in democracy, that she was by inclination and conviction a dictator," wrote JP from his prison cell. "As for the Congress party, I do not understand its spinelessness. Of course, quite a number of Congressmen are disguised communists. . . . Behind them is the Right CPI and behind it is Soviet Russia. Russia has backed Mrs. Gandhi to the hilt. Because the farther Mrs. Gandhi advances on her present course, the more powerful an influence will Russia have over this country. A time may come when, having squeezed the juice out of Mrs. Gandhi, the Russians through the CPI and their Trojan horses within the Congress will dump her on the garbage heap of history and install in her place their own man. . . . That India too should become another Pakistan or Bangladesh? What a shame that would be! These countries did not have a Gandhi, a Nehru. . . . Will all their work be reduced to ashes?"[2] Yet with as little warning as her proclamation of Emergency Rule, Indira called it off. On January 18, 1977, she announced release of her leading opponents, and called for national elections in March, suspending her ban of political parties. Did she think two months hardly time enough for her scattered, "demoralized" opposition to pull itself together and wage a winning campaign? Had reading her carefully controlled press convinced her of the "popularity" of her autocratic regime? Or was she, in fact, willing to stake her political future once more on the judgment of democratically enfranchised India? Some said her generals "advised" Prime Minister

<hr/>

[2] Jayaprakash Narayan, *Prison Diary, 1975*, ed. A. B. Shah (Bombay: Popular Prakashan, 1977), pp. 3–4.

Gandhi that it was time she took her party's platform to the people. Others thought she was just getting "tired" of "running the whole show alone."

Morarji and JP resurrected the Janata Morcha overnight. Two weeks later, Jagjivan Ram (b. 1908), Indira's former minister of defense and India's most powerful "ex-Untouchable," political leader of some 100 million Hindus at the bottom of the social pyramid, resigned from Mrs. Gandhi's cabinet and her Congress to form his own Congress for Democracy, and allied it to Janata. This unified opposition offered India's electorate, now totalling some 320 millions, the promise of "freedom and bread." JP declared the 1977 elections India's last chance to choose between "democracy or dictatorship" and "freedom or slavery." Insensitive lower echelon enforcement of government's strict policy of sterilization for men with two children or more, especially in crowded U.P. and Bihar towns, had alienated and terrified millions of peasants, who saw "Madam's Dictatorship" aimed at robbing them of potency and progeny. Sanjay's vigorous slum-clearance (without adequate resettlement) around old Delhi's Jama Masjid also made many lower-class Muslims, who traditionally supported Indira, lose faith in the wisdom or goodness of her administration. Over 200 million voters turned out in 1977, and after a week of polling a plurality of 43 percent of them voted Janata, while only 34 percent cast ballots for Indira's Congress. With a clear majority of Lok Sabha's 542 seats, the Janata Party thus ended 30 years of uninterrupted Congress power. Morarji Desai returned triumphant from provincial exile to New Delhi as India's fourth prime minister. Sixty-eight-year-old Jagjivan Ram had hoped to be elected premier by his party, but JP, who served as Janata's *guru*-mediator, convinced Jagjivan to allow his senior Brahman colleague to take his "turn" first at running the nation. Like Mahatma Gandhi, JP personally refused any office, preferring to serve as "conscience" to his colleagues, who would soon forget most of their selfless promises.

By April of 1977 it seemed that Indira Gandhi's career in politics had ended. She and Sanjay had both gone down to ignominious defeat in U.P. constituencies considered Nehru-Gandhi "pocketboroughs." The "people," after almost two years of silence and repressive suffering had spoken out, clearly for "freedom and democracy," and now all would be well, or so at least it may have seemed to those who did not understand the deeper currents of India's "Wounded Civilization," forgetting as V. S. Naipaul puts it that most of life in India "begins" with "the acceptance of distress as the condition of men."[3]

[3] V. S. Naipaul, *India: A Wounded Civilization* (New York: Alfred Knopf, 1977), p. 20.

TWENTY-FIVE

FROM JANATA RAJ TO RAJIV'S DEATH

Morarji Desai's Janata government remained so faction-ridden from its inception in 1977 that it could reach no consensus on how to resolve India's most compelling problems, dissolving its precious mandate in the acid of cabinet squabbles. Charan Singh (1902–87) leader of *Lok Dal*'s landlord faction, was home minister and later deputy prime minister, but his one ambition in life, as he often noted, was to be prime minister before he died. Though he had one heart attack while in Morarji's cabinet, his single-minded struggle for premier power did more to hasten the Janata's demise than his own. Jagjivan Ram, Morarji's "untouchable" minister of defense and later also deputy prime minister, had his heart and political sights set with equal firmness on the premiership, initially agreeing to step aside and wait only for the sake of party unity. How long, after all, could the octogenarian Morarji retain so exhausting a job as that of prime minister?

Morarji, however, seemed to grow younger in office. An austere orthodox Brahman, yoga-practicing Hindu, he ate no meat, drank no liquor, but daily imbibed some of his own urine as part of his personal "yoga-therapy," and remained at his desk long after younger colleagues went home. Much of his time there, however, seemed to be spent trying to devise and enforce laws prohibiting the use of intoxicating beverages or beef throughout a land plagued with poverty, rapidly rising unemployment and inflation, industrial stagnation, increasing urban violence, and inadequate housing. He considered American investments in outer space exploration wasted money since he claimed to know Indian yogis who "could go to the moon in just a few seconds." Like Indira, he had a son. Kantilal Desai (b. 1928) moved into the prime minister's residence, screened visitors vying for his father's ear, planned ambitious projects, closed deals, and intruded his presence so prominently into the highest circles of Indian politics that Charan Singh's ministry of home affairs finally decided to open an investigation into the rich and rapid rise of Kantilal. After learning of that unauthorized investigation's progress, Morarji dismissed his home minister, retaining full faith in his son. Charan now accused his prime minister's government of impotence and threatened to quit the Janata Party entirely.

With such unabated wrangling among the "old men" who ruled India, as they came to be popularly ridiculed, it was hardly surprising that ordinary people throughout the country lost confidence in Janata and its unkept promises. Or was the "freedom" they had restored merely to serve as license for increased economic exploitation, black marketeering, smuggling, tax evasion, and hoarding, all of which returned with a vengeance after the scrapping of Indira's emergency regulations? In four months of Janata rule the price of major foods rose almost 5 percent, and by the year's end inflation was back to its double-digit climb. Morarji's government swiftly used up Indira's unprecedented surplus of 18 million tons of foodgrain and in less than two years had expended its $3 billion of inherited foreign exchange reserves as well. Nor was the Indian government's perennial bureaucratic redundancy, indecision, and red tape eliminated or even diminished. Rhetoric remained exciting, promises soared, yet nothing was done. Small wonder that Indira rose from the ashes of her defeat within a year to win a new seat in a Lok Sabha by-election, this time in South India. Soon after she returned to Parliament, however, Morarji's government made the mistake of pressing formal charges against her within Lok Sabha itself, and though he commanded votes enough to oust her from that chamber, sending her to jail for a week in December 1978, by so doing he added the crown of martyrdom to her already battle-scarred brow. "Madam's" outraged young followers marched in noisy protest through India's major cities, stopping traffic, demanding justice, even highjacking an Indian Airlines plane, publicizing their total commitment to Indira.

Indians could also look at neighboring Pakistan at this time, where Prime Minister Bhutto, unlike Indira, had blatantly rigged his last elections, receiving more votes than the total electorate in some districts, while political opponents mysteriously disappeared, became "ill," suddenly died, or narrowly escaped assassination. Loss of public confidence in its political leadership led to so much unrest throughout Pakistan early in 1977 that army commander-in-chief General Zia ul-Haq (1924–88) (no relation to Bangladesh's General Zia) seized power on July 5th, leading a bloodless martial coup with the arrest of Bhutto. The latter, subsequently charged with ordering the assassination of several leaders of his political opposition, was tried and found "guilty" as charged and remained in jail till he was hanged before dawn on April 4, 1979. There was not only widespread shock and humane concern expressed throughout India about Bhutto's behavior and terrible fate, but an increased and inevitable comparison of his crimes and treatment with Indira's, putting in much milder perspective her era of "dictatorial misrule." If she was really so harsh an autocrat, after all, more and more Indians asked, why did she permit such free and fair elections? Most people began to believe that the political courage displayed by Mrs. Gandhi in defeat, her fortitude instead of despair, proved her a worthy heiress to her father's position, a true heroine of India's continuing freedom struggle.

By early summer of 1979 Morarji had lost the support of his party's left-wing. Socialist leaders Raj Narain (b. 1917) and George Fernandes (b. 1930) resigned from the Janata cabinet, taking so many members of Parliament with them that the government was left without a working majority in Lok Sabha. JP's idealistic lieutenants, secular-minded reformers eager to affect real changes in Indian society, union organizers and liberal journalists sick of Morarji's inertia, fed up with the Singh-Ram infighting and selfish bickering over loaves and fishes, worried about the growing power of orthodox Hindu Jan Sanghis within the Janata coalition, whose youthful leader, A. B. Vajpayee (b. 1926) was Morarji's foreign minister, all abandoned government's side. Still Morarji refused to step down, resigning only on the eve of a no-confidence motion, on July 19, 1979, when he knew positively that Janata was defeated. President Reddy first invited Congress opposition leader Y. B. Chavan to try to form a new government, but Chavan's efforts to rally a majority proved futile. Reddy turned next to Charan Singh (a Jat "peasant" by birth), who thus achieved his lifelong ambition, becoming India's first non-Brahman premier, though also its briefest one. Like Morarji, Charan resigned on the eve of a no-confidence motion within weeks of his investiture. Instead of turning next to Jagjivan Ram, new leader of the Janata, who might have been able to pull a majority coalition government together, India's Brahman president opted to dissolve Parliament, calling for new general elections in January 1980. Charan Singh remained caretaker prime minister as the curtain of history rang down on the Janata's sadly failed "last chance for democracy" in India. Just before the elections that would bury the party he brought to life, J. P. Narayan died of kidney failure in Bombay.

India's seventh general elections brought Indira Gandhi back to the center of Delhi power in January of 1980. Inexhaustible at 62, Indira toured the nation, campaigning on the slogan "Elect a Government that Works!" Her Congress-I (for Indira) won a commanding two-thirds majority in Lok Sabha, securing 351 of its 542 seats. Indira won both seats she contested, the Nehru family's traditional base of power at Rae Bareli in U.P. and South Indian Andhra's Medak constituency, proving her unique all-India voter appeal. She chose to retain the Andhra seat, reentering Lok Sabha as a representative of the "Dravidian" South, thus symbolically undermining any separatist aspirations from that region. Sanjay Gandhi was also elected, for the first time, as member for Amethi in U.P. and immediately reestablished his position as India's second most powerful person though he held no cabinet office. As head of the activist Youth Congress, whose cadres proved most important at getting out the vote, Sanjay personally "ticketed" almost half of the new Congress-I members, most of whom were elected to Lok Sabha for the first time. Sanjay, his mother's closest confidant and adviser, was viewed as her heir-apparent, potential third in the "Nehru-Gandhi Dynasty," which seemed a most appropriate First Family for independent India. Then, in June of 1980, Sanjay piloted his private stunt plane to his death over New Delhi. His shocking unexpected demise was mourned nationally, but Mrs. Gandhi carried on in the stoic spirit of her own prescribed cure for India's ills, with "austerity, discipline and hard work." Rajiv Gandhi (b. 1944) Indira's elder son and only other child, an excellent Indian Airlines pilot, who had never expressed any prior interest in politics, was urged to stand for his brother's seat. The unpretentious, well-liked Rajiv, whose name means "Ruler," was widely viewed as the ideal successor to Sanjay's leadership of the Youth Congress. Would he attain his younger brother's most coveted ambition as well, succeeding to his mother's high office?

Provincial unrest, with the potential for Bangladesh-like separatism, spread across north-eastern India's seven tribal states (Assam, Arunachal Pradesh, Nagaland, Manipur, Meghalaya, Mizoram, and Tripura) during 1979 and throughout 1980. Starting in sporadic non-violent local responses by indigenous inhabitants of these sparsely populated states to the mounting influx of Bengali immigrants from overcrowded, impoverished Bangladesh, the local protests became increasingly violent, popular, ideological, and integrated. An Assamese Liberation Army, initially inspired by Mahatma Gandhi's *Satyagraha* technique, won popular support in its struggle to oust an estimated five million Bengali migrants who had crossed the border and settled in Assam since 1947. That civil disobedience movement succeeded virtually in shutting down Assam's lucrative new oil industry, and also threatened the state's older tea and jute industries, both of which were important earners of foreign currency. Mrs. Gandhi's government sent troops into the troubled region, and she personally flew to Gauhati, trying without success to negotiate a political settlement. In Tripura, where Bengalis had entered in such numbers that they constituted a majority and controlled that state's "Maoist" CPI-M government, rampaging tribesmen murdered at least 382 "foreigners," unofficial estimates rising to thousands. Delhi imposed martial law, and all foreign newsmen were banished from the region as reports leaked out of a "Seven United Liberation Army," spearheaded by Naga, Mizo, and Assamese tribals, calling for "complete liberation of their homeland from imperialist invaders." Strategically situated as it is between the Sino-Burmese and Bangladesh borders with its Burmese and Mongoloid peoples, many of whom speak Sino-Tibetan and Tibeto-Burman languages and who are mostly Buddhists, Christians, or nature-worshipping "Animists," India's eastern wing is potentially its most vulnerable limb. If New Delhi's response to growing indigenous "unrest" in that region proves no more empathetic, imaginative, or enlightened than the British Raj's initial repressive response to Indian nationalism had been, the sporadic fires of 1980 could blaze on winds of discontent into terrifying conflagrations, depleting India's vital energies and precious resources for decades.

To the west, India faced even darker storm clouds rolling down from the Amu Daria (Oxus) River into Kabul in December of 1979. The Russian invasion of Afghanistan, which Moscow called mere "neighborly support" for a "beleaguered" friendly regime, was not initially condemned by Mrs. Gandhi's government. However, the massive, unrelentingly violent nature of the Soviet army's assertion of martial control over the Texas-size border-state to its south, soon became clear even to Indira, as Afghanistan's population of some 18 million desperately tried to reject the Russian invasion, as they had every previous one. Diplomatic appeals from New Delhi urging the Russians to withdraw their more than 85,000 regular army troops from occupied Afghanistan proved ineffectual. Russian helicopter gunships and heavy tanks moved their "mopping-up" raids to the Durand Line, driving over a million Afghan-Muslim refugees to seek shelter in Pakistan, at times firing across the border in hot pursuit of refugee raiders returning to their bases. Many Indians now recalled how virtually impossible it had so often been, historically, for South Asia's defenders to stop martial invaders, once they reached the Khyber, from pouring into the Indus Valley and then galloping across those flat plains toward the glittering lure of Delhi itself. Would the Russians be next in that series of predators, starting with nomadic Aryan tribes over 3,000 years ago, which subsequently included Alexander the Great, Timur the Lame, and Babur the Tiger among those who led such formidable legions? Merely to pose the question was enough to send shudders of apprehension through India's body politic.

For why, it was often asked, would Russia invest so much money in troops and equipment, sustaining so many losses (estimates varied from several to many thousands dead by the end of 1988), if it was simply to help "protect and secure" an unpopular Communist clique in Kabul? Was Afghanistan's barren, isolated plateau truly the *final* goal of Russia's latest imperial push-to-the-south? Did not the Soviet Union covet precisely what Tsarist Russia had long dreamed of attaining, warm-water access to the world at large? Why should the Kremlin stop at the Khyber when Peshawar was barely fifty miles east of that border pass and Lahore just two hundred miles of flat country east of that with Delhi less than 300 easy miles beyond? Why rest content with Kandahar when Baluchistan's capital of Quetta was scarcely 100 miles south and east of Afghanistan's second largest city. Then from Quetta to Karachi and the Arabian Sea was hardly more than a long day's push for any tank corps worth its firepower. On the other hand, the Russians had invested heavily in

Afghanistan's modernization since the mid-1950s; it was perhaps less surprising to find them eager to defend that investment than it would have been to expect them to write it off without a struggle when their communist puppets came under heavy attack from well-armed and organized Muslim fundamentalist guerrillas late in 1979. A return to tribal-dominated, orthodox, monarchic rule in Kabul would have meant, moreover, not simply the loss of Afghanistan to Moscow, but could send shock waves of discontent through the entire Muslim population of Russian Uzbekistan north of the Amu, and might even encourage "adventurism" by neighboring China! Nor were Afghan wool, fresh fruit, and Uranium ore, recently found in apparently substantial quantities in Afghanistan, insignificant assets from Moscow's point of view. Why surrender such valuable raw materials without at least a show of force? Perhaps the Russians expected that the mere sight of their formidable armor and death-dealing aircraft would stifle Afghan resistance. Any knowledge of British Indian history would, of course, have taught them that Afghans were among the world's most fiercely independence-loving people. Now faced as they were with Afghan hornets sniping at them from every rock and ridge were they really in any position to advance to the "conquest" of the rest of South Asia? And what on earth, one might ask, would the Russians *do* with India, or Pakistan, should they ever actually attempt or achieve such direct control over the subcontinent? Replay the history of the British Raj? Or that of the Great Mughals? Take up the "Communist-man's Burden"? Was that, in fact, the long-range strategy of Moscow's master planners? Could they be so stupid?

India's leaders, nonetheless, like their counterparts on Pakistan's front-line, slept less easy as the Bear inched closer. They relived the anxieties that had kept so many British statesmen awake nights at Whitehall in the nineteenth century, planning their next moves in the "Great Game," devising buffers to help secure their "spheres," drawing up contingencies, stashing railroad tracks in sheds along the frontier, bribing fierce Pathans to keep their rifles loaded, while manning beefed-up garrisons to their rear just in case they pointed them the wrong way! Yet the British had a *united* subcontinent to defend, giving them strategic advantages that neither India nor Pakistan possess today. For the latter two have pinned down the major muscle of their martial forces confronting one another, in

Kashmir, and on opposite banks of the river Sutlej, instead of joining forces to face a common threat from the north-west. Would Russia's presence on the Khyber help India and Pakistan to relearn the
value of subcontinental unity? Could that external catalyst possibly
achieve through the force of fear the miracle of Hindu-Muslim unity,
which Mr. Jinnah and Mahatma Gandhi both once called the key to
true national independence for India? Would New Delhi and Islamabad, and Dacca perhaps, one day awaken to appreciate that it was
in the best interests of all three nations to subordinate national pride
and communal prejudice to a coordinated plan for the protection and
defense of South Asia as a whole? Was such an "awakening" faintly
possible? Only the wildest of optimists could consider such reunification a forseeable prospect. If anything, the entire recent history of
South Asia, at least since partition, points the other way, toward fragmentation rather than unification.

The Islamic Republic of Pakistan under the leadership of President General Zia ul-Haq, moved by 1980 diplomatically closer to its
fellow-Muslim nations of the Middle East, rather than toward possible reunification with India. Zia hosted the multinational Islamic
Conference several times and was elected its chairman. Islamabad's
often-denied yet apparent continuing commitment to build a thermo-
nuclear bomb was seen as possibly a multinational Islamic effort to
secure such a bomb, perhaps for use against Israel; yet were Pakistan
ever to explode a nuclear weapon in Azad Kashmir, for example, India would naturally view such an act as a deliberate challenge to her
own security. By 1980 New Delhi had, moreover, stored enough nuclear fuel and acquired the technological capability of completing her
own intermediate missile delivery system, which could launch nuclear
warheads against Karachi, Lahore, or Islamabad more easily than at
any major center of Chinese power. India, like Pakistan, however,
continued to deny any intention of using nuclear power for other
than peaceful purposes. South Asia's escalating arms race, with Russia promising to pour billions into India's arsenal while Saudi Arabia
and Libya helped strengthen Pakistan, added a new dimension of terror and urgency to unresolved communal problems and potential
future conflicts that continued to haunt the subcontinent. With atomic
weapons at their command, either nation could trigger a war whose
fallout might prove global. The frustrations, fierce violence, and mer-

curial passions that have marred relations between India and Pakistan throughout the first three and a half decades of their independent existence can no longer be neatly contained within a two- or three-week clash yielding few casualties. Nehru's "No-War Zone" might, tragically enough, someday become the point of ignition for a global nuclear holocaust.

Or would the genius of Indian civilization's reconciliation of opposites, its ancient and enduring faith in nonviolence (*Ahimsa*) as Hinduism's highest religion, unique tolerance, long-suffering patience and trust in the healing powers of peace (*Shanti*), suffice to achieve the impossible, rebuilding—at least metaphorically—the roads and bridges destroyed by partition? Would history's wheel turn once again to a new Age of Unity? Forged in fear, hardened by fury and pain, enriched by the powers of science, could those ancient, impoverished lands leap from their semifeudal bullock-cart mire onto highroads of a peacefully applied nuclear-electronics epoch of growth, utilizing the biomedical and solar energy revolutions to overcome traditional handicaps of disease, overpopulation, mass ignorance, divisiveness, and poverty? Might not the lessons of so brilliant, rich, yet tragic a civilization still be mastered by its heirs before they fall victim to hatreds and fear that could only lead them into violent self-destruction?

"Before I ever knew anything of politics in my early youth, I dreamt the dream of communal unity of the heart," Mahatma Gandhi confessed, during his final fast in Delhi. "I shall jump in the evening of my life, like a child, to feel that the dream has been realized. Then we shall have real Swaraj. Then, though legally and geographically we may still be two States, in daily life no one will think that we were separate States. . . . I live and want to live for no lesser goal. Let the seekers from Pakistan help to come as near the goal as it is humanly possible. . . . In such paradise, whether it is in the Union or in Pakistan, there will be neither paupers nor beggars, nor high nor low, neither millionaire employers nor half-starved employees, nor intoxicating drinks or drugs. There will be the same respect for women as vouchsafed to men and the chastity and purity of men and women will be jealously guarded. . . . Where there will be no untouchability and where there will be equal respect for all faiths. They will be all proudly, joyously and voluntarily bread labourers. I hope everyone who listens to me or reads these lines will forgive me if stretched on my bed and basking in the sun, inhaling life-giving sunshine, I allow myself to indulge in this ecstasy. . . . It matters little if the ecstatic wishes of a fool like me are never realized and the fast is never broken."[1]

Two weeks after Gandhi uttered those words he was assassinated. Must his dream for India's future also die? Or will it, somehow, be nurtured to reality?

Prospering Punjab appeared to hold the best promise of translating the Mahatma's dream into reality. With its hard-working Sikh majority and fecund soil, the small state of modern Punjab (c. 20,000 square miles since 1966) enjoyed the highest per capita income in all of India, especially since the "Green Revolution" spur to agricultural production in the late 1960s. There were more motor cars and scooters, more tractors and TV sets, in tiny Punjab than in half a dozen of India's larger more backward states combined. The average Punjabi peasant family earned twice as much as the rest of India's peasant households. With only about 2 percent (c. 16 million) of India's 800 million population by 1987, talented Sikhs still held more than 10 percent of the higher officer posts in India's military services. Thanks to their beards and turbans, Sikhs remained the most visible of India's minority communities, and because so many "Sardarji's" drove their own taxi cabs, they were especially noticeable in Delhi. In 1982 Gyani Zail Singh (b. 1916), Mrs. Gandhi's Home Minister and close political ally, was the first Sikh elected to the office of president, where he remained until his retirement in 1987.

Since the final years of the British Raj, however, a vocal minority of Sikhs had called for the creation of their own "Nation," a demand first voiced by Master Tara Singh (1885–1967), who led the *Akali Dal* (Eternal Party), and periodically fasted for what came to be called *"Sikhistan"* or *"Khalistan"* ("Land of the Pure"). Much like the separatist Muslim demand for *"Pakistan"* (also meaning "Land of the Pure"), few took that Sikh cry seriously when it was first articulated, viewing it only as extremist political blackmail or British-

---

[1] M. K. Gandhi, *Delhi Diary*, January 14, 1948 (Ahmedabad: Navjivan Publishing House, 1948), pp. 341–3.

inspired duplicity. In 1966, however, Mrs. Gandhi's central government divided India's Punjab into three states, supposedly to "satisfy" Sikh demands for a *"Punjabi Suba,"* a separate Punjabi-majority Province, but in fact serving only to exacerbate Sikh alienation and anger by subdividing that previously partitioned state into tiny Punjab, with its Sikh majority, and neighboring Haryana to its southeast, with a Hindu majority, and Himachal Pradesh to the north, with its capital at Simla. Moreover, Punjab and Haryana were to share the centrally administered capital of Chandigarh, which had been built during Nehru's era to serve Punjab as a capital-symbol of modernity. The Akali Dal, under Prakash Singh Badal's and *Sant* ("Saint") Harchand Singh Longowal's (1928–85) leadership, agitated vigorously for the transfer of Chandigarh to Punjab alone, but Mrs. Gandhi initially refused to yield to that demand. Akali Dal gained popularity and power throughout the Punjab, and during the dark interlude of Mrs. Gandhi's "Emergency Raj" many thousands of Sikhs courted jail in outspoken opposition to that repressive central autocracy. From 1977 until 1980, in the aftermath of the Emergency, Akali Dal leadership was elected to govern Punjab, ousting and infuriating Congress both at the center and in the state. Sanjay Gandhi and Zail Singh sought to undermine Sant Longowal's popularity and power in Punjab by sponsoring a virtually unknown young fundamentalist opposing "Sant," Jarnail Singh Bhindranwale (1947–84).

By 1983, Bhindranwale had become far more extreme and dangerous than any previous Akali leader had ever been, directing a violent agitation for national autonomy for Punjab. By that date, Sanjay was dead, and though Zail Singh was President of India, he had no power over the "Sant," who heeded no Delhi Congress voice while emerging as the hero of Sikh fundamentalist youth and extremism. Early in 1984 Bhindranwale and his armed followers took control of the Golden Temple's *Akal Takht* ("Immortal Tower"), and resolved not to leave the sacred precincts of Amritsar's temple until New Delhi capitulated and granted Punjab complete autonomy. By May 1984, hundreds of moderate Sikh and Hindu lives were taken by extremist violence. Millions throughout India expressed their frustration at "Madam's" (Mrs. Gandhi's) apparent inability to take any "strong action" against Bhindranwale and terrorism. That month, however,

Indira met with her generals and opted for their "Devil's alternative," plan to impose martial law over Punjab and invade the Golden Temple. It was her fatal error of judgment.

"Operation Bluestar" was launched in early June 1984. Indian Army tanks rolled into the sacred Sikh temple grounds and fired their heavy weapons at Akal Takht until Bhindranwale and all his followers were either dead or had fled from that burning tower, once the sacrosanct seat of Sikh authority. Combat raged in the heart of Amritsar for two endless days and nights on June 5 and 6, leaving thousands of Sikhs dead inside the Golden Temple grounds, claiming more than 100 soldiers, as well as wounding many times that number, and turning to ashes the irreplaceable library of Sikh Scripture inside Harmandir. Khalistan had its first martyrs. Less than half a year later Mrs. Gandhi was to pay with her life for ordering the desecration of the Vatican or Mecca of the Sikh faith. Why had she opted for what clearly seems now to have been tantamount to signing her own death warrant?

Nationwide elections had to be held before the end of January 1985, as Mrs. Gandhi well knew. In a climate of continuing terrorist violence and central government inertia it was difficult to imagine Congress winning such elections. Martial success of any sort is usually an asset at the polls, and Mrs. Gandhi's "strong" action against the Sikh extremists certainly rallied India's Hindu majority behind Mrs. Gandhi's Congress Party by year's end. Perhaps she was inspired by memories of her glorious days in the aftermath of victory in the Bangladesh War of 1971. Moreover, since the death of Sanjay, she had become more of a fatalist, and may have lost much of her attachment to life, although she had always been courageous and never appeared to worry about threats to her personal safety. She might, of course, have hoped that "cleaning" Bhindranwale out of the Golden Temple grounds would prove a much easier job than it was to be. Such wishful thinking often appears to tempt powerful leaders the world over to unleash brutal martial force against political opponents. Whatever her reasons, the stunning, shocking violence of Operation Bluestar certainly served to put every minority in India on immediate notice of just how dangerous continued opposition to Delhi could be. Muslims as well as Sikhs got that message, and as if to underscore it, Mrs. Gandhi ordered her trusted governor of Kashmir to remove that

Muslim-majority state's popular chief minister, Farooq Abdullah, from his Srinagar seat of power in July 1984. The son of Kashmir's "Lion"-hero, Sheikh Abdullah, Farooq remained his state's most popular figure nonetheless, and Mrs. Gandhi's son felt obliged to bring him back to power with Congress-I coalition support early in 1987.

In the final months of her life, Mrs. Gandhi seemed to lose all tolerance for political opposition of any sort. A month after removing Farooq she ordered her trusted governor of Andhra to depose the popular Telugu movie idol chief minister of that state, N. T. Rama Rao (1924–96). "N. T. R.," as he was called, had lost none of his popularity, in fact, but as the leader of a non-Congress regional party, Telugu Desham, he was viewed by Mrs. Gandhi as a potential "threat" to her majority. Her final months were plagued by fears of political "enemies" lurking everywhere, though ironically enough it was two members of her own "security force" who would kill her. N. T. R. proved his continuing popularity by flying to Delhi with a majority of his legislative supporters, and remained at the helm of Andhra politics long after Mrs. Gandhi's demise, losing power a decade later, shortly before he died, to his own son-in-law!

Shortly after 9 A.M. on October 31, 1984, Prime Minister Indira Gandhi started her last walk, from her house inside a walled compound of New Delhi's best protected area, across her garden path, toward her office building. Two of her Sikh guards, Beant Singh and Satwant Singh, confronted her at the garden gate and emptied their weapons into her frail body. Some 30 bullets were later removed from Mrs. Gandhi's corpse. She was dead before reaching the hospital, where Delhi's best team of surgeons worked futilely for hours trying to revive her, pouring most of their supplies of blood into her unresponding form. "Khalistan" Sikhs in London and Texas cheered and were photographed drinking champagne to celebrate Mrs. Gandhi's assassination. But for Sikhs around Delhi and across North India's pain-filled plains the blood bath had only just begun.

Rampaging mobs of Hindu hoodlums took to Delhi's streets as word of Mrs. Gandhi's murder spread. Screaming "Blood for blood!" those criminal gangs poured kerosene over every Sikh they saw, setting fire to humans, cars, and stores or homes owned by Sikhs, wherever they could be found. For three days and nights, India's capital turned into the most lawless, terror-ridden place on earth as hired gangs of arsonists and killers ran free, stimulated by self-righteous "avengers" of the fallen prime minister. Police, meanwhile, turned a blind eye on all criminal assaults and India's Army slept. By official count more than one thousand Sikhs were murdered in Delhi alone during those first three days of November 1984, but unofficial observers saw many thousands of corpses in ravaged suburbs of Delhi, such as Trilokpuri, alone. The entire government of India seemed to have collapsed with the prime minister. President Zail Singh was abroad, in Yemen, when he received the dread news, and flew home. Rajiv Gandhi was in Calcutta, and he too flew back immediately to Delhi, where he was sworn in as India's new prime minister that Halloween night in the presidential palace. At 40, Rajiv was nine years younger than his mother had been when she ascended to the peak of Delhi's political power and 18 years younger than his grandfather, Jawaharlal Nehru, had been. "Captain" Rajiv was more famous as an Indian Airlines pilot than a politician, having embarked on his political career only after his younger brother's death four years earlier.

"She was mother not only to me but to the whole nation," Prime Minister Rajiv Gandhi said in his first broadcast to the nation. "We should remain calm and exercise maximum restraint. . . . Nothing would hurt the soul of our beloved Indira Gandhi more than the occurrence of violence in any part of the country." Yet even as he spoke, Delhi was burning. Not until Saturday, November 3, when Rajiv put the torch to his mother's funeral pyre on the banks of the Yamuna River, did that orgy of anti-Sikh violence in Delhi come to an end. The Army was finally called out and ordered to "shoot on sight" any gangs of killers or looters, and as the tanks rolled into Delhi's smoking streets, the cowardly gangs disappeared from sight. More than 50,000 terrified Sikh refugees sought shelter in improvised camps around Delhi, however, while property damage rose to an estimated $20 million.

Two weeks after the Delhi massacre ended, an impartial Citizens' Commission, headed by former Chief Justice S. M. Sikri, appealed to the government to appoint an official tribunal to investigate that criminal tragedy. Nothing was done for five months until April 1985, however, when Justice R. Misra was appointed to "inquire into the allegations in regard to the incidents of organised violence which took place in Delhi." The Misra Commission took another two years before producing a report that recommended no criminal action against any Delhi culprits, sounding more like an apologia for official inaction than a condemnation of what it termed "anti-social" behavior on Delhi's blood-soaked plain. India's People's Union for Democratic Rights and People's Union for Civil Liberties had by then long since published its own eye-witness account of the Delhi atrocities, called *Who Are The Guilty?*, which concluded that black November 1984 was the "outcome of a well-organized plan marked by acts of both deliberate commissions and omissions by important politicians of the Congress (I) at the top and by authorities in the administration."[1] Twelve years later, however, none of the culprits have as yet been put on trial.

"If it is possible, or it is perceived to be possible, for those in authority to escape the consequences of violating the law and the Constitution," warned a critique of the Misra Commission Report, entitled *Justice Denied,* "the very fabric of ordered social existence is liable to be irreparably damaged. . . . The schism between the Sikh community and the rest of country, especially the Hindu majority, will be further widened if the normal avenues to punish the guilty fail the victims of the riots. . . . Having been failed by the Commission, where will they go now?"[2] Sikh victims in Delhi continue to wait for Justice, but as yet it has remained denied to them.

In mid-November 1984, while popular sympathy for the orphaned young prime minister remained high, Rajiv wisely called for national elections to be held that December. The Congress (I) was to run on the sentimental slogan, "Remember Indira," and through her death Mrs. Gandhi helped her son and Party to muster a far greater victory at the polls than she could ever have achieved had she lived to be her own standard-bearer. Rajiv's own sister-in-law, Maneka Gandhi, Sanjay's widow, sought to challenge him in his Uttar Pradesh constitutency, but was easily defeated. On December 24, 1984, hundreds of millions of Indians started the trek toward tens of thousands of polling stations in what was to be the greatest landslide victory for Congress in independent India's history. Rajiv proved himself more

[1] *Who Are The Guilty?* published jointly by Gobinda Mukhoty and Rajni Kothari (Delhi: Sunny Graphica, 1984), p. 1.
[2] *Justice Denied:* A Critique of the Misra Commission Report on the Riots in November 1984, New Delhi, PUDC-PUCL, April 1987, p. 16.

popular as a campaigner than even his grandfather had been, his youthful energy and handsome demeanor on television and in person winning him the title of "Mister Clean." The new prime minister was more popular than any of Bombay's matinee idols. He led his party to a *Lok Sabha* victory of 400 out of 500 seats, and Congress (I) came closer to winning an absolute majority of the popular vote than it had in any previous election.

Dreadful tragedy struck central India's Bhopal just two weeks prior to those December 1984 elections. Deadly invisible gas from Union Carbide's insecticide storage tanks had escaped through defective valves to be blown by ill-winds over the slumbering bodies of thousands of poor innocents outside that giant plant's walls, less than 400 miles south of New Delhi. Within hours, 2,000 people were dead and hundreds of thousands of others had been injured by the worst industrial accident of recent history. Union Carbide was still American owned, one of the few multinationals in India that had not been taken over by a majority of Indian shareholders and management. The plant, which opened in 1977, produced more than 2,500 tons of pesticides each year, but since the December 1984 tragedy has become a symbol of how horribly dangerous to human life such modern chemical industries can prove to be. Though hundreds of billions of dollars in damages were quickly claimed in thousands of lawsuits filed both in India and America, virtually nothing was paid to any of the actual victims for seven full years after the tragedy devastated Bhopal. Legal snares and questions of jurisdiction plagued the process, serving to thwart justice instead of expediting its realization. Once again Indians learned an all too familiar lesson: death and suffering came swiftly, with justice and relief invariably lagging far behind in this ancient land plagued by modern sorrows.

Most Indians hoped that Prime Minister Rajiv Gandhi would lead them toward a better life, a happier horizon of a corruption-free society, national unity, and economic progress, without the pitfalls or pain of more Bhopals and Bluestars. After all, the new leader and his old party represented the richness of India's glorious past and promising future, linking the legacy of Mahatma Gandhi's nationalist ideal goal of *Ram Rajya* ("Rama's Rule"), an age of golden glory, to modernist faith in tomorrow's scientific technology, coordinated by computers. As the scion of India's First Family, Rajiv embodied all that was admired in the Nehru-Gandhi clan. He was also a daring new man—a pilot who had married the wife he chose for himself while studying in Cambridge, Italian-born Sonia Maino. He had even taught himself to use a personal computer.

"Our politics should be clean," India's youthful prime minister told his nation early in 1985, promising "speedy" development and "visible fruits" of economic growth for all his people. He retained enough cabinet portfolios in his own hands to expedite change in many fields and cut the red tape that had so long stymied Indian entrepreneurial initiative, abolishing licensing delays for the import of high-tech products, computers, color television sets, and VCRs—all the things Rajiv and his upwardly mobile generation of modern Indian managers coveted. He also cut taxes on wealth and inheritance, following the lead of Reaganomics in trying to lift India's lumbering economy out of its deep-rutted path toward Socialism and economic "equality" for all onto the smoother highroad of "trickle down" opportunities for faster economic growth at the top. He quickly sent most of his mother's top advisers packing and surrounded himself with his posh Doon School classmates, who shared his vision of a better India for the few as swiftly as possible. Mahatma Gandhi's dream faded faster, and the socialist rhetoric so eloquently articulated by Jawaharlal Nehru and Indira Gandhi was abandoned by the new team of modern technocrats who took control of India's cockpit.

Armed with his potent electoral mandate, Rajiv reached several impressive political accords with opposition parties in Punjab, Assam, and Mizoram, presaging what appeared to be the end of long-standing provincial conflicts. The Punjab Accord was signed by Rajiv and *Akali Dal* leader Sant Longowal in July 1985. It promised the transfer of Chandigarh to Punjab on January 26, 1986, some compensatory "Hindi-speaking areas" of Punjab to be transferred at that time to Haryana. Other points of the Accord included promises for which Sikhs had been agitating vigorously for many years, including retaining "merit" as the major criterion for army recruitment, consideration of an "All-India Gurudwara Bill," extension of the federal principle in more questions of Center-State relations and assurance to Punjab farmers of the continued use of river waters that rose and passed through their state. Moreover, the central government promised to take "steps" toward helping to "promote" the "Punjabi language." The Accord was hailed as marking an "end" to the period of "confrontation" that darkened relations between Delhi and Punjab, ushering in "an era of amity, goodwill and cooperation." Immediately after signing the Accord, Rajiv announced that elections would be held in Punjab. Less than a month later, Sant Longowal was assassinated by Sikh terrorists in his own village. Nonetheless, elections were held as scheduled, and Longowal's *Akali Dal* won a strong majority throughout the state. Surjeet Singh Barnala, Longowal's closest disciple, was chosen to serve as Punjab's chief minister. But on the eve of what was to have been the announced transfer of Chandigarh to Punjab in January 1986, lack of agreement between Delhi and Barnala left the Accord suspended in midimplementation. No further positive action was to be taken over the next six years, and as the bleak pall of inertia settled over Punjab, disillusionment followed, with bitter violence returning to shatter every vestige of hope for a swift or peaceful solution to Punjab's thorny problems.

Sikh terrorist violence was met with accelerating movement of tens of thousands of paramilitary forces into Punjab from Delhi's Central Reserve Police and crack Border Security Force. India's most famous police chief, Julio Ribeiro, was placed in command of Punjab's beefed-up security forces, the largest police force in all of India. Without proclaiming martial law, Punjab was turned into an armed camp, and Ribeiro himself spoke of his deadly struggle against terrorists as nothing less than "war." Leaders of the Sikh parties, and many Sikh reporters decried what they called "fake encounter" killings of young Sikhs by Punjab's police, while Chief Ribeiro insisted that "terrorist intruders" were daily crossing the border "from Pakistan," armed with weapons supplied by "foreign hands" to help Sikhs destroy India's Union. Rajiv also spoke of dangers from dark external forces trying to undermine India, and the spirit of Accord soon dissolved in the acid of communal violence, suspicion, and hatred. Chief Minister Barnala tried in vain to repair bridges of trust between Chandigarh and Delhi, but the more he turned toward Rajiv for help, the less his own Akali supporters trusted him. He finally lost the support of his own majority in Punjab's legislative assembly. Twenty-three members of Barnala's *Akali Dal* abandoned their chief minister's government before the end of 1986, following Prakash Singh Badal and Gurcharan Singh Tohra across the assembly floor, but not to prison cells. Instead of resigning his office, Barnala clutched desperately at the shifting sands of power, seeking to stay in charge of Punjab with Congress (I) support, thus becoming doubly dependent on Delhi's leadership. Soon, the once popular Akali leader was "excommunicated" from his own Sikh faith by leaders of the Golden Temple. He still refused to step down, appealing more urgently to Rajiv for visible support, such as the immediate transfer of Chandigarh, or a favorable decision on Punjabi demands for more water and more Sikh recruitment by the army. New Delhi did nothing to bolster the now impotent chief minister's position, however, and in May 1987 he was deposed and President's Rule was imposed on Punjab.

Governor Siddhartha Shankar Ray, Bengali-born barrister of Lincoln's Inn, and Director-General of Police Julio Ribeiro now ruled Punjab autocratically. Instead of reducing violence, the number of deaths more than doubled in the first year of President's Rule. New Draconian laws made it easier to arrest and detain, it appeared almost indefinitely, any Sikh youths who were deemed a "threat" to Punjab's "security." Reports varied and though it was impossible to secure accurate official information it was clear that hundreds of detainees, possibly thousands (some said as many as 5,000!), were jailed for many months, and in some cases years, without clear charges of any sort and with no prospect of coming to trial or being released from "preventive detention." Though India had signed the UN Declaration of Universal Human Rights, her legal system now clearly violated its basic provisions and left more and more of its citizens to languish behind prison bars without any stated cause for such action or real hope of freedom.

"The proximity of the Punjab to the frontier has enabled its administrators time and again to enforce their will," warned Congress President Motilal Nehru, Rajiv Gandhi's great grandfather, after World War One. "The bogey of the frontier is exploited to the uttermost . . . by the 'man on the spot'. . . . But repression and terrorism have never yet killed the life of a nation; they but increase the disaffection and drive it underground to pursue an unhealthy course. of . . . crimes of violence. And this brings further repression and so the vicious circle goes on."[1]

[1] Pandit Motilal Nehru, *The Voice of Freedom*, ed. by K. M. Panikkar and A. Pershad (London: Asia Publishing House, 1961), pp. 6–8.

While Punjab burned, Assam and Mizoram enjoyed a new interlude of nonviolence, thanks to accords that Rajiv reached with the
most powerful leaders of those eastern uprisings. Zenophobic passions
and fears had been roused in most of India's eastern wing by the millions of Bangladeshi Muslim migrants who had crossed the borders
into sparsely populated India, where they cleared jungle growth and
planted rice. Assamese, Mizo, and Naga tribals, as well as the more
urbanized Hindus and Christians of the region, decried the "Muslim
invasions" of their homeland and launched violent assaults on Muslim
farmers and peasant villages long-settled in remote areas of that rugged
eastern borderland. By negotiating directly with the popular leaders of
the *Asom Gana Parishad* ("Assam People's Party") as well as the
Mizo National Front, Rajiv reached agreements with both men, essentially granting them political control over their own provincial units
since they easily won the elections held there. Moreover, Mizoram
was raised to full statehood from its previous lesser union territory
status in 1987. The former "rebel agitators" thus emerged as chief
ministers of their states with political power and responsibility. Rajiv's
Congress (I) lost political power and patronage in the region, but India's army could withdraw from Assam and the eastern zone, and more
orderly and less expensive attempts were made to control illicit Muslim migration across India's porous borders. By early 1988, however,
both Accords in the east had started to unravel as more zenophobic
demands to "oust" all "foreigners" spread throughout the region. Renewed violence racked India's eastern horizon, and the Indian Army
returned to Assam in 1990.

As his much-touted political "accords" lost their initial luster,
Rajiv's once-glowing personal image began to tarnish. Finance minister V. P. Singh, one of the most highly respected members of the
cabinet, had launched attacks on some of India's wealthiest industrial
and commercial enterprises for tax evasion and other shady ventures
that had begun emerging by the end of 1986 under the full light of
public inquiry. In January 1987, Rajiv removed V. P. from finance and
shifted him overnight to the ministry of defense. Instead of lying low,
however, V. P. started a new inquiry into reported defense contract
"kickbacks" that were supposedly paid by the Swedish firm of Bofors
into a Swiss account code-named "Lotus," which had been opened by
the brother of one of Rajiv's movie-star, old school buddies. The

Bofors scandal became front-page news in Delhi, especially after V. P. was forced to resign from his second cabinet position in April 1987. The prime minister denied taking any money, but the *Indian Express*, under Arun Shourie's fearless editorial direction, ran editorials on its front page asking a number of embarrassingly pointed questions until that most popular newspaper was shut down by officials investigating its "property," upon finding a number of "violations" of building and fire codes. Few buildings in Delhi were without such minor infractions and before the end of December 1987 the *Express* was back on the streets, but the government's message could hardly be ignored or forgotten. Indeed, by year's end V. P. himself was under investigation for having "improperly" hired a foreign detective agency to investigate contract "kickbacks." The former minister of finance and defense became the foremost challenger to Rajiv's leadership within his own party and a possible white horse candidate of united opposition parties, something of a reincarnation of J. P. Narayan in the heyday of Janata Party opposition to Rajiv's mother.

Loss of popular confidence in Rajiv's Congress (I) leadership was reflected in a number of by-elections in 1987 as well. Haryana State was long considered a secure bastion of Congress (I) support and the western anchor of the party's Hindu-majority Hindi-belt that stretched across most of north and central India, including Uttar Pradesh, Rajasthan, Madhya Pradesh, and Bihar, but dropped out of the ruling party column in the mid-1987 elections. *Lok Dal* ("People's Party") leader Devi Lal scored a dramatic upset victory over Congress in every district of his predominantly Hindu peasant state, arguing that New Delhi's leadership "failed" Haryana farmers in the two things they required most: subsidized grain prices and adequate water supply. The populist chief minister promised to help his peasant backers in every way, including state assumption of their onerous land debts. Opponents of Devi Lal knew he would be much easier to defeat after a term in office, when so many of his promises could be raised against him. But for ousted local Congress politicians, 1989 seemed a long way off, and Rajiv's lightening helicopter campaign appearances on Haryana hustings did nothing to add any ghi to their chappattis.

In international affairs, Rajiv Gandhi achieved some success, which helped to counter his internal political losses. From his first visit to the United States in mid-1985, the handsome young prime minister won warm official and popular support for India in Washington. "The ties that bind our two peoples are many," Rajiv informed both houses of Congress in his address to them that June. "We share the conviction that democracy is the best guarantor of enduring development: that our people must live as free individuals." Agreement was reached at the White House and in the Pentagon on high-tech exports from Washington to India, supercomputers to assist India's economy in many ways, as well as in her martial capabilities. The total value of annual trade between the United States and India, however, was as yet only $4 billion, less than 5 percent of the value of the United States's trade with Japan. But during his first visit to Washington Rajiv inaugurated a yearlong Festival of India, which did help to introduce millions of Americans to the unique beauty and enduring fascination of Indic Civilization and its brilliant cultural and artistic contributions to world culture. Two years later Rajiv returned again briefly, stopping in Boston and Washington during his Commonwealth Conference visit to Canada, further advancing Indo-U.S. ties of friendship and trade, and sharing ideas as well as information.

India's intimate relations with the Soviet Union were not weakened by these growing bonds of friendship between New Delhi and Washington. The Treaty of Friendship concluded with Russia by Rajiv's mother remained the keystone of Indo-Soviet friendship and cooperation, and Rajiv's personal relations with General Secretary Mikhail Gorbachev were at least as cordial as were his meetings with President Ronald Reagan. Russia continued to supply India's armed forces with its most advanced, sophisticated weapons, including the new MIG-29 fighter-interceptor, Russia's counter to American F-16s shipped to Pakistan. Two nuclear-powered Russian submarines were shipped to India early in 1988, providing India's navy as well as her air force with as advanced an arsenal of deadly weapons as any found in South Asia or the Indian Ocean.

In December 1985 the first meeting of heads of seven states of South Asia was held in Bangladesh's Dhaka, marking the birth of the South Asian Association for Regional Cooperation (SAARC). By joining this most hopeful regional institution, Rajiv has helped defuse India's perennial tensions with Pakistan, as well as other explosive regional issues. The most dangerous issue with Pakistan in recent years remains the persistently reported "race" by Islamabad to complete a nuclear capability, which is often referred to as an "Islamic Bomb." Though Pakistan's president and other leading officials deny any intention of building such a bomb, Indian as well as U.S. intelligence reports indicate that Pakistan may have it already, thanks to Chinese as well as Libyan assistance. Shortly after SAARC was born, Rajiv met with Pakistan's President General Zia in Delhi to sign an agreement that neither nation would launch a first-strike attack against any of the other's nuclear plants or atomic facilities. Sporadic fighting has continued to plague the "cease-fire" line in Kashmir, and in 1986 India held such massive martial maneuvers along the Pakistani border in Punjab and Sind that major conflict seemed imminent. The Secretariat of SAARC, established in Nepal's capital of Kathmandu in 1987, now serves as a permanent South Asian information and communications center for reducing potential regional misunderstandings or conflicts. Like the European Common Market, SAARC's potential for economic exchange and regional growth as well as cultural and scientific cooperation are at least as great as its political utility.

India's importance to Sri Lanka, the tiny independent island suspended like a tear below Tamil Nadu, has always been enormous. Most of Sri Lanka's 20 million inhabitants are descendants of Indian migrants, the majority of whom are Sinhalese Buddhists, who first came to the region of Kandy and Colombo some two thousand years ago. A Tamil Hindu minority of almost five million, settled mostly around Jaffna in the north, are also indebted to India for their early roots and cultural heritage. Ethnic conflict between Sri Lanka's Sinhalese majority and her Tamil minority communities escalated in the wake of British Imperial withdrawal from Ceylon after World War Two. Colombo's "Sinhala-only" Official Language Act of 1956 incited violent Tamil opposition. By the early 1970s Tamils were well-organized politically, and most of them supported the Tamil United Liberation Front (TULF) party. After 1978, President J. R. Jayewardene's (b. 1906) United National Party, which won a clear victory in Sri Lanka's last general elections, enacted a Constitution that was viewed as anti-Tamil by TULF leaders. Tamil members of Parliament staged a boycott in protesting the Constitution and were subsequently "expelled" from Parliament by the ruling party. Violent Tamil-Sinhalese incidents flared in Colombo, Kandy, and Jaffna, and young

Tamils armed themselves in what soon became an incipient civil war. Half-a-dozen bands of Tamil "boys" launched guerrilla raids against Sri Lanka's armed forces or unarmed civilians, and by the end of the decade several of them called for creation of a separate Tamil "Nation" (*Eelam*) out of the northern and eastern districts of Sri Lanka. The fiercest and most militant of those Tamil separatist groups were the Liberation Tigers of Tamil Eelam (LTTE), called Tigers, led by their founder V. Prabakaran (b. 1954). Many of South India's 60 million Tamils sympathized with and supported the struggle of their Sri Lankan colinguists, offering food and money as well as arms and shelter to their "boys."

By 1986 the fighting in Sri Lanka had escalated so fiercely that India resolved to act to bring an end to its neighor's "civil war." Rajiv met with Jayewardene at the SAARC summit that year in Bangalore and had Prabakaran flown there as well in an Indian Air Force jet, trying to bring the Tigers to the peace table. The Tiger leader, however, adamantly refused to accept the chief ministership of a virtually autonomous Tamil-majority "province" of northern Sri Lanka, demanding full autonomy for both the northern and eastern districts. Since the Tamils lacked a majority in the east, Jayewardene could cogently argue that Prabakaran's demands were unreasonable and undemocratic. The Bangalore summit thus ended in failure, and by year's end fighting resumed with a vengeance around Jaffna, with Sri Lanka's air force moving in with rockets and bombs. India airlifted supplies that were parachuted over Jaffna from air force cargo planes. Soon after that, President Jayewardene proved his brilliant statesmanship by dramatically reversing himself and agreeing to sign a pact with Rajiv that was to put an "end" to Sri Lanka's War, permitting an Indian "Peacekeeping Force" to enter Jaffna and "disarm" the Tigers and other guerrilla bands, while the Sri Lankan Army withdrew to the south or remained inside their barracks. The Indo-Sri Lanka Peace Agreement was signed by Rajiv and Jayewardene in Colombo on July 29, 1987. India took over the thankless and bloody task of seeking first to disarm and then to subdue the intransigent Tigers shortly thereafter.

The Colombo Agreement had anticipated total Tamil guerrilla disarming by the end of August 1987, after which elections would be held throughout the island and a new Parliament convened by year's end. Such was the plan, to which India initially contributed an expeditionary force of under 15,000 troops. By year's end, however, that number had almost tripled, and the elusive goal of disarming the Tigers remained as far from completion as it had been in early August. The Indian Army was at first welcomed by Tamil civilians everywhere as "saviors" and "liberators," and several groups of "boys" readily surrendered their weapons, but not Prabakaran's Tigers. They preferred cyanide to captivity, and by the Fall of 1987 the Indian Army had taken upon itself the same bloody warfare that Sri Lanka's smaller force was most relieved to be able to abandon. By April 1990, when Delhi withdrew the last Indian troops from Sri Lanka, almost as many hundreds of Indian soldiers as Tigers had been killed or wounded in the bitter, futile struggle. A year later, Rajiv Gandhi himself fell victim to that still-unresolved conflict. Soon after President Chandrika Kumaratunga was elected to preside over Sri Lanka in 1994, she opened negotiations with Tamil leaders, hoping to reach a peaceful political settlement of the tragic civil war that has devastated that beautiful island state for more than two decades. Prabhakaran continued, however, to practice terror even when talking "peace" as deadly Tamil Tiger car bombs tore through Colombo's crowded bazaars, blowing up Sri Lanka's oil reserve tanks. Before the end of 1995 Sri Lanka's Sinhalese-led army captured the Tamil stronghold of Jaffna, consolidating its grip over the last great bastion of Tiger resistance. The cost of that final assault in Tamil lives was believed to be more than 10,000, yet Prabhakan himself escaped capture, vowing to "fight on" till his independent Tamil Eelam was achieved.

Sino-Indian relations remained tense as the Chinese moved more troops up to the frigid Himalayan border on India's northeast front. Fortunately, there was no actual fighting, but since the Chinese invasion of 1962 the mere movement of large numbers of Chinese troops sufficed to alert India's army and air force to potential danger of the highest order. In 1987 Indo-Nepalese relations were also strained as a growing Gurkha National Liberation Front (GNLF) movement pressed its demands for a separate "Gurkhaland" state to be carved out of West Bengal's Northern region around Darjeeling. Nepali-speaking Gurkhas naturally feel far more akin to their Nepalese Gurkha neighbors than they do to Calcutta's Bengali-speaking majority. Violence flared as Indian Border Security Force police sought to break up GNLF marches and protest rallies in high-altitude Darjeeling, from which Mt. Everest's peak was often visible. India's paramilitary forces crossed the Nepalese border in supposed hot pursuit of Gurkha "terrorists," but Kathmandu's foreign office insisted that such attacks by Indian "troops" were "unprovoked" and "totally unjustified," leaving many village homes burned to the ground in their rapid-fire wake.

India's unique linguistic and ethnic pluralism make that nation more vulnerable to such fragmenting demands than virtually any other state in the world. Since the Pandora's box of linguistic state redrawing of India's internal borders was opened by Prime Minister Nehru in the 1950s it has become more and more difficult to close. In Maharashtra, a Nag-Vidarbha separatist movement has long simmered round Nagpur, and since the end of 1987 a new "Uttarkhand" state separatist movement was born in the hills of Uttar Pradesh, bringing many thousands of its ardent advocates to New Delhi to demonstrate their solidarity. Similarly, in Orissa State, a long-standing "Jharkand" tribal separatist movement has gained momentum, while in Meghalaya a new demand by Garo hill tribals emerged in 1995 calling for the creation of a separate "Garoland."

The changes, however, that come most swiftly to modern India's ancient soil are the rising levels of choking pollution around Delhi, where fast-growing factory fumes, auto emissions, and the smoke of millions of kitchen fires threaten the lungs of all those who flock to India's capital plain. The air of Calcutta has long been as bad, and Bombay and Madras are not far behind Delhi's rush into the polluting perils of urban modernity. But it is Delhi's central power, fed by freeways and concentrically expanding ring roads, that attracts heavier diesel-belching traffic, while the pressures and demands from magnetized millions of migrants overtake marginal water as well as electric power supplies to India's megalopolis.

Now that all the "barbarisms" of civilization have come to India, as they earlier did to London, Manchester, New York, and Los Angeles, many urban-dwellers in the "best" neighborhoods of Delhi and Bombay keep armed guards posted outside the spiked gates of their posh homes around the clock and rarely drive anywhere without at least one private guard at the ready inside with them. Moreover, ancient Indian custom and religious violence has resurfaced of late, even in the heart of modern India's capital. Hundreds of "accidental deaths" of young Indian brides from burns by "kitchen fires" are now reported annually in the Delhi region alone. In fact, these dowry deaths are known to be maliciously inflicted murders, often planned or perpetrated by the dissatisfied husband and his parents, who feel that the former bride's family failed to pay the prearranged "price" promised during dowry negotiations. In Rajasthan, open revival and glorification of *sati* widow-burning has led to a number of arrests and vigorous official action in charging the culprits with murder. Thus, even though *sati*-immolations were outlawed by the British in 1829 and the offer or acceptance of a dowry of any size is against modern current Indian law, ancient Indian practice continues. The past clings to India's silken saried civilization and its flames continue to consume the bodies of young brides and widows, as they did for thousands of years, while the sacred mantras of Brahmans filled the air as Agni's flames "liberated" the immortal souls from fast-burning flesh.

India's peasant majority, her twice-born Hindu majority in cities as well as on the land, have not lost or abandoned their faith in the ancient Gods and the ritual sacrifices they require after all, despite the brief interlude of British rule with its liberal secular morality that left a patina of "Secular Democratic Socialism" over India's ancient Hindu body politic. India's modern computers are used to print out Hindu horoscopes more efficiently than ever before, just as wide and colored western film screens tell the epic tales of Rama and Sita and the Pandava brothers to more tens and hundreds of millions of enraptured Indian audiences than had earlier been possible.

Vishwanath Pratap Singh was sworn in as prime minister of India by President Ramaswamy Venkataraman in December of 1989. *Raja* V. P., heir to a small princely state in Uttar Pradesh, had only 145 Janata Party members in the *Lok Sabha*, yet thanks to backing from Lal Kishan (L. K.) Advani's eighty-eight Hindu-first Bharatiya Janata Party (BJP) supporters as well as from members of both Communist parties, V. P. managed to muster a *Lok Sabha* majority for his fragile coalition government. He promised to wield his "sword" of power on "behalf of the poor and the toiling masses," but fatally underestimated the strength of India's proud and traditionally pampered upper caste minority as he often repeated such populist appeals and later sponsored legislation designed to win nationwide majority support from India's impoverished "masses."

Prime Minister Singh launched his year at the helm of India's government with a courageous, inspiring visit to Amritsar, where he walked round the Golden Temple, trying to apply his "healing touch" to Punjab's wounded provincial polity. Sikhs have always admired courage, and most of them now believed that New Delhi was led by a man fearless enough to proclaim Chandigarh Punjab's own capital and to hold free elections throughout that police-run, violence-torn state. But as soon as V. P. returned to Delhi he was engulfed with so many other pressing problems that he forgot Punjab, failing to implement his brave promise. The equally disturbed State of Jammu and Kashmir became the most urgent focus of V. P.'s attention when the young daughter of Home Minister Mufti Mohamad Sayeed, himself a Kashmiri Muslim, was kidnapped by a Kashmiri separatist gang in Srinagar. The kidnappers demanded the release of all their recently arrested Jammu and Kashmir Liberation Front leaders for her life. Foreign Minister Inder Gujral flew to Srinagar to negotiate details of the release of his cabinet colleague's daughter, keeping in hourly touch with V. P. and the Mufti as the terrorists' deadline came and went without agreement. Happily, Miss Sayeed was released unharmed, in exchange for freeing all the incarcerated Liberation Front leaders. Home Minister Sayeed, in charge of India's police, could at last breathe easily again, as could the new prime minister and his trusted foreign minister. Yet some Indians feared that releasing prisoners for any hostage, no matter how important she was, had set a dangerous precedent that might provoke more kidnapping or escalating violence in Kashmir. Srinagar now ceased to be one of India's favorite and most lucrative tourist attractions, transformed instead into a front-line state, in constant conflict. Kashmir's best hotels were all used as barracks for Indian troops, more than 500,000 of whom were deployed around the thinly populated state, across whose "line-of-control" border in Azad Kashmir were some of the best units of Pakistan's regular army. By mid-1995 the death toll from fighting in Kashmir was much higher than in Punjab. India blamed "foreign" (that is, Pakistani) "infiltrators" for such rapidly escalating violence and refused to discuss holding a UN-run plebiscite or any other possible peaceful solution to Kashmir's conflict, either with Pakistan or any other nation. India insisted that Kashmir is purely an "internal" issue for New Delhi. Concerned human rights groups, including Amnesty International, Asia Watch, and International Alert, were repeatedly denied permission to visit Kashmir in order to investigate and report on a growing number of charges of brutal military as well as police violations of human rights. Indian officials either denied such accusations entirely, or insisted that they were "under investigation" by the "appropriate" Indian agencies.

The fall of V. P. Singh's Janata coalition government in less than a year, however, was precipitated by two issues that had nothing to do with Punjab or Kashmir. The first of those volatile matters flared up as soon as Prime Minister Singh announced his support for long-neglected or ignored affirmative action recommendations of the central government's musty Mandal Commission Report, which called for the reservation of almost 60 percent of all new civil service positions and higher-education admission slots for India's lower caste and ex-Untouchable progeny. Without consulting his own cabinet colleagues, V. P. decided that the Mandal Commission's radical therapy for India's ancient upper-caste-dominated society was precisely the strong medicine required to cure endemic Indian poverty as well as inequity. Opposition leaders, however, insisted that the wily prime minister merely hoped to win the support of most of India's electorate by associating himself with the championship of Mandal's "untouchable" recommendations. Many young upper-caste Hindus of Delhi, Haryana, and Uttar Pradesh feared implementation of the Mandal Commission proposals would be the deathblow to their ambitions for civil service or professional careers in medicine and law. Rather than face the bleak prospect of rejection from tightly competitive professional schools or civil service employment, tens of thousands of young Brahmans and Kashtriyas rallied and marched in angry loud protests against the prospective implementation of those hated Mandal Commission proposals. V. P. refused to back down, however, even after several frustrated, desperately depressed young men and women immolated themselves in fiery protests against his policy. But with each young death, many older high-caste Hindus announced their loss of faith in V. P.'s leadership, as did several of his own party stalwarts, who urged their prime minister to change course or lose his majority. V. P. did agree to refer the Mandal Commission's recommendations to India's Supreme Court for judicial review, yet never changed his opinion as to their socially salubrious value.

The most explosive issue leading directly to V. P.'s fall was the popular Hindu demand that a sixteenth-century mosque (*Babri Masjid*) erected by the Mughal emperor, Babur, in Ayodhya, be torn down and replaced with a new Hindu temple to mark the "Birthplace" and ancient "capital" of Lord Rama ("*Ram Janmabhoomi*"). Orthodox Hindus believe that Rama was born at the very spot on which Babur ordered his mosque built, first having desecrated, then torn down, an older Hindu temple there. As a Hindu sacred city, not only of Lord Rama's "birth" but also believed to be the capital of his "divine Raj," Ayodhya is visited annually by millions of Hindu pilgrims, who worship Rama and Sita in at least a hundred Hindu temples throughout the city. Rama's birthday is celebrated each spring and his coronation, following his epic battle with Lanka's demon Ravana, is hailed each fall, marking the dawn of Ayodhya's Golden Age under Rama's enlightened Raj. In October of 1990, therefore, BJP leader L. K. Advani mounted a golden chariot, which was dragged and pushed by thousands of his devout Hindu followers and brethren toward Ayodhya, where they vowed to rebuild the Hindu temple that Babur's "demon army" of Muslims had destroyed. Millions of "sacred bricks" were blessed by saffron-robed *sadhus* and Brahmans belonging to the orthodox *Vishwa Hindu Parishad* ("Universal Hindu Society"), which mobilized Hindus from every village of Uttar Pradesh and most other states in South as well as North India to join this pilgrimage to resanctify Lord Rama's "birthplace." Hundreds of thousands had joined the march before mid-October and by month's end millions were expected to converge on Ayodhya, each with a sacred brick or two in hand, ready to tear down the old mosque with their bare hands or potently sanctified brickbats, walking sticks, iron bars, knives, swords, and battering rams. V. P. appealed to Advani to "postpone" his pilgrimage, but the BJP leader had by October lost faith in his prime minister, whose climb to the top of Delhi's "slippery pole" would never have been achieved without the support of Advani's eighty-eight followers.

Sixty-three years earlier L. K. Advani had been born in a comfortable Hindu home in Karachi, the port city at the mouth of the Indus on the Arabian Sea. He was barely twenty-one when his entire family and all of their friends were forced to abandon the city that overnight became the capital of Muslim Pakistan. L. K., like most of the some one million enterprising Sindhi Hindus who, in the wake of 1947's partition, fled from Pakistan's Karachi to India's Bombay and Delhi, rebuilt his uprooted life and managed to prosper in his new Motherland. He thanked Lord Rama for the blessings of long life and good fortune he received, never forgetting that it was Muslims who forced him and his parents to flee their home. In dress and mild-mannered speech, L. K. was a thoroughly modern middle-class Indian political leader, who looked and sounded much like V. P. They might have almost been mistaken for brothers, the prime minister being just four years younger than his most powerful, erstwhile coalition partner. But on this one issue they could not agree. V. P. found it impossible to dissuade L. K. from undertaking the pilgrimage he considered the most important mission of his life. To Advani and tens, if not hundreds, of millions of Hindus who agreed with him, no public activity could be more valid, important, historically justified, divinely ordained, or urgently required than to reclaim and resanctify Lord Rama's birthplace. Delay was unnecessary, retreat inappropriate, cancellation impossible. Since almost 90 percent of all Indians are Hindus, and most of the ancestors of the some 10 percent who were Muslims had originally been Hindus or Buddhists, forced by Islamic swords or tax-free temptation to convert, politically astute Advani had reason to believe his pilgrimage was supported by India's vast majority.

On October 23, 1990, however, after all appeals to stop from V. P. and others in Delhi had failed, Home Minister Mufti Mahmood ordered the arrest of L. K. Advani, who was forced to descend his golden chariot and go off to jail. For India was, after all, a *secular* republic, was it not? For more than a century, the Indian National Congress and its leaders had stressed the importance of keeping independent India's body politic and administration separate from its religious pluralism, which had for millennia generated so many cataclysmic conflicts. The bedrock of Congress's argument against the Muslim League's "Two-Nation" theory and its demand for a separate *Muslim* Pakistan was that as a secular nation India would never discriminate against any religious minority. Jawaharlal Nehru was particularly proud of that claim and reiterated it whenever he spoke on questions even peripherally impinging upon religious pluralism. But many devout Hindus insisted that Congress and Nehru, as well as his daughter Indira and her son Rajiv, had much too long leaned over backward to "pamper" Muslims, and also Sikhs, always caring first and foremost for minorities, "robbing" the Hindu majority of resources and opportunities, ignoring Hindu demands, making light of Hindu fears or dreams. For V. P. Singh, however, there was never any question as to the primary value of secularism for India. No Hindu-first argument ever swayed him, though he must have realized in ordering Advani's arrest, in not permitting a bloody battle at Ayodhya, that BJP would withdraw its support and his government would fall. The vote of no-confidence came in two weeks, on November 7, 1990. V. P. spoke for several hours to a hostile House that voted him out by a resounding 356 to 151. He resigned the next day.

President Venkataraman initially invited Rajiv Gandhi, who led the opposition to V. P. in the *Lok Sabha*, to try to form a new government without returning to the electorate. But Rajiv refused the president's offer, never imagining, no doubt, that it would be his last chance of returning to power. He was still a youthful forty-six and thought he had plenty of time; he hoped that by waiting a bit the political climate would improve for him and his party. The economic situation was so bad, moreover, that he knew he would reap nothing but a harvest of blame for the nation's deepening depression and soaring inflation, as India found itself running out of hard currency and dipping into national reserves of gold treasure to avoid default on monthly international debt payments. Rajiv decided, therefore, to become India's "kingmaker" rather than the new "Raja of Delhi," choosing to back a Janata (S—for "Socialist") leader named Chandra Shekhar, who had a *Lok Sabha* bloc of only sixty followers, most of whom had splintered away from V. P.'s *Janata Dal* during the past year. With the support of Rajiv's 190 Congress (I) Party members, Chandra Shekhar could claim to "command" a *Lok Sabha* majority, but everyone understood that his government was, in fact, the shortest tail ever to wag the world's largest dog. Much to the surprise of many for whom Chandra Shekhar's lack of previous cabinet experience left him a totally unknown prospective premier, he managed for half a year at India's helm to launch several promising initiatives, including one attempt to resolve the growing conflict in Punjab by opening discussions with Sikh leaders of all factions, including some belonging to separatist pro-Khalistani groups. He could not even manage to win enough *Lok Sabha* support to pass a new budget, however, and early in 1991 it seemed obvious that Chandra Shekhar was no more than a stopgap prime minister. Fresh elections would have to be called soon, or else India's Constitution might have to be changed, converting it from its British parliamentary model into something closer to a French or U.S. presidential system. Rajiv cynically tried to remove himself from any share of blame for the state of political paralysis that gripped New Delhi, where nothing positive seemed to be accomplished and all of India's perennial problems merely festered or proliferated. Then suddenly, in March of 1991, Rajiv and his Congress followers walked out of *Lok Sabha*, charging Chandra Shekhar's government with "illegally tapping" Rajiv's home telephones. Instead of pleading with Rajiv to return, Prime Minister Shekhar resigned the same day and advised India's president to call new nationwide elections. *Lok Sabha* was dissolved, and new elections were set for three days in May 1991. Chandra Shekhar remained for the interim in charge of Delhi's sadly diminished skeletal administration.

India's most tragic, traumatic election started in what first seemed to be an almost evenly matched three-way race. Rajiv Gandhi's Congress was given a slight advantage by pollsters, thanks to its nationwide party machine and substantial financial backing. Rajiv attracted India's most modern young capitalists and upwardly mobile middle class, who liked his free enterprise approach to economic growth. V. P. Singh led a left-of-center coalition, appealing to India's lower castes and impoverished peasantry, promising to bring all of them more rapidly into the mainstream by implementing Mandal Commission reforms. V. P. won support from both Communist parties and from *Dalit* (former "Untouchable") groups in Bihar and Bengal as well as in Uttar Pradesh. After his disappointing performance as prime minister, however, V. P. was no longer credited by many of his previous supporters with the ability to deliver what he promised. Indeed, many of his former backers now abandoned V. P., and though he attracted large crowds in Bihar and Uttar Pradesh, he received little national publicity and fewer votes than anticipated. L. K. Advani's BJP was seen as the "dark horse" of the campaign. By appealing to Hindu consciousness, promising *Ram aur Roti* ("Rama and Bread") from campaign platforms all across the Hindi-speaking Hindu heartland, Advani and his revivalist party tapped the deepest emotional concerns of India's majority, drawing the largest crowds, eliciting passionate responses each time they promised to erect the appropriate temple over Lord Rama's "birthplace."

By late May, however, most election watchers predicted a Congress (I) victory as Rajiv's popularity seemed to grow whenever he appeared in public. Instead of remaining behind a phalanx of armed guards, Rajiv had adopted the warmer campaign strategy of moving right into crowds, much the way his grandfather and mother had done, throwing caution to the wind as his bandwagon gathered momentum. He flew to Madras for a tour of Tamil Nadu, but on the night of May 21, 1991, was blown up with a dozen others around him, including the female Tamil assassin who triggered a deadly explosion by bowing at the feet of Rajiv. She appears to have belonged to a Tamil "Tiger" terrorist team, whose one-eyed leader committed suicide when he was about to be caught in Bangalore several months later. Rajiv Gandhi was thus cut down on the eve of what would probably have been his triumphant return to power, halting India's elections as millions mourned the tragic act of terror. Congress Party leaders were so traumatized by the loss of their young standard-bearer that they turned in desperation to his Italian-born widow, Sonia, offering her the presidency of their party, appealing to this new Mrs. Gandhi to take up Rajiv's mantle and run in his place. Wisely, however, she refused to be lured by any prospect of political power into a post for which she was hardly prepared and from which it would have been too easy for her to become the next target of terrorist insanity. Sonia Gandhi bowed out of public life, withdrawing with her children soon after her husband's funeral to the calm of apolitical privacy.

TWENTY-SIX

INDIA TODAY

Congress turned in June 1991 to its seniormost leader, P. V. Nara-simha Rao (b. 1921), a lifelong disciple of Jawaharlal Nehru's "sec-ular line," who had served as foreign minister in the cabinets both of Indira Gandhi and Rajiv Gandhi, unanimously electing him its pres-ident. Narasimha Rao's humility, age, and recently impaired health (he'd had heart surgery in Texas less than a year earlier) made many wonder if he would be up to the strenuous task of leading his party to the successful completion of so traumatic a campaign. "The only way to exist in India is to co-exist," the new Congress president asserted, boldly arguing that "the alternative to the Nehru line in India is fas-cism." Narasimha Rao proved to be stronger, wiser, and more resilient than his critics, and many of his half-hearted supporters, had imagined. He led his party so deftly and effectively in the wake of Rajiv's death that Congress (I) won no less than 220 seats on its own, and though Narasimha Rao himself did not personally contest a *Lok Sabha* con-stituency in that general election, he was soon elected unanimously by his party's Parliamentary Board to serve as their prime minister, later easily won a seat from his home state of Andhra, and in April of 1992 was unanimously elected Congress (I) President as well. Tamil Nadu's most powerful party, the AIADMK, and the small but important Muslim League immediately joined forces with Congress in the *Lok Sabha*, assuring Narasimha Rao the majority he would need when the first test of his government's viability came a month after the elections. Chief Minister Jyoti Basu of W. Bengal, whose Communist "Left Front" followers won some forty seats in the *Lok Sabha*, also indicated his preference for Narasimha Rao as India's best possible premier. L. K. Advani, whose BJP had scored the most dramatic gains, jumping from little more than 80 seats to 120 in less than two years, took his seat as leader of the opposition in the *Lok Sabha*. V. P. Singh's *Janata Dal* was reduced to a mere sixty seats; V. P.

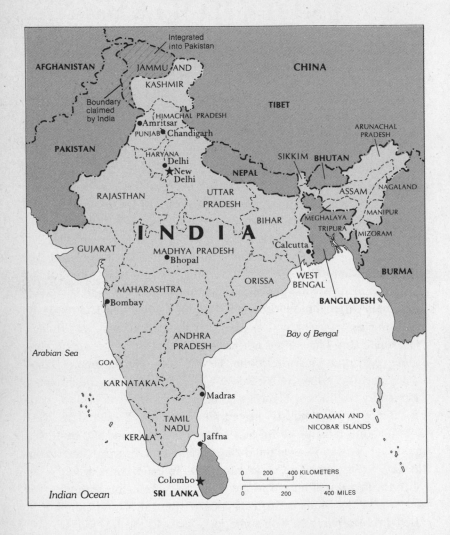

himself was challenged by several of his younger followers as having
proved too inept to continue to lead the party, but he survived such
attacks.

India's faltering economy suffered even more during the Gulf
War from rising oil prices and from the return of many thousands of
Indian laborers, mostly to Kerala and Maharashtra, who had not only
lost well-paid jobs in Kuwait and Iraq, but came home virtually without funds or hope of reemployment. In 1991 India was forced to sell
some gold treasure to keep from defaulting on payments of its international debt. Prime Minister Narasimha Rao and his party gave
highest priority, therefore, to preparing a lean budget that would restore confidence, internationally as well as internally, in India's ability
to develop at a brisker pace. Staggering population pressure, estimated
in some parts of India at nearly a 3 percent gain annually and overall
at about 2.5 percent, kept eating into economic growth, leaving the
nation unable to do more than sustain marginal sustenance for most
of its now some 960 million people. Prime Minister Rao chose as
finance minister Dr. Manmohan Singh, a Western-educated, free-
enterprise-oriented economist, who immediately proposed sharp cuts
in government spending and devaluation of the rupee, winning support both from the World Bank and IMF for India's requests for
several billion dollars in loans. Foreign investments were aggressively
courted, and nonresident Indians urged to pour their hard currency
savings into high interest-paying Indian banks and private enterprises.
India's bureaucratic redtape, which had held up the grant of any
economic permit for at least nine months, was stripped of most of its
discouraging requirements and negative regulations. By October of
1991 IBM and Ford Motor Company won official approval of joint-
venture agreements that would either have been denied or taken years
to process during the previous forty-four years of independent India's
history. Devaluation and budget-pruning led to a revival of rampant
inflation, however, and before year's end the cost of many vital necessities had risen 15 percent. More educated Indians found themselves
unemployed, as a capitalist free-enterprise economy started to replace
Socialism. Washington nodded warm approval of Manmohan Singh's
budget, but India's Communists and leftist socialist intellectuals predicted that runaway inflation and unemployment would suffice to
bring both Communist parties, V. P. Singh, and L. K. Advani back

together, giving the united opposition enough votes in the *Lok Sabha* to topple Narasimha Rao's Congress Party government. Prime Minister Narasimha Rao worked hard, however, at keeping his nation focused on the need to retain "secular development" and "social justice" as its principle national goals, urging his own party and others to stop quibbling over petty personal differences and devote more time and energy to problem solving in India's traditional spirit of selfless service and dedication to moral "duty" (*dharma*).

The popular appeal of rebuilding "Rama's birthplace" temple proved too great, however, to defer for more than a few months, even though the BJP, having won electoral control over Uttar Pradesh's state government in Lucknow, was itself responsible for maintaining law and order in that state. The *Vishwa Hindu Parishad* (VHP) that took the lead in vowing to "rebuild" Rama's "Birth Temple" promised to achieve its cherished goal during 1992. In late October 1991 V. P. Singh was arrested less than fifty miles from Ayodhya as he led a march of some 500 secularists opposed to Hindu efforts to tear down Ayodhya's *Babri Masjid*. V. P. was quickly released from detention but, in a surprisingly swift ripening of *karmic*-fruit, now had a bitter taste of the harsh medicine he had prescribed for L. K. Advani a year earlier. Nor would dark clouds of Hindu-Muslim conflict and violence of catastrophic dimension unseen in India since 1946–47 cease to hover over Ayodhya's doubly sacred soil. In less than nine hours on December 6, 1992 the massive mosque that had weathered more than 460 monsoons was destroyed by a mob of frenzied Hindu fanatics shouting "*Ram, Ram*" as they reduced *Babri Masjid* to rubble and choking dust, while Indian soldiers and police watched in smiling approval. Saffron-robed swamis and "*sadhus*" ignited that mob to demolition heat by their violent mantras, proclaiming this old Islamic monument nothing less than ancient demon Ravana's "dark palace" constructed over the "birthplace" of Lord Rama and Sita's "kitchen." Thus believing that they acted in the name of god and with divine approval, that army of VHP-led batterers triggered shock waves of terror across all of South Asia, leading to the murder of thousands of innocents in Bombay. In what equally misguided zealous Muslims considered appropriate retaliation, hundreds of Hindu temples were tragically destroyed in Pakistan and Bangladesh.

Hindu fundamentalist *Shiv Sena* ("Shiva's Army") fanatics of

Bombay (*Mumbai*), led by Bal Thackeray, warned Indian Muslims that if they wished to remain in "Hindustan" they would have to accept its *Hindutva* ("Hinduness") ethic, since some 90 percent of all Indians were Hindus. The alternative for Muslims who would not enjoy such domination was to emigrate to "their own country," Pakistan. From Ayodhya to Bombay's crowded slums, to the most remote villages of Maharashtra and Gujarat strident cries of "Hindustan for Hindus" now were heard. Initially, secular-minded Hindus as well as Muslims and other minorities were so repelled by such reactionary rhetoric that its BJP–*Shiv Sena* advocates lost ground in several provincial elections. Before the end of 1995, however, a tidal wave of support for the BJP swept over Uttar Pradesh's municipalities—while provincial victories in Maharashtra and Gujarat presaged the growing popularity of *Hindutva* across most of North India—anticipating its victory, in coalition with Punjab's Akali Dal and other key provincial allies in the south, over New Delhi's center of power in general elections held in April–May 1996. BJP President L. K. Advani boldly attacked Prime Minister Rao and his inflated cabinet with a blanket charge of "corruption," promising to restore "honesty" and "integrity" to India's centers of power. Then, on the eve of elections, Rao unleashed a long-suppressed federal police report of major "bribery" that tarred Advani himself, together with many members of the Congress cabinet, all of whom were indicted. Advani resigned his own seat in *Lok Sabha* immediately, vowing not to run again until he was cleared of all charges, which he denied, leaving his closest BJP coleaders, Atal Bihari Vajpayee (b. 1925) and Murli Manohar Joshi, to carry their party's campaign forward throughout the land. India's disillusioned electorate clearly found Congress (I) more "guilty" than the BJP of personal peculation and gross corruption. While election results did not give either party a majority of the seats in Lok Sabha, the BJP's control of over 194 seats proved just enough to permit them to form a fragile government that held power at the center for only twelve days in India's politically heated May of 1996.

Manmohan Singh's program of economic globalization, launched in June 1991, had opened India's long-protected socialist economy to stimulating winds of world market investment and dramatic change, fueling what many hailed as India's economic "miracle." In less than five years India's economy grew as much as it had during the previous forty. Over ten billion dollars worth of foreign capital pumped up India's long-stalled, stagnating economy, giving it a vigorous jump-start. Exports rocketed by more than 20 percent annually, industrial production jumped over 10 percent, and inflation fell to little more than 6 percent by the end of 1995. For the most affluent third of India's 960 million people globalization brought hitherto undreamed-of levels of prosperity and material comfort. Bombay's stock market boomed, and Bangalore looked more like California's Silicon Valley than a South Indian city of old, while New Delhi took on all the trappings of a great world capital. South Asia's long-sleeping giant was finally waking up! Import duties were slashed and western investors were wooed by brilliant Indian ministers of finance and commerce, who had studied at Oxford and Harvard, and now circled the globe urging the capitalist audiences they addressed to "Make our need *your* opportunity."

Pepsico led the way into Punjab, and soon IBM and Xerox followed, as did lesser enterprises. Kentucky Fried Chicken, which opened a restaurant in Delhi, was, however, forced to close down briefly by Indian "health" inspectors with many friends among India's countless tandoori chicken outlets who were "shocked" to find two flies in that KFC kitchen. A more serious deterrent to the ardor of western investors came early in 1995, when the newly elected BJP-*Shiv Sena* government of Maharashtra called a halt to Houston's huge ENRON electric power plant that had just begun to be built at Dabhol. The $2.8 billion power project had been contracted with Bombay's previous Congress government, whose "kickbacks" were reputedly no less than $300 million. The new Bombay government insisted, therefore, that ENRON cut the total cost of its project by at least that amount, which put all construction operations on hold for more than half a year, as both sides argued over the "fair price" of electricity. All agreed that by the next century West India would need three times its current power output to carry out the ambitious plans for economic development being drafted, yet now some investors fear political instability or a return to indigenous *Svadeshi* economics. If some American investors lost their enthusiasm for India's economic potential in the wake of ENRON wrangling, however, Singapore and Tokyo continued to pour money into India, and Ford Motor Company agreed jointly to produce a projected 100,000 cars annually in Madras with

Mahindra & Mahindra. India's rich pool of brilliant mathematicians and computer scientists helped to leapfrog its economy over the last stages of industrialization into orbits of futuristic global communication and electronically-guided satellite technology.

For India's impoverished 300 million landless peasants and urban slumdwellers bogged down in mud and squalor, at the mercy of monsoon rains bringing famine and flood, however, Manmohan's reforms brought little relief and less comfort. The daily drudgery of village India's bullock-cart economy remained as precarious as it had always been, while the wretched crowding of megalopolis slums in Bombay, Calcutta, and Delhi became more painful to those who labored to erect palaces of urban prosperity without earning enough to feed their families. Globalization increased disparities between the very rich and most poor, making those differences more disconcertingly glaring and harsh. Despite the significant economic advances of the early 1990s, therefore, Congress lost popular support as the mandatory date for holding general elections drew nearer. By mid-summer of 1995, when Prime Minister Rao announced his one billion dollar pre-poll package of free meals for poor children, maternity pay to poor mothers, and pensions for poor seniors, hoping thereby to win some 120 million impoverished votes with that generous gesture, most Indians had lost faith in the sincerity of Congress (I) and its leaders. Once again, India's electorate proved itself more sophisticated than many of New Delhi's smartest pandits and power brokers dreamed. Too much "cream," voters said, was being siphoned off the "milk" of foreign investment, pumped so lavishly into India's economy, by old politicians whose many children had much too expensive wasteful weddings, their young mistresses or new wives "needing" far too much for fine jewels and silken saris. Most Indians decided it was time for a change, hoping new brooms might sweep clean Delhi's Augean stables of corruption and ostentatious waste.

Late in September of 1995 word of a religious "miracle" sped across all of India in a matter of hours. "Lord Ganesha is drinking milk," excited Hindus reported, starting tens of thousands of parades of milk-bearing devotees to temples in Delhi, Bombay, Calcutta, and Madras, indeed, in every city where that elephant-headed younger son of Lord Shiva was to be found and worshipped. Reports of that miracle spread to London and Los Angeles as well, where icons of

Ganesh in wood, stone, and metal were "fed" that day by Hindus eager to help satisfy "God's thirst!" The entire milk supply of Delhi was sold out before sundown at record-breaking prices, millions of devout Hindus believing that Lord Shiva and his entire family had descended from on high, insatiably "thirsty" for milk, the nourishing gift of India's sacred cows. That "miracle" presaged the BJP-VHP election victory six months later, which many Hindus view as nothing less than the promised rebirth of an era of golden Hindu glory: the long-awaited *Ram Rajya* ("Rule of Rama") Mahatma Gandhi had predicted. VHP leader Ashok Singhal proclaimed 1996 "Cow-protection year," launching a massive movement spearheaded by saffron-clad disciples who drove everywhere in rose and marigold-garlanded jeeps hailed as "Cow-protection chariots," forcibly closing down cattle slaughter houses and butcher shops, intimidating those who worshipped neither Brahmans nor cows, and failed to appreciate the importance of *Ahimsa* ("Non-violence") to *Hindu Rashtra* ("Hindu Nationhood"). Singhal also promised to "liberate" ancient Hindu temples in Mathura, Lord Krishna's "birthplace," and Varanasi (Benaras), from Muslim mosques erected "on top of their dead bodies." He continued to vow, moreover, that Lord Rama's "birthplace" temple would be ressurected by the end of 1996 on the demolished site of *Babri Masjid* in Ayodhya.

Some 330 million Indians voted by mid-May of 1996 for several thousand candidates competing for the 545 seats of New Delhi's Lok Sabha. No single party secured a majority. The BJP won a plurality, however, and with firm support from several splinter parties commanded 194 seats. BJP's designated Prime Minister A. B. Vajpayee was, therefore, sworn into his high office by President S. D. Sharma on May 16, 1996, but had only till May 31 to prove his Lok Sabha majority. Mild-mannered Vajpayee, a Brahman born in U.P. whose constituency was Lucknow, was unable to lure any of East India's radical or Dalit leaders into his cabinet, and was shunned by most Muslim politicians and South Indian Dravidian bosses as well, all of whom felt more secure in the coalition of secular-socialist parties that came together shortly before these elections were called under a 13-party union known as the National Front-Left Front. Only twelve days after entering office, Vajpayee was obliged to resign, but vowed to return within the year at the head of a BJP-majority government.

The National Front, a loose coalition led by Bengali socialists and Bihari Dalits, was an expanded reincarnation of V. P. Singh's Janata Party, though V. P. himself did not feel healthy enough to run, and declined all offers to be the Front's prime minister designate. Bengal's old Communist Chief Minister Jyoti Basu hoped to be named to that post of potential power, but his comrades vetoed the idea. The Front contested elections, therefore, without naming its prime minister, but since it won only 117 seats, coming in third, behind Congress (with 138 seats) as well as the BJP, there seemed little prospect of its being called upon by India's president to form a new government. Former Prime Minister Rao opted to support the Front, however, from "outside" its cabinet, shrewdly positioning himself and his Congress minority in such a way as to keep his fingers on the pulse of power without further sullying them by responsibility in high office. Rao's opponents in the BJP had, moreover, threatened to "indict" him for major acts of "corruption," and India's fearless Barrister Ram Jethmalani, who was appointed minister of law and justice in Vajpayee's short-lived cabinet promised to bring "charges" against the deposed prime minister as soon as his new government secured its vote of confidence. That vote never came, however, and wily Narasimha Rao agreed, even before Vajpayee resigned, to back Karnataka's Chief Minister H. D. Deve Gowda (b. 1933) for prime minister.

On June 1, 1996 President Sharma swore in Prime Minister Deve Gowda, a non-Brahman Vokkaliga land-owner-businessman from Bangalore, as head of New Delhi's patchquilt coalition of more than fifteen parties from nearly every Indian state. Charismatic Lalu Prasad Yadev, former chief minister of Bihar, who had presided over the Front, refused to accept any cabinet post other than that of premier. Prime Minister Deve Gowda retained the powerful Home ministry and atomic energy in his own hands. Inder K. Gujral, the new government's most experienced diplomat, returned to lead the prestigious ministry of Foreign Affairs. Former Congress minister of commerce, Harvard-trained P. Chidambaram, was brought back to the Center as minister of finance, clearly indicating the new government's commitment to continuing economic reforms launched in 1991. Socialist opposition to greater globalization of India's economy, however, posed a potential rift in the coalition cabinet, whch might well lead to its breakdown before the full five year term expires. Should such an early "no-confidence" vote pass, moreover, India's president could invite Narasimha Rao or A. B. Vajpayee to try to form a more stable government, or dissolve Parliament and call for new elections, which might well give the BJP and its "Hindutva" a larger plurality or majority.

So India clings to ancient traditions as it hurtles toward a future of global modernity. But past and future are not always compatible. In dead of night a slow train stopped less than a mile from Firozabad to avoid mangling a slower cow crossing the tracks up ahead, while a poorly-paid flagman fell asleep at his switch, thus leaving a racing Delhi-bound express to crash into the wall of stalled steel, killing more than 350 voyagers, the luckiest of whom never awoke from their sleep to see or hear the crushing-screaming syncretism of ancient and modern India that killed them.

India's polity continued, moreover, to be jolted by terror as well as train wrecks and deadly fires. A "human-bomb" assassinated Punjab's Chief Minister Beant Singh (and fifteen others) in Chandigarh on August 31, 1995, reminding India and the world that, contrary to what Beant had so often said, Punjab had hardly "settled down." Beant's director general of police, K. P. S. Gill, had led the most ruthless campaign of official terror against suspected "Khalistani" Sikhs, "eliminating" everyone connected with such suspects, yet failing to save his own leader's life, hence his offer to retire was accepted with alacrity at year's end. Babbar Khalsa Khalistani terrorists claimed responsibility for Beant's assassination, redolent of the human-bomb killing of Rajiv Gandhi four years earlier. Since Operation Bluestar in 1984 Khalistani Sikhs and Tamil Tigers had helped one another in the planning and funding of such brutal acts of terror, if not in their actual execution.

Punjab, however, remained relatively peaceful compared to its northern neighbor state, Kashmir. With fewer than five million people, Jammu and Kashmir State was used as India's permanent cantonment for more than a half million regular army troops. India's central government understands too well by now how most Kashmiris feel about India's union, and about Pakistan: recent Srinagar polls indicate that three-quarters of all Kashmiris favor complete independence, or what India's National Congress used to call *Purna Swaraj*. Kashmir's economy has been battered by fighting, its once lucrative tourist trade vir-tully ended, except for climbers brave enough to risk terrorist attacks as well as glacial ice. In mid-summer of 1995 six such young tourists from Western Europe and America were taken hostage by a gang of Muslim terrorists who call themselves Al-Faran. One lucky American ran fast enough to escape, the other five apparently brutally murdered after an ordeal of frost and terror that lasted more than half a year.

Though Pakistan denied training and arming terrorist gangs like Al-Faran, since 1990 many such tough Muslim tribals have crossed the Line of Control (LOC) that still divides India's Jammu and Kashmir State, which includes Srinagar and its Vale, from Pakistan's western quarter, *Azad* ("Free") Kashmir. Those self-styled "freedom fighters," ready to risk their lives to "liberate" Kashmir from India's "occupation," are denigrated as "mercenaries" or "foreign agents" by India, New Delhi insisting that all of them are well-armed and covertly trained in the hills around Islamabad. One such force seized the town of Charar-e-Sharief in Kashmir, with its ancient sacred wooden shrine of Sheikh Nooruddin Noorani, one of Kashmir's great Sufi saints. In May of 1995 the Indian Army surrounded that mountain retreat and, after its captors refused to surrender, burned it to the ground, charging that the gang who held it committed that final act of terror. How many more such precious shrines must be destroyed, how many more lives lost before a peaceful solution is found to that dreadful conflict in Kashmir, to which all Kashmiris and the rest of South Asia remains hostage?

Since its birth in 1947 Pakistan has remained the primary focus of India's foreign policy, Kashmir its premier source of conflict. In 1965 Foreign Minister Zulfi Bhutto persuaded his martial president, Ayub Khan, to initiate a war in Kashmir, which ended in disaster for Pakistan as well as Ayub, but helped catapult Bhutto to premier power. Soon after he convinced Ayub's martial successor, Yahya Khan, to launch his "preemptive" strike in Dacca, triggering the birth of Bangladesh. Zulfi's fondest dream during his half decade at the top of Islamabad's "slippery pole" was to integrate all of Kashmir into Pakistan. His chosen general, Zia ul-Haq, hanged Bhutto, however, in 1979, and his political legacy was left to his elder daughter, Benazir. After over a decade of martial power General Zia and most of his high command were killed, together with U.S. Ambassador Arnold Raphel and an American general, in the still unexplained mid-air explosion of their plane in August 1988. Soon afterward free elections were held throughout Pakistan, and Benazir Bhutto was then invited by President Ghulam Ishaq Khan to lead Pakistan's central government in Islamabad. Benazir promised to inaugurate all the reforms her father had left unaccomplished, and seemed initially to lower tensions between India and Pakistan, meeting with Rajiv Gandhi, who, like Benazir, had inherited his prime ministership. Opposition leader Mian Nawaz Sharif, Punjab's chief minister, accused Benazir of being "too soft" on India, however, and in August 1990 President Khan dismissed her, acting with the support of his army. Before year's end fresh elections were held in Pakistan, and Nawaz Sharif was sworn in as prime minister, taking a tougher line on Kashmir, his own birthplace. Nawaz did nothing, however, to improve upon Pakistan's fast deteriorating economy, nor could he "liberate" Kashmir in the more than two years of power he enjoyed.

Benazir Bhutto was brought back to Pakistan's wobbly political pinnacle, after the World Bank's incorruptible caretaker, Prime Minister Moeen Qureshi, administered his stiff dose of fiscal medicine to Islamabad. Benazir's home province of Sind, however, was torn by virtual civil war, its crowded port city of Karachi brought to a standstill by violence in the streets that was to leave over 1,800 people dead in 1995 alone. Urdu-speaking *Muhajirs* ("Refugees"), whose MQM party and its leadership dominated Karachi's bustling business community, were long seen as "enemy intruders" by many members of Benazir's PPP and the *"Sindhi Desh"* provincial nationalists with whom they clashed daily. Karachi's murderous violence was intensified by Pakistan's booming traffic in drugs that kept rival drug-lords with small armies of their own in a constant state of armed conflict over those soaring "black money" profits. Only the army could restore any semblance of order to Karachi's streets, but knowing what the army had done to her father, Benazir was afraid to call them out for any length of time, fearing that the high command might decide she was no longer necessary. Her own brother, Mir Murtaza, and her mother, Begum Nusrat, also lost confidence in Benazir's leadership by early 1996, as had all of the opposition members in Parliament. Some Pakistanis feared that Sind might soon be "reabsorbed" by India if Karachi's violence continued much longer, the way East Pakistan had been turned, with Indira's help, into Bangladesh a quarter century earlier.

In Kashmir as well conflict escalated by early 1996 to the point at which it appeared to many observers that India was ready to trigger another nuclear explosion in Rajasthan. Both India and Pakistan continued officially to deny that they had nuclear weapons, of course, but India's secret arsenal was estimated to contain about thirty such warheads by 1996. India's long-range *Agni* (Vedic God of Fire) ballistic missile, capable of delivering nuclear payloads 5,000 kilometers from its Kerala State launch pad, could target Karachi's naval base and steel mills. India's medium-range Prithvi missiles could send nuclear warheads from 100 to 150 miles of their launch pads, while her huge Swedish Bofors artillery, mounted in Kashmir, could fire shells across the Line of Control to Muzaffarabad and Pakistan's Kohata nuclear plant on the outskirts of Islamabad. Both nations also had jet bombers capable of easy adaptation to carry nuclear bombs. Neither India nor Pakistan had signed the nuclear Non-Proliferation Treaty, nor would either agree to sign the new Comprehensive Test Ban Treaty. The world continued to hope, however, that neither would be so self-destructively foolish as to unleash any of the most deadly power they possessed.

The collapse of the Soviet Union brought an end to the superpower proxy war in Afghanistan by the close of 1991, but in and around Kabul warfare continued to take its toll of Afghan lives, leaving that capital city in rubble and ruin. Pakistan's proxy Taliban army of young Afghan tribals, many of whom had grown up in UN-supported camps around Peshawar and elsewhere along the frontier where more than a million Afghan refugees had lived throughout the Afghan War, tried in vain to capture Kabul for most of 1995. Pakistan hoped to reinstate old Zahir Shah on his Kabul throne, luring him back to Islamabad from his retirement villa outside Rome, but the youthful Taliban army could do no more than fire deadly mortar shells into Kabul from its trenches to the south, while Indian planes reinforced Kabul's Central Asian–backed government and president.

The more India and Indo-Pak relations changed, therefore, the more destined they seemed to revisit old trials and tribulations. A half century of freedom hardly sufficed to solve problems rooted millennia deep, whose *karmic* fruit of fears and hatreds, prejudice, inequality, greed, and violence continued daily to ripen, eliciting acts of retaliation and vengeance, sowing seeds of future *karmic* balances for India to confront in the twenty-first century.

Om!

BIBLIOGRAPHY

This select bibliography is arranged in priority ranking for each chapter for those interested in further reading.

I—*The Ecological Setting*

Schwartzberg, Joseph E., Shiva Bajpai et al., eds. *A Historical Atlas of South Asia.* Chicago, 1978.

Spate, O. H. K., and A. T. A. Learmonth. *India and Pakistan.* 3d ed. London, 1967.

Chatterjee, S. P., ed. *National Atlas of India.* Calcutta and Dehra Dun, 1957.

Davies, C. Collin. *An Historical Atlas of the Indian Peninsula.* 2d ed. Oxford, 1959.

Wadia, D. N. *Geology of India.* 3d ed. London, 1961.

Schwartzberg, Joseph E. "India: The Land and Natural Resources." In *The Encyclopedia Americana.* International ed. Vol. 14. New York, 1970.

Brown, W. Norman, ed. *India, Pakistan, Ceylon.* rev. ed. Philadelphia, 1960.

———. *Man in the Universe: Some Continuities in Indian Thought.* Berkeley and Los Angeles, 1966.

———. *The United States and India, Pakistan, Bangladesh.* 3d ed. of *The United States and India and Pakistan.* Cambridge, 1972.

Cohn, Bernard S. *India: The Social Anthropology of a Civilization.* Englewood Cliffs, 1971.

Srinivas, M. N. *Social Change in Modern India.* Berkeley and Los Angeles, 1966.

Singer, Milton. *When A Great Tradition Modernizes: An Anthropological Approach to Indian Civilization.* New York, 1972.

Crane, Robert I., ed. *Regions and Regionalism in South Asian Studies: An Exploratory Study.* Durham, 1967.

Metcalf, Thomas R. *Modern India: An Interpretive Anthology.* London, 1971.

Frykenberg, Robert Eric, ed. *Land Control and Social Structure in Indian History.* Madison, 1969.

II—*Indus Culture*

Wheeler, Sir Mortimer. *Civilizations of the Indus Valley and Beyond.* New York, 1966.

Allchin, Bridget and Raymond. *The Birth of Indian Civilization: India and Pakistan Before 500 B.C.* Harmondsworth, 1968.

Piggott, Stuart. *Prehistoric India to 1000 B.C.* 2d ed. London, 1962.

Fairservis, Walter A. *The Origin, Character, and Decline of an Early Civilization.* 2d rev. ed. New York, 1967.

Marshall, Sir John. *Mohenjo-Daro and the Indus Civilization.* 3 vols. London, 1931.

Possehl, Gregcry L., ed. *Ancient Cities of the Indus.* Durham, 1979.

Subbarao, Bendapudi. *The Personality of India.* Baroda, 1958.

Casal, Jean-Marie. *La Civilisation de l'Indus et ses enigmes.* Paris, 1969.

Dales, George F. "New Investigations at Mohenjo-daro." *Archaeology* 18 (1965): 145–50.

Dandekar, R. N. "Some Aspects of the Indo-Mediterranean Contacts." *Annals of the Bhandarkar Oriental Research Institute* (Poona) 50 (1969).

Raikes, R. L. "The end of the ancient cities of the Indus." *American Anthropologist* 66 (1964): 284–99.

Sankalia, H. D. "Early Man in India." *Journal of the Asiatic Society of Bombay,* 41–42 (1966–68): 173–81.

———. "The Philosophy of Archaeology in India, or Theoretical and Methodological Approaches in Archaeological Interpretation in India." *Journal of Indian History* vol. 48, pt. 1 (1970), pp. 27–42.

———. *Prehistory of India.* New Delhi, 1977.

———. *Pre-Historic Art in India.* Durham, 1978.

III—*The Aryan Age*

Basham, A. L. *The Wonder That Was India.* 3d ed. New York, 1967.

DeBary, Wm. T., ed. *Sources of Indian Tradition.* New York, 1958.

McNeill, Wm. H., and J. W. Sedlar, eds. *Classical India.* New York, 1969.

Thapar, Romila. *A History of India.* Vol. 1. Baltimore, 1969.

Brown, W. Norman. "Mythology of India." In *Mythologies of the Ancient World,* edited by S. N. Kramer. Garden City, N.Y., 1961.

———. "Agni, Sun, Sacrifice, and Vac: A Sacerdotal Ode by Dirghata-mas (Rig Veda 1.164)." *Journal of the American Oriental Society* 88 (1968): 199–218.

Bloomfield, Maurice. *The Religion of the Veda*. New York, 1908.

Childe, V. Gordon. *The Aryans*. 1926. Reprinted. New York, 1970.

Majumdar, R. C., ed. *History and Culture of the Indian People*. Vol. 1, *The Vedic Age*. London, 1951.

Bhargava, P. L. *India in the Vedic Age: A History of Aryan Expansion in India*. 2d rev. ed. Aminabad, Upper India, 1971.

Raychaudhuri, H. C. *The Political History of Ancient India*. 6th rev. ed. Calcutta, 1953.

Muller, F. Max, and H. Oldenberg, trans. *Vedic Hymns*. 2 vols. Sacred Books of the East, vols. 32 and 46. Oxford, 1891–97.

Muller, F. Max. *Heritage of India*. Calcutta, 1951.

Macdonell, A. A. *A Vedic Reader for Students*. London, 1960.

———. *History of Sanskrit Literature*. 5th impression. Delhi, 1962.

Gonda, Jan. "The Historical Background of the Name Satya Assigned to the Highest Being." *Annals of the Bhandarkar Oriental Research Institute* (Poona) 48–49 (1968): 83–93.

IV—*North Indian Conquest and Unification*

O'Flaherty, Wendy D. *Hindu Myths*. Harmondsworth, 1975.

Thomas, Paul. *Epics, Myths and Legends of India*. 12th ed. Bombay, 1961.

Coomaraswamy, A. K., and Sister Nivedita. *Myths of the Hindus and Buddhists*. New York, 1967.

Sengupta, Padmini. *Everyday Life in Ancient India*. 2d ed. London, 1957.

Keith, A. B., trans. *Aitareya Brahmana*. Harvard Oriental Series, vol. 25. Cambridge, 1920.

Eggeling, J., trans. *Satapatha Brahmana*. 5 vols. Sacred Books of the East, vols. 12, 26, 41, 43, and 44. Oxford, 1882–1900.

Mitra, R., ed. *Taittiriya Brahmana*. Calcutta, 1855–70.

Rajagopalachari, C., trans. *Mahabharata*. Bombay, 1955.

Griffith, R. T. H., trans. *The Ramayana*. Benares, 1915.

Dutt, R. C., trans. *The Mahabharata*. London, 1917.

———, trans. *The Ramayana*. London, 1917.

Pusalker, A. D. *Studies in the Epics and Puranas*. Bombay, 1963.

Narayan, R. K. *Gods, Demons, and Others*. New York, 1964.

Kosambi, D. D. "The Autochthonous Element in the Mahabharata." *Journal of the American Oriental Society* 84 (1964): 31–44.

Menon, Aubrey. *The Ramayana*. New York, 1954.

Hume, Robert E. *The Thirteen Principal Upanishads*. 2d rev. ed. New York, 1931.

Muller, F. Max, trans. *Upanishads*. 2 vols. 1962. Reprint. New York.

Conze, Edward. *Buddhism, its Essence and Development*. New York, 1959.

———. *Buddhist Scriptures*. Harmondsworth, 1959.

Thomas, E. J. *Early Buddhist Scriptures*. London, 1935.

Bapat, P. V., ed. *2500 Years of Buddhism*. Delhi, 1956.

Warren, Henry C. *Buddhism in Translations*. Harvard Oriental Series, vol. 3. Cambridge, 1922.

Eliot, Charles N. E. *Hinduism and Buddhism: An Historical Sketch*. 3 vols. New York, 1954.

Beane, Wendell C. *Myth, Cult and Symbols of Śākta Hinduism*. Leiden, 1977.

Rhys Davids, T. W. *Buddhism: Its History and Literature*. 3d ed. New York, 1918.

Asvaghosa. *Buddhacarita ("Life of the Buddha")*. Translated by E. B. Cowell, F. Max Muller, and J. Takakusu. Sacred Books of the East, vol. 49. Oxford, 1894.

Pali Text Society Translation Series. London, 1909–

Stevenson, Margaret Sinclair. *The Heart of Jainism*. London, 1915.

Jaini, J. L. *Outlines of Jainism*. Cambridge, 1940.

Nayanar, A. C. *The Religion of Ahimsa*. Bombay, 1957.

Jacobi, H. G., tr. *Jaina Sutras*. 2 vols. Oxford, 1884–95.

V—*India's First Imperial Unification*

Sastri, K. A. Nilakanta, ed. *A Comprehensive History of India*. Vol. 2, *The Mauryas & Satavahanos*. Bombay, 1957.

———. *The Age of the Nandas and Mauryas*. Banaras, 1952.

Majumdar, R. C., ed. *History and Culture of the Indian People*. Vol. 2: *The Age of Imperial Unity*. Bombay, 1951.

Mookerji, R. K. *Chandragupta Maurya and His Times*. 3d ed. Delhi, 1960.

———. *Asoka*. 2d rev. ed. London, 1955.

Thapar, Romila. *Asoka and the Decline of the Mauryas*. London, 1961.

Sen, A. C., trans. *Asoka's Edicts*. Calcutta, 1956.

Kautilya. *Arthashastra*. Translated by R. Shamasastry. 5th ed. Mysore, 1956.

Majumdar, R. C., ed. *The Classical Accounts of India*. Calcutta, 1960.

Woodcock, George. *The Greeks in India*. London, 1966.

McCrindle, John W. *Ancient India as Described by Megasthenes and Arrian*. London, 1877.

———. *The Invasion of India by Alexander the Great*. 2d ed. Westminster, 1896.

Brown, Percy. *Indian Architecture*. Vol. 1, *Buddhist and Hindu Periods*. Bombay, 1956.

Ling, Trevor. *The Buddha: Buddhist Civilization in India and Ceylon.* New York, 1973.

Jha, D. N. "Land Revenue in the Maurya and Gupta Periods." In *Land Revenue in India: Historical Studies,* edited by R. S. Sharma. Delhi, 1971.

VI—*Political Fragmentation and Economic and Cultural Enrichment*

Rosenfield, John. *The Dynastic Arts of the Kushans.* Berkeley and Los Angeles, 1967.

Schoff, Wilfred H., trans. *The Periplus of the Erythraen Sea.* New York, 1912.

McCrindle, John W. *Ancient India as Described by Ptolemy.* rev. ed. Calcutta, 1927.

Tarn, W. W. *The Greeks in Bactria and India.* rev. ed. Cambridge, 1951.

Narain, A. K. *The Indo-Greeks.* Oxford, 1957.

De Leeuw, Von Lohuizen. *The Scythian Period.* Leiden, 1949.

Chattopadhyaya, S. *The Sakas in India.* Santiniketan, 1955.

———. *Early History of North India (c. 200 B.C.–A.D. 650).* Calcutta, 1958.

McGovern, William M. *The Early Empires of Central Asia.* Chapel Hill, 1939.

Marshall, Sir John. *The Buddhist Art of Gandhara.* Cambridge, 1960.

Chaladar, Haran C. *Social Life in Ancient India.* 2d rev. ed. Calcutta, 1954.

Gandhara Sculpture from Pakistan Museum. New York, 1960.

VII—*The Classical Age*

Majumdar, R. C. *History and Culture of the Indian People.* Vol. 3, *The Classical Age.* Bombay, 1954.

Edgerton, Franklin, trans. *The Bhagavad Gita.* Cambridge, 1972.

Kalidasa, *Shakuntala and Other Writings.* Translated by Arthur W. Ryder. New York, 1959.

———. *Meghaduta: The Cloud Messenger.* Translated by S. K. De. New Delhi, 1957.

Monier-Williams, M. *Hinduism.* Calcutta, 1951.

Sen, K. M. *Hinduism.* Harmondsworth, 1961.

Embree, Ainslie T., ed. *The Hindu Tradition.* New York, 1966.

Radhakrishnan, S. *The Hindu View of Life.* London, 1954.

Zaehner, R. C. *Hinduism.* New York, 1966.

Keith, A. B. *Classical Sanskrit Literature.* 5th ed. Calcutta, 1947.

Beal, S., trans. *Si Yu Ki: Buddhist Records of the Western World.* 2 vols. London, 1883.

Sarma, D. S. *The Tales and Teachings of Hinduism.* Bombay, 1948.

Giles, H. A. *The Travels of Fa-hien.* Cambridge, 1923.

Mookerji, R. K. *The Gupta Empire.* 3d rev. ed. Bombay, 1959.

Dikshitar, V. R. Ramachandra. *Gupta Polity.* Madras, 1952.

Saletore, R. N. *Life in the Gupta Age.* Bombay, 1943.

Kramrisch, Stella. *The Art of India.* London, 1954.

———. *The Hindu Temple.* 2 vols. Calcutta, 1946.

Singer, Milton, ed. *Krishna: Myths, Rites, and Attitudes.* Honolulu, 1966.

Zaehner, R. C., trans. and comm. *The Bhagavad-Gita.* London, 1966.

Maity, S. K. *Economic Life of Northern India in the Gupta Period.* Calcutta, 1957.

Ghoshal, U. N. *The Agrarian System in Ancient India.* Calcutta, 1930.

Gokhale, B. G. *Indian Thought Through the Ages: A Study of Some Dominant Concepts.* London, 1961.

Das Gupta, S. N. *History of Indian Philosophy.* 4 vols. Cambridge, 1923–49.

Zimmer, Heinrich. *Philosophies of India.* Edited by Joseph Campbell. New York, 1951.

———. *Myths and Symbols in Indian Art and Civilization.* Edited by Joseph Campbell. New York, 1946.

Coomaraswamy A. K. *The Dance of Shiva.* rev. ed. New York, 1957.

Majumdar, R. C. *Hindu Colonies in the Far East.* Dacca, 1927.

Coedes, C. *L'État Hindouisé d'Indochine et d'Indonesie.* Paris, 1948.

Prasad, Beni. *Theory of Government in Ancient India (Post-Vedic).* Allahabad, 1927.

Mandelbaum, David G. *Society in India.* 2 vols. Berkeley, Los Angeles, and London, 1970.

Sastri, K. A. Nilakanta. *A History of South India.* London, 1955.

———. *The Cholas.* 2 vols., rev. ed. Madison, 1955.

Majumdar, A. K. *The Chalukyas of Gujarat.* Bombay, 1956.

Yazdani, Ghulam, ed. *The Early History of the Deccan.* 2 vols. London, 1960.

Bhandarkar, R. G. *Early History of the Deccan.* Calcutta, 1957.

Venkataramanayya, N. *The Eastern Chalukyas of Vengi.* Madras, 1950.

Dikshitar, V. R. Ramachandra. *Pre-Historic South India.* Madras, 1951.

Kanakasabhai, V. *The Tamils Eighteen Hundred Years Ago.* 2d rev. ed. Tirunelveli, 1956.

Derrett, J. Duncan M. *The Hoysalas, a Medieval Indian Royal Family.* London, 1957.

Panikkar, K. M. *A History of Kerala.* Annamalainagar, 1960.

Mudaliar, N. R. Balakrishna, trans. *The Golden Anthology of Ancient Tamil Literature.* 3 vols. Tirunelveli, 1959–60.

Gopalan, R. *History of the Pallavas of Kanchi.* Madras, 1928.

Karmarkar, A. P. *A Cultural History of Karnataka*. Dharwar, 1947.

Mahalingam, T. V. *South Indian Polity*. Madras, 1955.

Kingsbury, Francis, and G. D. Philips, eds. *Hymns of the Tamil Saivite Saints*. London, 1921.

O'Flaherty, Wendy D. *Asceticism and Eroticism in the Mythology of Siva*. London, 1973.

VIII—*The Impact of Islam*

Elliot, Henry M., and John Dowson, eds. *The History of India as Told by Its Own Historians: The Muhammadan Period*. 31 vols. Calcutta, 1952–59.

Majumdar, R. C., ed. *History and Culture of the Indian People*. Vol. 4, *The Age of Imperial Kanauj*. Bombay, 1955.

————, ed. Ibid. Vol. 5, *The Struggle for Empire*. 1957.

————, ed. Ibid. Vol. 6, *The Delhi Sultanate*. 1967.

Haig, Sir Wolseley, ed. *The Cambridge History of India*. Vol. 3, *Turks and Afghans*. New York, 1928.

Gibb, H. A. R. *Mohammedanism: An Historical Survey*. 2d ed. New York, 1955.

Rahman, Fazlur. *Islam*. Garden City, N.Y., 1968.

Ikram, S. M. *Muslim Civilization in India*. Edited by A. T. Embree. New York and London, 1964.

Chand, Tara. *Influence of Islam on Indian Culture*. Allahabad, 1954.

Sharma, S. R. *The Crescent in India: A Study in Medieval History*, rev. ed. Bombay, 1954.

Ibn Battuta. *Travels in Asia and Africa, 1325–1354*. Translated by H. A. R. Gibb. New York, 1929.

Ahmad, Aziz. *An Intellectual History of Islam in India*. Edinburgh, 1969.

————. *Studies in Islamic Culture in the Indian Environment*. Oxford, 1964.

Qureshi, I. H. *The Muslim Community of the Indo-Pakistan Subcontinent, 610–1947. A Brief Historical Analysis*. Mouton, 1962.

Habibullah, A. B. M. *The Foundation of Muslim Rule in India*. 2d rev. ed. Allahabad, 1961.

Misra, S. C. *The Rise of Muslim Power in Gujarat*. New York, 1963.

Habib, Muhammad, and A. U. S. Khan. *The Political Theory of the Delhi Sultanate*. Allahabad, 1961.

Sherwani, H. K. *The Bahmanis of the Deccan*. London, 1953.

Al Biruni. *Alberuni's India*. Translated by E. C. Sachau. 2 vols. London, 1910.

Tod, James. *Annals and Antiquities of Rajast'han*. 2 vols. London, 1957.

Qanungo, K. R. *Studies in Rajput History*. Delhi, 1960.

Hardy, Peter. *Historians of Medieval India*. London, 1960.

Babur. *Memoirs of Babur.* Translated by A. S. Beveridge. 2 vols. London, 1922.

Erskine, W. *Babur and Humayun.* 2 vols. London, 1854.

Prasad, Ishwari. *The Life and Times of Humayun.* Bombay, 1956.

Nizami, K. A. *Some Aspects of Religion and Politics in India During the Thirteenth Century.* Bombay, 1961.

Basham, A. L., ed. *A Cultural History of India.* Oxford, 1975.

IX—*Mughal Imperial Unification*

Abul Fazl. *Akbar Namah.* Translated by Henry Beveridge. 3 vols. Calcutta, 1897–1921.

———. *Ain-i-Akbari.* Translated by H. F. Blockman and H. S. Jarrett. 3 vols. Calcutta, 1873–96.

Burn, Sir Richard, ed. *The Cambridge History of India.* Vol. 4, *The Mughal Period.* New York, 1937.

Majumdar, R. C., ed. *History and Culture of the Indian People.* Vol. 8, *The Mughal Empire.* Bombay, 1969.

Smith, Vincent. *Akbar the Great Mogul.* Oxford, 1927.

Akbar, Muhammad. *The Punjab Under the Mughals.* Lahore, 1948.

Binyon, Lawrence. *Akbar.* London, 1932.

Raychaudhuri, Tapan. *Bengal Under Akbar and Jahangir: An Introductory Study in Social History.* Calcutta, 1953.

Sharma, Sri Ram. *Mughal Government and Administration.* Bombay, 1951.

Sarkar, Jadunath. *Mughal Administration.* 4th rev. ed. Calcutta, 1952.

Abdul Aziz. *The Mansabdari System and the Mughal Army.* London, 1946.

Roy Choudhury, M. *The Din-i-Ilahi.* Calcutta, 1941.

Moreland, William H. *India at the Death of Akbar.* London, 1920.

Habib, Irfan. *The Agrarian System of Mughal India, 1556–1707.* New York, 1963.

Monserrate, Antonio. *The Commentary of Father Monserrate.* Translated and edited by J. S. Hoyland and S. N. Banerji. London, 1922.

Irvine, William. *The Army of the Indian Moghuls.* Reprin. New Delhi, 1962.

DuJarric, Father Pierre. *Akbar and the Jesuits.* Translated by C. H. Payne. New York, 1926.

X—*Western Europe's Vanguard*

Hunter, W. W. *A History of British India.* Vol. 1. London, 1899.

Boxer, Charles R. *The Portuguese Seaborne Empire, 1415–1825.* London, 1969.

————. *The Dutch Seaborne Empire, 1600–1800*. London, 1965.

Danvers, Frederick C. *The Portuguese in India*. 2 vols. London, 1894.

da Gama, Vasco. *Journal of the First Voyage*. Translated by E. G. Ravenstein. London, 1898.

de Albuquerque, Alfonso. *The Commentaries of Albuquerque*. Translated by W. Birch. 4 vols. London, 1875–84.

Locke, J. C. *The First Englishmen in India: Letters and Narratives of Sundry Elizabethans*. London, 1930.

Spear, Percival. *A History of India*. Vol. 2. Harmondsworth, 1970.

Hakluyt, Richard, compiler. *The Principal Navigations, Voyages, Traffiques & Discoveries of the English Nation*. 10 vols. New York, 1927–28.

Foster, William, ed. *Early Travels in India*. London, 1921.

Panikkar, K. M. *Asia and Western Dominance*. Collier Books, 1969.

Raychaudhuri, Tapan. *Jan Company in Coromandel, 1605–1690*. The Hague, 1962.

Glamann, Kristof. *Dutch-Asiatic Trade*. Copenhagen, 1958.

Ross, Sir E. Denison. "The Portuguese in India." In *The Cambridge History of India*. Edited by H. H. Dodwell. Vol. 5, *British India, 1497–1858*. Cambridge, 1953.

Geyl, P. "The Dutch in India." In Ibid.

Van Leur, J. C. *Indonesian Trade and Society: Essays in Asian Social and Economic History*. The Hague, 1955.

XI—*Great Mughal Glory*

Hunter, W. W. *A History of British India*. Vol. 2. London, 1912.

Moreland, W. H. *From Akbar to Aurangzeb*. London, 1923.

————. *The Agrarian System of Moslem India*. Cambridge, 1929.

Jahangir. *Memoirs of Jahangir*. Translated by A. Rogers. Edited by H. Beveridge. 2 vols. London, 1909–14.

Roe, Sir Thomas. *The Embassy of Sir Thomas Roe to India*. Edited by W. Foster, rev. ed. London, 1926.

Mundy, Peter. *The Travels of Peter Mundy*. Edited by R. C. Temple. 5 vols. London, 1907–36.

Tavernier, Jean B. *Travels in India*. Translated by V. Ball. Edited by W. Cooke. 2 vols. 2d ed. London, 1925.

Bernier, Francois. *Travels in the Moghul Empire*. Translated by A. Constable. Edited by V. A. Smith. rev. ed. London, 1934.

Manucci, Niccolo. *Storia Do Mogor: or Mogul India*. Translated by W. Irvine. 4 vols. London, 1906–8.

————. *A Pepys of Moghul India*. Edited by M. L. Irvine. New York, 1913.

Krishna, Bal. *Commercial Relations Between India and England*. London, 1924.

Foster, William, ed. *English Factories in India, 1618–1669.* 13 vols. Oxford, 1906–1925.

Khan, Shafaat Ahmad. *The East India Trade in the 17th Century in Its Political and Economic Aspects.* London, 1923.

Sainsbury, Ethel B., ed. *Court Minutes of the East India Company, 1635–79.* 11 vols. Oxford, 1907–38.

Saksena, B. P. *History of Shah Jahan of Delhi.* Allahabad, 1962.

Sarkar, Sir Jadunath. *Anecdotes of Aurangzeb.* 2d ed. Calcutta, 1925.

―――. *A Short History of Aurangzeb.* 3d rev. ed. Calcutta, 1962.

―――. *Shivaji.* 5th rev. ed. Calcutta, 1952.

Sardesai, G. S. *New History of the Marathas.* 3 vols. Bombay, 1957.

Patwardhan, R. P., and H. G. Rawlinson. *Source Book of Maratha History.* Bombay, 1929.

Singh, Khushwant. *A History of the Sikhs.* Vol. 1, *1469–1839.* Princeton, 1963.

XII—*Twilight of the Mughal Empire*

Owen, Sidney J. *The Fall of the Mogul Empire.* 2d ed. Varanasi, 1960.

Sarkar, Sir Jadunath. *Fall of the Mughal Empire.* 2d rev. ed. 4 vols. Calcutta, 1949.

Sen, S. P. *The French in India.* Calcutta, 1958.

Malleson, George B. *History of the French in India.* 2d ed. London, 1893.

Dodwell, Henry H. *Dupleix and Clive.* London, 1920.

Pillai, Anandaranga. *Private Diary of Ananda Ranga Pillai.* Translated and edited by J. F. Price and H. H. Dodwell. 12 vols. Madras, 1904–28.

Hatalkar, V. G. *Relations Between the French and the Marathas, 1668–1851.* Bombay, 1958.

Mason [pseud. Woodruff], Philip. *The Men who Ruled India.* Vol. 1, *The Founders of Modern India.* New York, 1954.

Gupta, Brijen K. *Sirajuddaullah and the East India Company, 1756–1757.* Leiden, 1962.

Edwardes, Michael. *Plassey: The Founding of an Empire.* London, 1969.

Khan, Abdul Majed. *The Transition in Bengal, 1756–1775: A Study of Saiyid Muhammad Reza Khan.* London, 1969.

Dodwell, H. H., ed. *The Cambridge History of India.* Vol. 5, *British India.* Cambridge, 1929.

Roberts, Paul E. *History of British India.* 3d ed. London, 1952.

Muir, Ramsay. *The Making of British India, 1756–1858.* Manchester, 1923.

Gopal, Ram. *How the British Occupied Bengal.* New York, 1962.

Spear, T. G. P. *Twilight of the Mughuls.* Cambridge, 1951.

Hambly, Gavin. *Cities of Mughul India*. New York, 1968.

Chandra, Satish. *Parties and Politics at the Mughal Court*. Aligarh, 1959.

Raghuvanshi, V. P. S. *Indian Society in the Eighteenth Century*. New Delhi, 1969.

Chaudhuri, N. G. *Cartier, Governor of Bengal, 1769–1772*. Calcutta, 1970.

XIII—*John Company Raj*

Furber, Holden. *John Company at Work*. Cambridge, 1948.

———. *Henry Dundas: First Viscount Melville*. Oxford, 1931.

Philips, Cyril Henry. *The East India Company, 1784–1834*. 2d ed. Manchester, 1961.

———. *India*. New York, 1949.

Davies, A. M. *Strange Destiny: A Biography of Warren Hastings*. New York, 1935.

Nightingale, Pamela. *Trade and Empire in Western India, 1784–1806*. Cambridge, 1970.

Das Gupta, Anil C., ed. *The Days of John Company, 1824–1832*. Calcutta, 1959.

Mukherjee, S. N. *Sir William Jones: Eighteenth Century British Attitudes to India*. Cambridge, 1968.

Moon, Penderel. *Warren Hastings and British India*. New York, 1962.

Misra, B. B. *The Administration of the East India Company*. New York, 1960.

Sinha, Narendra K. *The Economic History of Bengal from Plassey to the Permanent Settlement*. 2 vols. Calcutta, 1961–62.

Ghosh, S. C. *The Social Condition of the British Community in Bengal, 1757–1800*. Leiden, 1970.

Spear, Percival. *India: A Modern History*. Ann Arbor, 1961.

Tripathi, Amales. *Trade and Finance in the Bengal Presidency, 1793–1833*. Bombay, 1956.

Thompson, Edward, and G. T. Garratt. *Rise and Fulfillment of British Rule in India*. Allahabad, 1958.

Aspinall, Arthur. *Cornwallis in Bengal*. Manchester, 1931.

Baden-Powell, B. H. *Land Systems of British India*. 3 vols. Oxford, 1892.

Chaudhuri, K. N. *The Trading World of Asia and the English East India Company, 1660–1760*. New York, 1978.

XIV—*The New Mughals*

Furber, Holden. *Indian Governor Generalship*. Cambridge, 1953.

Roberts, Paul E. *India Under Wellesley*. London, 1929.

Kopf, David. *British Orientalism and the Bengal Renaissance: The Dynamics of Indian Modernization, 1773–1834.* Berkeley and Los Angeles, 1969.

————. *The Brahmo Samaj and the Shaping of the Modern Indian Mind.* Princeton, 1979.

Embree, Ainslie T. *Charles Grant and British Rule in India.* New York, 1962.

Keith, A. B. *Speeches and Documents on Indian Policy, 1750–1921.* 2 vols. London, 1922.

O'Malley, Lewis S. S. *The Indian Civil Service, 1601–1930.* London, 1931.

Bearce, George D. *British Attitudes Towards India, 1784–1858.* London, 1961.

Ambirajan, S. *Classical Political Economy as British Policy in India.* Cambridge, 1978.

Stokes, Eric. *The English Utilitarians and India.* Oxford, 1959.

Frykenberg, Robert E. *Guntur District, 1788–1848: A History of Local Influence and Central Authority in South India.* Oxford, 1965.

Jones, Kenneth W. *Arya Dharm: Hindu Consciousness in Nineteenth Century Punjab.* Berkeley, 1976.

Pinney, Thomas, ed. *The Letters of Thomas Babington Macaulay.* vol. 3: Jan. 1834–August 1841. Cambridge, 1976.

Narain, V. A. *Jonathan Duncan and Varanasi.* Calcutta, 1959.

Ingram, Edward, ed. *Two Views of British India: The Private Correspondence of Mr. Dundas and Lord Wellesley, 1898–1901.* Bath, 1970.

Wellington, Arthur Wellesley. *A Selection from the Despatches, Memoranda, and Other Papers of the Marquess Wellesley Relating to India.* Edited by S. J. Owen. Oxford, 1877.

Ghosal, A. K. *Civil Service in India Under the East India Company.* Calcutta, 1944.

Bose, N. S. *The Indian Awakening and Bengal.* Calcutta, 1960.

Moorehouse, Geoffrey. *Calcutta.* New York, 1971.

Frykenberg, Robert E., ed. *Land Control and Social Structure in Indian History.* Madison, 1969.

Rammohun Roy. *The English Works of Raja Rammohun Roy.* Edited by K. Nag and D. Burman. Calcutta, 1945–51.

Singh, Iqbal. *Rammohun Roy.* New York, 1958.

Misra, B. B. *The Indian Middle Classes: Their Growth in Modern Times.* London, 1961.

Spear, Percival. *The Nabobs.* London, 1963.

Kincaid, Dennis C. *British Social Life in India, 1608–1937.* London, 1939.

Brown, Hilton. *The Sahibs.* London, 1948.

Busteed, Henry E. *Echoes from Old Calcutta*. London, 1908.

Collet, Sophia D. *The Life and Letters of Raja Rammohun Roy*. 3d ed. Edited by D. K. Biswas and P. C. Ganguli. Calcutta, 1962.

Poddar, Arabinda. *Renaissance in Bengal: Quests and Confrontations, 1800–1860*. Simla, 1970.

Ingham, K. *Reformers in India, 1793–1833*. Cambridge, 1956.

Cotton, H. E. A. *Calcutta; Old and New*. Calcutta, 1907.

Chand, Tara. *History of the Freedom Movement in India*. Vol. 1, *1750–1857*. Delhi, 1961.

Narain, V. A. *Social History of Modern India: Nineteenth Century*. Meerut, 1972.

Majumdar, R. C., ed. *History and Culture of the Indian People*. Pt. 1, vol. 9, *British Paramountcy and Indian Renaissance*. Bombay, 1963.

Kaye, Sir John W. *The Administration of the East India Company*. London, 1853.

————. *Lives of Indian Officers*. 2 vols. London, 1904.

Ballhatchet, Kenneth. *Social Policy and Social Change in Western India*. London, 1957.

Macaulay, Thomas B. *Lord Macaulay's Legislative Minutes*. Edited by C. D. Dharkar. London, 1946.

Galbraith, John S. "The 'Turbulent Frontier' as a Factor in British Expansion." *Comparative Studies in Society and History* 2 (1960): 150–68.

Banerjee, Anil C. *The Eastern Frontier of British India, 1784–1826*. 2d ed. Calcutta, 1946.

Bhanu, Dharma. *History of Administration of the North-Western Provinces, 1803–1858*. Agra, 1957.

Panigrahi, D. N. *Charles Metcalfe in India: Ideas and Administration, 1806–1835*. Delhi, 1968.

Thompson, Edward J. *Life of Charles, Lord Metcalfe*. London, 1937.

Sleeman, William H. *Rambles and Recollections of an Indian Official*. Edited by V. A. Smith. London, 1915.

Boulger, D. C. *Lord William Bentinck*. Oxford, 1897.

Imlah, A. H. *Lord Ellenborough*. London, 1939.

Huttenback, Robert A. *British Relations with Sind, 1799–1834: An Anatomy of Imperialism*. Berkeley, 1962.

Sale, Lady Floretia. *The First Afghan War*. Edited by Patrick MacRory. Harlow, 1969.

Kaye, John W. *History of the War in Afghanistan*. 3 vols. London, 1890.

Malcolm, Sir John. *Memoir of Central India*. 2 vols. London, 1832.

Elphinstone, Mountstuart. *Report on the Peshwa's Territories*. London, 1822.

Singh, Khushwant. *Ranjit Singh*. London, 1962.

XV—*Unification, Modernization, and Revolt*

Hunter, W. W. *The Marquess of Dalhousie*. London, 1905.

Das, M. N. *Studies in the Economic and Social Development of Modern India, 1848–56*. Calcutta, 1959.

Panikkar, K. M. *British Diplomacy in North India: A Study of the Delhi Residency, 1803–1857*. New Delhi, 1968.

Arnold, Edwin. *The Marquess of Dalhousie's Administration of British India*. 2 vols. London, 1862–65.

Edwardes, Michael. *The Necessary Hell*. London, 1958.

Baird, J. A., ed. *Dalhousie's Private Letters*. London, 1910.

Kaye, John W. *A History of the Sepoy War in India, 1857–1858*. 3 vols. 4th ed. London, 1880.

Sen, S. N. *Eighteen Fifty-Seven*. New Delhi, 1957.

Thompson, Edward. *The Other Side of the Medal*. London, 1925.

Khan, Sayyid Ahmad. *The Indian Revolt*. Benares, 1873.

Embree, Ainslie T. *1857 in India: Mutiny or War of Independence—* Boston, 1963.

Stokes, Eric. "Traditional Resistance Movements and Afro-Asian Nationalism: The Context of the 1857 Mutiny Rebellion in India." *Past and Present* 48 (1970): 100–117.

Chaudhuri, S. B. *Civil Rebellion in the Indian Mutinies*. Calcutta, 1957.

Joshi, P. C., ed. *Rebellion: 1857*. Bombay, 1957.

Palmer, J. A. B. *The Mutiny Outbreak at Meerut in 1857*. Cambridge, 1966.

XVI—*Crown Rule—A New Order*

Philips, C. H., ed. *The Evolution of India and Pakistan, 1858 to 1947. Select Documents*. London, 1962.

Dodwell, H. H., ed. *The Indian Empire, 1858–1918*. Delhi, 1964.

Sethi, R. R. *The Last Phase, 1919–1947*. Vol. 6 of *The Cambridge History of India*. Delhi, 1964.

Metcalf, Thomas R. *The Aftermath of Revolt*. Princeton, 1964.

Maclagan, Michael. *Clemency Canning*. London, 1962.

Mason [pseud. Woodruff], Philip. *The Men Who Ruled India*. Vol. 2, *The Guardians*. New York, 1954.

———. *Common Sense About Race Relations*. London, 1961.

———. *Prospero's Magic: Some Thoughts on Class and Race*. London, 1962.

Mannoni, O. *Prospero and Caliban: The Psychology of Colonization*. Translated by P. Powesland. New York, 1964.

Gopal, S. *British Policy in India, 1858–1905*. Cambridge, 1965.

Moore, R. J. *Sir Charles Wood's Indian Policy, 1853–1866.* Manchester, 1966.

Kling, Blair. *The Blue Mutiny.* Philadelphia, 1965.

Singh, S. N. *The Secretary of State and His Council, 1858–1919.* Delhi, 1962.

Pal, Dharma. *Administration of Sir John Lawrence in India, 1864–1869.* Simla, 1952.

Moulton, Edward C. *Lord Northbrook's Indian Administration, 1872–1876.* London, 1968.

Dutt, Romesh C. *The Economic History of India in the Victorian Age.* 7th ed. London, 1956.

Gadgil, D. R. *The Industrial Evolution of India in Recent Times.* London, 1944.

Buchanan, Daniel H. *The Development of Capitalist Enterprise in India.* New York, 1934.

Mehta, S. D. *The Cotton Mills of India, 1854 to 1954.* Bombay, 1954.

Balfour, Elizabeth. *History of Lord Lytton's Indian Administration, 1876–1880.* London, 1899.

Singh, Nagendra. *The Theory of Force and Organisation of Defence in Indian Constitutional History: From Earliest Times to 1947.* New York, 1969.

Hamilton, Lord George. *Parliamentary Reminiscenes and Reflections, 1868–1885.* Vol. 1. London, 1917.

Whitcombe, Elizabeth. *Agrarian Conditions in Northern India.* Vol. 1, *The United Provinces Under British Rule, 1860–1900.* Berkeley and Los Angeles, London, 1972.

Prasad, Bisheshwar. *The Foundations of India's Foreign Policy, 1860–1882.* Bombay, 1955.

Robinson, Francis. *Separatism Among Indian Muslims: The Politics of the United Provinces Muslims, 1860–1923.* Cambridge, 1974.

XVII—*Indian Nationalism—The First Movement*

Sisson, Richard, and Stanley Wolpert, eds. *Congress and Indian Nationalism,* Berkeley and Los Angeles, 1988.

Majumdar, R. C., ed. *History and Culture of the Indian People.* Pt. 2, vol. 10, *British Paramountcy and Indian Renaissance.* Bombay, 1965.

Seal, Anil. *The Emergence of Indian Nationalism.* Cambridge, 1968.

Copal, S. *The Viceroyalty of Lord Ripon, 1880–1884.* London, 1953.

Tinker, Hugh. *The Foundations of Local Self-Government in India, Pakistan and Burma.* London, 1954.

Wolpert, Stanley A. *Tilak and Gokhale.* Berkeley and Los Angeles, 1962.

Nanda, B. R. *Gokhale: The Indian Moderates and the British Raj.* Princeton, 1977.

Martin, Briton. *New India, 1885.* Berkeley and Los Angeles, 1969.

Banerjea, Surendra Nath. *A Nation in the Making.* London, 1925.

McCully, Bruce T. *English Education and the Origins of Indian Nationalism.* New York, 1940.

Sitaramayya, B. Pattabhi. *The History of the Indian National Congress.* 2 vols. Madras, 1935 and 1947.

Argov, Daniel. *Moderates and Extremists in the Indian Nationalist Movement, 1883–1920.* Bombay, 1967.

Rai, Lajpat. *Young India.* New York, 1916.

Naoroji, Dadabhai. *Poverty and Un-British Rule in India.* London, 1901.

Ganguli, B. N. *Dadabhai Naoroji and the Drain Theory.* Bombay, 1965.

Besant, Annie. *How India Wrought for Freedom.* Madras, 1915.

Karve, D. G. *Ranade: The Prophet of Liberated India.* Poona, 1942.

Parvate, T. V. *Mahadev Govind Ranade.* New York, 1963.

Mody, Homi. *Sir Pherozeshah Mehta.* New York, 1963.

Low, D. A., ed. *Soundings in Modern South Asian History.* Berkeley and Los Angeles, 1968.

Heimsath, C. H. *Indian Nationalism and Hindu Social Reform.* Princeton, 1964.

Bayly, C. A. *The Local Roots of Indian Politics: Allahabad, 1880–1920.* New York, 1975.

Rothermund, Dietmar. *The Phases of Indian Nationalism and Other Essays.* Bombay, 1970.

Mehrotra, S. R. *The Emergence of the Indian National Congress.* Delhi, 1971.

Hunter, W. W. *The Indian Musalmans.* London, 1871.

Hussain, M. Hadi. *Syed Ahmed Khan: Pioneer of Muslim Resurgence.* Lahore, 1970.

Smith, W. C. *Modern Islam in India.* London, 1946.

Khan, S. A. *An Account of the Loyal Mahomedans of India.* Meerut, 1860.

Troll, Christian W. *Sayyid Ahmad Khan.* New Delhi, 1978.

Malik, Hafeez. *Sir Sayyid Ahmad Khan as Muslim Modernization in India at Pakistan.* New York, 1980.

McLane, John R. *Indian Nationalism and the Early Congress.* Princeton, 1977.

Lelyveld, David. *Aligarh's First Generation: Muslim Solidarity in British India.* Princeton, 1978.

XVIII—*The Machine Solidifies*

Kaminsky, Arnold P. *The India Office, 1880–1910.* Westport, 1986.

Lewis, Martin D., ed. *The British in India: Imperialism or Trusteeship?* Boston, 1962.

Hamer, W. S. *The British Army: Civil Military Relations, 1885–1915.* Oxford, 1971.

Houghton, Walter E. *The Victorian Frame of Mind, 1830–1870.* New Haven, 1957.

Forster, E. M. *A Passage to India.* Harmondsworth, 1957.

Bolt, Christine. *Victorian Attitudes to Race.* London, 1971.

Hutchins, Francis G. *The Illusion of Permanence.* Princeton, 1967.

Parry, Benita. *Delusions and Discoveries: Studies on India in the British Imagination, 1880–1930.* Berkeley and Los Angeles, 1972.

Cornell, Louis L. *Kipling in India.* London, 1966.

Greenberger, Allen J. *The British Image of India.* London, 1969.

Curzon, Lord. *Speeches by Lord Curzon of Kedleston, 1898–1905.* 4 vols. Calcutta, 1900–06.

Dilks, David. *Curzon in India.* 2 vols. London, 1969–70.

Ronaldshay, Earl of. *The Life of Lord Curzon.* Vol. 2. London, 1928.

Lansdowne, Marquis of. *Speeches by the Marquis of Lansdowne, 1888–1894.* 2 vols. Calcutta, 1894.

Elgin, Earl of. *Speeches by the Earl of Elgin, 1894–1899.* Calcutta, 1899.

Reynolds, Reginald. *White Sahibs in India.* New York, 1973.

Griffiths, Percival J. *The British Impact on India.* London, 1952.

Gopal, Ram. *British Rule in India: An Assessment.* New York, 1963.

Lawrence, Sir Walter R. *The India We Served.* London, 1928.

Edwardes, Michael. *Bound to Exile: The Victorians in India.* London, 1969.

Rose, Kenneth. *Superior Person: A Portrait of Curzon and His Circle in Late Victorian England.* London, 1969.

XIX—*Revolt, Repression and Reform*

Broomfield, J. H. *Elite Conflict in a Plural Society.* Berkeley and Los Angeles, 1968.

Majumdar, R. C. *History of the Freedom Movement in India.* Vol. 2. Calcutta, 1963.

Darling, Sir Malcolm. *Apprentice to Power: India, 1904–1908.* London, 1966.

Tandon, Prakash. *Punjabi Century, 1857–1947.* Berkeley and Los Angeles, 1968.

Morley, John Viscount. *Recollections.* Vol. 2. New York, 1917.

Wolpert, Stanley. *Morley and India, 1906–1910.* Berkeley and Los Angeles, 1967.

Mary Minto. *India, Minto and Morley, 1905–1910.* London, 1934.

Das, M. N. *India Under Morley and Minto.* London, 1964.

Wasti, Syed Razi. *Lord Minto and the Indian Nationalist Movement, 1905–1910.* Oxford, 1964.

Chaudhuri, Nirad C. *Autobiography of an Unknown Indian.* New York, 1951.

Joshi, V. C., ed. *Sri Aurobindo: An Interpretation.* Delhi, 1973.

Chose, S. K. *The Poetry of Sri Aurobindo: A Short Survey.* Calcutta, 1969.

Pal, Bipin Chandra. *Writings & Speeches.* Vol. 1. Calcutta, 1958.

————. *The Soul of India.* 4th ed. Calcutta, 1958.

Hay, Stephen. *Asian Ideas of East and West: Tagore and His Critics in Japan, China, and India.* Cambridge, 1970.

Magnus, Philip. *Kitchener: Portrait of an Imperialist.* London, 1958.

Gilbert, Martin. *Servant of India: A Study of Imperial Rule from 1905 to 1910 as Told Through the Correspondence and Diaries of Sir James Dunlop Smith.* London, 1966.

Barrier, N. Gerald. *Banned: Controversial Literature and Political Control in British India, 1907–1947.* New York, 1974.

Leonard, Karen I. *Social History of an Indian Caste.* Berkeley, 1978.

Conlon, Frank F. *A Caste in a Changing World.* Berkeley, 1977.

XX–*The Impact of World War One*

Ellinwood, DeWitt C., and C. D. Pradhan, eds. *India and World War I.* Delhi, 1978.

Brown, Judith M. *Gandhi's Rise to Power: Indian Politics, 1915–1922.* Cambridge, 1972.

————. *Gandhi and Civil Disobedience.* London, 1977.

————. "War and the Colonial Relationship: Britain, India and the War of 1914–18." In *War and Society.* Edited by M. R. D. Foot. London, 1973.

Montagu, Edwin S. *An Indian Diary.* Edited by Venetia Montagu. London, 1930.

Hardinge, Lord of Penshurst. *My Indian Years, 1910–1916.* London, 1948.

Lucas, Sir C. *The Empire At War.* Vol. 5. London, 1920.

India's Contribution to the Great War. Calcutta, 1923.

Barker, A. J. *The Bastard War: The Mesopotamian Campaign of 1914–1918.* New York, 1967.

Busch, Briton Cooper. *Britain, India, and the Arabs, 1914–1921.* Berkeley and Los Angeles, 1971.

Tinker, Hugh. *A New System of Slavery: The Export of Indian Labour Overseas, 1830–1920.* London, 1974.

Hardy, Peter. *The Muslims of British India.* Cambridge, 1972.

Bolitho, Hector. *Jinnah: Creator of Pakistan.* London, 1954.

Gandhi, Mohandas K. *An Autobiography: The Story of My Experiments with Truth.* Boston, 1957.

Erikson, Erik H. *Gandhi's Truth: On the Origins of Militant Nonviolence.* New York, 1969.

Robbs, P. G. *The Government of India and Reform.* London, 1976.

Moore, R. J. *The Crisis of Indian Unity, 1917–1940.* Oxford, 1974.

Mehta, Ved. *Mahatma Gandhi and His Apostles.* New York, 1977.

India and the War. London, 1915.

Prasad, Rajendra. *Satyagraha in Champaran.* 2d rev. ed. Ahmedabad, 1949.

Deol, Gurdev Singh. *The Role of the Ghadar Party in the National Movement.* Delihi, 1969.

Chirol, Valentine. *India.* London, 1926.

Brown, Giles T. "The Hindu Conspiracy, 1914–1917." *Pacific Historical Review* 17 (1948): 299–310.

Dharmavira, ed. *Letters of Lala Har Dayal.* Ambala Cantt., 1970.

———. *Lala Har Dayal and Revolutionary Movements of His Times.* New Delhi, 1970.

Fein, Helen. *Imperial Crime and Punishment: The Massacre at Jallianwala Bagh and British Judgment 1919–1920.* Honolulu, 1977.

Furneaux, Rupert. *Massacre at Amritsar.* London, 1963.

Wolpert, Stanley. *An Error of Judgment.* Boston, 1970. Published as *Massacre at Jallianwala Bagh.* New Delhi, 1989.

XXI—*Toward Independence*

Gandhi, Mahatma. *Young India, 1919–1922.* New York, 1924.

Gandhi, Mohandas K. *Hind Swaraj [Indian Home Rule].* rev. ed. Ahmedabad, 1964.

———. *Satyagraha in South Africa.* Translated by V. G. Desai. 2d rev. ed. Ahmedabad, 1950.

Tendulkar, D. G. *Mahatma: Life of Mohandas Karamchand Gandhi.* 8 vols. Delhi, 1960–63.

Bondurant, Joan. *The Conquest of Violence: The Gandhian Philosophy of Conflict.* rev. ed. Berkeley and Los Angeles, 1969.

Gandhi, M. K. *The Collected Works of Mahatma Gandhi.* 50 vols. to date. Delhi, 1958–.

Iyer, Raghavan, *The Moral and Political Thought of Mahatma Gandhi.* New York, 1973.

Rudolph, Lloyd I., and Sysanne H. *The Modernity of Tradition: Political Development in India.* Chicago and London, 1967.

Curtis, Lionel. *Dyarchy.* London, 1920.

Morton, Eleanor. *The Women in Gandhi's Life.* New York, 1953.

Black, Jo Anne, et al. *Gandhi; the Man.* San Francisco, 1972.

Nehru, Jawaharlal. *Toward Freedom.* New York, 1941.

Brecher, Michael. *Nehru: A Political Biography.* London, 1959.

Dutt, R. P. *India, Today and Tomorrow.* Delhi, 1955.

Saiyid, M. H. *Mohammad Ali Jinnah*. Lahore, 1945.

Bose, Subhas Chandra. *The Indian Struggle*. Calcutta, 1952.

Parikh, N. D. *Sardar Vallabhbhai Pael*. 2 vols. Ahmedabad, 1953–56.

Gopal, S. *The Vice-Royalty of Lord Irwin, 1926–31*. Oxford, 1957.

Pandey, B. N. *Nehru*. London, 1976.

Nanda, B. R. *The Nehrus: Motilal and Jawaharlal*. New York, 1962.

Butler, Iris. *The Viceroy's Wife: Letters of Alice, Countess of Reading, from India, 1921–25*. London, 1969.

Ramusack, Barbara N. *The Princes of India in the Twilight of Empire*. Columbus, 1978.

Gopal, S. *Jawaharlal Nehru: A Biography*. 2 vols. London, 1975–79.

Pandey, Gyanendra. *The Ascendancy of the Congress in Uttar Pradesh, 1926–34: A Study in Imperfect Mobilization*. New York, 1978.

Low, D. A., ed. *Congress and the Raj: Facets of the Indian Struggle, 1917–47*. Columbia, Mo., 1977.

Nehru, Jawaharlal. *The Discovery of India*. New York, 1946.

Das Gupta, H. N. *Deshbandhu Chittaranjan Das*. 2d ed. New Delhi, 1969.

Wakefield, Sir Edward. *Past Imperative: My Life in India, 1927–1947*. London, 1966.

Das, Durga. *India from Curzon to Nehru and After*. New York, 1970.

Jones, Kenneth W. "Communalism in the Punjab: The Arya Samaj Contribution." In *Modern India*. Edited by T. R. Metcalf. London, 1971.

Krishna, Gopal. "The Development of the Indian National Congress as a Mass Organization." Ibid.

Gwyer, M., and A. Appodorai, eds. *Speeches and Documents on the Indian Constitution, 1921–1947*. 2 vols. London, 1957.

Coupland, Reginald. *The Indian Problem*. 3 vols. New York, 1944.

Majumdar, R. C. *History of the Freedom Movement in India*. Vol. 3. Calcutta, 1962.

Moon, Penderel. *Strangers in India*. New York, 1945.

Brailsford, Henry N. *Subject India*. New York, 1943.

Mehrotra, S. R. "The Congress and the Partition of India." In *The Partition of India: Policies and Perspectives, 1935–1947*. Edited by C. H. Philips and Mary D. Wainwright. London, 1970.

Zaidi, Z. H. "Aspects of the Development of Muslim League Policy, 1937–47." Ibid.

Zinkin, Maurice and Taya. "Impressions, 1938–47." Ibid.

Malik, Hafeez, ed. *Iqbal: Poet-Philosopher of Pakistan*. New York and London, 1971.

Singh, Iqbal. *The Ardent Pilgrim: An Introduction to the Life and Work of Mohammed Iqbal*. London, 1951.

Sayeed, Khalid B. *Pakistan: The Formative Phase, 1857–1948*. 2d ed. London, 1968.

Aziz, K. K. *Britain and Muslim India*. London, 1963.
Gopal, Ram. *Indian Muslims*. London, 1959.
Ambedkar, B. R. *What Congress and Gandhi Have Done to the Untouchables*. Bombay, 1945.
Keer, Dhanahjay. *Dr. Ambedkar: Life and Mission*. Bombay, 1954.

XXII—*The Impact of World War Two*

Wolpert, Stanley. *Jinnah of Pakistan*. New York, Oxford, 1984.
Tinker, Hugh. *India and Pakistan: A Political Analysis*. New York, 1962.
———. *Experiment with Freedom: India and Pakistan, 1947*. London, 1967.
Pandey, B. N. *The Break-up of British India*. New York, 1969.
Azad, A. K. *India Wins Freedom*. Bombay, 1959.
Ahmad, Jamil ud-Din. *Speeches and Writings of Mr. Jinnah*. 2 vols. Lahore, 1960 and 1964.
Allana, G. *Quaid-e-Azam Jinnah*. Karachi, 1967.
Brown, W. Norman. *The United States and India, Pakistan, Bangladesh*. Cambridge, 1972.
Bourke-White, Margaret. *Halfway to Freedom*. New York, 1949.
Coupland, R. *The Cripps Mission*. London, 1943.
———. *Indian Politics*. London, 1943.
Khaliquzzaman, Choudhry. *Pathway to Pakistan*. Lahore, 1961.
Venkataramani, M. S. *Bengal Famine of 1943*. Delhi, 1973.
Moon, Penderel, ed. *Wavell: The Viceroy's Journal*. Oxford, 1973.
———. *Divide and Quit*. Berkeley, 1962.
Panikkar, K. M. *An Autobiography of Madras*. Oxford, 1977.
Amery, L. S. *The Transfer of Power in India*. London, 1971.
Edwardes, Michael. *The Last Years of British India*. London, 1963.
Mosley, Leonard. *The Last Days of the British Raj*. London, 1962.
Menon, V. P. *The Transfer of Power in India*. Princeton, 1957.
Campbell-Johnson, Alan. *Mission With Mountbatten*. New York, 1953.
Mountbatten, Lord. *Time Only to Look Forward*. London, 1949.
Lumby, E. W. R. *The Transfer of Power in India*. London, 1954.
Tuker, Francis I. S. *While Memory Serves*. London, 1950.
Rajput, A. B. *The Cabinet Mission*. Lahore, 1946.
Prasad, Rajendra. *India Divided*. Bombay, 1946.
Lohia, Rammanohar. *Guilty Men of India's Partition*. Allahabad, 1960.
Vakil, C. N. *Economic Consequences of Divided India*. Bombay, 1950.
Wallbank, T. W., ed. *The Partition of India*. Boston, 1966.
Pyarelal. *Mahatma Gandhi: The Last Phase*. Vol. 1. Ahmedabad, 1956.
Singh, Khushwant. *Mano Majra*. New York, 1956.
Mansergh, N., and E. W. R. Lumby, eds. *The Transfer of Power, 1942-7*. Vol. 4. London, 1973.

Glendevon, John. *The Viceroy at Bay: Lord Linlithgow in India, 1939–1943.* London, 1971.

Collins, Larry and Dominique La Pierre. *Freedom At Midnight.* New York, 1975.

XXIII—*The Nehru Era*

Wolpert, Stanley. *Nehru: A Tryst With Destiny.* New York, 1996.

Sar Desai, D. R., and Anand Mohan, eds. *The Legacy of Nehru.* New Delhi, 1992.

Nehru, Jawaharlal. *Jawaharlal Nehru's Speeches, 1946–64.* 5 vols. Delhi, 1949–68.

————. *A Bunch of Old Letters.* rev. ed. Bombay, 1960.

————. *The Unity of India: Collected Writings, 1937–1940.* Edited by V. K. Krishna Menon. New York, 1948.

Norman, Dorothy, ed. *Nehru: The First Sixty Years.* 2 vols. London, 1965.

Crocker, Walter. *Nehru: A Contemporary Estimate.* London, 1966.

Das, M. N. *The Political Philosophy of Jawaharlal Nehru.* New York, 1961.

Palmer, Norman D. *The Indian Political System.* 2d ed. Boston, 1971.

Poplai, S. L., ed. *India, 1947–50.* 2 vols. Bombay, 1959.

Menon, V. P. *The Story of the Integration of the Indian States.* New York, 1956.

Misra, B. B. *The Indian Political Parties.* Delhi, 1976.

Frankel, Francine R. *India's Political Economy, 1947–1977: The Gradual Revolution.* Princeton, 1978.

Chandrasekhar, S. *Infant Mortality, Population Growth and Family Planning in India.* Chapel Hill, 1972.

Brecher, Michael. *The Struggle for Kashmir.* New York, 1953.

Korbel, Josef. *Danger in Kashmir.* Princeton, 1954.

Bazar, Prem Nath. *Kashmir in Crucible.* New Delhi, 1967.

Gupta, Sisir. *Kashmir.* New Delhi, 1966.

Datta, Kalikinkar. *Rajendra Prasad.* New Delhi, 1970.

Smith, Donald E. *India as a Secular State.* Princeton, 1963.

Srinivas, M. N. *Caste in Modern India and Other Essays.* Bombay, 1962.

Srinivas, M. N. et al., eds. *Dimensions of Social Change in India.* New Delhi, 1978.

Lengyel, Emil. *Krishna Menon.* New York, 1962.

Morris-Jones, W. H. *Parliament in India.* Philadelphia, 1957.

————. *The Government and Politics of India.* Garden City, N.Y., 1964.

Rau, B. N. *India's Constitution in the Making.* Edited by B. Shiva Rao. Bombay, 1960.

Park, Richard L., and Irene Tinker, eds. *Leadership and Political Institutions in India.* Princeton, 1959.

Nair, Kusum. *Blossoms in the Dust.* New York, 1962.

Chopra, P. N., ed. *The Collected Works of Sardar Vallabhbhai Patel.* Vol. I
    (1918–1925); Vol. II (1926–1929). Delhi, 1990–91.
——. *Maulana Abul Kalam Azad. Unfulfilled Dreams.* New Delhi, 1990.
Rosen, George. *Democracy and Economic Change in India.* 2d ed.
    Berkeley and Los Angeles, 1967.
Weiner, Myron. *The Politics of Scarcity.* Chicago, 1962.
Lewis, John P. *Quiet Crisis in India.* Garden City, N.Y., 1964.
Harrison, Selig S. *India: The Most Dangerous Decades.* Princeton, 1960.
Hanson, A. H. *The Process of Planning: A Study of India's Five Year
    Plans, 1950–1964.* London, 1966.
Narayana Rao, K. V. *The Emergence of Andhra Pradesh.* Bombay, 1973.
Narain, Iqbal, ed. *State Politics in India.* Meerut, 1968.
Aiyar, S. P., and R. Srinivasan, eds. *Studies in Indian Democracy.* Bom-
    bay, 1965.
Lamb, Alastair. *The China-India Border.* London, 1964.
Maxwell, Neville. *India's China War.* London, 1970.
Zinkin, Taya. *India Changes.* London, 1958.
Moraes, Frank. *India Today.* New York, 1960.
Lamb, Beatrice P. *India: A World in Transition.* New York, 1963.
Davis, Kingsley. *The Population of India and Pakistan.* Princeton, 1951.
Bowles, Chester. *Ambassador's Report.* New York, 1954.
——. *A View From New Delhi: Selected Speeches and Writings.* Bom-
    bay, 1969.
Bondurant, Joan V. *Regionalism Versus Provincialism: A Study in Prob-
    lems of Indian National Unity.* Berkeley and Los Angeles, 1958.
Government of India. *Report of the States Reorganisation Committee.*
    Delhi, 1955.
——. *Report of the Official Language Commission.* Delhi, 1957.
——. *The First Five Year Plan.* Delhi, 1953.
——. *The Second Five Year Plan.* Delhi, 1956.
——. *Third Five Year Plan: A Draft Outline.* Delhi, 1960.
Turner, Roy, ed. *India's Urban Future.* Berkeley and Los Angeles, 1962.
Tinberg, Thomas A. *The Marwaris: From Traders to Industrialists.* New
    Delhi, 1978.
Mathai, M. O. *Reminiscences of the Nehru Age.* New Delhi, 1978.
Myrdal, Jan. *India Waits.* Chicago, 1986.

XXIV—*From Collective Leadership to Indira Raj (1964–1977)*

Wolpert, Stanley. *Zulfi Bhutto of Pakistan.* New York, 1993.
Naipaul, V. S. *India: A Wounded Civilization.* New York, 1977.
Rao, V. K. R. V. *The Nehru Legacy.* Bombay, 1971.
Hagen, Welles. *After Nehru, Who?* New York, 1963.
Tennyson, Hallam. *India's Walking Saint: The Story of Vinoba Bhave.*
    Garden City, N.Y., 1955.

Narayan, Shriman. *Vinoba: His Life and Work*. Bombay, 1970.

Narayan, Jayaprakash. *A Plea for the Reconstruction of Indian Polity*. Kashi, 1959.

Béteille, André. *Caste, Class, and Power*. Berkeley and Los Angeles, 1969.

Vasudev, Uma. *Two Faces of Indira Gandhi*. Delhi, 1977.

Advani, Lal K. *A Prisoner's Scrap-Book*. New Delhi, 1978.

Malendaum, Wilfred. *Prospects for Indian Development*. London, 1961.

Mathur, M. V., and Iqbal Narain, eds. *Panchayati Raj, Planning and Democracy*. New York, 1969.

Maddick, Henry. *Panchayati Raj: A Study of Rural Local Government in India*. London, 1969.

Grover, V. P. *Panchayati Raj Administration in Rajasthan*. Agra, 1973.

Thorner, Daniel and Alice. *Land and Labour in India*. Bombay, 1962.

Wiser, William and Charlotte. *Behind Mud Walls, 1930–1960*. Berkeley, 1963.

Chandidas, R. et al., eds. *India Votes: A Source Book on Indian Elections*. New York, 1968.

Kothari, Rajni, ed. *Party Systems and Election Studies*. Bombay, 1967.

Speeches of Lal Bahadur Shastri. Delhi, 1965.

Overstreet, Gene D., and Marshall Windmiller. *Communism in India*. Berkeley and Los Angeles, 1959.

Erdman, Howard L. *The Swatantra Party and Indian Conservatism*. Cambridge, 1967.

Baxter, Craig. *The Jana Sangh: A Biography of an Indian Political Party*. Philadelphia, 1969.

Burger, Angela S. *Opposition in a Dominant Party System*. Berkeley, 1969.

Ram, Mohan. *Indian Communism: Split Within a Split*. Delhi, 1969.

Varma, S. P., and Iqbal Narain, eds. *Fourth General Elections in India*. Bombay, 1968.

Brecher, Michael. *India in World Politics*. New York, 1968.

Bandyopadhyaya, J. *The Making of India's Foreign Policy*. Bombay, 1970.

Brines, Russell. *The Indo-Pakistani Conflict*. London, 1968.

Gupta, Karunakar. *India in World Politics: A Period of Transition*. Calcutta, 1969.

Heimsath, Charles H., and Surjit Mansingh. *A Diplomatic History of Modern India*. Bombay, 1970.

Rao, P. V. R. *Defence Without Drift*. Bombay, 1970.

Ahluwalia, B. K., ed. *Zakir Husain: A Study*. New Delhi, 1970.

Ayer, Subbier A. *The Lone Sentinel: Morarji Desai*. Bombay, 1960.

Narayan, Jayaprakash. *Prison Diary, 1975*, ed. A. B. Shah. Bombay, 1977.

Hart, Henry, ed. *Indira Gandhi's India: A Political System Reappraised*. Boulder, 1976.

Ziring, Lawrence, Ralph Braibanti, W. H. Wriggins, eds., *Pakistan: The Long View*. Durham, 1977.

Masani, Zareer. *Indira Gandhi: A Biography*. London, 1975.

Khan, Ayub. *Friends Not Masters: A Political Autobiography*. New York, 1967.

Galbraith, John Kenneth. *Ambassador's Journal*. Boston, 1969.

Vasudevan, A. *The Strategy of Planning in India*. Meerut, 1970.

Sahgal, Nayantara. *Indira Gandhi's Emergence and Style*. Durham, 1979.

Mehta, Ved. *The New India*. New York, 1978.

Alexander, M. K. *Madame Gandhi: A Political Biography*. North Quincy, Mass., 1969.

Mohan, Anand. *Indira Gandhi: A Personal and Political Biography*. New York, 1967.

Sardahi, Ajit Singh. *Punjabi Suba: The Story of the Struggle*. Delhi, 1970.

Loshak, David. *Pakistan Crisis*. New York, 1971.

Kashyap, S. C., ed. *Bangladesh—Background and Perspectives*. New Delhi, 1971.

Chowdhury, Subrata Roy. *The Genesis of Bangladesh*. New York, 1972.

Kamal, K. A. *Sheikh Mujibur Rahman: Man and Politician*. Dacca, 1970.

Karim, Abdul. *Social History of the Muslims in Bengal*. Dacca, 1959.

Ali, Tariq. *Pakistan: Military Rule or People's Power*. New York, 1970.

Bhutto, Z. A. *The Myth of Independence*. London, 1969.

Brown, Lester R. *Seeds of Change: The Green Revolution and Development in the 1970's*. London, 1970.

Fox, Richard G. *Urban India: Society, Space and Image*. Durham, 1971.

Crane, Robert I., ed. *Transition in South Asia*. Durham, 1970.

Dandekar, K., and V. Bhate. *Prospects of Population Control*. New Delhi, 1971.

Patwardhan, Achyut. *Ideologies and the Perspective of Social Change in India*. Bombay, 1971.

Pathak, G. S. *Some Reflections on the Civil Revolution in India*. Delhi, 1972.

Sinha, Durganand. *The Mughal Syndrome: A Psychological Study of Inter-Generational Differences*. Bombay, 1972.

Jha, L. K. *Economic Development: Ends and Means*. Bombay, 1973.

Ghouse, Mohammad. *Secularism, Society, and Law in India*. Delhi, 1973.

Desai, A. R., and S. D. Pillai. *A Profile of an Indian Slum*. Bombay, 1972.

Singh, Nagendra. *India and International Law*. New Delhi, 1973.

India's Foreign Policy and Indo-Soviety Treaty. New Delhi, 1971.

The Fourth Plan Mid-term Appraisal: A Summary. New Delhi, 1972.

Mohan, Jag, ed. *Twenty-Five Years of Indian Independence*. Delhi, 1973.

Free India Forges Ahead: A Collection of Essays Depicting the Progress of India Since Independence. Secunderabad, 1972.

Subramaniam, C. *India of My Dreams*. New Delhi, 1972.

Costa, Benedict. *India's Socialist Princes and Garibi Hatao*. Ludhiana, 1973.

Somjee, A. H. *Democracy and Political Change in Village India: A Case Study*. New Delhi, 1971.

India: The Speeches and Reminiscences of Indira Gandhi. London, 1975.

XXV—*From Janata Raj to Rajiv's Death*

Akbar, M. J. *India: The Siege Within*. Harmondsworth, 1985.

Desai, Morarji. *The Story of My Life*, 2 vols. Delhi, 1974.

Bhargava, G. S. *Morarji Desai: Prime Minister of India*. Delhi, 1977.

Nargolkar, Vasant. *JP's Crusade for Revolution*. New Delhi, 1975.

Mankekar, D. R., and Kamla Mankekar. *Decline and Fall of Indira Gandhi*. New Delhi, 1977.

Prasad, Bimal, ed. *India's Foreign Policy*. New Delhi, 1979.

Freeman, James M. *Untouchable: An Indian Life History*. Stanford, 1979.

Weiner, Myron. *Sons of the Soil: Migration and Ethnic Conflict in India*. Princeton, 1979.

Ghose, S. K. *The Crusade and End of Indira Raj*. New Delhi, 1977.

Fishlock, Thevor. *Indira File*. London, 1983.

Advani, Lal K. *A Prisoner's Scrap-Book*. New Delhi, 1978.

XXVI—*India Today*

Naipaul, V. S. *India: A Million Mutinies Now*. New York, 1990.

Bouton, Marshall M., ed. *India Briefing, 1987–90*. Boulder, 1987–90.

Thapar, Romesh. *These Troubled Times . . .* Bombay, 1986.

Gupte, Pranay. *Vengeance: India After the Assassination of Indira Gandhi*. New York, 1985.

Tully, Mark, and Satish Jacob. *Amritsar, Mrs. Gandhi's Last Battle*. London, 1985.

Singh, Patwant, and Harji Malik. *Punjab: The Fatal Miscalculation*. New Delhi, 1985.

Kapur, Rajiv A. *Sikh Separatism: The Politics of Faith*. London, 1986.

Kalia, Ravi. *Chandigarh: In Search of an Identity*. Carbondale, 1987.

Mukhoty, Gobinda, and Rajni Kothari. *Who Are the Guilty?* New Delhi, 1984.

Lapierre, Dominique. *The City of Joy*. Garden City, 1985.

Souza, Alfred de, ed. *The Indian City: Poverty, Ecology and Urban Development*. New Delhi, 1978.

Johnson, B. L. C. *Development in South Asia*. Harmondsworth, 1983.

Nayar, Kuldip, and Khushwant Singh. *Tragedy of Punjab*. New Delhi, 1984.

Roach, James R. ed. *India 2000: The Next Fifteen Years*. Riverdale, 1986.

Hawley, John S., ed. *Sati, The Blessing and the Curse: The Burning of Wives in India*. New York, 1994.

Neale, C. *Developing Rural India: Policies, Politics, and Progress*. Riverdale, 1988.

Chopra, P. N. *Ocean of Wisdom: The Life of Dalai Lama XIV*. New Delhi, 1986.

Bhutto, Benazir. *Daughter of the East*. London, 1988.

Hardgrave, Robert L., and Stanley A. Kochanek. *India: Government and Politics in a Developing Nation*. 5th ed. San Diego, 1992.

Bouton, Marshall M., ed. *India Briefing, 1991*. Boulder, 1991.

Naipaul, V. S. *India: A Million Mutinies Now*. London, 1990.

Gupte, Pranay. *Mother India: A Political Biography of Indira Gandhi*. New York, 1992.

Thapar, Raj. *All These Years*. New Delhi, 1991.

Wolpert, Stanley. *India*. Berkeley and Los Angeles, 1992.

GLOSSARY

Agni	Vedic Aryan god of fire
Ahimsa	Nonviolence
Alim	Muslim scholar
Amir	Muslim prince or commander
Aparigraha	Hindu vow of poverty
Arthashastra	"Science of Material Gain"
Aryan	Indo-European Language Group of tribals who early invaded and conquered India
Arya-varta	Land of the Aryans
Asat	Unreal or untrue
Ashrama	Stage of Hindu life, or rural retreat
Ashva-medha	Vedic horse sacrifice
Atman	Vedic term originally for breath, later self, then soul
Avatara	Earthly emanation of Vishnu
Avidya	Ignorance
Azad	Free
Bagh	Garden
Bande Mataram	"Hail to Thee, Mother"; the first Indian national anthem
Bangladesh	Land of Bengal
Bania	A Hindu merchant caste, originally from Gujarat
Bhadralok	"Gentle people"; the intellectuals of Bengal
Bhagavad Gita	"Song of the Blessed One"
Bhakti	Hindu devotion
Bharat	India; possibly the name of the first great Vedic tribe or its chief
Bharatiya Janata	Indian People
Bhoodan	Gift of land
Brahmacharya	Celibate studenthood; the first stage of a twice-born (upper-class) Hindu's life
Brahman	Any sacred utterance; hence, he who controls such utterances, the class of Hindu priests called brahmans, first in the traditional four-fold hierarchy of Hindu classes, and finally the monistic principle of transcendental divinity so-called in Upanishadic philosophic scripture

Bodhisattva	He who has the essence of Buddhahood within; the savior of Mahayana Buddhism
Buddha	Enlightened One
Caliph	(also Khalifa) Deputy of god; ruler over the world community of orthodox Muslims
Chaitya	Sacred spot; a shrine of worship for Buddhists or Jains, and some Hindus
Chakra	Wheel
Chakravartin	Hindu universal emperor, for whom the wheel of the law will turn
Chatrapati	Ancient Hindu monarch over the "four quarters" of the universe
Dalit	ex-Untouchable
Dasa	Slave, initially pre-Aryan; lit. "dark-skinned one"
Deva	Aryan Vedic god, "shining one"
Deva-dasi	Slave of the god; used to describe Hindu temple dancer-prostitutes
Deva-nagari	Script in which Sanskrit and Hindi language are written; lit., "city of the gods"
Devi	The mother goddess
Dharma	Hindu religion, law, duty, responsibility, truth, etc.
Dharmashastra	A text (or code) of Hindu law
Dhimmis	Protected peoples of the book (non-Muslims, under Muslim rule)
Diwan	Muslim imperial (or royal) revenue collector; originally meant "court"
Diwani	Revenue-collection powers over a province granted by a Muslim emperor
Doab	Lit., "two rivers"; the strip of fertile land between two rivers
Dravidian	The family of languages spoken in South India; also generally used for people who speak those languages (Tamil, Telugu, Malayalam, and Kanarese)
Dukkha	Suffering or sorrow; the first of the Buddha's Four Noble Truths
Durga	The mother goddess
Feringi	Foreigner
Garibi	Poverty
Ghats	"Steps," either those leading from a Hindu temple to the water in which worshippers must bathe themselves, or the range of mountains in West India
Grama	Village
Guru	Teacher; Hindu preceptor
Hara	Shiva, one of the two great Hindu gods
Hari	Vishnu, the other great Hindu god

Harijan	Child of god (Vishnu); Mahatma Gandhi's name for untouchables
Hinayana	"Lesser vehicle"; a form of Buddhism
Indra	The Vedic war god
Jagir	A tax-free grant of land
Jagirdar	The Holder of a *jagir*
Jai	Victory
Jain	One who is a follower of Jainism
Jainism	A religion similar to Buddhism and Hinduism
Jajman	Patron
Jajmani system	The patronage system of barter on which most traditional Indian village economies were based
Jana	Vedic Aryan tribe
Janata Dal	People's Party
Jati	Lit., "birth"; hence kin group or "caste"
Jihad	Muslim holy war
Jina	The Conqueror; one title of the founder of the Jain religion
Jiva	Soul
Jizya	A head tax imposed on *dhimmis* by Muslim rulers
Kaffir	Infidel
Kali	The mother goddess
Kama	Love
Karma	Action or deed
Karma Yoga	The discipline of action; a method for attaining salvation expounded in the *Bhagavad Gita*
Khadi	Hand-spun and woven cotton cloth
Khalifa	See Caliph
"Khalistan"	Sikh "Land of the Pure"
Khalsa	Pure; the Sikh Army of the Pure
Krishna	Lit., "black"; a Hindu god, an *avatara* of Vishnu
Kshatriya	Hindu warrior; the second class in the Hindu hierarchy
Lok	People
Lok Sabha	House of the People; the lower house of the Republic of India's Parliament
Madhya Pradesh	Middle Province
Mahabharata	"Great *Bharata*"; the longer Aryan Hindu epic
Mahanadi	Great river
Maharaja	Great king
Maharashtra	Great country; the name of a state in western India
Mahatma	Great-Souled One; the honorific by which M. K. Gandhi was known
Mahavira	Great hero; one of the titles of the founder of Jainism
Mahayana	"Great vehicle"; the later form of Buddhist religion
Maheshvara	Great god; a term usually applied to Shiva
Malik	Muslim noble

Mandala	Circle
Mansab	Mughal bureaucratic office
Mansabdar	Holder of a *mansab*
Maya	Illusion
Mofussil	Rural hinterland
Moherjo-daro	"Mound of the Dead"; an ancient Indus city
Moksha	Lit., "release"; the goal of salvation to which orthodox Hindus aspire
Mulla	Muslim teacher or sage
Nabob	British corruption of *nawab*
Nagar	City
Nama Rupa	Name and form
Nataraja	"King of the Dance"; one of Shiva's names
Nawab	Muslim provincial governor or viceroy; lit. the plural of *naib,* "deputy"
Netaji	Leader
Nirvana	Lit., "the blowing out"; the goal of salvation to which Buddhists aspire
Nizamat	The administration of criminal justice and military matters under Muslim rule
Padishah	Mughal emperor
Pakistan	Land of the pure
Panchamas	"Fifths"; a term used for untouchables
Panch Shila	"Five principles"
Panchayat	Council of five; traditional ruling council of a Hindu village
Panchayati Raj	Government by councils of five
Parvati	The mother goddess
Peshwa	Maratha prime minister or premier
Pir	Muslim holy man of a Sufi order
Puja	Worshipful offering, usually flowers or food, to a Hindu god
Punjab	Land of the "Five Rivers"; a province in the northern Indus Valley
Pur	Fortified city
Puranas	Ancient tales; stories of the lives of Hindu gods
Purdah	Female seclusion, usually behind a dark veil
Purva Mimansa	"The Early Inquiry"; a school of Hindu philosophy
Raj	Rule, government, administration; royal
Raja	King
Ramayana	The shorter Hindu epic, story of Rama
Rishi	Vedic sage
Rita	Cosmic order; true
Ryot	Peasant
Sabha	Council or house

Samadhi	Deep meditation
Samiti	Ancient assembly
Samkhya	"The Count"; a school of Hindu philosophy
Samsara	Metempsychosis
Sangha	Monastic order in Buddhism
Sarkar	District; also later used for government as a whole under the British
Sarvodaya	"The Uplift of All"; Gandhian socialism
Sat, Satya	The real; hence true, the truth
Sati	"True one"; used for Hindu wives who immolated themselves on their husband's funeral pyres
Satyagraha	"Hold fast to the truth"; Mahatma Gandhi's method of nonviolent noncooperation
Sepoy (Sipahi)	Indian policeman; later any Indian soldier in a European army
Seth	Hindu banker or moneylender
Shakti	Female power; the active force in Hindu religion
Shastri	Teacher
Shi'ite	Heterodox Islamic sects, followers of Ali
Shiva	One of the two great gods of Hinduism
Shreni	Artisan or merchant guild
Shreshthin	Hindu banker or financier
Shruti	"Heard"; applied to sacred Vedic literature
Sikh	"Disciple"; a member of the syncretic North Indian religion founded by the gurus
Sikhism	The Indian religion started by Guru Nanak as a syncretism of Hinduism and Islam
Soma	Vedic god of divine nectar
Stupa	"Gathered"; a funerary mound to the Buddha's memory
Suba	A Mughal imperial province
Sudra	Menial worker; the fourth class in the Hindu social hierarchy
Sufi	Muslim mystic
Sufism	Muslim mysticism
Sunni	Islamic orthodoxy
Svadeshi	Indian-made goods; lit., "of our own country"
Svaraj	Self-rule; freedom; independence
Tamil	Classical Dravidian language
Tamil Nadu	Land of the Tamils; the South Indian state that was formerly most of Madras
Tapas	Heat generated by Yogic concentration
Tapasya	Suffering; a form of Hindu self-purification
Tat	"That one"; the monistic Brahman
Theravada	"Teachings of the Elders"; an early school of Buddhism

Thugi	Ritual murder by strangulation in the service of the mother goddess
Ulama	Islamic learned men; plural of *alim*
Umma	The Muslim universal brotherhood
Upanishad	"To sit down in front of"; hence the name given to esoteric Vedic texts consisting of philosophy and religion thus imparted
Vaishya	The merchant and landholding third class in the Hindu social hierarchy, originally "commoners"
Vaisheshika	"Individual characteristics" or "atomic" school of Hindu philosophy
Vakil	Regent
Vazir	Muslim premier
Varna	Hindu class; originally meant "color" or "skin covering"
Varuna	Vedic god of justice and order
Vedanta	Lit., "end of the vedas"; a school of Hindu philosophy and another name for the Upanishads
Vedas	Books of knowledge; the oldest Hindu scripture
Vihara	Multicell living quarters (caves), usually cut into mountainsides for Jain or Buddhist monks, especially in Bihar (Vihar)
Vira	Hero; manly
Vishnu	One of the two great gods of Hinduism, the solar deity
Vishwa Hindu Parishad	"Universal Hindu Society"
Yama	Vedic god of death
Zakat	Islamic income tax, imposed on Muslims only
Zamindar	Muslim revenue-collector overlord, confirmed by the British as landlord

INDEX

Abdullah, Farooq, 419
Abdullah, Sheikh Muhammad, 353–54
Abu-l Fazl, 128, 134
Achaemenid Empire, 55–56
Addiscombe, 216
Adoption and Maintenance Act, 367
Advani, Lal Kishan (L. K.), 433, 436ff, 439, 442–45
Adyar, 263
Afghan wars: first, 219ff; second, 254ff
Afghanistan, 129, 133, 219, 225, 270, 305, 411ff
Afghans, 172; invade India, 182ff
Afro-Asian Conference, in Bandung, 365
Aga Khan, 278, 279, 311, 319
Agartala, 404
Age of Consent Bill, 260
Age of marriage, 366–67
Agni (god of fire), 32, 33–34, 39, 274, 433, 453
Agra, 140, 150
Agrarian communalism, 309
Agrarian struggle, in Kashmir, 354
Agrarian unrest, 1857–58, 326ff
Agriculture: Aryan, 43ff; development after 1858, 243; expansion, 244; Guptan, 93
Ahimsa, 51, 53–54, 288
Ahmad Shah Abdali, see Ahmad Shah Durrani
Ahmad Shah Durrani, 181ff
Ahmed, Fakhruddin Ali (fifth president of India), 382
Ahmedabad, 119, 247, 275, 295; textile labor association of, 296
Aihole, 91
Aix-la-Chapelle, Treaty of, 177
Aiyar, G. Subramania, 262

Ajanta, 68, 79; art of, 91–92; caves of, 85–86
Ajatasatru (King of Magadha), 52, 55
Akal Takht, 417
Akali Dal, 379, 383, 416ff, 425, 445
Akbar (Mughal emperor), 158, 160, 165–66; birth of, 125; British description of capital cities of, 140; bureaucracy under, 130–31; chronicle of his reign, 128; conquest of Muslim sultanates of Deccan by, 132; conquest of Rajputs by, 127–28; consolidation of Empire by, 127ff; death of, 134; early years, 126–27; expansion of Mughal Empire under, 128ff; founder of syncretic faith, 132; invades South India, 132ff; Muslim opposition to, 132; political organization of, 129ff; patron of Sikhs, 161; tax system under, 130; toleration of Hindus, 127–28
Akbar, Prince, revolt against Alamgir by, 165–66
Aksai Chin, 364
Alam Shah, see Bahadur Shah I
Alamgir I (Mughal emperor), 153; appointed nawab of Deccan, 154; ascends throne, 157; conquers Deccan, 165–67; death of, 167; economy under, 159–60; kills Dara Shikoh, 157; orthodoxy of, 158–59; persecution of Hindus by, 158–59; reign of, 158ff; rivalry with brothers, 156; struggle to succeed Shah Jahan, 157–58
Alamgirpur, 37
Ala-ud-din Bahmani, 117
Albuquerque, Affonso d', 137ff
Alexander, A. V., 341

Alexander the Great, 56–57
Ali, Choudhry Rahmat, 317
Ali, Muhammad, 297
Ali, Shaukat, 297
Aligarh, 278; Muhammadan Anglo-Oriental College founded at, 264
Allahabad, 5; during War of 1857–58, 235; high court verdict on Indira Gandhi, 396; home of Nehrus, 303; home of Sapru, 316
Alptigin, founder of Samanid kingdom, 106
Amar, Das (Sikh guru), 161
Amaravati, 76; stupa at, 86
Ambedkar, Dr. B. R. (untouchable leader), 109, 310, 319, 356
Amboina, 142; massacre at, 144
Amery, L. S., 332
Amherst, Lord (governor-general), 217
Amir of Afghanistan, 255
Amitya Sabha, 211
Amnesty International, 435
Ampthill, Lord, 270
Amri, 11
Amrita Bazar Patrika, 261; *see also* Newspapers (Indian)
Amritsar, 161, 182; center of post-independence violence, 346; scene of Jallianwalla Bagh Massacre, 299
Amu Daria (Oxus), 411, 413
Anand Marg, 396
Anandamath (novel), 262
Andaman Islands, 283, 337
Andhra, rebirth of as a state, 368
Andhra dynasty (Satavahana), 75
Angad (Sikh guru), 161
Angkor Wat (temple), 100
Anglo-Burmese War: first, 216ff; second, 225; third, 259
Anglo-Indian Association, 257
Anglo-Indian War of 1857–58, 230, 234ff; end of, 238; legacy of, 239ff; "national" struggle in Oudh, 236; racial hatred left by, 244
Anglo-Russian Convention, 1907, 270
Anglo-Tibetan relations, 271
Annadurai, C. N., 369
Antiochus I, 61
Antwerp, 141
Anwar-ud-din, 175ff
Aparigraha, 51
Arab, *see* Islam
Archbold, W. A. J., 278
Archeological sites, earliest, 10

Archeology, director-general of appointed, 269
Architecture, Dravidian, 100–101; Hindu, 91
Arcot, *nawab* of, 193
Arjun (Sikh guru), 161
Arms race in South Asia, 414
Army: coup in Bangladesh, 401; East India Company's, 232; General Service Enlistment Act, 233; reorganization after 1858, 241; *see also* British-Indian Army; Indian Army; Pakistani Army
Art, Indus Valley, 17
Arthashastra, 57ff, 61, 62
Arunachal Pradesh, 391, 411
Arya Samaj, 263, 277, 280, 306, 309
Aryan civilization, 24ff; agriculture, 43ff; culture, 25; economy, 25, 31; invasion of South India, 41; invasions, 14, 25; iron technology, 37–38; optimism replaced by Upanishadic pessimism, 47ff; religion, 27, 32ff, 35; rice cultivation, 39; Ritual sacrifices, 38–39, 43, 274; social organization, 28–29
Asat, 48
Ashoka (Mauryan emperor), 17, 61, 351; administration of, 63ff; art of his reign, 67; and Buddhism, 62, 67; conquers Kalinga, 62; edicts of, 62ff; pillars of, 61–62
Ashramas, 83ff
Ashvaghosha, 74
Ashvamedha, 38–39, 89
Asia Watch, 435
Asian Relations Conference, 364
Asian unity, 364
Asom Gana Parishad, 426
Assam, 95, 268, 411; British conquest of, 216; "Clearance Act," 247; tea cultivation in, 247
Assamese Liberation Army, 411
Astrology, ancient, 86–87
Astronomy, Vedic, 86
Atharva Veda, *see* Vedas
Atlantic Charter, 333
Atman, concept of in Upanishads, 45–46
Atomic energy, 357; development of under Indira Gandhi, 414; India's first explosion of, 395
Atrocities, Lord Hunter's Parliamentary Commission of Inquiry into, 299

Attlee, Clement (prime minister), 311, 322, 339, 341, 345–46
Attock, 56, 216, 354
Auchinleck, Field Marshal, 354
Auckland, Lord, see Eden, George
Aungier, Gerald, 148
Aurangabad, Treaty of, 191
Aurangzeb, see Alamgir I
Automobile industry, 394, 402
Avatara, 71, 80
Awami League, 384; victory of, in Bangladesh, 386
Ayodhya, 39, 437–38, 444, 448
Ayurvedic medicine, 86
Azad, Maulana Abul Kalam, 305, 325, 331, 339, 343
Azad Kashmir, government of, 354, 436
Azes I, 73

Babri Masjid, 436, 444
Babur, 119, 121ff; early victories of, 123–24
Bactria, 70; coins of, 70; conquest of northwestern India, 71
Badal, Prakash Singh, 426
Badami, 101
Baghdad, 106
Bahadur Shah I (Prince Muazzam; Shah Alam), 168–69, 185, 191
Bahadur Shah II, 228, 234, 236; exiled to Burma, 238
Bahmani dynasty, 117
Baji Rao, 172, 178; defeats Nizam at Bhopal, 173
Baji Rao II, 203
Baker, Sir Herbert, 285
Baksar, see Buxar
Balaji Rao, 178
Balaji Vishvanath, 170
Balban, Sultan, 110–11
Baldwin, Stanley, 311
Balfour, Arthur, 272
Bali, 92
Baluchi Hills, 3
Baluchistan, 9–10
Banaras, 5, 49
Banda Bahadur, 169
Bande Mataram, 262, 325; see also Newspapers (Indian)
Bandung Conference, 364
Banerjea, Surendranath ("Surrender Not"), 274; Bengali political leader and editor, 261; convenes national conference of his Indian Association, 258; denounces official-dom, 267; elected president of Indian National Congress, 263; joins and is expelled from Indian Civil Service, 251
Bangalore, 430, 446, 449
Bangladesh, 5, 13; anticipated by Lahore Resolution, 330; becomes People's Republic of, 390; birth-pangs of, 388ff; during Bengal's first partition, 278; Nationalist Party (BNP), 401; representative government abolished in, 400; Six Point Program for, 384, war in, 387ff
Bank of India, nationalized, 392
Bankers, in British India, 255
Bantam, 142
Barabudur, 100
Baramula Road, 354
Barnala, Surjeet Singh, 424
Barooah, Dev Kant, 399
Barrackpur, 234; sepoy mutiny at, 217
Barwell, Richard, 190
Basic Democrats, Ayub Khan's, 384
Bassein, Treaty of, 204
Basu, Jyoti, 442, 449
Bay of Bengal, 6
Beas River (Hyphasis), 56
Beck, Theodore, 264
Bengal, 6, 13, 95, 96, 155, 178, 196, 262, 326, 328, 330, 337, 340, 346; administration of, passes from Company to Crown, 194; Awami League in, 384; British army in, 234; British oust French from, 180; British rule destroys economy of, 187ff; communalism in, 331; conquered by Turko-Afghans, 108–9; elections of 1970, 386; famine of 1943, 336–37; first partition of, 273, 275, 278; flooded with refugees from Bangladesh, 388ff; governor of, becomes governor-general, 190; growth of, 335; Islam in, 117ff; jute industry of, 248; Muslims of, 273, 331; National College of, 276; Naxalite movement in, 386; radical youth of, 276; reunification of, 285–86; sepoys of, 238; size and administration of, in 1905, 273; terrorist revival in, 333
Bengalee, The, 261; see also Newspapers (Indian)
Benn, William Wedgewood, 314
Bentinck, Lord William (governor-general), 212ff, 217
Berlin, 333

Besant, Annie, 263, 290
Best, Captain, 143
Bhabha, Dr. Homi, 376
Bhadralok (of Calcutta), 251, 273
Bhagavad Gita, 71, 81, 156, 261
Bhakti, 81, 95, 98ff, 120–21
Bharat Mata, 303
Bharata, 27
Bharatiya Jan Sangh ("Indian People's Party"), 362
Bharatiya Janata Party (BJP), 434, 442, 444ff, 448–50
Bharhut, 75
Bhattacharya, Abinash, 281
Bhave, Vinoba, 333, 359
Bhilai, 362
Bhima, 38
Bhindranwaje, Jarnail Singh, 417–18
Bhonsle clan, 191
Bhonsle of Nagpur, 198
Bhoodan movement, 359
Bhopal tragedy, 422
Bhutto, Benazir, 444–45, 451–52
Bhutto, Zulfkar Ali (prime minister of Pakistan), 373ff, 451; arrested by Zia, 409; as Ayub Khan's protegee, 373; becomes prime minister, 390; elections, 408; hanged, 451; heads Pakistan People's Party, 384; rejects Mujib as Prime Minister, 387
Bihar, 30; student movement in, 395
Bijapur, 163; conquered by Alamgir, 166
Bilgrami, S. H., 283
Bimbisara, 49
Bindusara, 61
al-Biruni, 107
"Black Hole" of Calcutta, 179, 237
Blavatsky, Madame, 263
Blue Mutiny, 246
Board of Control, 194
Bodhisattva, 72
Bofors, 428
Bokharo, 362
Bolan Pass, 255
Bombay (*Mumbai*), 258, 438, 445; acquired by East India Company, 145; under British rule, 148; cotton mill industry of, 247–48, 276; comes under rule of Bengal, 190; emerges as East India Company's west coast headquarters, 174; home of *Arya Samaj,* 263; home of Parsis, 254; linguistic conflict in 369–70; naval mutiny at, 340–41; partition of, 370; presidency army of, 236

Bonnerjee, Womesh C., 258
Border Security Force, 365
Bose, Khudiram, 282
Bose, Sarat Chandra, 328
Bose, Subhas Chandra (nationalist leader), contests Mahatma Gandhi's control of Congress, 327; dies in plane crash, 337; disciple of Das, 309; elected president of Indian National Congress, 326; escapes house arrest, 333; followers of, tried for treason, 340; in Germany, 333; heads Free Indian Legion, 337; helps Nehru organize Socialist League, 313; jailed, 318; joins nationalist movement, 304; opposes Gandhi, 322; in Singapore, 337
Boundary, between British India and Afghanistan, 268
Boundary Commission, 348
Bourbon Islands, 175
Bourdonnais, Mahé de la, 175
Bowles, Chester, 388
Boycott movement, 274–75; discontinued, 317; extended to liquor shops in Poona, 281; major Congress weapon, 277; stimulates *svadeshi* movement, 302
Brahmacharya, 51, 83
Brahmadeya, 99–100
Brahman, 46
Brahmana Age, late Aryan era, 43
Brahmanas, 26, 38; *Satapatha,* 39
Brahmanic ritual, disenchantment with in eighth century, 44ff
Brahmans, 26, 28, 41; importance in later Vedic era, 43; revolt against, 45ff
Brahmaputra River, 5
Brahmi: alphabet, 18; script, 62
Brahmo Samaj, 211ff, 253
Brihadratha, 69
Britain: attempts to restore Raghunath as *peshwa*, 192; bureaucracy of 256, 265; captures Ormuz, 145; cloth of, ruins Bengal textile industry, 214; district and provincial government of, under crown rule, 246; early contacts with India, 139ff; education of, impact on India, 209–10; and expansion of Indian trade, 144ff; home charges" of, 267; imperial policy of racial exclusion, 293; introduces social reforms, 212ff; invests in public works, 244; judicial system of, 199; and major

imports from India, 144–45; and marriage habits of India after 1858, 245; paternalism of, 246, 259; paramountcy of, ceases, 349; and relations with Indian princes, 203; repression by, 275, 280, 282, 335; taxation of peasants by, 207; trade with Chinese, 200; at war with France, 175

British Club, 245

British East Indian Company, see East India Company

British Indian Association, 242, 251, 262

British parliament: asserts control over East India Company's affairs, 194; concerned over Company rapacity in Bengal, 188; forces reforms on Company, 189

British-Indian Army, 216–17; in Bengal, 234; civilian domination of, 272; destroyed in retreat from Kabul, 221; growth during World War II, 336; impact of, on World War I, 292; invasion of Kashmir, 353; role of Muslims in, 332; Indians' trials for treason during World War II, 340; under Kitchener's command, 271; role in World War I, 289ff

British-Portuguese rivalry, 143

British-Russian rivalry, in Afghanistan, 218ff

Brodrick, St. John, 272

Bubonic plague, 267

Buddha, 10; as an avatara of Vishnu, 71; birth of, 49; death of, 52; development of image of, 84ff; first sermon, 50; images of, 92

Buddha Charita, 74

Buddhism, 49ff; basic concepts of, 50ff; demise in India, 108–9; development of art of, 84ff; development of monastic order, 51ff; four noble truths of, 50ff; Fourth Council of, 74; in Guptan era, 90; under Harsha, 95; Mahayana, 72; Tantric, 96; Theravada, 72; Third Council of, 67; Vajrayana, 96; see also Stupa

Buddhist Pali Canon, 49

Bull, worship of the, 10

Bureaucracy, Mauryan, 58–59

Bureaucratic indifference, 266ff; sabotage of reforms, 280

Burke, Edmund, 190

Burma, 216; final conquest of, 268; made separate colony, 322; second Anglo-Burmese War, 225; third Anglo-Burmese War, 259

Burnes, Alexander, 218

Bussy, Marquis de, 176, 184

Butler, William Harcourt, 278

Buxar, Battle of, 185

Cabinet mission (1946): sent by Attlee to resolve constitutional problems, 341; proposals, 342

Cabot, Sebastian, 139

Cabral, Pedro Alvarez, 136

Cairo Declaration, 381

Calcutta, 179, 258, 285; bhadralok of, 251; bureaucratic center of power, 265; capital of British India, 210ff; foundation of, 174; "Great Killing" in, 344; Indian National Congress session in, 277; municipal and provincial politics of, 309ff; Naxalite terror in, 386; reaction to first Partition of Bengal, 273–74; reactions to "mutiny," 236; refugee camps around, 402; University of, 257; under Warren Hastings' administration, 189; see also "Black Hole" of Calcutta

Calendar, ancient, 86

Calico, 144

Calicut, 135; destruction of, by Vasco da Gama, 136

Cambodia, 100

Campbell, Sir Colin, 238

Campbell-Bannerman, Sir Henry (prime minister), 275

Canal irrigation, 335

Canning, Paul, 143

Cape of Good Hope, 140

Capital investment, after 1858, 243

Carey, William, 207–8

Carnatic, nawab of, 228

Cartwright, Ralph, 146

Caste, and political power, 367

Caste Disabilities Act, 233

Caste system: origins and development, 32, 41ff; relation to Hinduism, 82

Castes, "martial," 242

Cauvery, 6

Cavagnari, Louis, 285

Cave art: influence of, in temples, 68; paintings, 91–92; sculpture, 68ff

Cawnpore (Kanpur), 232, 234–35

Census, of India, 267

CENTO, 373
Central Legislative Council, 245–46, 307
Central Provinces, revenue collectors, 243
Central Reserve Police, 397
Chaitanya, 121
Chaitya, 68; at Karla, 85
Chakravartin, 63
Chalukyas, 101
Chamberlain, Sir Austen, 292
Chanda Sahib, 176–77
Chandarnagar, 174
Chandigarh, 379, 434
Chanhu-daro, 21
Charaka Samhita, 86
Charar-e-Sharief, 451
Chariar, M. Viraraghava, 262
Charka, 310, 325
Charles I, 146
Charles II, 147
Charnock, Job, 174
Charter Acts: 1813, 208; 1833, 213; 1853, 239
Chatterji, Bankim Chandra, 261–62, 281
Chauri Chaura, 307
Chavan, Y. B., 372, 409
Chelmsford, Lord (viceroy), 295, 299
Chidambaram, 112–13
Chidambaram, P., 449
Child, Sir John, 173–74
Child, Sir Josia, 173–74
Chilianwala, Battle of, 223
China: and Indian opium trade with, 200; invades India (1962), 365; and trade with Guptan Empire, 92
Chiplunkar, Vishnu Hari, 254
Chishti, 117
Chitor, seige of, 127–28
Chitpavin Brahmans, 205, 252, 357
Chitral, 269–70
Chola dynasty, 63, 76ff, 99, 112ff; art of, 112ff
Choudhury, Abu Sayeed, 390
Christianity in India, 73, 84; *see also* Missionaries, British
Chumbi Valley, 271
Church, Senator Frank, 388
Churchill, Winston, 316, 322, 333, 334
Civil disobedience, *see* Gandhi, Mohandas Karamchand; *Satyagraha*
Civil liberties, suspension of, 280, 409
Civil lines (suburbs), 244
Civil Service, *see* Indian Civil Service

Civil War of Succession (1946–47), 344ff
Clavering, General Sir John, 190
Clive, Sir Robert, 175, 180; buys control of East India Company, 181; captures Arcot, 177; defeats Siraj at Plassey, 180; governor of Bengal, 187ff; *mansabdar* of the Mughal Empire, 180; rule in Bengal, 185–86; seeks Hindu support, 180; suicide, 188
Coal: mining of, 243; production of, 361; transport by rail of, 229
Coen, Jan Pieterszoon, 144
Coffee cultivation, 247
Coins, *see* Bactria; Kushans
Colbert, Jean-Baptiste, 174
Collective leadership, interlude of, 371ff; ended, 382
Colombo, 430; Agreement, 430–31
Commerce and Industry, Department of, 269
Communal Award, 319, 323
Communal electorates, *see* Separate electorates
Communalism, 305, 330, 414; in countryside, 309; in Punjab, 346; rioting, 306, 311
Communist parties, support Indira Gandhi, 403
Communist Party of India, 305, 312, 362; "left" or pro-Maoist Party, 378; splits into several parties, 389; victory in Kerala, 378
Community development program, 360
Competitive examination system, 239
Congress Party (from 1885 to 1947), *see* Indian National Congress
Congress Party (post-1947): adopts "socialist pattern of society" goal, 362; business support of, 362; criticisms of, 381; elects Sastri prime minister, 371; fourth general election, 380; launches attack against princely privilege, 386; "old guard" expels Indira Gandhi, 383; split in 1969, 383; under Kamaraj's leadership, 372; "Requisition" Party wins 1971 elections, 389; ruled by Indira and Sanjay Gandhi, 410; run by the "Syndicate," 372; "spinelessness" during Emergency Raj, 403; traumatic 1991 election, 441–42; victory of, in 1951, 357

Congress Socialist Party, 359
Congress-League deadlock, 341
Constituent Assembly, 345
Constitution of Bangladesh, 390, 400
Constitution of India, 356ff; emergency clause to suspend, 356; fundamental rights of threatened, 399
Constitution of Pakistan, 372–73
Coolie labor, 287
Cooperatives, agricultural, 386
Coote, Sir Eyre, 184, 193
Cornwallis, Lord (governor-general): administrative code, 199; becomes governor-general, 195; British supremacist, 198; moves against Mysore, 198; Permanent Settlement, 199, 240
Coromandel coast, 8
Cottage industries, destroyed by British imports, 248, 276
Cotton, 229; excise tax on abolished, 308; imports, 275; industry, 247–48, 321, 362; origins of, 19
Council of India, 240; first Indian members of, 283
Council of State, 307
Courten, Sir William, 146
Cremation, 44
Crewe, Lord (secretary of state), 287
Crimean War, 229, 231, 248
Criminal Jurisdiction Bill of 1883 (Ilbert Bill), 257
Criminal prosecution, of political activists, 276
Cripps, Sir Stafford, 334, 339, 341, 345
Cromwell, Oliver, 147
Crown Raj, 263, 264; last year of, 344ff; manners and mores of, 245
Cunning, Lord Charles John: first viceroy of the Crown, 240; last governor-general of East India Company, 233; taluqdars and, 242
Cunningham, Sir Alexander, 14
Curzon, Lord George Nathaniel (viceroy), 265, 268; administrative reforms, 269; conflict with Kitchener, 271–72; foreign policy of, 270; and Ibbetson, 280; loan to nawab of Dacca, 279; military reorganization problems of, 271; partition of Bengal, 272ff; resignation, 272; strategic use of railroads, 270
Curzon-Wyllie, Sir William, 283

Dabhol, 446
Dacca, 277–78, 387, 401; birthplace of Muslim League, 279; capital of Bangladesh, 390; coup in, 400–401; held by West Pakistan Army, 387; nawab of, 286
Dalhousie, Marquess of (governor-general), 223ff; annexes Oudh, 232; and doctrines of "lapse" and "paramountcy," 226–27; egalitarian policy of reversed, 242; modernizes communications, 228ff, 243
Dalit, 439, 449
Dane, Sir Louis, 270–71
Dange, Shripat Amrit, 305
Dara Shikoh, 153ff; murdered by Alamgir, 157; supported by Sikhs, 161
Darjeeling, 247, 432
Das, Chittaranjan (nationalist leader of Bengal), 301, 303, 304; death of, 309; founds Swarajist Party with Motilal Nehru, 307–8
Das, Ram Charan, 255
Dasas, 25, 27, 32, 33, 38
Datta, Bhupendra Nath, 281
Daulatabad, 115
Day, Francis, 145
Dayal, Har, 293
Deccan, 6, 8, 101; conquered by Aurangzeb, 166–67; medieval kingdoms of, 111ff; sultanates, 139
Deccan Education Society, 259
Defense of India Act, 298
Delhi, 181, 235–36, 263, 340; capture of, by Ghurids, 108; conquered by Tamerlane, 119; durbar in, 285; falls to Persians, 173; massacres, 419ff; recaptured by British in 1857, 238; Shah Jahan's capital, 154ff
Delhi Sultanate, 109ff; decline of, 119
Demetrius, 71
Deogarh, 91
"Depressed classes," 319; see also Untouchables
Derozio, Henry Louis, 212
Desai, Bhulabhai, 340
Desai, Kantilal, 407–8
Desai, Morarji (nationalist leader), 371–72, 376–77, 391, 395–96, 403–4, 409; arrested, 397; defeats Indira's Congress in Gujarat, 395–96; goes on a fast unto-death, 395; industrial support for, 377; joins In-

Desai (*cont.*)
  dira's Cabinet, 380; political defeat,
  409; released from jail, 403–4; re-
  signs from Indira's Cabinet, 382;
  returns as fourth prime minister,
  404ff
Devadasis, 77; *see also* Women: in
  ancient India
Devagiri, 111, 114
Devaluation of rupee, 387
Devangari script, 62
Devicandragupta, 89
Dharma, 40, 66–67, 82, 415
Dharma chakra ("wheel of the law"),
  67
Dharmaraja, 38
Dharmashastras, 83
Dhimmis, 105
Dhingra, Madan Lal, 283
Diesel engines, 364
Din-i-Ilahi, 132
Direct Action Day, 344
Disraeli, Benjamin (prime minister),
  249, 255, 256
District government boards, 256
Divide et impera ("divide and rule"),
  273–74, 319
Divorce, 366
Diwan, 129, 287
Dogra Rajputs, 353
Dominion of India, 347ff
Dominion of Pakistan, 347ff
Dominion status, 314
Dost Muhammad Khan (amir of
  Afghanistan), 219, 221, 255
Drake, Roger, 179
Draupadi, 38
Dravida Kazhagam, 369
Dravida Munnetra Kazhagam
  (DMK), 369, 402
Dravidian, 8, 263, 369
Dravidian consciousness, 369; *see also*
  Tamil
Drought, 268
Dufferin, Lord (viceroy), 259, 265
Dundas, Henry, 194ff, 201
Dunlop Smith, J. R., 278
Dupleix, Joseph François, 175–77
Durand Line, 268, 412
Durga, 82, 281
Durgapur, 362
Dutch, in South India, 145; massacre
  British at Amboina, 144; publish
  navigational maps, 141; trade with
  India, 141ff
Dutch-Spanish rivalry, 140ff

Dutt, Romesh Chandra, 284
Dyarchy, 297
Dyer, Brigadier R. E. H., 299

East India Company, 140, 147, 174,
  194, 200; appeals for parliamentary
  support, 146; chartered by the
  Crown, 142; decline of, 147; demise
  of, 238; "diwan" of Bengal, Bihar,
  and Orissa, 186ff; establishes rule,
  187ff; first expedition of, 142; first
  factory at Surat, 144; recruit training, 215; regulated by Parliament,
  190; salt monopoly of, 200; seizes
  Salsette and Bassein, 192; trade
  monopoly abolished, 213, 246
East Pakistan, 331, 384–85, 387;
  "colony" of West Pakistan, 383;
  renamed Bangladesh, 387
Eastern Bengal and Assam, 273, 277,
  284
Economic depression, 320
Economic development, under Dalhousie, 229ff
Economic planning, 358; *see also*
  Five-Year-Plan
Economic reforms, Twenty Point Pro-
  gram, 398ff
Eden, George (Lord Auckland), 219
Education, 367; birth of Presidency
  University, 241; British missionary,
  241; compulsory (elementary), 367;
  Muslim modernists, 264; national,
  281, 303, 367; reforms, 285
Education Despatch of 1854, 241
Edward VII, King (emperor of In-
  dia), 272, 283
Egalitarian ideals, 356
Elections, India: 1937, 323; 1945–46,
  340; 1967, 381–82; 1971, 387; 1976
  "postponed," 402; 1977, 404; 1980,
  410
Elections, Pakistan: 1964, 384; 1970,
  386
Elective principle, 266, 284
Electric telegraph, 228ff, 244; politi-
  cal and commercial impact of, 230;
  use in War of 1857–58, 243; value
  to Postal Service, 231
Elgin, Lord (viceroy), 265, 268
Elizabeth I, Queen, 142
Ellenborough, Lord (governor-
  general), 219ff
Ellora, caves and temple, 68, 79, 86,
  103

Elphinstone, Mountstuart, 204ff, 209
Emergency clause suspending Constitution: included in 1949 Constitution, 356; invoked by Indira Gandhi, 397
Emergency ordinance (raj), British, 280
Enfield rifle, 233
Enfranchisement, for Indians in 1910, 284
England, see Britain
ENRON, 446
Eucratides, 71
European community, non-official, 257
Euthydemus, 71
Evans, Colonel Sir De Lacy, 218

"Factory" system, establishment of, 136
Fa-hsien, 89–90
Family planning, 362
Famine, 362; in 1895, 267; in Maharashtra (1879), 248
Farrukh-siyar, 169
Fast, as a political weapon, 296; see also Gandhi, Mohandas Karamchand
Fatehpur Sikri, 128, 133, 140
Fazlul Huq, Abul Kasem, 311, 326, 330
Female education, 367–68
Feudalism, in India, 111ff
Firdawsi, 107, 117
Firingis, 281
Firozshah, Battle of, 223
Firuz, Tughluq Sultan, 118–19
Fiscal autonomy, 308
Fitch, Ralph, 140
Five-Year Plan: First, 358–59; Second, 361–62; Third, 362–63; Fourth, 378, 385; Fifth, 394–95
Flogging, 282
Ford Foundation, 360
Ford Motor Company, 446
Foreign Department, 245
Fort St. George, see Madras
Fort Williams, 174, 179
Forward Bloc Party, 328, 333
France: controls Hyderabad, 177; defeats nawab's forces, 176; forces driven from Bengal by Clive, 180; and rivalry with Britain, 175ff; and trade with India, 175
Francis, Philip, 190
Fraser, Sir Andrew, 274

French East India Company, 174ff
Fu-nan (South Vietnam), 92

Gaekwar of Baroda, 178
Galbraith, John K., 388
Ganapati festivals, 260
Gandhara, 5, 55, 71; Buddhist art of, 71–72, 85
Gandhi, Firoze, 371
Gandhi, Indira (prime minister), 300, 371–72, 376ff, 389, 394ff, 401ff, 410ff; abolishes Maharajas' privy purses, 386; assassination of, 419ff; on atomic explosion, 395; backs Giri for president, 381–82; becomes prime minister, 377; devalues rupee, 378; defeated, 404; and "disciplined democracy," 401; "expelled" from Congress Party, 383; and fourth general elections, 381; imposes Emergency Rule, 397ff; invites Desai to be deputy prime minister, 380; leads national "left-wing" coalition, 383; overwhelms opposition, 387; in Shastri's government, 372; signs Treaty with Soviet Union, 389; struggle with Morarji Desai, 376–77; takes Finance Ministry from Desai, 382; unveils Twenty-Point Program, 398ff; victorious in 1980, 410; visits Washington, 378
Gandhi, Kasturbai (Mahatma Gandhi's wife), 316
Gandhi, Maneka, 421
Gandhi, Mohandas Karamchand ("Mahatma"), 294–95, 351, 371, 414ff; on ahimsa, 302; assassination of, 355; Bardoli satyagraha, 306; broadcast to U.S.A., 318; calls off satyagraha, 307; Champaran victory, 295–96; fasts of, 319, 337, 355; first nationwide satyagraha, 301; and Hindu-Muslim unity, 309; and Jinnah, 338; at London Round Table Conference, 319; march protesting salt tax, 315; and Mountbatten, 346ff; one-man "border patrol," 349; opposes Rowlatt Acts, 298; and "Pakistan" demand, 331–32; and Patel, 352; "Quit India" satyagraha, 335; reaction to Attlee's cabinet mission plan, 343; "saintly style" of politics, 295; satyagrahas, 288, 296, 315, 333; and S. C. Bose, 327–28; in South Africa, 287; sup-

Gandhi (*cont.*)
 ports Nehru for Congress president,
 313; and untouchables, 310; and
 vocational education, 367; and
 women, 366; and World War I,
 289–90; and World War II, 329
Gandhi, Rajiv, 410, 420ff, 439ff; as
 prime minister, 423ff; Assam Ac-
 cord, 426; assassination of, 441–42;
 Punjab Accord, 425–26; and
 SAARC, 430; and Sri Lanka, 431
Gandhi, Sanjay, 394, 401–2, 404, 410
Gandhi, Sonia, 423, 440
Gandhi-Irwin pact, 317–18
Ganesh, 82, 260, 447–48
Ganga River (Ganges), 5, 39, 370;
 goddess, 37
Ganga-Yamuna plain (Gangetic
 plain), 12
Ganges, *see* Ganga River
Ganges Canal, 230
Gangu, Hasan, 116
"Garoland," 432
Gauhati, 411
Gautama, *see* Buddha
George V, King, 285
Germany, 289
Ghadr Party, 293
Ghazni, 106–7
Ghengiz Khan, 109
Ghose, Dr. Rash Behari, 281
Ghosh, Arabinda, 261, 276, 281
Ghosh, Motilal, 261
Ghosh, Shishir, 261
Ghulam Qadir, 198
Ghur, Muhammad, 108–9
Gilgit, 268
Gill, K. P. S., 450
Giri, Varahagiri V. (president),
 381–82
Gladstone, William E., 256
GNLF (Gurkha National Liberation
 Front), 433
"Gurkhaland," 431
Goa, 140; inquisition in, 138; libera-
 tion from Portuguese rule, 363; mis-
 sionaries in, 138; Portuguese con-
 quest of, 137
Goan Liberation Committee, 363
Gobind Rai, Sikh Guru, 162
Godavari River, 6
Godse, Naturam V., 355
Gokhale, Gopal Krishna, 253, 284;
 character of, 294; Congress presi-
 dent, 273; death of, 290; Elemen-
 tary Education Bill, author of, 285;

impact on Morley, 277; on Imperial
 Legislative Council, 266; and Jin-
 nah, 287; and Surat Congress, 281;
 and Tilak, 259
Golconda, 166
Golden Temple, 161, 434
Gondophernes, 73
Gorbachev, Mikhail, 428
Gough, Lord, 223
Government of India Acts: 1858, 239;
 1921, 307; 1935, 322
Governor-General's Council, 190; first
 Indian member of, 283
Gowda, H. D. Deve, 449
Gracey, General, 354
Gramdan, 359
Grand Trunk Road, 243
Greased cartridges, 233
"Great Game," 413
"Great Revolt," *see* Anglo-Indian War
 of 1857–58
Greco-Bactrian kingdoms, 70ff
Green Revolution, 385; in Punjab, 416
Gryhastha, 83
Gujarat, 11, 119–20; 395; birth of
 State of, 370
Gujral, Inder, 434, 449
Gujrat, Battle of, 223
Gupta, Chandra, I, 88ff
Gupta, Chandra, II, 89ff
Gupta, K. G., 283
Gupta, Kumara, 94–95
Gupta, Samudra, 88
Gupta, Skanda, 94
Gupta Empire, 88ff; agriculture in, 93;
 art of, 91ff; commerce with China
 and Southeast Asia, 92ff; decline of,
 94–95; extent of, 88–89; royal mines
 and lands, 93–94; tax system of, 93
"Gurkhaland," 431
Guru, massacre at, 271
Gurumukhi script, 379
Gwalior, 109
Gyantze, 271

Habibullah (amir of Afghanistan),
 270–71
Haidar Ali Khan, 193
Haileybury College, 215
Han dynasty, 72
Handicrafts and cottage industries,
 248
Harappa, 12, 14ff, 15; burial practices
 of, 23
Hardinge, Lord (viceroy), 285–86,
 289–90

Hardwar, 5
Hare, David, 212
Hargobind, Sikh Guru, 161
Harihara, I, 116–17
Harijans, 310, 320, 365; *see also*
  Untouchables
Harmika, 68
Harsha Vardhana, 88, 95
Haryana, 380, 425, 429
Hastinapura, 12, 30, 37ff
Hastings, Marquess of (governor-
  general), 210ff
Hastings, Warren, 192–93; appointed
  governor of Bengal, 189; resigna-
  tion and retirement, 194; rule in
  Bengal, 189ff; stops Company's an-
  nual payment to Mughal Emperor,
  189
Hathigumpha inscription, 76
Hawkins, Captain William, 142–43
Heliodorus, 71
Hemu, 126
Harmaeus, 72
Herodotus, on India, 56
Hijrat, 104
Hill, Sir Rowland, 231
Himachal Pradesh, 391
Himalaya Mountains, 3, 5
Hinayana Buddhism, 50
Hindi language, 370
Hindu, 262; *see also* Newspapers
  (Indian)
Hindu caste society, 32
Hindu College, 212
Hindu communial Party, 355
Hindu Kush Mountains, 3, 5
Hindu *Mahasabha* Party, 306, 309,
  312, 355
Hindu Marriage Validating Act, 366
Hindu orthodoxy, undermined, 233
Hindu Rashtra ("Hindu Nation-
  hood"), 448
Hindu Renaissance, 210ff, 263, 277
Hindu Succession Act, 367
Hindu temple architecture, 91
Hinduism: basic tenets of, 82ff; *bhakti*
  sect of, 98ff; classical schools of,
  96ff; compared with Islam, 104;
  emergence of, 80; gods of, 82; ori-
  gins of, 44; response to Christian
  missionaries, 211; salvation in, 81–
  82; social structure of, 82ff; in
  South India, 98–99; traditionalism
  of, 260
Hindu-Muslim conflicts, 302, 309,
  311, 326, 338, 437–38, 444

Hindu-Muslim-Sikh rioting, 346
Hindu-Muslim syncretism, 120ff
Hindu-Muslim unity, 306, 414
Hindutva ("Hinduness"), 445, 447–
  48, 450
Hittites, 24, 27
Hoare, Sir Samuel, 322
Hobhouse, Sir John (Lord Brough-
  ton), 219, 227
Holkar of Indore, 178, 191
Holwell, J. Z., 179
Home industry, of India, 247–48
Horse sacrifice, *see Ashvamedha*
Houtman, Cornelius de, 141
Hoysalas, 116
Hsuan Tsang, 95, 101
Hughli River, 148, 174, 248
Human rights violations, 299, 419ff,
  425, 436, 446
Humayun, 124–25
Huns, 94
Husain, Dr. Zakir (president), 381
Husain Ali (Sayyid), 170
Hussain, Mushtaq, *see* Viqur-ul-Mulk
Hyderabad: emerges as independent
  kingdom, 171; land reform revolt
  in, 359; seeks independence of In-
  dia, 352; surrounded by British,
  227; and transfer of power, under
  British control, 203; "Usmanistan"
  separatism, 330

Ibbetson, Sir Denzil, 280
Ibn Battuta, 115
Ignatius of Loyola, 138
Ilbert, Sir Courtney, 257
Ilbert Bill, 257
Iletmish (Iltutmish), Shams-ud-din,
  109
Imad-ul-Mulk, 182–83
Imam, Sayyid Ali, 284
Imperial Legislative Assembly, 307
Imphal, 268, 337
Indentured labor system, 287
Independence Day, 314
India: Dominion of, created, 347; Re-
  public of, born, 356; social reform
  in, 365ff; supports Bangladesh, 388ff
India Act (1784), 194
India Office, 267, 277, 282
Indian Army: Border Security Force
  of, 365; compared with Chinese
  Army, 365; compared with Paki-
  stani Army, 374; conquers Hydera-
  bad, 352; and Goa, 363; helps lib-

Indian Army (*cont.*)
    erate Bangladesh, 390; in Kashmir,
    354, 374; represses Nagas, 411
Indian Association, 252, 258
Indian Civil Service, 265; competitive
    examinations for, 239; few Indians
    admitted, 251; simultaneous exami-
    nations for India and England, 266,
    308
Indian Councils Act: of 1861, 245; of
    1892, 246, 266; of 1909, 284
Indian languages, 87
Indian Muslim leaders, 263–64
Indian National Army (INA), 334,
    337; trial of officers, 340
Indian National Congress: Amritsar
    session, 300; anthem and flag of,
    325; birth of New Party within,
    277; under Bose, 327–28; British
    Committee of, discontinued, 304,
    conflict with Muslim League, 338–
    39; demands independence, 343;
    during World War II, 329ff; election
    victories of, in 1937, 323; first meet-
    ing of, 258; first woman president
    of, 263; under Gokhale, 273; impact
    of first partition of Bengal upon,
    274ff; inspiration for, 257; Lucknow
    reunification of, 293; Muslim mem-
    bership in, 258, 326; Nagpur ses-
    sion, 314; in Punjab, 280; under
    Sinha, 284; Surat split, 281; trans-
    formed under Gandhi's leadership,
    301; *see also* Congress Party
Indian National Movement, *see* In-
    dian National Congress; Nation-
    alism
Indian Ocean: navigation of, 141;
    Portuguese domination of, 137;
    trade in Roman era, 78
Indian Statutory Commission, 311
Indigo industry, 246, 296
Indira Raj, 383ff; *see also* Gandhi,
    Indira
Indo-Afghan relations, 270–71, 411ff
Indo-American friendship, poisoned
    by Nixonian "tilt," 388ff
Indo-Aryans, *see* Aryans
Indo-Europeans, 8, 24
Indo-Iranians, *see* Aryan civilization
Indo-Pakistani relations: clash over
    Rann of Kutch, 374; War of 1965,
    374–75; war over Bangladesh, 389ff
Indo-Russian alliance, 389
Indra, 32–33, 35
Indus River, 3–5

Indus Valley civilization, 10–11; ani-
    mals of, 11; art of, 17; burial rites
    of, 21, 23; decline of, 21–22; deities
    of, 18–19; economy of, 20; extent
    of, 19; iconography of, 17; pottery
    of, 17; seals of, 16–17; trade of,
    19–20
Industrial unrest, 321
Industrialization, 276, 362–63, 378,
    394; in Nehru's era, 361ff
Internal unification, 232
International Alert, 435
Iqbal, Muhammad, 316, 319, 324
Iron, in Aryan Age, 30
Iron and Steel industry, 276, 296
Irrigation canals, 226, 244
Irwin, Lord (Edward Wood), viceroy,
    316–17
Islam: basic tenets of, 104–5; dynas-
    ties of, in North India, 109; early
    invasions by adherents of, in India,
    104–9; expansion of, 105
Islamic socialism, 395

Jaffna, 430–31
Jagirs, 112, 197
Jahandar Shah, 169
Jahangir (Mughal emperor), 142, 161;
    assumes throne, 134; death of, 152;
    Hindu influence on, 149; Persian
    influences on, 149–50
Jainism, 52ff; basic concepts, 53ff;
    division into sects, 84
Jallianwala Begh Massacre, 299–300
James I, 142
Jammu, and Kasmir, 353–54, 434–36,
    445
Jammu and Kashmir National Con-
    ference, 353
Jamshedpur, 296, 336
Jan Sangh, 362, 383, 396, 409
Janata Dal, 436, 440, 443
Janata Morcha, 396–97, 404
Japanese attacks, 334
Java, 92, 100; Dutch arrival in, 141
Jayakar, Dr. Mukund Ramrao, 316
Jayewardene, J. R., 429–30
Jessore, 389
Jesuits: in India, 138; at Jahangir's
    Court, 143
Jethmalani, Ram, 449
Jewish community in India, 84
Jhangar culture, 23
Jhansi, Rani of, 227, 238
"Jharkand," 432
Jhelum (Hydaspes) River, 56

Jhukar culture, 21–22
Jihad, 306
Jinnah, Fatima, 384
Jinnah, Muhammad Ali (*Qa'id-i-
Azam*), 329, 331, 332, 343, 345,
349, 372, 414; at All-Parties Con-
ference, 312; calls for Direct Action
by League, 344; during World War
II, 335; early life and National
Congress career, 286ff; on eve of
partition, 347; on Kashmir, 345; at
London Round Table Conference,
319; loses national leadership to
Gandhi, 303; and Lucknow Pact,
293ff; and Mountbatten, 346; op-
poses Rowlatt Acts, 298; president
of Muslim League, 309; revitalizes
League, 324ff; spokesman for Mus-
lim India, 323; talks with Gandhi,
338–39; wins mandate of Muslim
India, 340
Jiva, 53
Jivan-dan, 359
Jodhpur, 165
John Company, *see* East India Com-
pany
Jones, Sir William, 25, 209
Joshi, Murli Manohar, 445
Junagadh, 349, 352
Jute, 229, 248

Kabir, 120
Kabul, 255, 268, 270, 333; British
capture of, 220
Kafur, Malik, 114
Kailasanatha temple, 103
Kalat, 268
Kali (goddess), 82, 174, 281
Kalibangan, 12, 17, 21
Kalidasa, 89; works of, 90ff
Kalinga, 62, 76
Kalyani, 103
Kama, 36
Kama Sutra, 83, 92
Kamaraj, Kumaraswami (Nadar),
371ff, 376–77, 380–81
Kamaraj Plan, 371
Kanarese, 371
Kanauj, 95, 107
Kanchipuram (Conjeevaram), 99, 112
Kandahar, 255
Kanishka (king of Kushans), 74–75
Kanvayana dynasty, 75
Kapilavastu, 49
Karachi, 221, 279, 349, 352, 438
Karla, cave temples of, 85

Karma yoga, 81
Karoshthi script, 62
Kartikeya, 82, 94
Kashmir: All-Jammu and Kashmir
Muslim Conference, 353; Bhutto-
Singh talks about, 372; escalating
conflict in, 434–36, 445, 450ff; first
undeclared war over, 354; invasion
of, 353–54; and Russian "danger,"
268; second Indo-Pak war over,
374–75; Sikh conquest of, 216; and
Sino-Indian border conflict, 364;
surrendered to British, 223; UN
cease-fire in, 354
Kasi (Varanasi), *see* Banaras
Kathiwar, 352
Kausambi, 37
Kautilya, 57ff
Keane, Sir John, 220–21
Kennedy, Senator Edward, 388
Kentucky Fried Chicken, 446
Kerala: kingdom of, 63; people of,
76ff
Kesari, 282; *see also* Newspapers
(Indian)
Khadi, 307
Khalistan, 416ff; Khalistani, 440, 450
Khalji, Ala-ud-din, 111, 113ff
Khalji, Jalal-ud-din Firuz, 111
Khalji dynasty, 111–14
Khalsa, 162, 346
Khan, Abdul Ghaffar, 324, 346
Khan, Afzal, 164, 261
Khan, Aga Muhammad Yahya, 383–
84, 451; dictatorship of, supported
by Nixon, 385; as Nixon's middle-
man in negotiations with China,
388; resigns, 390
Khan, Ali Verdi, *nawab* of Bengal,
178
Khan, Asaf, 149, 152
Khan, Ayub, 383, 384, 451
Khan, Bayram, 126–27
Khan, Dilavar, 119
Khan, Dost Muhammad, *see* Dost
Muhammad Khan
Khan, Ghulam Ishaq, 452
Khan, Khizr Hayat, 346
Khan, Liaquat Ali, 323–24, 344–45,
372–73
Khan, Lieutenant General Tikka, 387
Khan, Muhammad Ayub, 373, 375–
76, 383, 451
Khan, Sir Sayyid Ahmad, 263–64,
279
Khan, Sir Shafatt Ahmed, 344

Khan, Sir Sikander Hyat, 326
Khan, Yakub, 255
Khan, Zafar, 119
Khan, Zulfiqar, 169
Khan Sahib, Dr., 346
Khanua, Battle of, 122–24
Kharavela (king of Kalinga), 76, 78
Kheda, 296, 351
Khilafat Movement, 297, 301, 303, 305, 309
Khojas, 286
Khusrau, 152
Khyber Pass, 5, 106, 255, 412
Kingsford, Douglas, 282
Kingship, Indian ideal of, 57–58
Kipling, Rudyard, 244
Kisan Sabhas, 304–5
Kissinger, Henry, 388–89
Kitchener, Lord, 271, 275, 283, 290
Kitchlu, Dr., 298
Komagata Maru, 292
Kosala, 48–49
Kosygin, Aleksei, 375
Kot Diji, 11, 17
Krishak Praja Party, 311
Krishna, 6, 71, 80–81
Kshatriyas, 29, 41, 45
Kujala Kadphises, 74
Kulli culture, 10
Kumar Barindra, 281
Kumaratunga, Chandrika, 431
Kumbakonam, 113
Kurukshetra, 38; refugee camp at, 355
Kushans, 72, 73, 75, 89; coins of, 74
Kuvera, Queen, 90

Lahore: captured by Ghurids, 108; center of partition carnage, 346; focus of British repression, 280; within range of Indian tanks, 375
Lahore Resolution, 330
Lakshanas, 85
Lakshmi, 82
Lal, Devi, 429
Lally, Count de, 184
Lama, Dalai, 365
Lancashire, 248
Lancaster, James, 142
Land ownership, system transformed, 197
Land revenue, as political weapon, 296
Land settlement: permanent *zamindari*, 196ff; *ryotwari*, in Madras, 207; with *taluqdars*, 235; in Uttar Pradesh, 242; *see also Zamindars*

Landless labor, growth of, 321, 362
Landlords, of Oudh, 235; "props" of British rule, 204
Landsdowne, Lord (viceroy), 265–66
Language, earliest written, 8, 18
Lapse, doctrine of, 226–27; rejection of, 240
Lawrence, Henry, 224, 235
Lawrence, John, Lord (viceroy), 224, 237, 246, 255
Le Corbusier, 379
Legislative Council, 266
Lhasa, 271, 366
Li Peng, 446
Liberal Federation, 316
Liberal government, 277
Libya, 414
Linguistic provinces, 368
Linguistic separatism, 8, 381, 368–89
Linlithgow, Lord (viceroy), 323, 329, 332
Liquor boycott, 281, 316
Literacy, 369; *see also* Education
"Little Clay Cart," 91
Local self-government, 256
Lodi dynasty, 120
Lodi sultans: Buhlul, 120; Ibrahim, 121; Mahmud, 124; Sikandar, 120
Lok Dal, 407, 429
Lok Sabha ("House of the People"), 357, 383, 439, 442; women elected to, 366–67
London Company, first expedition of, 139
Longju, 365
Longowal, Sant Harchand Singh, 417, 424–25
Lothal, 17, 23; graves at, 21
LTTE ("Liberation Tigers of Tamil Eelam"), 431, 440
Lucknow, 184, 203, 232, 234–35, 242
Lucknow Pact, 294, 309
Luytens, Sir Edward, 285
Lytton, Lord (viceroy), 248ff, 256; and Indian nationalism, 252; Second Anglo-Afghan War, 255; Vernacular Press Act, 254

Macaulay, Thomas Babington, 208ff, 214–15; education, 215
MacDonald, Ramsay, 314; and the Communal Award, 319, 323
McMahon, Sir Arthur Henry, 364
Macnaghten, Sir William, 219
Madhav Rao, Peshwa, 191, 192

Madho, Rana Beni, 237
Madras, 8, 193, 236, 262, 369; attacked by *nawab* of Carnatic, 176; captured by French, 175; comes under Bengal's control, 190; center of Anglo-French struggle, 174; divided linguistically, 368; founding of, 145; Native Association founded, 262; restored to British, 177
Madura (Madurai), 77; Sultanate of, 116
Magadha, kingdom of, 48–49
Maha Gujarat Parishad, 369
Mahabalipuram, 100
Mahabat Khan, 152
Mahabharata, 27, 31, 37–38, 71
Mahanadi, 6
Maharajas ("Great King"), 39, 397; *ex-maharajas*, 406
Maharashtra ("Great Country"), 163, 281, 283; famine in, 248; modern state, 369
Mahatma ("Great Soul") Gandhi, *see* Gandhi, Mohandas Karamchand
Mahavira, Vardhamana, 52ff
Mahayana Buddhism, 50, 72
Mahindra & Mahindra, 447
Mahmud of Ghazni, 106–7
Maithuna, 96
Malabar, 6; coast of, 198
Malaviya, Pandit Madan Mohan, 306
Malayalam, 371
Malcolm, John, 204
Maldive Islands, 112
Malguzars, 243
Malwa, 119–20
Mamdot, Khan of, 346
Mamluk ("Slave") dynasty, 109ff
Mamluks, 106
Manava Dharmashastra ("Law Code of Manu"), 80
Manchester, imports from, 276
Mandal Commission Report, 435ff, 441
Mandalay prison (Burma), 280, 282
Manekshaw, General Sam, 390
Manipur, 268, 337, 411
Mansabdari system, 129ff, 153
Mansel, Charles, 224
Marathas: alliance with Nizam, 172; in Bengal, 178; brought within Mughal imperial system, 170; capture Delhi, 191; civil war, 191, 203; defeated at Panipat, 183–84; expansion of, 170–71, 178, 182; fight against Nizam, 191; final defeat of

by British, 204–5; and Hindu power, 159–60, 162ff; tax system of, 170; treaty with British, 203
Marriage: intercaste, 366; reforms in, 260
Marshall, Sir John, 14
Marshman, Joshua, 207
Martial "law," 298ff; imposed by Indira Gandhi, 397
Martin, François, 174
Marwari merchants, 275, 301, 362
Massacre, in Bangladesh, 387–88; at Jallianwala Bagh, 299–300
Mata Hari, 293
Mathura, Buddhist Art of, 84–85
Maues, King, 73
Mauritius, 175
Maurya, Chandragupta (emperor), 55ff, 61
Mauryan dynasty and Empire, 57ff; administration of, 59–60; art of, 67; economic organization of, 60; expansion of, 59–60; decline of, 68–69; political organization of, 57–58; social structure within, 60ff; under Ashoka's rule, 63ff
Maya, 48
Mecca, 104
Medicine, ancient Indian, 86
Medieval Hindu kingdoms, 111ff
Medina, 104
Meerut, 222, 234–35, 237, 280; conspiracy trial at, 313
Megasthenes, 58
Meghaduta, 91
Meghalaya, 391, 411
Mehta, Pherozeshah ("Uncrowned King of Bombay"), 254, 266, 277, 290
Memons, 287
Menander, 71
Menon, V. K. Krishna, 365
Menon, V. P., 339, 347, 352
Mesolithic Age, 8
Mesopotamian war, 292
Metcalfe, Charles, 204ff
Mewar, 165
Middle class, 259
Mihirakula, 94
Milindapanho, 71
Military: and China, 364–65; expenditures, 259; operations in Kashmir, 353–54, 374; organization of Bengal Army, 234; in Bombay presidency, 236; in Madras presidency, 236; in Pakistan, 373ff, 384ff;

Military (*cont.*)
  seizes power in Bangladesh, 401;
  gets support from Russia, 402; *see
  also* Army; British-Indian Army;
  Indian Army
Mill, James, 208
Mill, John Stuart, 208, 239
Mills, cotton, 247–48
Minto, Fourth Earl of (viceroy), 275,
  278, 280, 283–84
Mir Jafar, 180ff, 187
Mir Kasim, 181, 184
Mirza Abdulla (Shah Alam), 183–84
Mirza Beg, 149
Mishra, L. N., 396
Misra, Justice R., 420
Missionaries, British: reform efforts
  of, 233; under the Company, 207ff;
  and education, 241; under the
  Crown, 241ff
Mizo rebellion, 377; uprisings, 411
Mizoram, 391, 411
Modernization of India: under British,
  226ff, 232; after 1858, 243ff, 263;
  after World War I, 304; secular, 358
Modh bania caste, 287
Mody, Piloo, 397
Mohenjo-daro, 11, 14ff, 36, 42, 67
Mohsin-ul Mulk, 278–79
Moksha, 45ff, 48, 81, 84, 435
Monasticism, 51ff
Monetary reform, under Tughluqs,
  115–16
Money-lending castes (urban), 236
Mongol invasions, 114
Monism, origins of, 35; in Upani-
  shads, 46
Monson, Colonel, 190
Monsoons, 6, 361, 385
Montagu, Edwin Samuel (secretary of
  state), 285, 294, 299
Montagu declaration, 314
Montagu-Chelmsford Reforms, 297
Moplahs, 306
Morison, Theodore, 274
Morley, John (secretary of state),
  275, 277–78, 280, 283–84, 285
Moscow, 333
Mother goddess, 10, 19, 82; worship,
  262
Mountbatten, Edwina, 345
Mountbatten, Lord Louis (viceroy),
  345ff, 349, 354
Mrichakatika, 91
Mudki, Battle of, 223
Mudras, 85

Mughal Empire, 173, 185, 238, 263;
  under Alamgir (Aurangzeb), 157ff;
  birth under Babur, 112ff; in decline
  after Alamgir's death, 168ff; developed by Humayun, 124–25; expands
  under Akbar, 126ff; *mansabdari*
  system of administration, 129–30;
  Persian influence on, 133ff; power
  seized by Jahangir, 134; pursued
  by British enterprise, 142ff; Shah
  Jahan's reign, 151ff
Muhajirs ("Refugees"), 452
Muhammad, 104
Muhammad Ali, 177
Muhammad Shah, 171, 172, 181
Muhammadan Anglo-Oriental Col-
  lege, 264
Mujibur Rahman, Sheikh ("*Banga-
  bandhu*"), 386ff; arrested, 387; as-
  sassinated, 401; becomes prime
  minister of Bangladesh, 390;
  changes constitution, 400; Six-Point
  Program of, 384; victory of, in
  1970, 386
Mukti Bahini, 389
Muller, Friedrich Max, 24, 26
Mumbai, see Bombay
Mumtaz Mahal, 149, 152–53
Munda tribe, 8
Municipal corporations, 266
Municipal government boards, 256
Munro, Hector, 185
Munro, Thomas, 204, 296–97
Murad, Prince, 155ff
Murshidabad, 178, 180–81, 184
Muscovy Company, 139
Mushtaq Hussain, *see* Viqur-ul-Mulk
Music, 30–31
Muslim League, 264, 278, 324, 335,
  347; election results for, in 1937,
  323; founded in Dacca, 279, 286;
  growing popularity of, 325; under
  Iqbal, 316; under Jinnah, 293; La-
  hore session, 330; Punjab organiza-
  tion of, 346; response to cabinet
  mission plan, 342–43; role in gene-
  sis of Pakistan, 339ff; turns to Di-
  rect Action, 344
Muslim separatism, 302; political
  consciousness of, 277; role in elections, 284; *see also* Separate electorates
Muslims, 342; of Bengal, 273; and
  Congress rule, 329
"Mutiny" of 1857–58, *see* Anglo-
  Indian War of 1857–58

Mutiny of Royal Indian Air Force and Navy, 340
Muzaffar Jang, 176
Muzaffarabad, 354
Mysore, 193, 201; coffee plantations, 247

"Nabobism," 175
Nadir Shah (Nadir Quli), 172–73, 181
Nagaland (state), 377; revolt in, 411
Nagasena, 71
Nagpur, 227, 275–76
Nag-Vidarbha, 432
Naicker, E. V. Ramaswami, 380
Naidu, Sarojini, 315, 368
Najib-ud-Dawla, 184
Nalanda, 108
Nama rupa, 48
Namboodripad, E. M. S., 378, 389
Nana Phadnis, 191–92, 203
Nana Sahib, 234–35, 238, 278
Nanak, Sikh Guru, 121, 161
Nanda, Gulzarilal, 375, 384
Nandi, 10
Naoroji, Dadabhai ("Grand Old Man" of Congress), 254, 259, 286
Napier, Sir Charles, 221ff
Napoleon, 184
Narain, Raj, 409
Narayan, Jagat, 255
Narayan, Jayaprakash (JP) (leader of *Janata* opposition to Indira's Raj), 359, 395ff, 404, 427; arrested, 397; dies, 410; on Indira Gandhi, 403
Narayan Rao, 192
Narmada River, 9
Nasik, 260
Nasir Jang, 173, 176
Natal, 287
Nataraja (Shiva), 82
National anthem, 262
National Front-Left Front, 448–49
National language controversy, 368
National Social Conference, 253
Nationalism: in Bombay, 254; consciousness of, emerges, 250ff; and English education, 251; and Hindu reform, 253; leadership of, 256ff; Maharashtra as breeding ground of, 253–54; provincial roots of, 252; role of the press in, 262
Nationalist movement, *see* Indian National Congress
Nationalization of banks, 382
"*Nawab*" of Dacca, 279
Naxalbari district, 386

Naxalites, 386
Nehru-Gandhi "dynasty," 410
Nehru-Kaul clan, 353
Nehru, Jawaharlal (nationalist leader and prime minister), 303ff, 351ff, 439; advice to his daughter, Indira, 400; and Allahabad politics, 309; barrister for INA "rebels," 340; and Bose, 328; and cabinet mission plan, 343; and Cripps, 334; death of, 370; economic policy of, 358ff; first prime minister, 351ff; foreign policy of, 363ff; and Gandhi, 306; after Gandhi's assassination, 355; jailed, 333; and Jinnah, 326; Kamaraj Plan, 371; and Kashmir, 352ff; and linguistic provinces, 368–69; and Mountbatten, 345; and Pakistan, 347; and Patel, 351; presidential address to Congress, 323; Republic Day address, 358; and secularism, 356; and social policy, 365ff; testament of, 370; *see also Panditji*
Nehru, Motilal, 300–301, 427; proposes "Commonwealth of India," 312–13; resigns from assembly, 315; and Swarajists, 307–8
Neolithic Age, 9–10
Netaji ("Leader"), *see* Bose, Subhas Chandra
New Delhi, 285, 341, 349; *see also* Delhi
New India, 258
New Party, 281–82
New towns, 244–45
Newspaper, vernacular, 262
Newspapers (Indian): *Amrita Bazar Patrika,* 261; *Bande Mataram,* 276; *Bengalee, The,* 261; *Hindu,* 262; *Kesari,* 282; *Pioneer,* 282; *Punjabee,* 280; *Swadeshamitram,* 262; *Yugantar,* 281
Niazi, General, 403
Nicholson, John, 237–38
Nijalingappa, S., 375, 390, 393
Nilgiri hills, tea cultivation in, 247
Nirvana, 51
Nixon White House, supports Yahya Khan's dictatorship, 383ff
Nizam of Hyderabad, 198; allied with British Raj, 203; reluctant to join Dominion of India, 353
Nizam-ul-Mulk, 171ff, 176
Noorani, Sheikh Nooruddin, 451
North, Lord, ministry passes Regulating Act, 190

North India, kingdoms of, in sixth
  century, 48ff
Northeast Frontier Agency (NEFA),
  365
Northeastern India, reorganized, 390–
  91
Northwest frontier fighting, 268
Northwest Frontier Province, 269,
  323–24, 246
Northwestern Provinces and Oudh,
  land settlement in, 242
Nuclear arms, 385, 414; Comprehen-
  sive Test Ban Treaty, 453; Non-
  proliferation Treaty, India's refusal
  to sign, 395, 453
Nur Jahan, 149–50, 152
Nutrition, improved, 385
Nyaya, 97

Occupational segregation, 42–43
O'Dwyer, Sir Michael, 299
Oil, 407
One-nation theory, 331
Operation Bluestar, 418
Opium trade, 200
Opposition, against Indira Raj, 395ff;
  leadership arrested, 397
Ordinance Raj, 319
Orissa, 6, 95, 146, 322, 405, 432
O'Shaughnessy, Dr. William, 230
Oudh, 182, 190–91, 203, 234–35; an-
  nexation of, 232; taluqdars of, 235,
  242
"Outcaste" communities, 310; see also
  Harijans; Panchamas; Untouch-
  ables
Outram, Sir James, 232

Pakistan, 5, 224, 264, 279, 352, 372ff,
  408–9, 412, 414, 430; and Assam,
  340; and Bengalis, 311; and cabinet
  mission plan, 342–43; civil war,
  388ff; creation of, "inevitable," 347;
  Dominion of, born, 347–48; genesis
  of, 325, 338; hopelessly divided,
  383–85; "Islamic Bomb," 430; and
  Lahore resolution, 330; national
  election in, 384; nuclear explosion,
  414; and partition, 348; and the
  Punjab, 346; and Russians in Afghanistan, 412ff; in SAARC, 430;
  surrenders Bangladesh, 390; undeclared war with India over Kashmir,
  353ff; see also Bangladesh
Pakistani Army: in Dacca, 387ff;
  in Kashmir, 354; see also Army

Pal, Bipin Chandra, 276, 279, 282
Paleolithic Age, 6
Pallavas, 78, 99ff
Panch shila, 364
Panchamas, 42; see also Harijans;
  "Outcaste" communities; Untouch-
  ables
Panchayat, 360–61
Panchayati Raj, 360ff
Panchayati Samiti, 361
Panditji, 351; see also Nehru, Jawa-
  harlal
Pandits, 353
Pandyas, 63, 76ff
Panini, 87
Panipat, Battle of, 183–84
Pant, Pandit G. B., 328
Paramountcy, doctrine of, 227–28
Parantaka I, 112
Parliamentary government, 356–67
Parsis, 254
Parthians, 73
Partition: of Bengal, 273, 275; of
  Kashmir, 354; of subcontinent,
  347ff
Parvati, 82
Pataliputra, 58, 67, 88, 90
Patanjali, 87
Patel, Chimanbhai, 408
Patel, Vallabhbhai ("Sardar"), 296,
  350, 355; as first deputy prime
  minister, 351ff; and Indian princess,
  347ff; and Mahatma Gandhi,
  327–28; and Nehru, 313
Pathans, 255, 270, 324, 413
Patna, 184
Peasants: awakening, 379; conditions
  in Rajputana, 236; revolt in
  Alamgir's reign, 159–60; revolt in
  Kashmir, 353; revolt in Poonch,
  353
Peerbhoy, Sir Adamjee, 279
Peninsular and Orient Steamship
  Company, 245
Pensions, lapse of, 228
Pentagon (U.S.), supports Pakistan,
  388
Pepsico, 446
Periplus of the Erythrean Sea, 78
Periyanadu, 99–100
Periyar, 389
Permanent Zamindari Settlement, 196;
  see also Land settlement; Zamindars
Persia, 270; culture of, 133ff
Persian Gulf, 291–92
Peshawar, 70, 108, 270

Peshwa, 170, 191, 198, 228
Pethick-Lawrence, Lord (secretary of state), 339, 341, 345–46
Phadke, Vasudeo Balwant, 253–54
Phallic worship, 18
Pheruman, Darshan Singh, 380
Pillay, Chidambaram, 282
Pitt, William, 194
Plassey, 178, 180
Polarization of Indian politics, 383
Police: Act of 1861, 280; Curzon's reforms, 269
Political arrests in India, 397ff
Political opposition to Indira's Congress, 383, 395ff, 404
Political pressure, 257
Political significance of caste, 365–6
Pondicherry, 173, 175, 184, 262
Poona, 171, 178, 191, 258, 260, 277, 281
Poona Sarvajanik Sabha, 252, 259
Poonch, 353
Population: mounting pressure due to increase in, 321, 335, 358–59, 361, 378, 385, 394; in third century B.C., 59
Porbandar, 287
Porter, Endymion, 146
Portugal: Dutch defeat of, in Indian Ocean, 142; first contacts, 135ff; secures toehold in India, 121; trade, 136ff, 139
Postal service: Office Act passed for, 231; revolutionary impact of penny, 228ff, 231; service linked to electric telegraph, 231
Poverty, attacked, 382, 394
Prabakaran, V., 430–31
Praja Socialist Party, 359
Prajapati, 35
Prathana Samaj, 253
Prasad, Dr. Rajendra (first President), 308, 328, 356
Presidential powers, 358
Press, censorship: by British, 283; by Indians, 399, 411
Preventive detention, 357, 397
Price controls, 398
Prime minister's elite guard, 397
Princes, 227; accession to India by, 349; British wooing of, 240
Prinsep, James, 61
Profits, industrial, 406
Prohibition, nationalist, 281; *see also* Liquor boycott
Property, inheritance of, 233

Provincial autonomy, 322; divisiveness, 402; pluralism, 264, 369; unrest, 411
Provisional Government of *Azad* ("Free") India, 337
Public works, 244
Puja, 95
Pulakeshin I, 101
Pulakeshin II, 101
Pulicat, 145
Punjab, the, 4, 8, 29, 70, 182, 216, 268, 292, 326, 340, 342, 450; Accord, 425; atrocities in, 298ff; British conquest of, 222ff, 225–26; communal rioting in, 346; government of, 1907, 280; Khalistan demand in, 416ff; linguistic separatism within, 380; "preventive detention" in, 427, repression in, 280; revolts against Mughal rule, 160ff; Sikh *sirdars* of, 242; system of administration adopted elsewhere in British India, 246; *suba* ("state"), demand for, 379; terrorist violence in, 425ff
Puranas, 80
Purna svaraj, 314, 318, 329
Purva-mimansa, 97–98

Qa'id-i-Azam, see Jinnah, Muhammad Ali
Quetta, 10, 412
Qureshi, Moeen, 452
Qutb-ud-din Aybak, 108–9
Qutb-ud-din Mubarak, 115

Race, 244, 250, 283, 288; and Anglo-Indian struggle, 257; and British theories of "martial" and "nonmartial," 241, 282–83; and violence, 237
Radcliffe, Sir Cyril, 348
Radha, 80
Radhakrishnan, Dr. Sarvepalli (second president), 356
Raghunath Bhonsle, 177–78
Raghunathrao, Peshwa, 182, 191–92
Rahim, Sir Abdur, 269–70, 311
Rahman, Zia-ur, 401
Rai, Lala Lajpat (nationalist leader), 277ff; and *Arya Samaj,* 280; exiled in New York, 282; president of Trade Union Congress, 304
Railroads: commercial, political, and sociocultural impact of, 229–30; construction begun, 228ff; expansion of network, 243–44, 269; strategic

Railroads (*cont.*)
value of, 229; used to combat
famine, 267
Raj Ghat, 370
Raja Bharmal, 127
Raja Birbal, 133
Raja Ram, 166
Rajagopalachari, Chakravarti, 308
Rajaraja, 112
Rajasthan, 5–6, 11, 12, 73, 95; eco-
nomic planning in, 360ff; nuclear
explosion in, 395; *sati* continues
in, 433
Rajendra I, 112
Rajput: confederacy, 122; dynasties,
108; people, 73, 108, 158
Rajputana, *see* Rajasthan
Rajya Sabba ("Council of State"),
357
Ram, Jagjivan, 404–5, 410
Ram Das, Sikh Guru, 161
Ram Janmabhoomi, 436, 444
Ram Rajya, 423, 448
Rama, 39, 71
Ramananda, 120–21
Ramanuja, 98
Ramayana, 39ff, 133
Rana Sanga, 121ff
Ranade, Mahadev Govind, 252, 258
Rann of Kutch, clash over, 374
Rao, N. T. Rama, 419
Rao, P. V. Narasimha, 442–44, 447,
449–50
Raphel, Arnold, 451
Rashtrakutas, 103
Rashtriya Loktantrik Dal, 396
Rashtriya Svayamsevak Sangh, 355
Ravana, 40
Ravi River, 14
Ray, A. N., 396
Ray, Siddhartha Shankar, 425
Raya, Krishna Deva, 139
Raziyya, 109–10
Reading, Lord (Rufus Isaacs), vice-
roy, 307–8, 316
Reagan, Ronald, 430
Red Fort, 240
Reddy, Sanjiva, 372, 382, 409
Refugees: from East Pakistan, 388;
flood of, due to creation of Pakistan,
348
Reincarnation, *see* Samsara
Religious separatism, *see* Communal-
ism
Repression, *see* Britain, repression by
Republic Day, 356

Republic of India, 356ff
Reunification of Bengal, 285–86
Revenue collectors of Nagpur, 243
Revolution: nationalist movement
gives rise to, 261; potential of
Panchayati Raj for, 360–61
Ribeiro, Julio, 425
Rice, 385
Ripon, Lord (viceroy), 256ff
Rishis, 29, 403
Risley, H. H., 274
Roe, Sir Thomas, 143–44
Rohilla Afghans, 190
Rome, trade with, 92
Roosevelt, Franklin D., 333
Rose, Sir Hugh, 238
Round Table Conference: first, 316;
second, 318; third, 322
Rourkela, 362
Rowlatt, Justice, heads committee,
297–98
Rowlatt Acts, 298
Roy, Manabendra Nath, 305
Roy, Ram Mohun, 210ff, 250
Royal Indian Air Force, mutiny of,
340
Royal Indian Navy, mutiny of, 340
Rudrasena II, 90
Rupee, 267
Rural areas: development planning,
359ff; Extension Program for
(1952), 360; and leaders of, 237;
polarization of, 361; rebellions in,
236–37; revitalization of, 385; revo-
lution in, 395
Russia: influence of, 389; interference
of, 270, 281; invasion of Afghan-
istan, 411ff; and rivalry with British
in Afghanistan, 218ff
Ryotwari settlement, 207; *see also*
Land settlement

SAARC (South Asian Association for
Regional Cooperation), 428–29
Sabarmati, 315
Sadashivrao, 183ff
Sadhu, 289
Safavid dynasty, 172
Safdar Jung, 182
Sahibs, 250; society of, 244
Sakyamuni, *see* Buddha
Salabat Jang, 177
Salbai, Treaty of, 193
Salimullah Khan, *see* "*Nawab*" of
Dacca
Salisbury, Lord Robert, 255

Salt tax, 308, 315
Samanid kingdom, 106
Sambhaji, 165–66
Samkhya, 96
Samsara, 44, 47ff
Samyukta Maharashtra Samiti, 369;
    see also Maharashtra
Sanchi, 68, 75, 85
Sangams, 77
Sangha, 51
Sannyasin, 84
Sanskrit grammar, 87
Sapru, Sir Tej Bahadur, 316, 339, 340
Sarabhai, Ambalal, 296
Sarasvati (goddess), 82
Saraswati, Dayananda, 263
Sarnath, 50, 67
Sarvodaya, 346, 359
Sat, 48
Satarat, 178; raja of, 227
Satavahana, *see* Andhra dynasty
Sati, 21, 82; abolition of, 212; impact
    of abolition, 233; revival of, 433
Satpura range, 6, 9
Satya, 288
Satyagraha, 238, 295, 317; *see also*
    Gandhi, Mohandas Karamchand
Satyapal, Dr., 298
Satya-pir, 121
Saudi Arabia, 414
Savarkar, Ganesh Damodar ("Veer"),
    283, 355
Sayeed, Mufti M., 434–36, 439
Sayyid brothers, 169, 171
Sayyid dynasty, 120
Science, early development of, 86
Scoble, Sir Andrew, 260
Scott, David, 217
Scythia: invasions by, 72ff, 89; peoples
    of, 72
Sea route, 243
Seals, iconography, 18
SEATO, 373
Secretary of state for India, 239–40,
    272, 275
Secular ideals, of Republic of India,
    356
Seeds, high-yield, 385
Seleucus Nikator, 59
Self-government, local, 256
Separate electorates, 284, 319, 323
Sepoys, 138, 233–34
Serampore, 207
Seringapatam, 201
Seth, Jagat, 180
Seven United Liberation Army, 411

Seven Years' War, 178, 184
Shah, Ahsan, 116
Shah, Khwaja Yusuf, 278
Shah, Zahir, 453
Shah Alam, *see* Bahadur Shah I
Shah Jahan (Khurram), 151ff, 163;
    construction of Delhi under, 154;
    court life during reign of, 155–56;
    defeat of Deccan sultanates by, 153;
    struggle for power by, 152; succes-
    sion struggle among sons of, 156ff
Shah Nama, 107
Shah Shuja (Durrani), 219
Shahji, 162–63
Shahu, 170
Shakas, *see* Scythia
Shakti, 95
Shakuntala, 90ff
Shankara, 98
Shanti Ghat, 372
Sharif, Mian Nawaz, 452
Sharma, President S. D., 448
Shastri, Lal Bahadur (prime minis-
    ter), 371ff; charges Pakistan with
    aggression in Kashmir, 375; death
    of, 376; meets Ayub Khan in
    Tashkent, 375–76; succession
    struggle following, 376ff
Shekhar, Chandra, 438–39
Sher Ali, 255
Shiva, 10, 18–19, 31, 48, 80, 82, 260,
    447ff; as *Nataraja,* 113; as *Pashu-
    pati,* 82; temple at Ellora, 103;
    temple at Tanjore, 113
Shivaji Bhonsle, 162, 163ff; birth cele-
    brated by modern festival, 260–61;
    conquests in Deccan, 165; crowned
    Chatrapati at Rajgarh, 165; death
    of, 165; defeats Afzal Khan, 164;
    escapes from Alamgir's court, 164–
    65
Shore, Sir John, 196, 201, 208
Shreni, 60–61, 79; in Guptan Empire,
    93
Shri Aurobindo, *see* Ghosh, Arabinda
Shuddhi, 306
Shudraka, 91
Shudras, 32, 41–42
Shuja, Prince, 155, 156, 157–58
Shuja-ud-Daula, 182–84, 185, 190
Shunga, Pushyamitra, 69
Shunga dynasty, 75
Siddhartha Gautama, *see* Buddha
Sikhistan, 416
Sikhs, 121, 159ff, 416ff, 424–25, 434;
    and assassination of Mrs. Gandhi,

Sikhs (*cont.*)
    419; Delhi massacre of, 420–21; de-
    mand separate state, 346, 379; dur-
    ing War of 1857–58, 238; emergence
    as martial force, 162; impact of Op-
    eration Bluestar on, 418; and
    Khalistan, 416ff; perils of partition
    for, 342; first war with British,
    222ff; political party of, 379, 416;
    Punjabi Subha, demand of, 417;
    ravaged by partition of Punjab,
    348; represented by Tara Singh,
    319; revolt against Bahadur Shah,
    169; revolutionaries, 292; *sirdars,*
    242; and struggle against Mughal
    power, 182; surrender to British,
    223; second war with British,
    223–24; under Ranjit Singh, 216
Sikkim, 391
Sino-Indian relationship, 432
Silappadikaram, 78
Silver, world slump in, 267
Simla, 265, 268, 271, 278, 332, 391;
    first Conference at, 339; second
    Conference at, 341
Simon, Sir John, heads Indian Statu-
    tory Commission, 311, 316
Sind, 106, 218–20, 221ff, 322
Sindia, Dattaji, 182–83
Sindia, Mahadji, 191, 198–99, 201
Singapore, 334, 446
Singh, Ajit, 280
Singh, Beant, 450
Singh, Charan, 407–8, 410
Singh, Devi, 237
Singh, Gulab, 223
Singh, Hari, 352–53
Singh, Kadam, 237
Singh, Manmohan, 443, 445–47
Singh, Master Tara, 319, 346, 380
Singh, Ranjit, 219, 222
Singh, Sant Fateh, 380
Singh, Satwant, 419
Singh, Sawran, 372
Singh, V. P., 426–27, 433–39, 441,
    443–44
Singh, Zail, 416–17, 420
Singhal, Ashok, 448
Sinha, Jag Mohan Lal, 396
Sinha, Sir Satyendra P., 283
Sinhagarh, 163
Sinkiang, 364
Sino-Indian relations: border conflict,
    364–65; friendship, 364, 446; war,
    365
Sino-Pakistani treaty, 374

Siraj-ud-daula, 178ff
Sirdars, 242
Sita, 39, 437
Sitaramayya, Pattabhi, 327
Six-Point Program, 384
Slums, urban, 321
Smythe, Thomas, 142
Soan River, 4, 6
Sobraon, Battle of, 223
Social change: class conflicts, 264;
    reform, 240, 253, 365ff; under
    Dalhousie, 229ff
Socialist Independence for India
    League, 313
Socialist Party, Congress, 359
Soma, 31, 32, 34; sacrifice, 43, 98
Somnath, 107
Sonora, 64, 385
South Asia: area of, 3; climate of, 4;
    early man in, 4–5; geography of, 4;
    languages and linguistic families
    within, 8; neolithic age in, 9–10;
    topography of, 4
South Asian Association for Regional
    Cooperation (SAARC), 429–30
South India, 6; social and political
    institutions of, 99–100
Southeast Asia, 92
Soviet Union, relation with, 429
Spain, rivalry with Dutch, 140ff
Spice Islands, 145
Spice trade, 137–38, 142
Sravasti, 49
Sri Lanka, 430ff; civil war in, 430;
    Peace Agreement with India, 431,
    432
Sri Vijaya, 112
Srinagar, 354, 419, 434–36, 445
Sriramalu, Potti, 368
State assemblies: Congress-dominated,
    367; women elected to, 369
Steam transport, to India, 243
Steel factories and production, 362
Stone Age, 9
Street-dwellers, 321
Strikes: in Ahmedabad, 296; elim-
    inated by emergency rule, 399;
    Indian labor's first, 246–47;
    widespread, in 1928–29, 321
Stupa, 16, 68, 85
Suburban development, 244
Suez Canal, 243, 245
Sufism: in Bengal, 118; orders of,
    117ff, 132
Suhrawardy, Huseyn Shaheed, 311,
    344

Sumatra, 92
Sumeria, 19
Supreme Court, 412
Sur, Sher Khan, 124–25
Surat, 142–43, 145, 174, 203, 281;
    Treaty of, 192
Surya, 33
Sutkagen Dor, 19
Svadeshi, 274, 275ff, 303, 321, 446;
    movement, 276, 317
Svaraj, 163, 253, 261, 282, 297, 304,
    306, 311, 313
Swarajist Party, 308–9, 313
Swatantra Party, 362
Syndicate, of Congress, 389, 405
Szulc, Tad, 401

Tagore, Dwarkanath, 211, 214
Tagore, Rabindranath, 262, 274, 300
Taj Mahal, 150
Taliban army, 453
Taluqdars, 235, 242, 250, 316; see also
    Anglo-Indian War of 1857–58
Tamerlane (Timur the Lame), 119
Tamil: culture, 77; kingdoms, 77ff;
    laborers, 288; Tamil Nadu, 369,
    402; see also Dravidian
    consciousness
Tamil Gelam (Nation), 430
Tamil Tigers (LTTE), 430–31, 440
Tamil United Liberation Front
    (TULF), 430
Tamilnad, 77
Tanjore, 112, 203
Tantia Topi, 238
Tantrism, 95–96
Tapas, 35–36
Tapasya, 302
Tapti River, 6
Tara Bai, 166
Taras, 96
Tariff Board, 308
Tashkent: summit meetings at, 375–76
Tat tvam asi, 47
Tata, Jamshed N., 276, 296
Tata, Ratan, 276
Tata Iron and Steel Company, 276,
    336
Taxes: collection of, 170; evasion of,
    410; system, under Mauryan rule,
    58; and system for under Muslim
    sultanate, 114
Taxila, 55, 71
Tea: plantations, 247; planters, 268
Tegh Bahadur (Sikh guru), 161–62

Temple Entry Bill, Dr. Subbaroyan's,
    321
Terrorist attacks, 261, 281, 282, 314
Textiles, 131, 214
Thakurs, 237
Theosophical Society, 263
Theravada Buddhism, 50, 72
Thugi, 212
Tibet, 271, 364–65
Tibetan refugee protests, 446
Tilak, Balwantrao Gangadhar (Loka-
    manya), nationalist leader, 254,
    259–60, 262, 279; compared with
    Gokhale and Gandhi, 294; con-
    victed of "sedition," 282; dies, 301;
    on legislative council, 266; and
    national education, 281; supports
    svadeshi movement, 277; and World
    War I, 290
Tipu Sultan, 198, 201
Tirthankaras, 53
Titles, lapse of, 218
Todar Mal, 129
Tohra, S. G., 425
Tojo, 337
Toramana, 94
Tordesillas, Treaty of, 136–37, 139
Trade: between India and Rome, 78–
    79; "triangular," 145; via overland
    route, 79
Trade Union Congress, All-India, 304,
    312
Transfer of power, 345ff
Travancore, 310
Treaty of Paris, 184
Trichinopoly, 177
Tripura, 391, 411
Tughluq, Ghiyas-un-din, 115
Tughluq, Muhammad, 115ff
Tughluq dynasty, 115ff
Tulsi Das, 133
Turko-Afghan Muslims, 107ff
Twenty-Point Program, 398ff
Two-nation theory, 331
Two-race theory, 369

Ujjain, 109
UN Declaration of Universal Human
    Rights, violation of, 426
Underemployment, 321
Unemployment, 362–63, 387
Unification, by British of India, 226ff
Unionist Party, 326, 340, 346
United East India Company, Dutch
    (Vereenigde Oostindische Com-
    pagnie), 141–42

United Provinces, 244; *see also* Uttar
  Pradesh
United States: aid to India, 363, 378,
  389, 395, 400; Civil War, impact of,
  in India, 229, 247; relations with
  India, 428–29; supplies arms to
  Pakistan, 373, 389
Universal franchise, impact of, 367
University student repression, 276
Untouchability: abolished, 358; Abo-
  lition Bill, 321; Abolition Week,
  320; (Offences) Act, 365
Untouchables, 32, 90; and Ambedkar,
  310; and Communal Award, 319;
  exuntouchables, 358, 367, 436, 441;
  and Mahatma Gandhi, 310; origins
  of, 42; *see also Harijans*
Upanishads, 26, 44ff, 211; concept of
  time in, 48; mystic authors of, 44ff;
  revolt of, 44ff
Urban magnates, 254
Urbanization, 304, 321; and its prob-
  lems, 433
Urdu, 383
Uri, 354; and Poonch "bulge," 375
Ushas, 34
Utbi, 107
Uttar Pradesh, 242, 432; *see also*
  United Provinces
"Uttarakhand" State Movement, 432
Uttara-mimansa, see Vedanta
Uzbekistan, Russian, 413

Vaisesika, 97
Vaishyas, 41
Vajpayee, A. B., 409, 445, 448, 450
Vakatakas, 101
Vakil, 129
Valabhi, 94
Valmiki, 39
Van Linschoten, Jan Huygen, 141
Vanaprastha, 83–84
Varman I, Mahendra Vikrama, 100
Varman I, Narasimha, 100
Varna, 29, 32, 41–42, 48
Varuna, 32–33
Vasco da Gama, 135ff
Vedanta, 97–98; basic principles of,
  45ff; pessimism of, 47ff
Vedas, 25ff; Atharva, 26, 30; Rig,
  25ff, 29, 31–32, 35–36, 45; Sama,
  26, 30; Yajur, 26
Vedic society: burial practices of, 44;
  pantheon of, 27; proselytizing of,

  263; sacrifices of, 26; social organ-
  ization of, 41ff
Vedic religion, *see* Aryan civilization,
  religion
Venkataraman, President Rama-
  swamy, 433, 438
Vernacular Press Act, 254
Viceroy's Executive Council, 245
Victoria, Queen (empress of India),
  240, 248
Victoria's Proclamation, 240
Vidarbha, Nag-, movement, 32
Viharas, 68, 85
Vijayalaya, 112
Vijayanagar, 116–17; defeat of, 138–
  39
Villages: revitalization of, 377; society
  in, 362
Vindhya mountains, 6, 9, 40
Viqur-ul-Mulk, 279, 286
Vishnu, 33, 71; *avataras* of, 80
Vishwa Hindu Parishad (VHP), 436,
  444, 448
Vritra, 33, 48

Wajid Ali, 232, 234
Wali-ullah, Shah, 182
Wandiwash, Battle of, 184
Ward, William, 207
Wardha scheme of education, 367
Wavell, Field Marshal Archibald
  (viceroy), 336ff, 339, 344–45
Wedderburn, William, 258, 290
Wellesley, Sir Arthur (Duke of
  Wellington), 201
Wellesley, Richard Colley (governor-
  general), 201ff, 204
Western assistance to India, 363
Western Ghats, 6
Western idealism, impact on India,
  358
Wheeler, Sir Hugh, 235
White House, support for Yahya
  Khan, 400
White Man's Burden, 244
Whitehall, 240, 259, 283
Wholesale food-grain markets, 394
Willingdon, Lord (viceroy), 318
Willoughby, Sir Hugh, 139
Wilson, Horace H., 209, 212
Wima, Kushan king, 74
Women: in ancient India, 77; elected
  to *Lok Sabha,* 367; enfranchised,
  366; Hindu widows, 233; liberation
  struggle by, 366ff; in politics, 316,

367; rights of, 260; suicide rates of, 366; traditional legal status of, 366
Wood, Sir Charles, 241
Wood, Edward, *see* Irwin, Lord
Workers and Peasants Party, All-India, 312
World War I, 287ff; aftermath of, 301; Congress-League scheme developed during, 294; economic impact of, 291–92; Indian responses to, 289; Indian terrorism in, 292; industrial development during, 296; legacy of, 329; Madras attacked during, 290–91; "Mespot" disaster, 292; and Muslims, 291
World War II, 329ff; economic impact of, 335; Indian response to, 335; industrial development during, 336; political impact of, 337ff; results of, 340ff

Xavier, Francis, 138

Yadavas, 111
Yadev, Lalu Prasad, 449
Yahya Khan, A. M., *see* Khan, Aga Muhammad Yahya
Yajnavalkya, 45
Yamuna, 373; Ganga, 5
Yashodhara, 91
Yoga, 36, 96–97; early forms of, 36; powers of, 288, 302, 413
Younghusband, Major Francis, 269, 271
Yudhishthira, 38
Yueh-chih, *see* Kushans

Zaman, Aga Muhammad, 146
Zamindars, 196, 242, 251, 258, 274, 316
Zhob River, 10
Zia ul-Haq, 409, 414, 429, 451
Zila Parishad, 361
Zinkin, Maurice, 326
Zoroastrians, 106, 254; *see also* Parsis